# Exercises and Solutions in Biostatistical Theory

# CHAPMAN & HALL/CRC
## Texts in Statistical Science Series

Series Editors

Bradley P. Carlin, *University of Minnesota, USA*
Julian J. Faraway, *University of Bath, UK*
Martin Tanner, *Northwestern University, USA*
Jim Zidek, *University of British Columbia, Canada*

**Analysis of Failure and Survival Data**
P. J. Smith

**The Analysis of Time Series — An Introduction, Sixth Edition**
C. Chatfield

**Applied Bayesian Forecasting and Time Series Analysis**
A. Pole, M. West and J. Harrison

**Applied Nonparametric Statistical Methods, Fourth Edition**
P. Sprent and N.C. Smeeton

**Applied Statistics — Handbook of GENSTAT Analysis**
E.J. Snell and H. Simpson

**Applied Statistics — Principles and Examples**
D.R. Cox and E.J. Snell

**Applied Stochastic Modelling, Second Edition**
B.J.T. Morgan

**Bayesian Data Analysis, Second Edition**
A. Gelman, J.B. Carlin, H.S. Stern and D.B. Rubin

**Bayesian Ideas and Data Analysis: An Introduction for Scientists and Statisticians**
R. Christensen, W. Johnson, A. Branscum, and T.E. Hanson

**Bayesian Methods for Data Analysis, Third Edition**
B.P. Carlin and T.A. Louis

**Beyond ANOVA — Basics of Applied Statistics**
R.G. Miller, Jr.

**Computer-Aided Multivariate Analysis, Fourth Edition**
A.A. Afifi and V.A. Clark

**A Course in Categorical Data Analysis**
T. Leonard

**A Course in Large Sample Theory**
T.S. Ferguson

**Data Driven Statistical Methods**
P. Sprent

**Decision Analysis — A Bayesian Approach**
J.Q. Smith

**Design and Analysis of Experiment with SAS**
J. Lawson

**Elementary Applications of Probability Theory, Second Edition**
H.C. Tuckwell

**Elements of Simulation**
B.J.T. Morgan

**Epidemiology — Study Design and Data Analysis, Second Edition**
M. Woodward

**Essential Statistics, Fourth Edition**
D.A.G. Rees

**Exercises and Solutions in Biostatistical Theory**
L.L. Kupper, B.H. Neelon, and S.M. O'Brien

**Extending the Linear Model with R — Generalized Linear, Mixed Effects and Nonparametric Regression Models**
J.J. Faraway

**A First Course in Linear Model Theory**
N. Ravishanker and D.K. Dey

**Generalized Additive Models: An Introduction with R**
S. Wood

**Graphics for Statistics and Data Analysis with R**
K.J. Keen

**Interpreting Data — A First Course in Statistics**
A.J.B. Anderson

**Introduction to General and Generalized Linear Models**
H. Madsen and P. Thyregod

**An Introduction to Generalized Linear Models, Third Edition**
A.J. Dobson and A.G. Barnett

**Introduction to Multivariate Analysis**
C. Chatfield and A.J. Collins

**Introduction to Optimization Methods and Their Applications in Statistics**
B.S. Everitt

**Introduction to Probability with R**
K. Baclawski

**Introduction to Randomized Controlled Clinical Trials, Second Edition**
J.N.S. Matthews

**Introduction to Statistical Inference and Its Applications with R**
M.W. Trosset

**Introduction to Statistical Methods for Clinical Trials**
T.D. Cook and D.L. DeMets

**Large Sample Methods in Statistics**
P.K. Sen and J. da Motta Singer

**Linear Models with R**
J.J. Faraway

**Logistic Regression Models**
J.M. Hilbe

**Markov Chain Monte Carlo — Stochastic Simulation for Bayesian Inference, Second Edition**
D. Gamerman and H.F. Lopes

**Mathematical Statistics**
K. Knight

**Modeling and Analysis of Stochastic Systems, Second Edition**
V.G. Kulkarni

**Modelling Binary Data, Second Edition**
D. Collett

**Modelling Survival Data in Medical Research, Second Edition**
D. Collett

**Multivariate Analysis of Variance and Repeated Measures — A Practical Approach for Behavioural Scientists**
D.J. Hand and C.C. Taylor

**Multivariate Statistics — A Practical Approach**
B. Flury and H. Riedwyl

**Pólya Urn Models**
H. Mahmoud

**Practical Data Analysis for Designed Experiments**
B.S. Yandell

**Practical Longitudinal Data Analysis**
D.J. Hand and M. Crowder

**Practical Statistics for Medical Research**
D.G. Altman

**A Primer on Linear Models**
J.F. Monahan

**Probability — Methods and Measurement**
A. O'Hagan

**Problem Solving — A Statistician's Guide, Second Edition**
C. Chatfield

**Randomization, Bootstrap and Monte Carlo Methods in Biology, Third Edition**
B.F.J. Manly

**Readings in Decision Analysis**
S. French

**Sampling Methodologies with Applications**
P.S.R.S. Rao

**Statistical Analysis of Reliability Data**
M.J. Crowder, A.C. Kimber, T.J. Sweeting, and R.L. Smith

**Statistical Methods for Spatial Data Analysis**
O. Schabenberger and C.A. Gotway

**Statistical Methods for SPC and TQM**
D. Bissell

**Statistical Methods in Agriculture and Experimental Biology, Second Edition**
R. Mead, R.N. Curnow, and A.M. Hasted

**Statistical Process Control — Theory and Practice, Third Edition**
G.B. Wetherill and D.W. Brown

**Statistical Theory, Fourth Edition**
B.W. Lindgren

**Statistics for Accountants**
S. Letchford

**Statistics for Epidemiology**
N.P. Jewell

**Statistics for Technology — A Course in Applied Statistics, Third Edition**
C. Chatfield

**Statistics in Engineering — A Practical Approach**
A.V. Metcalfe

**Statistics in Research and Development, Second Edition**
R. Caulcutt

**Stochastic Processes: An Introduction, Second Edition**
P.W. Jones and P. Smith

**Survival Analysis Using S — Analysis of Time-to-Event Data**
M. Tableman and J.S. Kim

**The Theory of Linear Models**
B. Jørgensen

**Time Series Analysis**
H. Madsen

**Time Series: Modeling, Computation, and Inference**
R. Prado and M. West

Texts in Statistical Science

# Exercises and Solutions in Biostatistical Theory

Lawrence L. Kupper

University of North Carolina
Chapel Hill, North Carolina, USA

Brian H. Neelon

Duke University
Durham, North Carolina, USA

Sean M. O'Brien

Duke University Medical Center
Durham, North Carolina, USA

CRC Press
Taylor & Francis Group
Boca Raton London New York

CRC Press is an imprint of the
Taylor & Francis Group, an **informa** business
A CHAPMAN & HALL BOOK

Chapman & Hall/CRC
Taylor & Francis Group
6000 Broken Sound Parkway NW, Suite 300
Boca Raton, FL 33487-2742



ISBN 13: 978-1-58488-722-5 (pbk)



**Library of Congress Cataloging-in-Publication Data**

Kupper, Lawrence L.
Exercises and solutions in biostatistical theory / Lawrence L. Kupper, Sean M. O'Brien, Brian H. Neelon.
p. cm. -- (Chapman & Hall/CRC texts in statistical science series)
Includes bibliographical references and index.
ISBN 978-1-58488-722-5 (pbk. : alk. paper)
1. Biometry--Problems, exercises, etc. I. O'Brien, Sean M. II. Neelon, Brian H. III. Title.

QH323.5.K87 2010
570.1'5195--dc22 2010032496

**Visit the Taylor & Francis Web site at**
**http://www.taylorandfrancis.com**

**and the CRC Press Web site at**
**http://www.crcpress.com**

*To my wonderful wife Sandy, to the hundreds of students who have taken my courses in biostatistical theory, and to the many students and colleagues who have collaborated with me on publications involving both theoretical and applied biostatistical research.*

**Lawrence L. Kupper**

*To Sara, Oscar, and my parents for their unwavering support, and to Larry, a true mentor.*

**Brian H. Neelon**

*To Sarah and Avery, for support and inspiration.*

**Sean M. O'Brien**

# Contents

# Preface

This exercises-and-solutions book contains exercises and their detailed solutions covering statistical theory (from basic probability theory through the theory of statistical inference) that is taught in courses taken by advanced undergraduate students, and first-year and second-year graduate students, in many quantitative disciplines (e.g., statistics, biostatistics, mathematics, engineering, physics, computer science, psychometrics, epidemiology, etc.).

The motivation for, and the contents of this book, stem mainly from the classroom teaching experiences of author Lawrence L. Kupper, who has taught graduate-level courses in biostatistical theory for almost four decades as a faculty member with the University of North Carolina Department of Biostatistics. These courses have been uniformly and widely praised by students for their rigor, clarity, and use of real-life settings to illustrate the practical utility of the theoretical concepts being taught. Several exercises in this book have been motivated by actual biostatistical collaborative research experiences (including those of the three authors), where theoretical biostatistical principles have been used to address complicated research design and analysis issues (especially in fields related to the health sciences). The authors strongly believe that the best way to obtain an in-depth understanding of the principles of biostatistical theory is to work through exercises whose solutions require nontrivial and illustrative utilization of relevant theoretical concepts. The exercises in this book have been prepared with this belief in mind. Mastery of the theoretical statistical strategies needed to solve the exercises in this book will prepare the reader for successful study of even higher-level statistical theory.

The exercises and their detailed solutions are divided into five chapters: Basic Probability Theory; Univariate Distribution Theory; Multivariate Distribution Theory; Estimation Theory; and Hypothesis Testing Theory. The chapters are arranged sequentially in the sense that a good understanding of basic probability theory is needed for exercises dealing with univariate distribution theory, and univariate distribution theory provides the basis for the extensions to multivariate distribution theory. The material in the first three chapters is needed for the exercises on statistical inference that constitute the last two chapters of the book. The exercises in each chapter vary in level of difficulty from fairly basic to challenging, with more difficult exercises identified with an asterisk. Each of the five chapters begins with a detailed introduction summarizing the statistical concepts needed to help solve the exercises in that chapter of the book. The book also contains a brief summary of some useful mathematical results (see Appendix A).

The main mathematical prerequisite for this book is an excellent working knowledge of multivariable calculus, along with some basic knowledge about matrices (e.g., matrix multiplication, the inverse of a matrix, etc.).

This exercises-and-solutions book is not meant to be used as the main textbook for a course on statistical theory. Some examples of excellent main textbooks on statistical theory include Casella and Berger (2002), Hogg, Craig, and McKean (2005), Kalbfleish (1985), Ross (2006), and Wackerly, Mendenhall III, and Scheaffer (2008). Rather, our book should serve as a supplemental source of a wide variety of exercises and their detailed solutions both for advanced undergraduate and graduate students who take such courses in statistical theory and for the instructors of such courses. In addition, our book will be useful to individuals who are interested in enhancing and/or refreshing their own theoretical statistical skills. The solutions to all exercises are presented in sufficient detail so that users of the book can see how the relevant statistical theory is used in a logical manner to address important statistical questions in a wide variety of settings.

**Lawrence L. Kupper**
**Brian H. Neelon**
**Sean M. O'Brien**

# *Acknowledgments*

Lawrence L. Kupper acknowledges the hundreds of students who have taken his classes in biostatistical theory. Many of these students have provided valuable feedback on the lectures, homework sets, and examinations that make up most of the material for this book. In fact, two of these excellent former students are coauthors of this book (Brian H. Neelon and Sean M. O'Brien). The authors want to personally thank Dr. Susan Reade-Christopher for helping with the construction of some exercises and solutions, and they want to thank the reviewers of this book for their helpful suggestions. Finally, the authors acknowledge the fact that some exercises may overlap in concept with exercises found in other statistical theory books; such conceptual overlap is unavoidable given the breadth of material being covered.

# Authors

**Lawrence L. Kupper, PhD,** is emeritus alumni distinguished professor of biostatistics, School of Public Health, University of North Carolina (UNC), Chapel Hill, North Carolina. Dr. Kupper is a fellow of the American Statistical Association (ASA), and he received a Distinguished Achievement Medal from the ASA's Environmental Statistics Section for his research, teaching, and service contributions. During his 40 academic years at UNC, Dr. Kupper has won several classroom teaching and student mentoring awards. He has coauthored over 160 papers in peer-reviewed journals, and he has published several coauthored book chapters. Dr. Kupper has also coauthored three textbooks, namely, *Epidemiologic Research—Principles and Quantitative Methods*, *Applied Regression Analysis and Other Multivariable Methods* (four editions), and *Quantitative Exposure Assessment*. The contents of this exercises-and-solutions book come mainly from course materials developed and used by Dr. Kupper for his graduate-level courses in biostatistical theory, taught over a period of more than three decades.

**Brian H. Neelon, PhD**, is a research statistician with the Children's Environmental Health Initiative in the Nicholas School of the Environment at Duke University. He obtained his doctorate from the University of North Carolina, Chapel Hill, where he received the Kupper Dissertation Award for outstanding dissertation-based publication. Before arriving at Duke University, Dr. Neelon was a postdoctoral research fellow in the Department of Health Care Policy at Harvard University. His research interests include Bayesian methods, longitudinal data analysis, health policy statistics, and environmental health.

**Sean M. O'Brien, PhD,** is an assistant professor in the Department of Biostatistics & Bioinformatics at the Duke University School of Medicine. He works primarily on studies of cardiovascular interventions using large multicenter clinical registries. He is currently statistical director of the Society of Thoracic Surgeons National Data Warehouse at Duke Clinical Research Institute. His methodological contributions are in the areas of healthcare provider performance evaluation, development of multidimensional composite measures, and clinical risk adjustment. Before joining Duke University, he was a research fellow at the National Institute of Environmental Health Sciences. He received his PhD in biostatistics from the University of North Carolina at Chapel Hill in 2002.

# 1

# Basic Probability Theory

## 1.1 Concepts and Notation

### 1.1.1 Counting Formulas

#### 1.1.1.1 N-tuples

With sets $\{a_1, a_2, \ldots, a_q\}$ and $\{b_1, b_2, \ldots, b_s\}$ containing $q$ and $s$ distinct items, respectively, it is possible to form $qs$ distinct pairs (or 2-tuples) of the form $(a_i, b_j), i = 1, 2, \ldots, q$ and $j = 1, 2, \ldots, s$. Adding a third set $\{c_1, c_2, \ldots, c_t\}$ containing $t$ distinct items, it is possible to form $qst$ distinct triplets (or 3-tuples) of the form $(a_i, b_j, c_k), i = 1, 2, \ldots, q, j = 1, 2, \ldots, s,$ and $k = 1, 2, \ldots, t$. Extensions to more than three sets of distinct items are straightforward.

#### 1.1.1.2 Permutations

A *permutation* is defined to be an ordered arrangement of $r$ distinct items. The number of distinct ways of arranging $n$ distinct items using $r$ at a time is denoted $\mathrm{P}_r^n$ and is computed as

$$\mathrm{P}_r^n = \frac{n!}{(n-r)!},$$

where $n! = n(n-1)(n-2)\cdots(3)(2)(1)$ and where $0! \equiv 1$. If the $n$ items are not distinct, then the number of distinct permutations is less than $\mathrm{P}_r^n$.

#### 1.1.1.3 Combinations

The number of ways of dividing $n$ distinct items into $k$ distinct groups with the $i$th group containing $n_i$ items, where $n = \sum_{i=1}^{k} n_i$, is equal to

$$\frac{n!}{n_1! n_2! \cdots n_k!} = \frac{n!}{\left(\prod_{i=1}^{k} n_i!\right)}.$$

The above expression appears in the *multinomial* expansion

$$(x_1 + x_2 + \cdots + x_k)^n = \sum^* \frac{n!}{\left(\prod_{i=1}^{k} n_i!\right)} x_1^{n_1} x_2^{n_2} \cdots x_k^{n_k},$$

where the summation symbol $\sum^*$ indicates summation over all possible values of $n_1, n_2, \ldots, n_k$ with $n_i, i = 1, 2, \ldots, k$, taking the set of possible values $\{0, 1, \ldots, n\}$ subject to the restriction $\sum_{i=1}^{k} n_i = n$.

With $x_1 = x_2 = \cdots = x_k = 1$, it follows that

$$\sum^* \frac{n!}{\left(\prod_{i=1}^{k} n_i!\right)} = k^n.$$

As an important special case, when $k = 2$, then

$$\frac{n!}{n_1!n_2!} = \frac{n!}{n_1!(n - n_1)!} = C_{n_1}^{n},$$

which is also the number of ways of selecting *without replacement* $n_1$ items from a set of $n$ distinct items (i.e., the number of *combinations* of $n$ distinct items selected $n_1$ at a time).

The above combinational expression appears in the *binomial* expansion

$$(x_1 + x_2)^n = \sum^* \frac{n!}{n_1!n_2!} x_1^{n_1} x_2^{n_2} = \sum_{n_1=0}^{n} C_{n_1}^{n} x_1^{n_1} x_2^{n-n_1}.$$

When $x_1 = x_2 = 1$, it follows that

$$\sum_{n_1=0}^{n} C_{n_1}^{n} = 2^n.$$

**Example**

As a simple example using the above counting formulas, if 5 cards are dealt from a well-shuffled standard deck of 52 playing cards, the number of ways in which such a 5-card hand would contain exactly 2 aces is equal to $qs = C_2^4 C_3^{48} = 103,776$, where $q = C_2^4 = 6$ is the number of ways of selecting 2 of the 4 aces and where $s = C_3^{48} = 17,296$ is the number of ways of selecting 3 of the remaining 48 cards.

#### 1.1.1.4 *Pascal's Identity*

$$C_k^n = C_{k-1}^{n-1} + C_k^{n-1}$$

for any positive integers $n$ and $k$ such that $C_k^n \equiv 0$ if $k > n$.

#### 1.1.1.5 Vandermonde's Identity

$$C_r^{m+n} = \sum_{k=0}^{r} C_{r-k}^{m} C_k^{n},$$

where $m, n$, and $r$ are nonnegative integers satisfying $r \leq \min\{m, n\}$.

### 1.1.2 Probability Formulas

#### 1.1.2.1 Definitions

Let an *experiment* be any process via which an observation or measurement is made. An experiment can range from a very controlled experimental situation to an uncontrolled observational situation. An example of the former situation would be a laboratory experiment where chosen amounts of different chemicals are mixed together to produce a certain chemical product. An example of the latter situation would be an epidemiological study where subjects are randomly selected and interviewed about their smoking and physical activity habits.

Let $A_1, A_2, \ldots, A_p$ be $p(\geq 2)$ possible events (or outcomes) that could occur when an experiment is conducted. Then:

1. For $i = 1, 2, \ldots, p$, the *complement* of the event $A_i$, denoted $\overline{A}_i$, is the event that $A_i$ does *not* occur when the experiment is conducted.
2. The *union* of the events $A_1, A_2, \ldots, A_p$, denoted $\cup_{i=1}^{p} A_i$, is the event that *at least* one of the events $A_1, A_2, \ldots, A_p$ occurs when the experiment is conducted.
3. The *intersection* of the events $A_1, A_2, \ldots, A_p$, denoted $\cap_{i=1}^{p} A_i$, is the event that *all* of the events $A_1, A_2, \ldots, A_p$ occur when the experiment is conducted.

Given these definitions, we have the following probabilistic results, where $\text{pr}(A_i), 0 \leq \text{pr}(A_i) \leq 1$, denotes the *probability* that event $A_i$ occurs when the experiment is conducted:

(i) $\text{pr}(\overline{A}_i) = 1 - \text{pr}(A_i)$. More generally,

$$\text{pr}\left(\overline{\cup_{i=1}^{p} A_i}\right) = 1 - \text{pr}\left(\cup_{i=1}^{p} A_i\right) = \text{pr}\left(\cap_{i=1}^{p} \overline{A}_i\right)$$

and

$$\text{pr}\left(\overline{\cap_{i=1}^{p} A_i}\right) = 1 - \text{pr}\left(\cap_{i=1}^{p} A_i\right) = \text{pr}\left(\cup_{i=1}^{p} \overline{A}_i\right).$$

(ii) The probability of the union of $p$ events is given by:

$$\begin{aligned}
\text{pr}\left(\cup_{i=1}^{p} \text{A}_i\right) &= \sum_{i=1}^{p} \text{pr}(\text{A}_i) - \sum_{i=1}^{p-1} \sum_{j=i+1}^{p} \text{pr}(\text{A}_i \cap \text{A}_j) \\
&\quad + \sum_{i=1}^{p-2} \sum_{j=i+1}^{p-1} \sum_{k=j+1}^{p} \text{pr}(\text{A}_i \cap \text{A}_j \cap \text{A}_k) - \cdots \\
&\quad + (-1)^{p-1} \text{pr}\left(\cap_{i=1}^{p} \text{A}_i\right).
\end{aligned}$$

As important special cases, we have, for $p = 2$,

$$\text{pr}(\text{A}_1 \cup \text{A}_2) = \text{pr}(\text{A}_1) + \text{pr}(\text{A}_2) - \text{pr}(\text{A}_1 \cap \text{A}_2)$$

and, for $p = 3$,

$$\begin{aligned}
\text{pr}(\text{A}_1 \cup \text{A}_2 \cup \text{A}_3) &= \text{pr}(\text{A}_1) + \text{pr}(\text{A}_2) + \text{pr}(\text{A}_3) \\
&\quad - \text{pr}(\text{A}_1 \cap \text{A}_2) - \text{pr}(\text{A}_1 \cap \text{A}_3) - \text{pr}(\text{A}_2 \cap \text{A}_3) \\
&\quad + \text{pr}(\text{A}_1 \cap \text{A}_2 \cap \text{A}_3).
\end{aligned}$$

#### 1.1.2.2 Mutually Exclusive Events

For $i \neq j$, two events $\text{A}_i$ and $\text{A}_j$ are said to be *mutually exclusive* if these two events cannot both occur (i.e., cannot occur together) when the experiment is conducted; equivalently, the events $\text{A}_i$ and $\text{A}_j$ are mutually exclusive when $\text{pr}(\text{A}_i \cap \text{A}_j) = 0$. If the $p$ events $\text{A}_1, \text{A}_2, \ldots, \text{A}_p$ are *pairwise* mutually exclusive, that is, if $\text{pr}(\text{A}_i \cap \text{A}_j) = 0$ for every $i \neq j$, then

$$\text{pr}\left(\cup_{i=1}^{p} \text{A}_i\right) = \sum_{i=1}^{p} \text{pr}(\text{A}_i),$$

since pairwise mutual exclusivity implies that any intersection involving more than two events must necessarily have probability zero of occurring.

#### 1.1.2.3 Conditional Probability

For $i \neq j$, the *conditional* probability that event $\text{A}_i$ occurs *given that* (or conditional on the fact that) event $\text{A}_j$ occurs when the experiment is conducted, denoted $\text{pr}(\text{A}_i|\text{A}_j)$, is given by the expression

$$\text{pr}(\text{A}_i|\text{A}_j) = \frac{\text{pr}(\text{A}_i \cap \text{A}_j)}{\text{pr}(\text{A}_j)}, \quad \text{pr}(\text{A}_j) > 0.$$

Using the above definition, we then have:

$$\begin{aligned}
\text{pr}\left(\cap_{i=1}^{p} A_i\right) &= \text{pr}\left(A_p | \cap_{i=1}^{p-1} A_i\right) \text{pr}\left(\cap_{i=1}^{p-1} A_i\right) \\
&= \text{pr}\left(A_p | \cap_{i=1}^{p-1} A_i\right) \text{pr}\left(A_{p-1} | \cap_{i=1}^{p-2} A_i\right) \text{pr}\left(\cap_{i=1}^{p-2} A_i\right) \\
&\vdots \\
&= \text{pr}\left(A_p | \cap_{i=1}^{p-1} A_i\right) \text{pr}\left(A_{p-1} | \cap_{i=1}^{p-2} A_i\right) \cdots \text{pr}(A_2|A_1)\text{pr}(A_1).
\end{aligned}$$

Note that there would be $p!$ ways of writing the above product of $p$ probabilities. For example, when $p = 3$, we have

$$\begin{aligned}
\text{pr}(A_1 \cap A_2 \cap A_3) &= \text{pr}(A_3|A_1 \cap A_2)\text{pr}(A_2|A_1)\text{pr}(A_1) \\
&= \text{pr}(A_2|A_1 \cap A_3)\text{pr}(A_1|A_3)\text{pr}(A_3) \\
&= \text{pr}(A_1|A_2 \cap A_3)\text{pr}(A_3|A_2)\text{pr}(A_2), \text{ and so on.}
\end{aligned}$$

#### 1.1.2.4 Independence

The events $A_i$ and $A_j$ are said to be *independent* events if and only if the following equivalent probability statements are true:

1. $\text{pr}(A_i|A_j) = \text{pr}(A_i)$;
2. $\text{pr}(A_j|A_i) = \text{pr}(A_j)$;
3. $\text{pr}(A_i \cap A_j) = \text{pr}(A_i)\text{pr}(A_j)$.

When the events $A_1, A_2, \ldots, A_p$ are *mutually independent*, so that the conditional probability of any event is equal to the unconditional probability of that same event, then

$$\text{pr}\left(\cap_{i=1}^{p} A_i\right) = \prod_{i=1}^{p} \text{pr}(A_i).$$

#### 1.1.2.5 Partitions and Bayes' Theorem

When $\text{pr}\left(\cup_{i=1}^{p} A_i\right) = 1$, and when the events $A_1, A_2, \ldots, A_p$ are pairwise mutually exclusive, then the events $A_1, A_2, \ldots, A_p$ are said to constitute a *partition* of the experimental outcomes; in other words, when the experiment is conducted, exactly one and only one of the events $A_1, A_2, \ldots, A_p$ must occur. If B is any event and $A_1, A_2, \ldots, A_p$ constitute a partition, it follows that

$$\text{pr}(B) = \text{pr}\left[B \cap \left(\cup_{i=1}^{p} A_i\right)\right] = \text{pr}\left[\cup_{i=1}^{p}(B \cap A_i)\right]$$
$$= \sum_{i=1}^{p} \text{pr}(B \cap A_i) = \sum_{i=1}^{p} \text{pr}(B|A_i)\text{pr}(A_i).$$

As an illustration of the use of the above formula, if the events $A_1, A_2, \ldots, A_p$ represent an exhaustive list of all $p$ possible causes of some observed outcome B, where $\text{pr}(B) > 0$, then, given values for $\text{pr}(A_i)$ and $\text{pr}(B|A_i)$ for all $i = 1, 2, \ldots, p$, one can employ *Bayes' Theorem* to compute the probability that $A_i$ was the cause of the observed outcome B, namely,

$$\text{pr}(A_i|B) = \frac{\text{pr}(A_i \cap B)}{\text{pr}(B)} = \frac{\text{pr}(B|A_i)\text{pr}(A_i)}{\sum_{j=1}^{p} \text{pr}(B|A_j)\text{pr}(A_j)}, \quad i = 1, 2, \ldots, p.$$

Note that $\sum_{i=1}^{p} \text{pr}(A_i|B) = 1$.

As an important special case, suppose that the events $A_1, A_2, \ldots, A_p$ constituting a partition are elementary events in the sense that none of these $p$ events can be further decomposed into smaller events (i.e., for $i = 1, 2, \ldots, p$, the event $A_i$ cannot be written as a union of mutually exclusive events each having a smaller probability than $A_i$ of occurring when the experiment is conducted). Then, any more complex event B (sometimes called a *compound* event) must be able to be represented as the union of two or more of the elementary events $A_1, A_2, \ldots, A_p$. In particular, with $2 \leq m \leq p$,

$$\text{if } B = \cup_{j=1}^{m} A_{i_j},$$

where the set of positive integers $\{i_1, i_2, \ldots, i_m\}$ is a subset of the set of positive integers $\{1, 2, \ldots, p\}$, then

$$\text{pr}(B) = \sum_{j=1}^{m} \text{pr}(A_{i_j}).$$

In the very special case when the elementary events $A_1, A_2, \ldots, A_p$ are *equally likely* to occur, so that $\text{pr}(A_i) = \frac{1}{p}$ for $i = 1, 2, \ldots, p$, then $\text{pr}(B) = \frac{m}{p}$.

**Example**

To continue an earlier example, there would be $p = C_5^{52} = 2{,}598{,}960$ possible 5-card hands that could be dealt from a well-shuffled standard deck of 52 playing cards. Thus, each such 5-card hand has probability $\frac{1}{2{,}598{,}960}$ of occurring. If B is the event that a 5-card hand contains exactly two aces, then

$$\text{pr}(B) = \frac{m}{p} = \frac{103{,}776}{2{,}598{,}960} = 0.0399.$$

## EXERCISES

**Exercise 1.1.** Suppose that a pair of balanced dice is tossed. Let $E_x$ be the event that the sum of the two numbers obtained is equal to $x, x = 2, 3, \ldots, 12$.

(a) Develop an explicit expression for $\text{pr}(E_x)$.

(b) Let A be the event that "$x$ is divisible by 4," let B be the event that "$x$ is greater than 9," and let C be the event that "$x$ is *not* a prime number." Find the numerical values of the following probabilities: pr(A), pr(B), pr(C), $\text{pr}(A \cap B)$, $\text{pr}(A \cap C)$, $\text{pr}(B \cap C)$, $\text{pr}(A \cap B \cap C)$, $\text{pr}(A \cup B \cup C)$, $\text{pr}(A \cup B | C)$, and $\text{pr}(A | B \cup \bar{C})$.

**Exercise 1.2.** For any family in the United States, suppose that the probability of any child being male is equal to 0.50, and that the gender status of any child in a family is unaffected by the gender status of any other child in that same family. What is the minimum number, say $n^*$, of children that any U.S. couple needs to have so that the probability is no smaller than 0.90 of having at least one male child *and* at least one female child?

**Exercise 1.3.** Suppose that there are three urns. Urn 1 contains three white balls and four black balls. Urn 2 contains two white balls and three black balls. And, Urn 3 contains four white balls and two black balls. One ball is randomly selected from Urn 1 and is put into Urn 2. Then, one ball is randomly selected from Urn 2 and is put into Urn 3. Then, two balls are simultaneously selected from Urn 3. Find the exact numerical value of the probability that both balls selected from Urn 3 are white.

**Exercise 1.4.** In the National Scrabble Contest, suppose that the two players in the final match (say, Player A and Player B) play consecutive games, with the national champion being that player who is the first to win five games. Assuming that no game can end in a tie, the two finalists must necessarily play at least 5 games but no more than 9 games. Further, assume (probably somewhat unrealistically) that the outcomes of the games are mutually independent of one another, and also assume that $\pi$ is the probability that Player A wins any particular game.

(a) Find an explicit expression for the probability that the final match between Player A and Player B lasts exactly 6 games.

(b) Given that Player A wins the first two games, find an explicit expression for the probability that Player A wins the final match in exactly 7 games.

(c) Find an explicit expression for the probability that Player B wins the final match.

**Exercise 1.5.** Suppose that there are two different diagnostic tests (say, Test A and Test B) for a particular disease of interest. In a certain large population, suppose that the prevalence of this disease is 1%. Among all those people who have this disease in this large population, 10% will incorrectly test negatively for the presence of the disease when given Test A; and, independently of any results based on Test A, 5% of these diseased people will incorrectly test negatively when given Test B. Among all those people who do *not* have the disease in this large population, 6% will incorrectly test positively when given Test A; and, independently of any results based on Test A, 8% of these nondiseased people will incorrectly test positively when given Test B.

(a) Given that both Tests A and B are positive when administered to a person selected randomly from this population, what is the numerical value of the probability that this person actually has the disease in question?

(b) Given that Test A is positive when administered to a person randomly selected from this population, what is the numerical value of the probability that Test B will also be positive?

(c) Given that a person selected randomly from this population actually has the disease in question, what is the numerical value of the probability that at least one of the two different diagnostic tests given to this particular person will be positive?

**Exercise 1.6.** A certain medical laboratory uses three machines (denoted $M_1$, $M_2$, and $M_3$, respectively) to measure prostate-specific antigen (PSA) levels in blood samples selected from adult males; high PSA levels have been shown to be associated with the presence of prostate cancer. Assume that machine $M_1$ has probability 0.01 of providing an *incorrect* PSA level, that machine $M_2$ has probability 0.02 of providing an *incorrect* PSA level, and that machine $M_3$ has probability 0.03 of providing an *incorrect* PSA level. Further, assume that machine $M_1$ performs 20% of the PSA analyses done by this medical laboratory, that machine $M_2$ performs 50% of the PSA analyses, and that machine $M_3$ performs 30% of the PSA analyses.

(a) Find the numerical value of the probability that a PSA analysis performed by this medical laboratory will be done *correctly*.

(b) Given that a particular PSA analysis is found to be done incorrectly, what is the numerical value of the probability that this PSA analysis was performed either by machine $M_1$ or by machine $M_2$?

(c) Given that two *independent* PSA analyses are performed and that exactly one of these two PSA analyses is found to be correct, find the numerical value of the probability that machine $M_2$ did not perform both of these PSA analyses.

**Exercise 1.7.** Suppose that two medical doctors, denoted Doctor #1 and Doctor #2, each examine a person randomly chosen from a certain population to check for the presence or absence of a particular disease. Let $C_1$ be the event that Doctor #1 makes the correct diagnosis, let $C_2$ be the event that Doctor #2 makes the correct diagnosis, and let D be the event that the randomly chosen patient actually has the disease in question; further, assume that the events $C_1$ and $C_2$ are independent *conditional* on disease status. Finally, let the *prevalence* of the disease in the population be $\theta = \text{pr}(D)$, let $\pi_1 = \text{pr}(C_1|D) = \text{pr}(C_2|D)$, and let $\pi_0 = \text{pr}(C_1|\overline{D}) = \text{pr}(C_2|\overline{D})$.

(a) Develop an explicit expression for $\text{pr}(C_2|C_1)$. Are the events $C_1$ and $C_2$ *unconditionally* independent? Comment on the more general implications of this particular example.

(b) For this particular example, determine specific conditions involving $\theta$, $\pi_0$, and $\pi_1$ such that $\text{pr}(C_2|C_1) = \text{pr}(C_2)$.

**Exercise 1.8.** For a certain state lottery, 5 balls are drawn each day randomly *without replacement* from an urn containing 40 balls numbered individually from 1 to 40. Suppose that there are $k$ (>1) consecutive days of such drawings. Develop an expression for the probability $\pi_k$ that there is *at least* one matching set of 5 numbers in those $k$ drawings.

**Exercise 1.9.** In a certain small city in the United States, suppose that there are $n$ ($\geq 2$) dental offices listed in that city's phone book. Further, suppose that $k$ ($2 \leq k \leq n$) people each independently and randomly call one of these $n$ dental offices for an appointment.

(a) Find the probability $\alpha$ that none of these $k$ people call the same dental office, and then find the numerical value of $\alpha$ when $n = 7$ and $k = 4$.

(b) Find the probability $\beta$ that all of these $k$ people call the same dental office, and then find the numerical value of $\beta$ when $n = 7$ and $k = 4$.

**Exercise 1.10.** Suppose that the positive integers $1, 2, \ldots, k, k \geq 3$, are arranged randomly in a horizontal line, thus occupying $k$ slots. Assume that all arrangements of these $k$ integers are equally likely. For $j = 0, 1, \ldots, (k-2)$, develop an explicit expression for the probability $\theta_j$ that there are exactly $j$ integers between the integers 1 and $k$.

**Exercise 1.11.** Suppose that a balanced die is rolled $n$ ($\geq 6$) times. Find an explicit expression for the probability $\theta_n$ that each of the six numbers $1, 2, \ldots, 6$ appears at least once during the $n$ rolls. Find the numerical value of $\theta_n$ when $n = 10$.

**Exercise 1.12.** An urn contains $N$ balls numbered $1, 2, 3, \ldots, (N-1), N$. A sample of $n$ ($2 \leq n < N$) balls is selected at random *with replacement* from this urn, and the $n$ numbers obtained in this sample are recorded. Derive an explicit expression for the probability that the $n$ numbers obtained in this sample of size $n$ are all different from one another (i.e., no two or more of these $n$ numbers are the same). If $N = 10$ and $n = 4$, what is the numerical value of this probability?

**Exercise 1.13.** Suppose that an urn contains $N$ ($N > 1$) balls, each individually labeled with a number from 1 to $N$, where $N$ is an unknown positive integer.

(a) If $n$ ($2 \leq n < N$) balls are selected one-at-a-time *with replacement* from this urn, find an explicit expression for the probability $\theta_{wr}$ that the ball labelled with the number $N$ is selected.

(b) If $n$ ($2 \leq n < N$) balls are selected one-at-a-time *without replacement* from this urn, find an explicit expression for the probability $\theta_{wor}$ that the ball labelled with the number $N$ is selected.

(c) Use a proof by induction to determine which method of sampling has the higher probability of selecting the ball labeled with the number $N$.

**Exercise 1.14.** A midwestern U.S. city has a traffic system designed to move morning rushhour traffic from the suburbs into this city's downtown area via three tunnels. During any weekday, there is a probability $\theta$ ($0 < \theta < 1$) that there will be inclement weather. Because of the need for periodic maintenance, tunnel $i$ ($i = 1, 2, 3$) has probability $\pi_i$ ($0 < \pi_i < 1$) of being closed to traffic on any weekday. Periodic maintenance activities for any particular tunnel occur independently of periodic maintenance activities for any other tunnel, and all periodic maintenance activities for these three tunnels are performed independently of weather conditions.

The rate of rushhour traffic flow into the downtown area on any weekday is considered to be *excellent* if there is no inclement weather and if all three tunnels are open to traffic. The rate of traffic flow is considered to be *poor* if either: (i) more than one tunnel is closed to traffic; or, (ii) there is inclement weather and exactly one tunnel is closed to traffic. Otherwise, the rate of traffic flow is considered to be *marginal*.

(a) Develop an explicit expression for the probability that exactly one tunnel is closed to traffic.

(b) Develop explicit expressions for the probability that the rate of traffic flow is excellent, for the probability that the rate of traffic flow is marginal, and for the probability that the rate of traffic flow is poor.

(c) Given that a particular weekday has a marginal rate of traffic flow, develop an explicit expression for the conditional probability that this particular weekday of marginal flow is due to inclement weather and not to a tunnel being closed to traffic.

**Exercise 1.15.** Bonnie and Clyde each independently toss the same unbalanced coin and count the number of tosses that it takes each of them to obtain the first head. Assume that the probability of obtaining a head with this unbalanced coin is equal to $\pi, 0 < \pi < 1$, with $\pi \neq \frac{1}{2}$.

(a) Find the probability that Bonnie and Clyde each require the same number of tosses of this unbalanced coin to obtain the first head.

(b) Find the probability that Bonnie will require more tosses of this unbalanced coin than Clyde to obtain the first head.

**Exercise 1.16.** Suppose that 15 senior math majors, 7 males and 8 females, at a major public university in the United States each take the same Graduate Record Examination (GRE) in advanced mathematics. Further, suppose that each of these 15 students has probability $\pi, 0 < \pi < 1$, of obtaining a score that exceeds the 80-th percentile for all scores recorded for that particular examination. Given that exactly 5 of these 15 students scored higher than the 80-th percentile, what is the numerical value of the probability $\theta$ that at least 3 of these 5 students were female?

**Exercise 1.17.** In the popular card game *bridge*, each of four players is dealt a *hand* of 13 cards from a well-shuffled deck of 52 standard playing cards. Find the numerical value of the probability that any randomly dealt hand of 13 cards contains all three face cards of the same suit, where a face card is a jack, a queen, or a king; note that it is possible for a hand of 13 cards to contain all three face cards in at least two different suits.

**Exercise 1.18*.** In the game known as "craps," a dice game played in casinos all around the world, a player competes against the casino (called "the house") according to the following rules. If the player (called "the shooter" when rolling the dice) rolls either a 7 or an 11 on the first roll of the pair of dice, the player wins the game (and the house, of course, loses the game); if the player rolls either 2, 3, or 12 on the first roll, the player loses the game (and the house, of course, wins the game). If the player rolls any of the remaining numbers 4, 5, 6, 8, 9, or 10 on the first roll (such a number is called "the point"), the player keeps rolling the pair of dice until either the point is rolled again or until a 7 is rolled. If the point (e.g., 4) is rolled before a 7 is rolled, the player wins the game; if a 7 is rolled before the point (e.g., 4) is rolled, the player loses the game. Find the exact numerical value of the probability that the player wins the game.

**Exercise 1.19*.** In a certain chemical industry, suppose that a proportion $\pi_h$ $(0 < \pi_h < 1)$ of all workers is exposed to a high daily concentration level of a certain potential carcinogen, that a proportion $\pi_m$ $(0 < \pi_m < 1)$ of all workers is exposed to a moderate daily concentration level, that a proportion $\pi_l$ $(0 < \pi_l < 1)$ of all workers is exposed to a low daily concentration level, and that a proportion $\pi_o$ $(0 < \pi_o < 1)$ of all workers receives no exposure to this potential carcinogen. Note that $(\pi_h + \pi_m + \pi_l + \pi_o) = 1$. Suppose that $n$ workers in this chemical industry are randomly selected. Let $\theta_n$ be the probability that an *even number* of highly exposed workers is included in this randomly selected sample of $n$ workers, where 0 is considered to be an even number.

(a) Find a *difference equation* of the form $\theta_n = f(\pi_h, \theta_{n-1})$ that expresses $\theta_n$ as a function of $\pi_h$ and $\theta_{n-1}$, where $\theta_0 \equiv 1$.

(b) Assuming that a solution to this difference equation is of the form $\theta_n = \alpha + \beta\gamma^n$, find an explicit solution for this difference equation (i.e., find specific values for $\alpha, \beta$, and $\gamma$), and then compute the numerical value of $\theta_{50}$ when $\pi_h = 0.05$.

**Exercise 1.20*.** In epidemiological research, a *follow-up study* involves enrolling randomly selected disease-free subjects with different sets of values of known or suspected risk factors for a certain disease of interest and then following these subjects for a specified period of time to investigate how these risk factors are related to the risk of disease development (i.e., to the probability of developing the disease of interest).

A model often used to relate a (row) vector of $k$ risk factors $\boldsymbol{x}' = (x_1, x_2, \ldots, x_k)$ to the probability of developing the disease of interest, where D is the event that a person develops the disease of interest, is the *logistic model*

$$\text{pr}(\text{D}|\boldsymbol{x}) = \left[1 + \text{e}^{-(\beta_0 + \sum_{j=1}^{k} \beta_j x_j)}\right]^{-1} = \left[1 + \text{e}^{-(\beta_0 + \boldsymbol{\beta}'\boldsymbol{x})}\right]^{-1} = \frac{\text{e}^{\beta_0 + \boldsymbol{\beta}'\boldsymbol{x}}}{1 + \text{e}^{\beta_0 + \boldsymbol{\beta}'\boldsymbol{x}}},$$

where the intercept $\beta_0$ and the (row) vector $\boldsymbol{\beta}' = (\beta_1, \beta_2, \ldots, \beta_k)$ constitute a set of $(k+1)$ regression coefficients.

For certain rare chronic diseases like cancer, a follow-up study can take many years to yield valid and precise statistical conclusions because of the length to time required for sufficient numbers of disease-free subjects to develop the disease. Because of this limitation of follow-up studies for studying the potential causes of rare chronic diseases, epidemiologists developed the *case–control study*. In a case–control study, random samples of *cases* (i.e., subjects who have the disease of interest) and *controls* (i.e., subjects who do not have the disease of interest) are asked to provide information about their values of the risk factors $x_1, x_2, \ldots, x_k$. One problem with this *outcome-dependent* sampling design is that statistical models for the risk of disease will now depend on the probabilities of selection into the study for both cases and controls.

More specifically, let S be the event that a subject is selected to participate in a case–control study. Then, let

$$\pi_1 = \text{pr}(\text{S}|\text{D}, \boldsymbol{x}) = \text{pr}(\text{S}|\text{D}) \quad \text{and} \quad \pi_0 = \text{pr}(\text{S}|\bar{\text{D}}, \boldsymbol{x}) = \text{pr}(\text{S}|\bar{\text{D}})$$

be the probabilities of selection into the study for cases and controls, respectively, where it is assumed that these selection probabilities do *not* depend on $\boldsymbol{x}$.

(a) Assuming the logistic model for $\text{pr}(\text{D}|\boldsymbol{x})$ given above, show that the risk of disease development for a case–control study, namely $\text{pr}(\text{D}|\text{S}, \boldsymbol{x})$, can be written as a logistic model, but with an intercept that functionally depends on $\pi_1$ and $\pi_0$. Comment on this finding with regard to using a case–control study to estimate disease risk as a function of $\boldsymbol{x}$.

(b) The *risk odds ratio* comparing the odds of disease for a subject with the set of risk factors $(\boldsymbol{x}^*)' = (x_1^*, x_2^*, \ldots, x_k^*)$ to the odds of disease for a subject with the set of risk factors $\boldsymbol{x}' = (x_1, x_2, \ldots, x_k)$ is defined as

$$\theta_{\text{r}} = \frac{\text{pr}(\text{D}|\boldsymbol{x}^*)/\text{pr}(\bar{\text{D}}|\boldsymbol{x}^*)}{\text{pr}(\text{D}|\boldsymbol{x})/\text{pr}(\bar{\text{D}}|\boldsymbol{x})}.$$

Show that

$$\theta_{\text{r}} = \text{e}^{\boldsymbol{\beta}'(\boldsymbol{x}^* - \boldsymbol{x})},$$

and then show that the risk odds ratio expression for a case–control study, namely,

$$\theta_{\text{c}} = \frac{\text{pr}(\text{D}|\text{S}, \boldsymbol{x}^*)/\text{pr}(\bar{\text{D}}|\text{S}, \boldsymbol{x}^*)}{\text{pr}(\text{D}|\text{S}, \boldsymbol{x})/\text{pr}(\bar{\text{D}}|\text{S}, \boldsymbol{x})},$$

is also equal to $\text{e}^{\boldsymbol{\beta}'(\boldsymbol{x}^* - \boldsymbol{x})}$. Finally, interpret these results with regard to the utility of case–control studies for epidemiological research.

**Exercise 1.21*.** In a certain population of adults, the prevalence of inflammatory bowl disease (IBD) is $\theta, 0 < \theta < 1$. Suppose that three medical doctors each independently examine the same adult (randomly selected from this population) to determine whether or not this adult has IBD. Further, given that this adult has IBD, suppose that each of the three doctors has probability $\pi_1, 0 < \pi_1 < 1$, of making the correct diagnosis that this adult does have IBD; and, given that this adult does not have IBD, suppose that each of the three doctors has probability $\pi_0, 0 < \pi_0 < 1$, of making the correct diagnosis that this adult does not have IBD.

Consider the following two diagnostic strategies:

*Diagnostic Strategy* #1: The diagnosis is based on the *majority* opinion of the three doctors;

*Diagnostic Strategy* #2: One of the three doctors is randomly chosen and the diagnosis is based on the opinion of just that one doctor.

(a) Find ranges of values for $\pi_1$ and $\pi_0$ that jointly represent a sufficient condition for which Diagnostic Strategy #1 has a higher probability than Diagnostic Strategy #2 of providing the correct diagnosis. Comment on your findings.

(b) Under the stated assumptions, suppose a fourth doctor's opinion is solicited. Would it be better to make a diagnosis based on the *majority* opinion of four doctors (call this Diagnostic Strategy #3) rather than on the *majority* opinion of three doctors (i.e., Diagnostic Strategy #1)? Under Diagnostic Strategy #3, note that no diagnosis will be made if two doctors claim that the adult has IBD and the other two doctors claim that the adult does not have IBD.

**Exercise 1.22*.** Consider the following three events:

D: an individual has Alzheimer's Disease;

E: an individual has diabetes;

M: an individual is male.

And, consider the following list of conditional probabilities:

$$\pi_{11} = \text{pr}(\text{D}|\text{E} \cap \text{M}), \pi_{10} = \text{pr}(\text{D}|\text{E} \cap \bar{\text{M}}), \pi_{01} = \text{pr}(\text{D}|\bar{\text{E}} \cap \text{M}),$$

$$\pi_{00} = \text{pr}(\text{D}|\bar{\text{E}} \cap \bar{\text{M}}), \pi_{1} = \text{pr}(\text{D}|\text{E}), \text{ and } \pi_{0} = \text{pr}(\text{D}|\bar{\text{E}}).$$

The *risk ratio* comparing the risk of Alzheimer's Disease for a diabetic to that for a nondiabetic among males is equal to

$$RR_1 = \frac{\pi_{11}}{\pi_{01}};$$

the *risk ratio* comparing the risk of Alzheimer's Disease for a diabetic to that for a nondiabetic among females is equal to

$$RR_0 = \frac{\pi_{10}}{\pi_{00}};$$

and, the *crude* risk ratio *ignoring gender status* that compares the risk of Alzheimer's Disease for a diabetic to that for a nondiabetic is equal to

$$RR_c = \frac{\pi_1}{\pi_0}.$$

Assuming that $RR_1 = RR_0 = RR$ (i.e., there is *homogeneity* [or equality] of the risk ratio across gender groups), then gender status is said to be a *confounder* of the true association between diabetes and Alzheimer's Disease when $RR_c \neq RR$.

Under this homogeneity assumption, find two *sufficient* conditions for which gender status will *not* be a confounder of the true association between diabetes and Alzheimer's Disease; that is, find two sufficient conditions for which $RR_c = RR$.

**Exercise 1.23*.** Consider a diagnostic test which is being used to diagnose the presence ($\text{D}_1$) or absence ($\bar{\text{D}}_1$) of some particular disease in a population with pretest probability (or prevalence) of this particular disease equal to $\text{pr}(\text{D}_1) = \pi_1$; also, let $\text{pr}(\bar{\text{D}}_1) = 1 - \pi_1 = \pi_2$. Further, let $\theta_1 = \text{pr}(\text{T}^+|\text{D}_1)$ and $\theta_2 = \text{pr}(\text{T}^+|\bar{\text{D}}_1)$ where $\text{T}^+$ denotes the event that the diagnostic test is positive (i.e., the diagnostic test indicates the presence of the disease in question).

(a) Given that the diagnostic test is positive, prove that the posttest odds of an individual having, versus not having, the disease in question is given by the formula

$$\frac{\text{pr}(D_1|T^+)}{\text{pr}(\bar{D}_1|T^+)} = \left(\frac{\theta_1}{\theta_2}\right)\left(\frac{\pi_1}{\pi_2}\right) = LR_{12}\left(\frac{\pi_1}{\pi_2}\right)$$

where $LR_{12} = \theta_1/\theta_2$ is the so-called *likelihood ratio* for the diagnostic test and where $\pi_1/\pi_2$ is the pretest odds of the individual having, versus not having, the disease in question. Hence, knowledge of the likelihood ratio for a diagnostic test permits a simple conversion from pretest odds to posttest odds (Birkett, 1988).

(b) Now, suppose we wish to diagnose an individual as having one of three mutually exclusive diseases (i.e., the patient is assumed to have exactly one, but only one, of the three diseases in question). Thus, generalizing the notation in part (a), we have $\sum_{i=1}^{3} \pi_i = 1$, where $\text{pr}(D_i) = \pi_i$ is the pretest probability of having disease $i$, $i = 1, 2, 3$. With $\theta_i = \text{pr}(T^+|D_i)$, $i = 1, 2, 3$, prove that

$$\frac{\text{pr}(D_1|T^+)}{\text{pr}(\bar{D}_1|T^+)} = \left[\sum_{i=2}^{3}\left(\frac{\pi_1}{\pi_i}LR_{1i}\right)^{-1}\right]^{-1},$$

where $LR_{1i} = \theta_1/\theta_i$, $i = 2, 3$. Further, prove that the posttest probability of having disease 1 is

$$\text{pr}(D_1|T^+) = \left[1 + \sum_{i=2}^{3}\left(\frac{\pi_1}{\pi_i}LR_{1i}\right)^{-1}\right]^{-1}.$$

(c) As a numerical example, consider an emergency room physician attending a patient presenting with acute abdominal pain. This physician is considering the use of a new diagnostic test which will be employed to classify patients into one of three mutually exclusive categories: non-specific abdominal pain (NS), appendicitis (A), or cholecystitis (C). The published paper describing this new diagnostic test reports that a positive test result gives a likelihood ratio for diagnosing NS versus A of 0.30, namely, $\text{pr}(T^+|NS)/\text{pr}(T^+|A) = 0.30$. Also, the likelihood ratio for diagnosing NS versus C is 0.50, and the likelihood ratio for diagnosing A versus C is 1.67. In addition, a study of a very large number of patients seen in emergency rooms revealed that the pre-test probabilities for the three diseases were $\text{pr}(NS) = 0.57$, $\text{pr}(A) = 0.33$, and $\text{pr}(C) = 0.10$. Using all this information, calculate for each of these three diseases, the posttest odds and the posttest probability of disease. Based on your numerical results, what is the most likely diagnosis (NS, A, or C) for an emergency room patient with a positive test result based on the use of this particular diagnostic test?

**Exercise 1.24***. In medicine, it is often of interest to assess whether two distinct diseases (say, disease A and disease B) tend to occur together. The odds ratio parameter $\psi$ is defined as

$$\psi = \frac{\text{pr}(A|B)/\text{pr}(\bar{A}|B)}{\text{pr}(A|\bar{B})/\text{pr}(\bar{A}|\bar{B})},$$

and serves as one statistical measure of the tendency for diseases A and B to occur together. An observed value of $\psi$ significantly greater than 1 may suggest that diseases

A and B have a common etiology, which could lead to better understanding of disease processes and, ultimately, to prevention.

Suppose, however, that the diagnosis of the presence of diseases A and B involves the presence of a third factor (say, C). An example would be where a person with an abnormally high cholesterol level would then be evaluated more closely for evidence of both ischemic heart disease and hypothyroidism. In such a situation, one is actually considering the odds ratio

$$\psi_C = \frac{\text{pr}(A \cap C|B \cap C)/\text{pr}(\bar{A} \cap C|B \cap C)}{\text{pr}(A \cap C|\bar{B} \cap C)/\text{pr}(\bar{A} \cap C|\bar{B} \cap C)},$$

which is a measure of the association between diseases A and B each observed *simultaneously* with factor C.

(a) Show that $\psi$ and $\psi_C$ are related by the equation

$$\psi_C = \psi \left[\frac{\text{pr}(C|A \cap B)\text{pr}(C|\bar{A} \cap \bar{B})}{\text{pr}(C|\bar{A} \cap B)\text{pr}(C|A \cap \bar{B})}\right]\left[1 + \frac{\text{pr}(\bar{C})}{\text{pr}(\bar{A} \cap \bar{B} \cap C)}\right].$$

(b) If A, B, and C actually occur completely independently of one another, how are $\psi$ and $\psi_C$ related? Comment on the direction of the bias when using $\psi_C$ instead of $\psi$ as the measure of association between diseases A and B.

**Exercise 1.25*.** Suppose that a certain process generates a sequence of $(s + t)$ outcomes of two types, say, $s$ successes (denoted as S's) and $t$ failures (denoted as F's). A *run* is a *subsequence* of outcomes of the same type which is both preceded and succeeded by outcomes of the opposite type or by the beginning or by the end of the complete sequence. For example, consider the sequence

SSFSSSFSFSFFS

of $s = 8$ successes and $t = 5$ failures. When rewritten as

SS|F|SSS|F|S|F|S|FF|S,

it is clear that this particular sequence contains a *total* of nine runs, namely, five S runs (three of length 1, one of length 2, and one of length 3) and four F runs (three of length 1 and one of length 2).

Since the S runs and F runs alternate in occurrence, the number of S runs differs by at most one from the number of F runs.

(a) Assuming that all possible sequences of $s$ successes and $t$ failures are equally likely to occur, derive an expression for the probability $\pi_x$ that any sequence contains a *total* of exactly $x$ runs. HINT: Consider separately the two situations where $x$ is an even positive integer and where $x$ is an odd positive integer.

(b) For each year over a 7-year period of time, a certain cancer treatment center recorded the percentage of pancreatic cancer patients who survived at least 5 years following treatment involving both surgery and chemotherapy. For each of the seven years, let the event S be the event that the survival percentage is at least 20%, and let the event $F = \bar{S}$. Suppose that the following sequence (ordered chronologically) is observed:

FFSFSSS.

Does this observed sequence provide evidence of a nonrandom pattern of 5-year survival percentages over the 7-year period of time?
For additional information about the theory of runs, see Feller (1968).

**Exercise 1.26***. Consider the following experiment designed to examine whether a human subject has extra-sensory perception (ESP). A set of $R$ ($R > 2$) chips, numbered individually from 1 to $R$, is arranged in random order by an examiner, and this random order cannot be seen by the subject under study. Then, the subject is given an identical set of $R$ chips and is asked to arrange them in exactly the same order as the random order constructed by the experimenter.

(a) Develop an expression for the probability $\theta(0, R)$ that the subject has no chips in their correct positions (i.e., in positions corresponding to the chip positions constructed by the experimenter). Also, find the limiting value of $\theta(0, R)$ as $R \to \infty$, and then comment on your finding.

(b) For $r = 0, 1, 2, \ldots, R$, use the result in part (a) to develop an expression for the probability $\theta(r, R)$ that the subject has exactly $r$ out of $R$ chips in their correct positions.

(c) Assuming that $R = 5$, what is the probability that the subject places at least 3 chips in their correct positions?

**Exercise 1.27***. Suppose that two players (denoted Player A and Player B) play a game where they alternate flipping a balanced coin, with the winner of the game being the first player to obtain $k$ heads (where $k$ is a known positive integer).

(a) With $a$ and $b$ being positive integers, let $(a, b, \text{A})$ denote that specific game where Player A needs $a$ heads to win, where Player B needs $b$ heads to win, and where it is Player A's turn to flip the balanced coin. Similarly, let $(a, b, \text{B})$ denote that specific game where Player A needs $a$ heads to win, where Player B needs $b$ heads to win, and where it is Player B's turn to flip the balanced coin. Also, let $\pi(a, b, \text{A})$ be the probability that Player A wins game $(a, b, \text{A})$, and let $\pi(a, b, \text{B})$ be the probability that Player A wins game $(a, b, \text{B})$. Show that

$$\pi(a, b, \text{A}) = \left(\frac{2}{3}\right) \pi(a-1, b, \text{B}) + \left(\frac{1}{3}\right) \pi(a, b-1, \text{A}),$$

and that

$$\pi(a, b, \text{B}) = \left(\frac{1}{3}\right) \pi(a-1, b, \text{B}) + \left(\frac{2}{3}\right) \pi(a, b-1, \text{A}).$$

(b) Assuming that Player A goes first in any game, find the exact numerical values of the probabilities that A wins the game when $k = 2$ and when $k = 3$. In other words, find the exact numerical values of $\pi(2, 2, \text{A})$ and $\pi(3, 3, \text{A})$.

**Exercise 1.28*.** The first author (LLK) has been a University of North Carolina (UNC) Tar Heel basketball fan for close to 50 years. This exercise is dedicated to LLK's alltime favorite Tar Heel basketball player, Tyler Hansbrough; Tyler is the epitome of a student-athlete, and he led the Tar Heels to the 2009 NCAA Division I men's basketball national championship. During his 4-year career, Tyler also set numerous UNC, ACC, and NCAA individual records.

In the questions to follow, assume that Tyler has a fixed probability $\pi, 0 < \pi < 1$, of making any particular free throw, and also assume that the outcome (i.e., either a make or a miss) for any one free throw is independent of the outcome for any other free throw.

(a) Given that Tyler starts shooting free throws, derive a general expression (as a function of $\pi, a$, and $b$) for the probability $\theta(\pi, a, b)$ that Tyler makes $a$ consecutive free throws before he misses $b$ consecutive free throws, where $a$ and $b$ are positive integers. For his 4-year career at UNC, Tyler's value of $\pi$ was 0.791; using this value of $\pi$, compute the numerical value of the probability that Tyler makes 10 consecutive free throws before he misses two consecutive free throws.

HINT: Let $A_{ab}$ be the event that Tyler makes $a$ consecutive free throws before he misses $b$ consecutive free throws, let $B_a$ be the event that Tyler makes the first $a$ free throws that he attempts, and let $C_b$ be the event that Tyler misses the first $b$ free throws that he attempts. Express $\alpha = \text{pr}(A_{ab}|B_1)$ as a function of both $\pi$ and $\beta = \text{pr}(A_{ab}|\bar{B}_1)$, express $\beta$ as a function of both $\pi$ and $\alpha$, and then use the fact that $\theta(\pi, a, b) = \text{pr}(A_{ab}) = \pi\alpha + (1 - \pi)\beta$.

(b) Find the value of $\theta(\pi, a, b)$ when both $\pi = 0.50$ and $a = b$; also, find the value of $\theta(\pi, a, b)$ when $a = b = 1$. For these two special cases, do these answers make sense? Also, comment on the reasonableness of any assumptions underlying the development of the expression for $\theta(\pi, a, b)$.

(c) If Tyler continues to shoot free throws indefinitely, show that he must eventually either make $a$ consecutive free throws or miss $b$ consecutive free throws.

## SOLUTIONS

### Solution 1.1

(a) Let $D_{ij}$ be the event that die #1 shows the number $i$ *and* that die #2 shows the number $j, i = 1, 2, \ldots, 6$ and $j = 1, 2, \ldots, 6$. Clearly, these 36 events form the finest partition of the set of possible experimental outcomes. Thus, it follows that $\text{pr}(E_x) = \sum^* \text{pr}(D_{ij})$, where $\text{pr}(D_{ij}) = \frac{1}{36}$ for all $i$ and $j$, and where $\sum^*$ indicates summation over all $(i, j)$ pairs for which $(i + j) = x$.

For example,

$$\text{pr}(E_6) = \text{pr}(D_{15}) + \text{pr}(D_{51}) + \text{pr}(D_{24}) + \text{pr}(D_{42}) + \text{pr}(D_{33}) = \frac{5}{36}.$$

In general,

$$\text{pr}(E_x) = \frac{\min\{(x-1), (13-x)\}}{36}, \quad x = 2, 3, \ldots, 12.$$

(b) Note that

$$A = E_4 \cup E_8 \cup E_{12}, \quad B = E_{10} \cup E_{11} \cup E_{12}, \quad \text{and}$$
$$C = E_4 \cup E_6 \cup E_8 \cup E_9 \cup E_{10} \cup E_{12},$$

so that $\bar{C} = E_2 \cup E_3 \cup E_5 \cup E_7 \cup E_{11}$.

So, it follows directly that $\text{pr}(A) = \frac{1}{4}, \text{pr}(B) = \frac{1}{6}$, and $\text{pr}(C) = \frac{7}{12}$. Also, $A \cap B = E_{12}$, so that $\text{pr}(A \cap B) = \frac{1}{36}$; $A \cap C = E_4 \cup E_8 \cup E_{12}$, so that $\text{pr}(A \cap C) = \frac{1}{4}$; $B \cap C = E_{10} \cup E_{12}$, so that $\text{pr}(B \cap C) = \frac{1}{9}$; $A \cap B \cap C = E_{12}$, so that $\text{pr}(A \cap B \cap C) = \frac{1}{36}$; and, $A \cup B \cup C = E_4 \cup E_6 \cup E_8 \cup E_9 \cup E_{10} \cup E_{11} \cup E_{12}$, so that $\text{pr}(A \cup B \cup C) = \frac{23}{36}$.

Also,

$$\begin{aligned}\text{pr}(A \cup B|C) &= \text{pr}(A|C) + \text{pr}(B|C) - \text{pr}(A \cap B|C) \\ &= \frac{\text{pr}(A \cap C)}{\text{pr}(C)} + \frac{\text{pr}(B \cap C)}{\text{pr}(C)} - \frac{\text{pr}(A \cap B \cap C)}{\text{pr}(C)} \\ &= \frac{\frac{1}{4}}{\frac{7}{12}} + \frac{\frac{1}{9}}{\frac{7}{12}} - \frac{\frac{1}{36}}{\frac{7}{12}} = \frac{4}{7}.\end{aligned}$$

Finally,

$$\begin{aligned}\text{pr}(A|B \cup \bar{C}) &= \frac{\text{pr}[A \cap (B \cup \bar{C})]}{\text{pr}(B \cup \bar{C})} = \frac{\text{pr}[(A \cap B) \cup (A \cap \bar{C})]}{\text{pr}(B \cup \bar{C})} \\ &= \frac{\text{pr}(A \cap B) + \text{pr}(A \cap \bar{C}) - \text{pr}(A \cap B \cap \bar{C})}{\text{pr}(B) + \text{pr}(\bar{C}) - \text{pr}(B \cap \bar{C})}.\end{aligned}$$

Since $\text{pr}(A \cap \bar{C}) = \text{pr}(A \cap B \cap \bar{C}) = 0$ and since $\text{pr}(B \cap \bar{C}) = \text{pr}(E_{11}) = \frac{2}{36}$, we obtain

$$\text{pr}(A|B \cup \bar{C}) = \frac{\text{pr}(A \cap B)}{\text{pr}(B) + \text{pr}(\bar{C}) - \text{pr}(B \cap \bar{C})} = \frac{\frac{1}{36}}{\left(\frac{1}{6} + \frac{5}{12} - \frac{2}{36}\right)} = \frac{1}{19}.$$

**Solution 1.2.** Let $\theta_n$ be the probability that a family with $n$ children has at least one male child *and* at least one female child among these $n$ children. Further, let $M_n$ be the event that all $n$ children are male, and let $F_n$ be the event that all $n$ children are female. And, note that the events $M_n$ and $F_n$ are mutually exclusive. Then,

$$\begin{aligned}\theta_n &= 1 - \text{pr}(M_n \cup F_n) = 1 - \text{pr}(M_n) - \text{pr}(F_n) \\ &= 1 - \left(\frac{1}{2}\right)^n - \left(\frac{1}{2}\right)^n = 1 - \left(\frac{1}{2}\right)^{n-1}.\end{aligned}$$

So, we need to find the smallest value of $n$, say $n^*$, such that

$$\theta_n = 1 - \left(\frac{1}{2}\right)^{n-1} \geq 0.90.$$

It then follows that $n^* = 5$.

**Solution 1.3.** Define the following events:

$W_1$: a white ball is selected from Urn 1;

$W_2$: a white ball is selected from Urn 2;

$W_3$: two white balls are selected from Urn 3;

$B_1$: a black ball is selected from Urn 1;

$B_2$: a black ball is selected from Urn 2.

Then,

$$\begin{aligned}
\text{pr}(W_3) &= \text{pr}(W_1 \cap W_2 \cap W_3) + \text{pr}(W_1 \cap B_2 \cap W_3) \\
&\quad + \text{pr}(B_1 \cap W_2 \cap W_3) + \text{pr}(B_1 \cap B_2 \cap W_3) \\
&= \text{pr}(W_1)\text{pr}(W_2|W_1)\text{pr}(W_3|W_1 \cap W_2) + \text{pr}(W_1)\text{pr}(B_2|W_1) \\
&\quad \times \text{pr}(W_3|W_1 \cap B_2) + \text{pr}(B_1)\text{pr}(W_2|B_1)\text{pr}(W_3|B_1 \cap W_2) \\
&\quad + \text{pr}(B_1)\text{pr}(B_2|B_1)\text{pr}(W_3|B_1 \cap B_2) \\
&= (3/7)(3/6)[(5/7)(4/6)] + (3/7)(3/6)[(4/7)(3/6)] \\
&\quad + (4/7)(2/6)[(5/7)(4/6)] + (4/7)(4/6)[(4/7)(3/6)] = 0.3628.
\end{aligned}$$

**Solution 1.4**

(a) pr(final match lasts exactly 6 games) = pr[(Player A wins 4 of first 5 games) $\cap$ (Player A wins sixth game)] + pr[(Player B wins 4 of first 5 games) $\cap$ (Player B wins sixth game)] = $\left[C_4^5 \pi^4 (1-\pi)\right](\pi) + \left[C_4^5 (1-\pi)^4 \pi\right](1-\pi)$.

(b) pr[(Player A wins match in 7 games)|(Player A wins first 2 games)]

$$= \frac{\text{pr[(Player A wins match in 7 games)} \cap \text{(Player A wins first 2 games)]}}{\text{pr(Player A wins first 2 games)}}$$

Since

$$\begin{aligned}
&\text{pr[(Player A wins match in 7 games)} \cap \text{(Player A wins first 2 games)]} \\
&\quad = \text{pr[(Player A wins two of games \#3 through \#6)} \\
&\qquad \cap \text{(Player A wins game \#7)} \cap \text{(Player A wins first 2 games)]} \\
&\quad = \text{pr(Player A wins two of games \#3 through \#6)} \\
&\qquad \times \text{pr(Player A wins game \#7)pr(Player A wins first 2 games),}
\end{aligned}$$

it follows that

$$\begin{aligned}
&\text{pr[(Player A wins match in 7 games)|(Player A wins first 2 games)]} \\
&\quad = \text{pr(Player A wins two of games \#3 through \#6)} \\
&\qquad \times \text{pr(Player A wins game \#7)} \\
&\quad = \left[C_2^4 \pi^2 (1-\pi)^2\right](\pi) = C_2^4 \pi^3 (1-\pi)^2.
\end{aligned}$$

(c)

$$\begin{aligned}
&\text{pr(Player B wins final match)}\\
&= \text{pr}[\cup_{j=5}^{9}(\text{Player B wins match in } j \text{ games})]\\
&= \sum_{j=5}^{9} \text{pr}[(\text{Player B wins 4 of first}(j-1)\text{games})\\
&\quad \cap (\text{Player B wins } j\text{th game})]\\
&= \sum_{j=5}^{9} [C_4^{j-1}(1-\pi)^4\pi^{j-5}](1-\pi)\\
&= \left(\frac{1-\pi}{\pi}\right)^5 \sum_{j=5}^{9} C_4^{j-1}\pi^j.
\end{aligned}$$

**Solution 1.5**

(a) Define the following events:

D: "a person has the disease of interest"

$A^+$: "Test A is positive"

$B^+$: "Test B is positive"

Then,

$$\begin{aligned}
\text{pr}(D) &= 0.01\\
\text{pr}(A^+|D) &= 1 - 0.10 = 0.90,\\
\text{pr}(B^+|D) &= 1 - 0.05 = 0.95,\\
\text{pr}(A^+|\bar{D}) &= 0.06,
\end{aligned}$$

and

$$\text{pr}(B^+|\bar{D}) = 0.08.$$

So,

$$\begin{aligned}
&\text{pr}(D|A^+ \cap B^+)\\
&= \frac{\text{pr}(D \cap A^+ \cap B^+)}{\text{pr}(A^+ \cap B^+)}\\
&= \frac{\text{pr}(A^+ \cap B^+|D)\text{pr}(D)}{\text{pr}(A^+ \cap B^+|D)\text{pr}(D) + \text{pr}(A^+ \cap B^+|\bar{D})\text{pr}(\bar{D})}
\end{aligned}$$

$$= \frac{\text{pr}(A^+|D)\text{pr}(B^+|D)\text{pr}(D)}{\text{pr}(A^+|D)\text{pr}(B^+|D)\text{pr}(D) + \text{pr}(A^+|\bar{D})\text{pr}(B^+|\bar{D})\text{pr}(\bar{D})}$$
$$= \frac{(0.90)(0.95)(0.01)}{(0.90)(0.95)(0.01) + (0.06)(0.08)(0.99)}$$
$$= \frac{0.0086}{0.0086 + 0.0048}$$
$$= 0.6418.$$

(b)

$$\text{pr}(B^+|A^+) = \frac{\text{pr}(A^+ \cap B^+)}{\text{pr}(A^+)}$$
$$= \frac{\text{pr}(A^+ \cap B^+ \cap D) + \text{pr}(A^+ \cap B^+ \cap \bar{D})}{\text{pr}(A^+ \cap D) + \text{pr}(A^+ \cap \bar{D})}$$
$$= \frac{\text{pr}(A^+|D)\text{pr}(B^+|D)\text{pr}(D) + \text{pr}(A^+|\bar{D})\text{pr}(B^+|\bar{D})\text{pr}(\bar{D})}{\text{pr}(A^+|D)\text{pr}(D) + \text{pr}(A^+|\bar{D})\text{pr}(\bar{D})}$$
$$= \frac{(0.90)(0.95)(0.01) + (0.06)(0.08)(0.99)}{(0.90)(0.01) + (0.06)(0.99)}$$
$$= \frac{0.0086 + 0.0048}{0.0090 + 0.0594} = 0.1959.$$

(c)

$$\text{pr}(A^+ \cup B^+|D)$$
$$= \frac{\text{pr}[(A^+ \cup B^+) \cap D]}{\text{pr}(D)}$$
$$= \frac{\text{pr}[(A^+ \cap D) \cup (B^+ \cap D)]}{\text{pr}(D)}$$
$$= \frac{\text{pr}(A^+ \cap D) + \text{pr}(B^+ \cap D) - \text{pr}(A^+ \cap B^+ \cap D)}{\text{pr}(D)}$$
$$= \frac{\text{pr}(A^+|D)\text{pr}(D) + \text{pr}(B^+|D)\text{pr}(D) - \text{pr}(A^+|D)\text{pr}(B^+|D)\text{pr}(D)}{\text{pr}(D)}$$
$$= \frac{(0.90)(0.01) + (0.95)(0.01) - (0.90)(0.95)(0.01)}{0.01}$$
$$= \frac{0.0090 + 0.0095 - 0.0086}{0.01} = \frac{0.0099}{0.01} = 0.9900.$$

**Solution 1.6**

(a) For $i = 1, 2, 3$, let $M_i$ be the event that "machine $M_i$ performs the PSA analysis"; and, let C be the event that "the PSA analysis is done correctly." Then,

$$\text{pr}(C) = \text{pr}(C \cap M_1) + \text{pr}(C \cap M_2) + \text{pr}(C \cap M_3)$$

$$= \text{pr}(\text{C}|\text{M}_1)\text{pr}(\text{M}_1) + \text{pr}(\text{C}|\text{M}_2)\text{pr}(\text{M}_2) + \text{pr}(\text{C}|\text{M}_3)\text{pr}(\text{M}_3)$$
$$= 0.99(0.20) + (0.98)(0.50) + 0.97(0.30) = 0.979.$$

(b)

$$\text{pr}(\text{M}_1 \cup \text{M}_2|\bar{\text{C}}) = \frac{\text{pr}[(\text{M}_1 \cup \text{M}_2) \cap \bar{\text{C}}]}{\text{pr}(\bar{\text{C}})} = \frac{\text{pr}[(\text{M}_1 \cap \bar{\text{C}}) \cup (\text{M}_2 \cap \bar{\text{C}})]}{1 - \text{pr}(\text{C})}$$
$$= \frac{\text{pr}(\text{M}_1 \cap \bar{\text{C}}) + \text{pr}(\text{M}_2 \cap \bar{\text{C}})}{1 - \text{pr}(\text{C})}$$
$$= \frac{\text{pr}(\bar{\text{C}}|\text{M}_1)\text{pr}(\text{M}_1) + \text{pr}(\bar{\text{C}}|\text{M}_2)\text{pr}(\text{M}_2)}{1 - \text{pr}(\text{C})}$$
$$= \frac{(0.01)(0.20) + (0.02)(0.50)}{1 - 0.979} = 0.5714.$$

Equivalently,

$$\text{pr}(\text{M}_1 \cup \text{M}_2|\bar{\text{C}}) = 1 - \text{pr}(\text{M}_3|\bar{\text{C}}) = 1 - \frac{\text{pr}(\bar{\text{C}}|\text{M}_3)\text{pr}(\text{M}_3)}{\text{pr}(\bar{\text{C}})}$$
$$= 1 - \frac{(0.03)(0.30)}{0.021} = 0.5714.$$

(c)

$$\text{pr}(1 \text{ of } 2 \text{ PSA analyses is correct}) = \text{C}_1^2\,\text{pr}(\text{C})\text{pr}(\bar{\text{C}})$$
$$= 2(0.979)(0.021) = 0.0411.$$

Now, pr(machine $\text{M}_2$ did *not* perform both PSA analyses|1 of 2 PSA analyses is correct)

$$= 1 - \text{pr}(\text{machine M}_2 \text{ performed both PSA analyses}|1 \text{ of } 2 \text{ PSA analyses is correct})$$
$$= 1 - \frac{\left[\text{C}_1^2\,\text{pr}(\text{C}|\text{M}_2)\text{pr}(\bar{\text{C}}|\text{M}_2)\right][\text{pr}(\text{M}_2)]^2}{0.0411}$$
$$= 1 - \frac{2(0.98)(0.02)(0.50)^2}{0.0411} = 0.7616.$$

**Solution 1.7**

(a) First, $\text{pr}(\text{C}_2|\text{C}_1) = \text{pr}(\text{C}_1 \cap \text{C}_2)/\text{pr}(\text{C}_1)$. Now,

$$\text{pr}(\text{C}_1) = \text{pr}(\text{C}_2) = \text{pr}(\text{C}_2 \cap \text{D}) + \text{pr}(\text{C}_2 \cap \bar{\text{D}})$$
$$= \text{pr}(\text{C}_2|\text{D})\text{pr}(\text{D}) + \text{pr}(\text{C}_2|\bar{\text{D}})\text{pr}(\bar{\text{D}})$$
$$= \pi_1\theta + \pi_0(1 - \theta) = \theta(\pi_1 - \pi_0) + \pi_0.$$

And, appealing to the conditional independence of the events $C_1$ and $C_2$ given disease status, we have

$$\begin{aligned}\text{pr}(C_1 \cap C_2) &= \text{pr}(C_1 \cap C_2|D)\text{pr}(D) + \text{pr}(C_1 \cap C_2|\bar{D})\text{pr}(\bar{D}) \\ &= \text{pr}(C_1|D)\text{pr}(C_2|D)\text{pr}(D) + \text{pr}(C_1|\bar{D})\text{pr}(C_2|\bar{D})\text{pr}(\bar{D}) \\ &= \pi_1^2\theta + \pi_0^2(1-\theta) = \theta(\pi_1^2 - \pi_0^2) + \pi_0^2.\end{aligned}$$

Finally,

$$\text{pr}(C_2|C_1) = \frac{\theta(\pi_1^2 - \pi_0^2) + \pi_0^2}{\theta(\pi_1 - \pi_0) + \pi_0},$$

so that, in this example, $\text{pr}(C_2|C_1) \neq \text{pr}(C_2)$. More generally, this particular example illustrates the general principle that conditional independence between two events does not allow one to conclude that they are also unconditionally independent.

(b) Now,

$$\text{pr}(C_2|C_1) = \text{pr}(C_2) \Leftrightarrow \theta(\pi_1^2 - \pi_0^2) + \pi_0^2 = [\theta(\pi_1 - \pi_0) + \pi_0]^2,$$

which is equivalent to the condition

$$\theta(1-\theta)(\pi_1 - \pi_0)^2 = 0.$$

So, $\text{pr}(C_2|C_1) = \text{pr}(C_2)$ when either $\theta = 0$ (i.e., the prevalence of the disease in the population is equal to zero, so that nobody in the population has the disease), $\theta = 1$ (i.e., the prevalence of the disease in the population is equal to one, so that everybody in the population has the disease), or the probability of a correct diagnosis does not depend on disease status [i.e., since $\text{pr}(C_1) = \text{pr}(C_2) = \theta(\pi_1 - \pi_0) + \pi_0$, the condition $\pi_1 = \pi_0$ gives $\text{pr}(C_1) = \text{pr}(C_2) = \pi_1 = \pi_0$].

**Solution 1.8.** Now, $\pi_k = 1 - \text{pr}(\text{no matching sets of 5 numbers in } k \text{ drawings})$, so that

$$\begin{aligned}\pi_k &= 1 - \left(\frac{C_5^{40} - 1}{C_5^{40}}\right)\left(\frac{C_5^{40} - 2}{C_5^{40}}\right)\cdots\left(\frac{C_5^{40} - (k-1)}{C_5^{40}}\right) \\ &= 1 - \frac{\prod_{j=1}^{k-1}\left(C_5^{40} - j\right)}{\left(C_5^{40}\right)^{(k-1)}}.\end{aligned}$$

**Solution 1.9**

(a) For $i = 1, 2, \ldots, k$, let $A_i$ be the event that the $i$th person calls a dental office that is different from the dental offices called by the preceding $(i-1)$ people. Then,

$$\alpha = \text{pr}\left(\cap_{i=1}^{k} A_i\right)$$

$$= \left(\frac{n}{n}\right)\left(\frac{n-1}{n}\right)\left(\frac{n-2}{n}\right)\cdots\left[\frac{n-(k-1)}{n}\right]$$

$$= \frac{[n!/(n-k)!]}{n^k}.$$

When $n = 7$ and $k = 4$, then $\alpha = 0.350$.

(b) For $j = 1, 2, \ldots, n$, let $B_j$ be the event that all $k$ people call the $j$th dental office. Then,

$$\beta = \text{pr}\left(\cup_{j=1}^{n} B_j\right) = \sum_{j=1}^{n} \text{pr}(B_j)$$

$$= \sum_{j=1}^{n} \left(\frac{1}{n}\right)^k = \frac{1}{n^{k-1}}.$$

When $n = 7$ and $k = 4$, then $\beta = 0.003$.

**Solution 1.10.** For $j = 0, 1, \ldots, (k-2)$, there are exactly $(k-j-1)$ pairs of slots for which the integer 1 precedes the integer $k$ and for which there are exactly $j$ integers between the integers 1 and $k$. Also, the integer $k$ can precede the integer 1, and the other $(k-2)$ integers can be arranged in the remaining $(k-2)$ slots in $(k-2)!$ ways. So,

$$\theta_j = \frac{2(k-j-1)[(k-2)!]}{k!} = \frac{2(k-j-1)}{k(k-1)}, \quad j = 0, 1, \ldots, (k-2).$$

**Solution 1.11.** For $i = 1, 2, \ldots, 6$, let $A_i$ be the event that the number $i$ does *not* appear in $n$ rolls of this balanced die. Then, $\theta_n = 1 - \text{pr}\left(\cup_{i=1}^{6} A_i\right)$, where $\text{pr}\left(\cup_{i=1}^{6} A_i\right)$ may be calculated using Result (ii) on page 4. By symmetry, $\text{pr}(A_1) = \text{pr}(A_2) = \cdots = \text{pr}(A_6)$ and $\text{pr}(A_{i_1} \cap A_{i_2} \cap \cdots \cap A_{i_k}) = \text{pr}(\cap_{i=1}^{k} A_i)$, $(1 \le i_1 < i_2 < \cdots < i_k \le 6)$. Thus:

$$\text{pr}\left(\cup_{i=1}^{6} A_i\right) = C_1^6[\text{pr}(A_1)] - C_2^6[\text{pr}(\cap_{i=1}^{2} A_i)] + C_3^6[\text{pr}(\cap_{i=1}^{3} A_i)]$$

$$- C_4^6[\text{pr}(\cap_{i=1}^{4} A_i)] + C_5^6[\text{pr}(\cap_{i=1}^{5} A_i)]$$

$$= 6\left(\frac{5}{6}\right)^n - 15\left(\frac{4}{6}\right)^n + 20\left(\frac{3}{6}\right)^n - 15\left(\frac{2}{6}\right)^n + 6\left(\frac{1}{6}\right)^n.$$

When $n = 10$, $\theta_{10} \approx (1 - 0.73) = 0.27$.

**Solution 1.12.** Let $A_i$ denote the event that the first $i$ numbers selected are different from one another, $i = 2, 3, \ldots, n$. Note that $A_n \subset A_{n-1} \subset \cdots \subset A_3 \subset A_2$. So,

$$\begin{aligned}
\text{pr}(A_n) &= \text{pr(first } n \text{ numbers selected are different from one another)} \\
&= \text{pr}\left[\bigcap_{i=2}^{n} A_i\right] \\
&= \text{pr}(A_2)\text{pr}(A_3|A_2)\text{pr}(A_4|A_2 \cap A_3) \cdots \text{pr}\left[A_n \,\middle|\, \bigcap_{i=2}^{n-1} A_i\right] \\
&= \left(1 - \frac{1}{N}\right)\left(1 - \frac{2}{N}\right)\left(1 - \frac{3}{N}\right)\cdots\left[1 - \frac{(n-1)}{N}\right] \\
&= \prod_{j=1}^{(n-1)}\left(1 - \frac{j}{N}\right) = \left(\frac{N-1}{N}\right)\left(\frac{N-2}{N}\right)\cdots\left[\frac{N-(n-1)}{N}\right] \\
&= \frac{N!}{(N-n)!\,N^n}.
\end{aligned}$$

For $N = 10$ and $n = 4$,

$$\begin{aligned}
\text{pr}(A_4) &= \prod_{j=1}^{3}\left(1 - \frac{j}{10}\right) = \left(1 - \frac{1}{10}\right)\left(1 - \frac{2}{10}\right)\left(1 - \frac{3}{10}\right) \\
&= \frac{10!}{(10-4)!(10)^4} = 0.504.
\end{aligned}$$

**Solution 1.13**

(a) We have

$$\begin{aligned}
\theta_{wr} &= 1 - \text{pr(all } n \text{ numbers have values less than } N) \\
&= 1 - \left(\frac{N-1}{N}\right)^n.
\end{aligned}$$

(b) We have

$$\begin{aligned}
\theta_{wor} &= 1 - \text{pr(all } n \text{ numbers have values less than } N) \\
&= 1 - \frac{C_n^{N-1} C_0^1}{C_n^N} \\
&= 1 - \frac{(N-n)}{N} = \frac{n}{N}.
\end{aligned}$$

(c) First, note that

$$\delta_n = (\theta_{wor} - \theta_{wr}) = \frac{n}{N} - \left[1 - \left(\frac{N-1}{N}\right)^n\right]$$
$$= \left(\frac{N-1}{N}\right)^n - \left(\frac{N-n}{N}\right).$$

Now, for $n = 2$,

$$\delta_2 = \left(\frac{N-1}{N}\right)^2 - \left(\frac{N-2}{N}\right)$$
$$= \frac{1}{N^2}\left[(N-1)^2 - N(N-2)\right] = \frac{1}{N^2} > 0.$$

Then, assuming $\delta_n > 0$, we have

$$\delta_{n+1} = \left(\frac{N-1}{N}\right)^{n+1} - \left[\frac{N-(n+1)}{N}\right]$$
$$= \left(\frac{N-1}{N}\right)^n \left(\frac{N-1}{N}\right) - \left(\frac{N-n}{N}\right) + \frac{1}{N}$$
$$= \left(\frac{N-1}{N}\right)^n - \left(\frac{N-1}{N}\right)^n \left(\frac{1}{N}\right) - \left(\frac{N-n}{N}\right) + \frac{1}{N}$$
$$= \left[\left(\frac{N-1}{N}\right)^n - \left(\frac{N-n}{N}\right)\right] + \frac{1}{N}\left[1 - \left(\frac{N-1}{N}\right)^n\right]$$
$$= \delta_n + \left(\frac{1}{N}\right)\left[1 - \left(\frac{N-1}{N}\right)^n\right] > 0,$$

which completes the proof by induction.

Therefore, sampling without replacement has a higher probability than sampling with replacement of selecting the ball labeled with the number $N$.

**Solution 1.14**

(a) Let $T_1$ be the event that tunnel 1 is closed to traffic, let $T_2$ be the event that tunnel 2 is closed to traffic, and let $T_3$ be the event that tunnel 3 is closed to traffic. If $\alpha$ is the probability that exactly one tunnel is closed to traffic, then, since the events $T_1, T_2$, and $T_3$ are mutually independent, it follows that

$$\alpha = \text{pr}(T_1)\text{pr}(\bar{T}_2)\text{pr}(\bar{T}_3) + \text{pr}(\bar{T}_1)\text{pr}(T_2)\text{pr}(\bar{T}_3)$$
$$+ \text{pr}(\bar{T}_1)\text{pr}(\bar{T}_2)\text{pr}(T_3)$$
$$= \pi_1(1-\pi_2)(1-\pi_3) + (1-\pi_1)\pi_2(1-\pi_3)$$
$$+ (1-\pi_1)(1-\pi_2)\pi_3.$$

(b) Define the following five mutually exclusive events:

A: no inclement weather and all tunnels open to traffic

B: inclement weather and all tunnels are open to traffic

C: inclement weather and at least one tunnel is closed to traffic

D: no inclement weather and exactly one tunnel is closed to traffic

E: no inclement weather and at least two tunnels are closed to traffic

If $\beta$ is the probability of an excellent traffic flow rate, then

$$\beta = \text{pr}(A) = (1-\theta)(1-\pi_1)(1-\pi_2)(1-\pi_3).$$

And, if $\gamma$ is the probability of a marginal traffic flow rate, then

$$\gamma = \text{pr}(B) + \text{pr}(D) = \theta(1-\pi_1)(1-\pi_2)(1-\pi_3) + (1-\theta)\alpha.$$

Finally, if $\delta$ is the probability of a poor traffic flow rate, then

$$\begin{aligned}\delta = \text{pr}(C) + \text{pr}(E) &= \theta[1-(1-\pi_1)(1-\pi_2)(1-\pi_3)] \\ &+ (1-\theta)[1-(1-\pi_1)(1-\pi_2)(1-\pi_3)-\alpha] = 1-\beta-\gamma.\end{aligned}$$

(c) Using the event definitions given in part (b), we have

$$\begin{aligned}\text{pr}(B|B\cup D) &= \frac{\text{pr}[B\cap(B\cup D)]}{\text{pr}(B\cup D)} \\ &= \frac{\text{pr}[B\cup(B\cap D)]}{\gamma} = \frac{\text{pr}(B)}{\gamma} \\ &= \frac{\theta(1-\pi_1)(1-\pi_2)(1-\pi_3)}{\gamma}.\end{aligned}$$

**Solution 1.15**

(a) Let $A_x$ be the event that it takes $x$ tosses of this unbalanced coin to obtain the first head. Then,

$$\begin{aligned}\text{pr}(A_x) &= \text{pr}\{[\text{first } (x-1) \text{ tosses are tails}] \cap [x\text{th toss is a head}]\} \\ &= (1-\pi)^{x-1}\pi, \quad x = 1,2,\ldots,\infty.\end{aligned}$$

Now, letting $\theta$ = pr(Bonnie and Clyde each require the same number of tosses to obtain the first head), we have

$$\begin{aligned}\theta &= \text{pr}\{\cup_{x=1}^{\infty}[(\text{Bonnie requires } x \text{ tosses}) \cap (\text{Clyde requires } x \text{ tosses})]\} \\ &= \sum_{x=1}^{\infty}[\text{pr}(A_x)]^2 = \sum_{x=1}^{\infty}[(1-\pi)^{x-1}\pi]^2\end{aligned}$$

$$= \left(\frac{\pi}{1-\pi}\right)^2 \sum_{x=1}^{\infty}[(1-\pi)^2]^x = \left(\frac{\pi}{1-\pi}\right)^2 \left[\frac{(1-\pi)^2}{1-(1-\pi)^2}\right]$$

$$= \frac{\pi}{(2-\pi)}.$$

(b) By symmetry, pr(Bonnie requires more tosses than Clyde to obtain the first head) = pr(Clyde requires more tosses than Bonnie to obtain the first head) = $\gamma$, say. Thus, since $(2\gamma + \theta) = 1$, it follows that

$$\gamma = \frac{(1-\theta)}{2} = \frac{1 - \left(\frac{\pi}{2-\pi}\right)}{2} = \frac{(1-\pi)}{(2-\pi)}.$$

To illustrate a more complicated approach,

$$\gamma = \sum_{x=1}^{\infty} \sum_{y=x+1}^{\infty} \text{pr}[(\text{Clyde needs } x \text{ tosses for first head}) \cap (\text{Bonnie needs } y \text{ tosses for first head})]$$

$$= \sum_{x=1}^{\infty}[(1-\pi)^{x-1}\pi] \sum_{y=x+1}^{\infty} [(1-\pi)^{y-1}\pi]$$

$$= \frac{\pi^2}{(1-\pi)} \sum_{x=1}^{\infty}(1-\pi)^x \left[\frac{(1-\pi)^x}{1-(1-\pi)}\right]$$

$$= \frac{\pi}{(1-\pi)} \sum_{x=1}^{\infty}[(1-\pi)^2]^x = \frac{\pi}{(1-\pi)}\left[\frac{(1-\pi)^2}{1-(1-\pi)^2}\right]$$

$$= \frac{(1-\pi)}{(2-\pi)}.$$

**Solution 1.16.** For $x = 0, 1, \ldots, 5$, let A be the event that exactly 5 of these 15 students scored higher than the 80-th percentile, and let $\text{B}_x$ be the event that exactly $x$ females and exactly $(5-x)$ males scored higher than the 80-th percentile.

So,

$$\text{pr}(\text{B}_x|\text{A}) = \frac{\text{pr}(\text{A} \cap \text{B}_x)}{\text{pr}(\text{A})} = \frac{\text{pr}(\text{B}_x)}{\text{pr}(\text{A})}$$

$$= \frac{\left\{\left[\text{C}_x^8 \pi^x (1-\pi)^{8-x}\right]\left[\text{C}_{5-x}^7 \pi^{5-x}(1-\pi)^{2+x}\right]\right\}}{\text{C}_5^{15}\pi^5(1-\pi)^{10}}$$

$$= \frac{\text{C}_x^8 \text{C}_{5-x}^7}{\text{C}_5^{15}}, \quad x = 0, 1, \ldots, 5.$$

Thus,

$$\theta = \sum_{x=3}^{5} \frac{\text{C}_x^8 \text{C}_{5-x}^7}{\text{C}_5^{15}} = \frac{(1176 + 490 + 56)}{3003} = 0.5734.$$

**Solution 1.17.** For $i = 1, 2, 3, 4$, let the event $A_i$ be the event that a hand contains all three face cards of the $i$th suit. Note that the event of interest is $\cup_{i=1}^{4} A_i$. So,

$$\text{pr}(A_i) = \frac{C_3^3 C_{10}^{49}}{C_{13}^{52}}, \quad i = 1, 2, 3, 4.$$

For $i \neq j$,

$$\text{pr}(A_i \cap A_j) = \frac{C_6^6 C_7^{46}}{C_{13}^{52}};$$

for $i \neq j \neq k$,

$$\text{pr}(A_i \cap A_j \cap A_k) = \frac{C_9^9 C_4^{43}}{C_{13}^{52}};$$

and, for $i \neq j \neq k \neq l$,

$$\text{pr}(A_i \cap A_j \cap A_k \cap A_l) = \frac{C_{12}^{12} C_1^{40}}{C_{13}^{52}}.$$

Then, using Result (ii) on page 4, we have

$$\text{pr}(A) = \text{pr}\left(\cup_{i=1}^{4} A_i\right) = \frac{\sum_{m=1}^{4} C_m^4 (-1)^{m-1} C_{13-3m}^{52-3m}}{C_{13}^{52}},$$

which is equal to 0.0513.

**Solution 1.18*.** Let W denote the event that the player wins the game and let $X$ denote the number obtained on the first roll. So,

$$\begin{aligned}
\text{pr}(W) &= \sum_{x=2}^{12} \text{pr}(W|X = x)\text{pr}(X = x) \\
&= \sum_{x=2}^{12} \text{pr}(W|X = x) \left[\frac{\min(x - 1, 13 - x)}{36}\right] \\
&= (0)\frac{1}{36} + (0)\frac{2}{36} + \sum_{x=4}^{6} \text{pr}(W|X = x)\frac{(x - 1)}{36} \\
&\quad + (1)\frac{6}{36} + \sum_{x=8}^{10} \text{pr}(W|X = x)\frac{(13 - x)}{36} + (1)\frac{2}{36} + (0)\frac{1}{36} \\
&= \frac{2}{9} + 2\sum_{x=4}^{6} \text{pr}(W|X = x)\frac{(x - 1)}{36},
\end{aligned}$$

since the pairs of numbers "4 and 10," "5 and 9," and "6 and 8" lead to the same result.

So, for $x = 4, 5,$ or 6, let $\pi_x = \text{pr(number } x \text{ is rolled before number 7 is rolled)}$. Then,

$$\pi_x = \sum_{j=1}^{\infty} [\text{pr(any number but } x \text{ or 7 is rolled)}]^{(j-1)} \text{pr(number } x \text{ is rolled)}$$

$$= \sum_{j=1}^{\infty} \left[1 - \frac{(x-1)}{36} - \frac{6}{36}\right]^{(j-1)} \frac{(x-1)}{36}$$

$$= \frac{(x-1)}{36} \sum_{j=1}^{\infty} \left(\frac{31-x}{36}\right)^{(j-1)}$$

$$= \frac{(x-1)}{36} \left[1 - \frac{(31-x)}{36}\right]^{-1} = \frac{(x-1)}{(x+5)}, \quad x = 4, 5, 6.$$

So,

$$\text{pr(W)} = = \frac{2}{9} + \frac{1}{18} \sum_{x=4}^{6} \pi_x (x-1) = \frac{2}{9} + \frac{1}{18} \sum_{x=4}^{6} \frac{(x-1)^2}{(x+5)}$$

$$= \frac{2}{9} + \frac{1}{18}\left(\frac{9}{9} + \frac{16}{10} + \frac{25}{11}\right) = 0.4931.$$

Thus, the probability of the house winning the game is (1−0.4931)=0.5069; so, as expected with any casino game, the house always has the advantage. However, relative to many other casino games (e.g., blackjack, roulette, slot machines), the house advantage of $(0.5069 - 0.4931) = 0.0138$ is relatively small.

**Solution 1.19***

(a) Let E be the event that the first worker randomly selected is a highly exposed worker. Then,

$$\theta_n = (1 - \theta_{n-1})\text{pr(E)} + \theta_{n-1}[1 - \text{pr(E)}]$$
$$= (1 - \theta_{n-1})\pi_h + \theta_{n-1}(1 - \pi_h)$$
$$= \pi_h + \theta_{n-1}(1 - 2\pi_h), \text{ with } \theta_0 \equiv 1.$$

(b) Now, assuming that $\theta_n = \alpha + \beta\gamma^n$ and using the result in part (a), we have

$$\alpha + \beta\gamma^n = \pi_h + (\alpha + \beta\gamma^{n-1})(1 - 2\pi_h)$$
$$= \pi_h + (1 - 2\pi_h)\alpha + (1 - 2\pi_h)\beta\gamma^{n-1},$$

with the restriction that $(\alpha + \beta) = 1$ since $\theta_0 \equiv 1$.

Thus, we must have $\alpha = \beta = \frac{1}{2}$ and $\gamma = (1 - 2\pi_h)$, giving

$$\theta_n = \frac{1}{2} + \frac{1}{2}(1 - 2\pi_h)^n, n = 1, 2, \ldots, \infty.$$

Finally, when $\pi_h = 0.05$, $\theta_{50} = \frac{1}{2} + \frac{1}{2}[1 - 2(0.05)]^{50} = 0.5026$.

**Solution 1.20***

(a) We have

$$\text{pr}(\text{D}|\text{S}, \boldsymbol{x}) = \frac{\text{pr}(\text{D} \cap \text{S}|\boldsymbol{x})}{\text{pr}(\text{S}|\boldsymbol{x})} = \frac{\text{pr}(\text{S}|\text{D}, \boldsymbol{x})\text{pr}(\text{D}|\boldsymbol{x})}{\text{pr}(\text{S}|\bar{\text{D}}, \boldsymbol{x})\text{pr}(\bar{\text{D}}|\boldsymbol{x}) + \text{pr}(\text{S}|\text{D}, \boldsymbol{x})\text{pr}(\text{D}|\boldsymbol{x})}$$

$$= \frac{\pi_1 \left[\frac{e^{\beta_0 + \boldsymbol{\beta}'\boldsymbol{x}}}{1+e^{\beta_0 + \boldsymbol{\beta}'\boldsymbol{x}}}\right]}{\pi_0 \left[\frac{1}{1+e^{\beta_0 + \boldsymbol{\beta}'\boldsymbol{x}}}\right] + \pi_1 \left[\frac{e^{\beta_0 + \boldsymbol{\beta}'\boldsymbol{x}}}{1+e^{\beta_0 + \boldsymbol{\beta}'\boldsymbol{x}}}\right]}$$

$$= \frac{\pi_1 e^{\beta_0 + \boldsymbol{\beta}'\boldsymbol{x}}}{\pi_0 + \pi_1 e^{\beta_0 + \boldsymbol{\beta}'\boldsymbol{x}}}$$

$$= \frac{\left(\frac{\pi_1}{\pi_0}\right) e^{\beta_0 + \boldsymbol{\beta}'\boldsymbol{x}}}{1 + \left(\frac{\pi_1}{\pi_0}\right) e^{\beta_0 + \boldsymbol{\beta}'\boldsymbol{x}}}$$

$$= \frac{e^{\beta_0^* + \boldsymbol{\beta}'\boldsymbol{x}}}{1 + e^{\beta_0^* + \boldsymbol{\beta}'\boldsymbol{x}}},$$

where $\beta_0^* = \beta_0 + \ln(\pi_1/\pi_0)$.

So, for a case–control study, since $\beta_0 = \beta_0^* - \ln(\pi_1/\pi_0)$, to estimate the risk $\text{pr}(\text{D}|\boldsymbol{x})$ of disease using logistic regression would necessitate either knowing (or being able to estimate) the ratio of selection probabilities, namely, the ratio $\pi_1/\pi_0$.

(b) Since

$$\frac{\text{pr}(\text{D}|\boldsymbol{x})}{\text{pr}(\bar{\text{D}}|\boldsymbol{x})} = e^{\beta_0 + \boldsymbol{\beta}'\boldsymbol{x}},$$

it follows directly that

$$\theta_r = \frac{e^{\beta_0 + \boldsymbol{\beta}'\boldsymbol{x}^*}}{e^{\beta_0 + \boldsymbol{\beta}'\boldsymbol{x}}} = e^{\boldsymbol{\beta}'(\boldsymbol{x}^* - \boldsymbol{x})}.$$

Analogously, since

$$\frac{\text{pr}(\text{D}|\text{S}, \boldsymbol{x})}{\text{pr}(\bar{\text{D}}|\text{S}, \boldsymbol{x})} = e^{\beta_0^* + \boldsymbol{\beta}'\boldsymbol{x}},$$

it follows directly that

$$\theta_c = \frac{e^{\beta_0^* + \boldsymbol{\beta}'\boldsymbol{x}^*}}{e^{\beta_0^* + \boldsymbol{\beta}'\boldsymbol{x}}} = e^{\boldsymbol{\beta}'(\boldsymbol{x}^* - \boldsymbol{x})} = \theta_r.$$

Hence, we can, at least theoretically, use case–control study data to estimate risk odds ratios via logistic regression, even though we cannot estimate the risk (or probability) of disease directly without information about the quantity $\pi_1/\pi_0$.

There are other potential problems with the use of case–control studies in epidemiologic research. For further discussion about such issues, see Breslow and Day (1980) and Kleinbaum, Kupper, and Morgenstern (1982).

**Solution 1.21***

(a) Let A be the event that Diagnostic Strategy #1 provides the correct diagnosis, let B be the event that Diagnostic Strategy #2 provides the correct diagnosis, and let D be the event that the adult has IBD. Then,

$$\begin{aligned}
\text{pr}(A) &= \text{pr}(A \cap D) + \text{pr}(A \cap \bar{D}) = \text{pr}(A|D)\text{pr}(D) + \text{pr}(A|\bar{D})\text{pr}(\bar{D}) \\
&= [3\pi_1^2(1-\pi_1) + \pi_1^3]\theta + [3\pi_0^2(1-\pi_0) + \pi_0^3](1-\theta) \\
&= (3\pi_1^2 - 2\pi_1^3)\theta + (3\pi_0^2 - 2\pi_0^3)(1-\theta).
\end{aligned}$$

And,

$$\text{pr}(B) = \text{pr}(B|D)\text{pr}(D) + \text{pr}(B|\bar{D})\text{pr}(\bar{D}) = \pi_1\theta + \pi_0(1-\theta).$$

Now,

$$\begin{aligned}
\text{pr}(A) - \text{pr}(B) &= [(3\pi_1^2 - 2\pi_1^3)\theta + (3\pi_0^2 - 2\pi_0^3)(1-\theta)] - [\pi_1\theta + \pi_0(1-\theta)] \\
&= (3\pi_1^2 - 2\pi_1^3 - \pi_1)\theta + (3\pi_0^2 - 2\pi_0^3 - \pi_0)(1-\theta) \\
&= \pi_1(1-\pi_1)(2\pi_1 - 1)\theta + \pi_0(1-\pi_0)(2\pi_0 - 1)(1-\theta).
\end{aligned}$$

So, a sufficient condition for the ranges of $\pi_1$ and $\pi_0$ so that pr(A)>pr(B) is

$$\frac{1}{2} < \pi_1 < 1 \quad \text{and} \quad \frac{1}{2} < \pi_0 < 1.$$

In other words, if each doctor has a better than 50% chance of making the correct diagnosis conditional on disease status, then Diagnostic Strategy #1 is preferable to Diagnostic Strategy #2.

(b) Let C be the event that Diagnostic Strategy #3 provides the correct diagnosis. Then,

$$\begin{aligned}
\text{pr}(C) &= \text{pr}(C|D)\text{pr}(D) + \text{pr}(C|\bar{D})\text{pr}(\bar{D}) \\
&= [4\pi_1^3(1-\pi_1) + \pi_1^4]\theta + [4\pi_0^3(1-\pi_0) + \pi_0^4](1-\theta) \\
&= (4\pi_1^3 - 3\pi_1^4)\theta + (4\pi_0^3 - 3\pi_0^4)(1-\theta).
\end{aligned}$$

Since

$$\begin{aligned}
\text{pr}(A) - \text{pr}(C) &= [(3\pi_1^2 - 2\pi_1^3) - (4\pi_1^3 - 3\pi_1^4)]\theta \\
&\quad + [3\pi_0^2 - 2\pi_0^3) - (4\pi_0^3 - 3\pi_0^4)](1-\theta) \\
&= 3\pi_1^2(1-\pi_1)^2\theta + 3\pi_0^2(1-\pi_0)^2(1-\theta) > 0,
\end{aligned}$$

Diagnostic Strategy #1 (using the majority opinion of three doctors) has a higher probability than Diagnostic Strategy #3 (using the majority opinion of four doctors) of making the correct diagnosis.

**Solution 1.22***. First,

$$\begin{aligned}\pi_1 = \text{pr}(\text{D}|\text{E}) &= \frac{\text{pr}(\text{D} \cap \text{E})}{\text{pr}(\text{E})} \\ &= \frac{\text{pr}(\text{D} \cap \text{E} \cap \text{M}) + \text{pr}(\text{D} \cap \text{E} \cap \bar{\text{M}})}{\text{pr}(\text{E})} \\ &= \frac{\pi_{11}\text{pr}(\text{E} \cap \text{M}) + \pi_{10}\text{pr}(\text{E} \cap \bar{\text{M}})}{\text{pr}(\text{E})} \\ &= \text{pr}(\text{M}|\text{E})\pi_{11} + \text{pr}(\bar{\text{M}}|\text{E})\pi_{10}.\end{aligned}$$

Similarly,

$$\pi_0 = \text{pr}(\text{M}|\bar{\text{E}})\pi_{01} + \text{pr}(\bar{\text{M}}|\bar{\text{E}})\pi_{00}.$$

So,

$$\begin{aligned}RR_c = \frac{\pi_1}{\pi_0} &= \frac{\text{pr}(\text{M}|\text{E})\pi_{11} + \text{pr}(\bar{\text{M}}|\text{E})\pi_{10}}{\text{pr}(\text{M}|\bar{\text{E}})\pi_{01} + \text{pr}(\bar{\text{M}}|\bar{\text{E}})\pi_{00}} \\ &= \frac{\text{pr}(\text{M}|\text{E})\pi_{01}RR_1 + \text{pr}(\bar{\text{M}}|\text{E})\pi_{00}RR_0}{\text{pr}(\text{M}|\bar{\text{E}})\pi_{01} + \text{pr}(\bar{\text{M}}|\bar{\text{E}})\pi_{00}} \\ &= RR\left[\frac{\text{pr}(\text{M}|\text{E})\pi_{01} + \text{pr}(\bar{\text{M}}|\text{E})\pi_{00}}{\text{pr}(\text{M}|\bar{\text{E}})\pi_{01} + \text{pr}(\bar{\text{M}}|\bar{\text{E}})\pi_{00}}\right].\end{aligned}$$

Thus, a sufficient condition for $RR_c = RR$ is

$$\text{pr}(\text{M}|\text{E})\pi_{01} + \text{pr}(\bar{\text{M}}|\text{E})\pi_{00} = \text{pr}(\text{M}|\bar{\text{E}})\pi_{01} + \text{pr}(\bar{\text{M}}|\bar{\text{E}})\pi_{00},$$

or equivalently,

$$[\text{pr}(\text{M}|\text{E}) - \text{pr}(\text{M}|\bar{\text{E}})]\pi_{01} + [\text{pr}(\bar{\text{M}}|\text{E}) - \text{pr}(\bar{\text{M}}|\bar{\text{E}})]\pi_{00} = 0.$$

Using the relationships $\text{pr}(\bar{\text{M}}|\text{E}) = 1 - \text{pr}(\text{M}|\text{E})$ and $\text{pr}(\bar{\text{M}}|\bar{\text{E}}) = 1 - \text{pr}(\text{M}|\bar{\text{E}})$ in the above expression, it follows that $RR_c = RR$ when

$$[\text{pr}(\text{M}|\text{E}) - \text{pr}(\text{M}|\bar{\text{E}})](\pi_{01} - \pi_{00}) = 0.$$

Thus, the two sufficient conditions for *no confounding* are

$$\text{pr}(\text{M}|\text{E}) = \text{pr}(\text{M}|\bar{\text{E}}) \quad \text{and} \quad \pi_{01} = \pi_{00}.$$

Further, since

$$\text{pr}(\text{M}) = \text{pr}(\text{M}|\text{E})\text{pr}(\text{E}) + \text{pr}(\text{M}|\bar{\text{E}})\text{pr}(\bar{\text{E}}),$$

the condition $\text{pr}(\text{M}|\text{E}) = \text{pr}(\text{M}|\bar{\text{E}})$ means that $\text{pr}(\text{M}) = \text{pr}(\text{M}|\text{E})$, or equivalently, that the events E and M are independent events.

Finally, the two *no confounding* conditions are:

(i) The events E and M are independent events;

(ii) $\text{pr}(\text{D}|\bar{\text{E}} \cap \text{M}) = \text{pr}(\text{D}|\bar{\text{E}} \cap \bar{\text{M}})$.

**Solution 1.23***

(a) First,

$$\text{pr}(\text{D}_1|\text{T}^+) = \frac{\text{pr}(\text{D}_1 \cap \text{T}^+)}{\text{pr}(\text{T}^+)} = \frac{\text{pr}(\text{T}^+|\text{D}_1)\text{pr}(\text{D}_1)}{\text{pr}(\text{T}^+ \cap \text{D}_1) + \text{pr}(\text{T}^+ \cap \bar{\text{D}}_1)}$$
$$= \frac{\text{pr}(\text{T}^+|\text{D}_1)\text{pr}(\text{D}_1)}{\text{pr}(\text{T}^+|\text{D}_1)\text{pr}(\text{D}_1) + \text{pr}(\text{T}^+|\bar{\text{D}}_1)\text{pr}(\bar{\text{D}}_1)}$$
$$= \frac{\theta_1\pi_1}{\theta_1\pi_1 + \theta_2\pi_2}.$$

And,

$$\text{pr}(\bar{\text{D}}_1|\text{T}^+) = 1 - \text{pr}(\text{D}_1|\text{T}^+) = \frac{\theta_2\pi_2}{\theta_1\pi_1 + \theta_2\pi_2}.$$

Finally,

$$\frac{\text{pr}(\text{D}_1|\text{T}^+)}{\text{pr}(\bar{\text{D}}_1|\text{T}^+)} = \frac{\theta_1\pi_1}{\theta_2\pi_2} = LR_{12}\left(\frac{\pi_1}{\pi_2}\right).$$

(b) First,

$$\text{pr}(\text{D}_1|\text{T}^+) = \frac{\text{pr}(\text{T}^+|\text{D}_1)\text{pr}(\text{D}_1)}{\sum_{i=1}^{3}\text{pr}(\text{T}^+|\text{D}_i)\text{pr}(\text{D}_i)} = \frac{\theta_1\pi_1}{\sum_{i=1}^{3}\theta_i\pi_i},$$

and so

$$\text{pr}(\bar{\text{D}}_1|\text{T}^+) = 1 - \text{pr}(\text{D}_1|\text{T}^+) = \frac{\sum_{i=2}^{3}\theta_i\pi_i}{\sum_{i=1}^{3}\theta_i\pi_i}.$$

Finally,

$$\frac{\text{pr}(\text{D}_1|\text{T}^+)}{\text{pr}(\bar{\text{D}}_1|\text{T}^+)} = \frac{\theta_1\pi_1}{\theta_2\pi_2 + \theta_3\pi_3} = \frac{1}{\frac{\theta_2\pi_2}{\theta_1\pi_1} + \frac{\theta_3\pi_3}{\theta_1\pi_1}}$$
$$= \left[\sum_{i=2}^{3}\left(\frac{\pi_1}{\pi_i}LR_{1i}\right)^{-1}\right]^{-1}.$$

And,

$$\text{pr}(D_1|T^+) = \frac{\theta_1\pi_1}{\sum_{i=1}^{3}\theta_i\pi_i} = \frac{1}{1+\left(\frac{\theta_1\pi_1}{\theta_2\pi_2}\right)^{-1}+\left(\frac{\theta_1\pi_1}{\theta_3\pi_3}\right)^{-1}}$$

$$= \left[1+\sum_{i=2}^{3}\left(\frac{\pi_1}{\pi_i}LR_{1i}\right)^{-1}\right]^{-1}.$$

(c) For notational convenience, let $\pi_1$=pr(NS)=0.57, $\pi_2$=pr(A)=0.33, $\pi_3$ = pr (C) = 0.10, $LR_{12}$=pr($T^+$|NS)/pr($T^+$|A)=0.30, $LR_{13}$=pr($T^+$|NS)/pr($T^+$|C) = 0.50, and $LR_{23}$=pr($T^+$|A)/pr($T^+$|C)=1.67.

Following the developments given in part (b), it then follows directly that

$$\frac{\text{pr}(\text{NS}|T^+)}{\text{pr}(\overline{\text{NS}}|T^+)} = 0.4385 \quad \text{and} \quad \text{pr}(\text{NS}|T^+) = 0.3048,$$

$$\frac{\text{pr}(\text{A}|T^+)}{\text{pr}(\bar{\text{A}}|T^+)} = 1.4293 \quad \text{and} \quad \text{pr}(\text{A}|T^+) = 0.5883,$$

and

$$\frac{\text{pr}(\text{C}|T^+)}{\text{pr}(\bar{\text{C}}|T^+)} = 0.1196 \quad \text{and} \quad \text{pr}(\text{C}|T^+) = 0.1069.$$

Thus, based on this particular diagnostic test, the most likely diagnosis is *appendicitis* for an emergency room patient with a positive test result.

**Solution 1.24***

(a) The four probabilities appearing in the expression for $\psi_c$ can be rewritten as follows:

$$\text{pr}(A\cap C|B\cap C) = \frac{\text{pr}(A\cap B\cap C)}{\text{pr}(B\cap C)} = \frac{\text{pr}(C|A\cap B)\text{pr}(A|B)\text{pr}(B)}{\text{pr}(B\cap C)}$$

$$\text{pr}(\overline{A\cap C}|B\cap C) = \frac{\text{pr}[(\bar{A}\cup\bar{C})\cap(B\cap C)]}{\text{pr}(B\cap C)} = \frac{\text{pr}(\bar{A}\cap B\cap C)}{\text{pr}(B\cap C)}$$

$$= \frac{\text{pr}(C|\bar{A}\cap B)\text{pr}(\bar{A}|B)\text{pr}(B)}{\text{pr}(B\cap C)}$$

$$\text{pr}(A\cap C|\overline{B\cap C}) = \frac{\text{pr}[(A\cap C)\cap(\bar{B}\cup\bar{C})]}{\text{pr}(\overline{B\cap C})} = \frac{\text{pr}(A\cap\bar{B}\cap C)}{\text{pr}(\overline{B\cap C})}$$

$$= \frac{\text{pr}(C|A\cap\bar{B})\text{pr}(A|\bar{B})\text{pr}(\bar{B})}{\text{pr}(\overline{B\cap C})}$$

and

$$\mathrm{pr}(\overline{A\cap C}|\overline{B\cap C}) = \frac{\mathrm{pr}[(\bar{A}\cup\bar{C})\cap(\bar{B}\cup\bar{C})]}{\mathrm{pr}(\overline{B\cap C})}$$
$$= \frac{\mathrm{pr}[(\bar{A}\cap\bar{B})\cup(\bar{A}\cap\bar{C})\cup(\bar{B}\cap\bar{C})\cup\bar{C}]}{\mathrm{pr}(\overline{B\cap C})}$$
$$= \frac{\mathrm{pr}(\bar{A}\cap\bar{B})+\mathrm{pr}(\bar{C})-\mathrm{pr}(\bar{A}\cap\bar{B}\cap\bar{C})}{\mathrm{pr}(\overline{B\cap C})},$$

since

$$\mathrm{pr}[(\bar{A}\cap\bar{B})\cup(\bar{A}\cap\bar{C})\cup(\bar{B}\cap\bar{C})\cup\bar{C}]$$
$$= \mathrm{pr}(\bar{A}\cap\bar{B})+\mathrm{pr}(\bar{C})-\mathrm{pr}(\bar{A}\cap\bar{B}\cap\bar{C})$$

via use of the general formula for the union of four events.

Then, inserting these four expansions into the formula for $\psi_c$ and simplifying gives the desired result, since $\mathrm{pr}(\bar{A}\cap\bar{B})+\mathrm{pr}(\bar{C})-\mathrm{pr}(\bar{A}\cap\bar{B}\cap\bar{C})$ can be rewritten as

$$\mathrm{pr}[(\bar{A}\cap\bar{B})\cup\bar{C}] = \mathrm{pr}[(\bar{A}\cup\bar{C})\cap(\bar{B}\cup\bar{C})]$$
$$= \mathrm{pr}[(\bar{A}\cup\bar{C})\cap(\bar{B}\cup\bar{C})\cap(C\cup\bar{C})]$$
$$= \mathrm{pr}[(\bar{A}\cap\bar{B}\cap C)\cup\bar{C}] = \mathrm{pr}(\bar{A}\cap\bar{B}\cap C)+\mathrm{pr}(\bar{C}).$$

(b) If events A, B, and C occur completely independently of one another, then $\psi = 1, \mathrm{pr}(C|A\cap B) = \mathrm{pr}(C)$, so on, so that

$$\psi_c = (1)(1)\left[1+\frac{\mathrm{pr}(\bar{C})}{\mathrm{pr}(\bar{A})\mathrm{pr}(\bar{B})\mathrm{pr}(C)}\right] > 1.$$

Thus, using $\psi_c$ instead of $\psi$ introduces a *positive* bias. So, using $\psi_c$ could lead to the false conclusion that diseases A and B are related when, in fact, they are not related at all (i.e., $\psi = 1$).

**Solution 1.25***

(a) First, given that there is a total of $(s+t)$ available positions in a sequence, then $s$ of these $(s+t)$ positions can be filled with the letter S in $C_s^{s+t}$ ways, leaving the remaining positions to be filled by the letter F. Under the assumption of randomness, each of these $C_s^{s+t}$ sequences is equally likely to occur, so that each random sequence has probability $1/C_s^{s+t}$ of occurring.

Now, for $x$ an even positive integer, let $x = 2y$. Since the S and F runs alternate, there will be exactly $y$ S runs and exactly $y$ F runs, where $y = 1, 2, \ldots, \min(s,t)$. The number of ways of dividing the $s$ available S letters into $y$ S runs is equal to $C_{y-1}^{s-1}$, which is simply the number of ways of choosing $(y-1)$ spaces from the $(s-1)$ spaces between the $s$ available S letters. Analogously, the $t$ available F letters can be divided into $y$ runs in $C_{y-1}^{t-1}$ ways. Thus, since the first run in the sequence can be either an S run or an F run, the *total* number of sequences that each contain exactly $2y$ runs is equal to $2C_{y-1}^{s-1}C_{y-1}^{t-1}$. Hence, under the assumption that all sequences containing exactly $s$ successes (the letter S) and $t$ failures (the letter F) are equally likely to occur, the probability of observing a sequence containing a *total* of exactly $x = 2y$ runs is equal to

$$\pi_{2y} = \frac{2C_{y-1}^{s-1}C_{y-1}^{t-1}}{C_s^{s+t}}, \quad y = 1, 2, \ldots, \min(s, t).$$

Now, for $x$ an odd positive integer, let $x = (2y + 1)$. Either there will be $(y + 1)$ S runs and $y$ F runs, or there will be $y$ S runs and $(y + 1)$ F runs, where $y = \min(s, t)$. In the former case, since the complete sequence must begin with an S run, the total number of runs will be $C_y^{s-1}C_{y-1}^{t-1}$; analogously, in the latter case, the total number of runs will be $C_{y-1}^{s-1}C_y^{t-1}$. Hence, under the assumption that all sequences containing exactly $s$ successes (the letter S) and $t$ failures (the letter F) are equally likely to occur, the probability of observing a sequence containing a *total* of exactly $x = (2y + 1)$ runs is equal to

$$\pi_{2y+1} = \frac{C_y^{s-1}C_{y-1}^{t-1} + C_{y-1}^{s-1}C_y^{t-1}}{C_s^{s+t}}, \quad y = 1, 2, \ldots, \min(s, t),$$

where

$$C_y^{s-1} \equiv 0 \text{ when } y = s \quad \text{and} \quad C_y^{t-1} \equiv 0 \text{ when } y = t.$$

(b) For the observed sequence, $s = 4$ and $t = 3$. Also, the observed total number of runs $x$ is equal to 4; in particular, there are two S runs, one of length 1 and one of length 3, and there are two F runs, one of length 1 and one of length 2. Using the formula $\pi_{2y}$ with $y = 2$ gives

$$\pi_4 = \frac{2C_1^3C_1^2}{C_4^7} = \frac{12}{35} = 0.343.$$

Since this probability is fairly large, there is no statistical evidence that the observed sequence represents a deviation from randomness.

**Solution 1.26***

(a) For $i = 1, 2 \ldots, R$, let $A_i$ be the event that the subject's $i$th chip is in its correct position. Then,

$$\theta(0, R) = 1 - \text{pr}\left(\cup_{i=1}^{R} A_i\right) = 1 - \sum_{i=1}^{R} \text{pr}(A_i) + \sum_{i=1}^{R-1}\sum_{j=i+1}^{R} \text{pr}(A_i \cap A_j)$$

$$- \sum_{i=1}^{R-2}\sum_{j=i+1}^{R-1}\sum_{k=j+1}^{R} \text{pr}(A_i \cap A_j \cap A_k) + \cdots + (-1)^R \text{pr}\left(\cap_{i=1}^{R} A_i\right).$$

Now, for all $i$, $\text{pr}(A_i) = 1/R = (R-1)!/R!$. For all $i < j$,

$$\text{pr}(A_i \cap A_j) = \text{pr}(A_i)\text{pr}(A_j|A_i) = \left(\frac{1}{R}\right)\left(\frac{1}{R-1}\right) = \frac{(R-2)!}{R!}.$$

And, for $i < j < k$,

$$\begin{aligned}\text{pr}(A_i \cap A_j \cap A_k) &= \text{pr}(A_i)\text{pr}(A_j|A_i)\text{pr}(A_k|A_i \cap A_j) \\ &= \left(\frac{1}{R}\right)\left(\frac{1}{R-1}\right)\left(\frac{1}{R-2}\right) = \frac{(R-3)!}{R!}.\end{aligned}$$

In general, for $r = 1, 2, \ldots, R$, the probability of the intersection of any subset of $r$ of the $R$ events $A_1, A_2, \ldots, A_R$ is equal to $(R-r)!/R!$. Thus, we have

$$\begin{aligned}\theta(0, R) &= 1 - C_1^R \frac{(R-1)!}{R!} + C_2^R \frac{(R-2)!}{R!} \\ &\quad - C_3^R \frac{(R-3)!}{R!} + \cdots + \frac{(-1)^R}{R!} \\ &= 1 - 1 + \frac{1}{2!} - \frac{1}{3!} + \cdots + \frac{(-1)^R}{R!} \\ &= \sum_{l=0}^{R} \frac{(-1)^l}{l!}.\end{aligned}$$

So,

$$\begin{aligned}\lim_{R\to\infty}\theta(0, R) &= \lim_{R\to\infty} \sum_{l=0}^{R} \frac{(-1)^l}{l!} \\ &= \sum_{l=0}^{\infty} \frac{(-1)^l}{l!} = e^{-1} \approx 0.368,\end{aligned}$$

which is a somewhat counterintuitive answer.

(b) For a particular set of $r$ chips, let the event $B_r$ be the event that these $r$ chips are all in their correct positions, and let $C_{R-r}$ be the event that none of the remaining $(R-r)$ chips are in their correct positions. Then,

$$\begin{aligned}\text{pr}(B_r \cap C_{R-r}) &= \text{pr}(B_r)\text{pr}(C_{R-r}) \\ &= \left\{\left(\frac{1}{R}\right)\left(\frac{1}{R-1}\right)\cdots\left[\frac{1}{R-(r-1)}\right]\right\}\theta(0, R-r) \\ &= \left[\frac{(R-r)!}{R!}\right]\sum_{l=0}^{R-r}\frac{(-1)^l}{l!}.\end{aligned}$$

Finally, since there are $C_r^R$ ways of choosing a particular set of $r$ chips from a total of $R$ chips, it follows directly that

$$\theta(r,R) = C_r^R\left[\frac{(R-r)!}{R!}\right]\sum_{l=0}^{R-r}\frac{(-1)^l}{l!}$$
$$= \frac{\sum_{l=0}^{R-r}(-1)^l/l!}{r!}, \quad r = 0,1,\ldots,R.$$

(c) The probability of interest is

$$\theta(3,5)+\theta(4,5)+\theta(5,5) = \sum_{r=3}^{5}\sum_{l=0}^{5-r}\frac{(-1)^l/l!}{r!}$$
$$= \left(\frac{1}{3!}\right)\left(\frac{1}{2!}\right)+0+\left(\frac{1}{5!}\right)(1)$$
$$= \frac{1}{12}+\frac{1}{120} = \frac{11}{120} = 0.0917.$$

Note, in general, that $\theta(r-1,R) \equiv 0$ and that $\sum_{r=0}^{R}\theta(r,R) = 1$.

**Solution 1.27***

(a) First, let $H_A$ be the event that Player A obtains a head before Player B when it is Player A's turn to flip the balanced coin. In particular, if H is the event that a head is obtained when the balanced coin is flipped, and if T is the event that a tail is obtained, then

$$\text{pr}(H_A) = \text{pr}(H) + \text{pr}(T\cap T\cap H) + \text{pr}(T\cap T\cap T\cap T\cap H) + \cdots$$
$$= \frac{1}{2}+\frac{1}{8}+\frac{1}{32}+\cdots = \frac{2}{3}.$$

And, if $H_B$ is the event that Player A obtains a head before Player B when it is Player B's turn to flip the balanced coin, then

$$\text{pr}(H_B) = \text{pr}(T\cap H) + \text{pr}(T\cap T\cap T\cap H)$$
$$+ \text{pr}(T\cap T\cap T\cap T\cap T\cap H) + \cdots$$
$$= \frac{1}{4}+\frac{1}{16}+\frac{1}{64}+\cdots = \frac{1}{3}.$$

Then, we move from game $(a,b,\text{A})$ to game $(a-1,b,\text{B})$ if Player A obtains the next head before Player B (an event that occurs with probability 2/3); and, we move from game $(a,b,\text{A})$ to game $(a,b-1,\text{A})$ if Player B obtains the next head before Player A (an event that occurs with probability 1/3).
Thus, we have

$$\pi(a,b,\text{A}) = \left(\frac{2}{3}\right)\pi(a-1,b,\text{B}) + \left(\frac{1}{3}\right)\pi(a,b-1,\text{A}).$$

Using analogous arguments, we obtain

$$\pi(a, b, \mathrm{B}) = \left(\frac{1}{3}\right) \pi(a-1, b, \mathrm{B}) + \left(\frac{2}{3}\right) \pi(a, b-1, \mathrm{A}).$$

(b) First, note that the following boundary conditions hold:

$$\pi(0, b, \mathrm{A}) = \pi(0, b, \mathrm{B}) = 1, b = 1, 2, \ldots, \infty$$

and

$$\pi(a, 0, \mathrm{A}) = \pi(a, 0, \mathrm{B}) = 0, a = 1, 2, \ldots, \infty.$$

From part (a), we know that

$$\pi(1, 1, \mathrm{A}) = \frac{2}{3} \quad \text{and} \quad \pi(1, 1, \mathrm{B}) = \tfrac{1}{3}.$$

Now,

$$\pi(2, 2, \mathrm{A}) = \left(\frac{2}{3}\right) \pi(1, 2, \mathrm{B}) + \left(\frac{1}{3}\right) \pi(2, 1, \mathrm{A}),$$

so that we need to know the numerical values of $\pi(1, 2, \mathrm{B})$ and $\pi(2, 1, \mathrm{A})$.

So,

$$\begin{aligned}\pi(1, 2, \mathrm{B}) &= \left(\frac{1}{3}\right) \pi(0, 2, \mathrm{B}) + \left(\frac{2}{3}\right) \pi(1, 1, \mathrm{A}) \\ &= \left(\frac{1}{3}\right)(1) + \left(\frac{2}{3}\right)\left(\frac{2}{3}\right) = \frac{7}{9};\end{aligned}$$

and,

$$\begin{aligned}\pi(2, 1, \mathrm{A}) &= \left(\frac{2}{3}\right) \pi(1, 1, \mathrm{B}) + \left(\frac{1}{3}\right) \pi(2, 0, \mathrm{A}) \\ &= \left(\frac{2}{3}\right)\left(\frac{1}{3}\right) + \left(\frac{1}{3}\right)(0) = \frac{2}{9}.\end{aligned}$$

Finally,

$$\pi(2, 2, \mathrm{A}) = \left(\frac{2}{3}\right)\left(\frac{7}{9}\right) + \left(\frac{1}{3}\right)\left(\frac{2}{9}\right) = \frac{16}{27} = 0.593.$$

Now,

$$\pi(3, 3, \mathrm{A}) = \left(\frac{2}{3}\right) \pi(2, 3, \mathrm{B}) + \left(\frac{1}{3}\right) \pi(3, 2, \mathrm{A}),$$

where

$$\begin{aligned}\pi(2, 3, \mathrm{B}) &= \left(\frac{1}{3}\right) \pi(1, 3, \mathrm{B}) + \left(\frac{2}{3}\right) \pi(2, 2, \mathrm{A}) \\ &= \left(\frac{1}{3}\right) \pi(1, 3, \mathrm{B}) + \left(\frac{2}{3}\right)\left(\frac{16}{27}\right)\end{aligned}$$

and

$$\pi(3,2,\mathrm{A}) = \left(\frac{2}{3}\right)\pi(2,2,\mathrm{B}) + \left(\frac{1}{3}\right)\pi(3,1,\mathrm{A}).$$

Now,

$$\begin{aligned}\pi(2,2,\mathrm{B}) &= \left(\frac{1}{3}\right)\pi(1,2,\mathrm{B}) + \left(\frac{2}{3}\right)\pi(2,1,\mathrm{A})\\ &= \left(\frac{1}{3}\right)\left(\frac{7}{9}\right) + \left(\frac{2}{3}\right)\left(\frac{2}{9}\right) = \frac{11}{27};\end{aligned}$$

so,

$$\pi(3,2,\mathrm{A}) = \left(\frac{2}{3}\right)\left(\frac{11}{27}\right) + \left(\frac{1}{3}\right)\pi(3,1,\mathrm{A}).$$

Since

$$\begin{aligned}\pi(2,1,\mathrm{B}) &= \left(\frac{1}{3}\right)\pi(1,1,\mathrm{B}) + \left(\frac{2}{3}\right)\pi(2,0,\mathrm{A})\\ &= \left(\frac{1}{3}\right)\left(\frac{1}{3}\right) + \left(\frac{2}{3}\right)(0) = \frac{1}{9},\end{aligned}$$

we have

$$\begin{aligned}\pi(3,1,\mathrm{A}) &= \left(\frac{2}{3}\right)\pi(2,1,\mathrm{B}) + \left(\frac{1}{3}\right)\pi(3,0,\mathrm{A})\\ &= \left(\frac{2}{3}\right)\left(\frac{1}{9}\right) + \left(\frac{1}{3}\right)(0) = \frac{2}{27}.\end{aligned}$$

And, since

$$\begin{aligned}\pi(1,2,\mathrm{A}) &= \left(\frac{2}{3}\right)\pi(0,2,\mathrm{B}) + \left(\frac{1}{3}\right)\pi(1,1,\mathrm{A})\\ &= \left(\frac{2}{3}\right)(1) + \left(\frac{1}{3}\right)\left(\frac{2}{3}\right) = \frac{8}{9},\end{aligned}$$

we have

$$\begin{aligned}\pi(1,3,\mathrm{B}) &= \left(\frac{1}{3}\right)\pi(0,3,\mathrm{B}) + \left(\frac{2}{3}\right)\pi(1,2,\mathrm{A})\\ &= \left(\frac{1}{3}\right)(1) + \left(\frac{2}{3}\right)\left(\frac{8}{9}\right) = \frac{25}{27}.\end{aligned}$$

Finally, since

$$\pi(2,3,\mathrm{B}) = \left(\frac{1}{3}\right)\left(\frac{25}{27}\right) + \left(\frac{2}{3}\right)\left(\frac{16}{27}\right) = \frac{57}{81}$$

and

$$\pi(3,2,\text{A}) = \left(\frac{2}{3}\right)\left(\frac{11}{27}\right) + \left(\frac{1}{3}\right)\left(\frac{2}{27}\right) = \frac{24}{81},$$

we have

$$\pi(3,3,\text{A}) = \left(\frac{2}{3}\right)\left(\frac{57}{81}\right) + \left(\frac{1}{3}\right)\left(\frac{24}{81}\right) = \frac{46}{81} = 0.568.$$

Clearly, this procedure can be programed to produce the numerical value of $\pi(k,k,\text{A})$ for any positive integer $k$. For example, the reader can verify that $\pi(4,4,\text{A}) = 0.556$ and that $\pi(5,5,\text{A}) = 0.549$. In general, $\pi(k,k,\text{A})$ monotonically decreases toward the value $1/2$ as $k$ becomes large, but the rate of decrease is relatively slow.

**Solution 1.28***

(a) Now,

$$\begin{aligned}
\alpha = \text{pr}(\text{A}_{ab}|\text{B}_1) &= \text{pr}(\text{A}_{ab} \cap \text{B}_a|\text{B}_1) + \text{pr}(\text{A}_{ab} \cap \bar{\text{B}}_a|\text{B}_1) \\
&= \text{pr}(\text{A}_{ab}|\text{B}_a \cap \text{B}_1)\text{pr}(\text{B}_a|\text{B}_1) + \text{pr}(\text{A}_{ab}|\bar{\text{B}}_a \cap \text{B}_1)\text{pr}(\bar{\text{B}}_a|\text{B}_1) \\
&= (1)\pi^{a-1} + \beta[1 - \pi^{a-1}] \\
&= \pi^{a-1} + \beta[1 - \pi^{a-1}],
\end{aligned}$$

since the event "$\text{A}_{ab}$ given $\bar{\text{B}}_a \cap \text{B}_1$" is equivalent to the event "$\text{A}_{ab}$ given $\bar{\text{B}}_1$." More specifically, the event "$\bar{\text{B}}_a \cap \text{B}_1$" means that the first free throw is made and that there is at least one missed free throw among the next $(a-1)$ free throws. And, when such a miss occurs, it renders irrelevant all the previous makes, and so the scenario becomes exactly that of starting with a missed free throw (namely, the event "$\bar{\text{B}}_1$").

Similarly,

$$\begin{aligned}
\beta = \text{pr}(\text{A}_{ab}|\bar{\text{B}}_1) &= \text{pr}(\text{A}_{ab} \cap \text{C}_b|\bar{\text{B}}_1) + \text{pr}(\text{A}_{ab} \cap \bar{\text{C}}_b|\bar{\text{B}}_1) \\
&= \text{pr}(\text{A}_{ab}|\text{C}_b \cap \bar{\text{B}}_1)\text{pr}(\text{C}_b|\bar{\text{B}}_1) + \text{pr}(\text{A}_{ab}|\bar{\text{C}}_b \cap \bar{\text{B}}_1)\text{pr}(\bar{\text{C}}_b|\bar{\text{B}}_1) \\
&= (0)(1-\pi)^{b-1} + \alpha[1 - (1-\pi)^{b-1}] \\
&= \alpha[1 - (1-\pi)^{b-1}],
\end{aligned}$$

since the event "$\text{A}_{ab}$ given $\bar{\text{C}}_b \cap \bar{\text{B}}_1$" is equivalent to the event "$\text{A}_{ab}$ given $\text{B}_1$."

Solving these two equations simultaneously, we have

$$\begin{aligned}
\alpha &= \pi^{a-1} + \beta[1 - \pi^{a-1}] \\
&= \pi^{a-1} + \{\alpha[1 - (1-\pi)^{b-1}]\}[1 - \pi^{a-1}],
\end{aligned}$$

giving

$$\alpha = \frac{\pi^{a-1}}{\pi^{a-1} + (1-\pi)^{b-1} - \pi^{a-1}(1-\pi)^{b-1}}$$

and

$$\beta = \frac{\pi^{a-1}[1-(1-\pi)^{b-1}]}{\pi^{a-1}+(1-\pi)^{b-1}-\pi^{a-1}(1-\pi)^{b-1}}.$$

Finally, it follows directly that

$$\theta(\pi, a, b) = \pi\alpha + (1-\pi)\beta = \frac{\pi^{a-1}[1-(1-\pi)^{b}]}{\pi^{a-1}+(1-\pi)^{b-1}-\pi^{a-1}(1-\pi)^{b-1}}.$$

When $\pi = 0.791, a = 10$, and $b = 2$, then $\theta(0.791, 10, 2) = 0.38$.

(b) When both $\pi = 0.50$ and $a = b$, then $\theta(0.50, a, a) = \theta(0.50, b, b) = 0.50$; this answer makes sense because runs of makes and misses of the same length are equally likely when $\pi = 0.50$. When $a = b = 1$, then $\theta(\pi, 1, 1) = \pi$; this answer also makes sense because the event $A_{11}$ (i.e., the event that the first free throw is made) occurs with probability $\pi$. Finally, once several consecutive free throws are made, the pressure to continue the run of made free throws will increase; as a result, the assumption of mutual independence among the outcomes of consecutive free throws is probably not valid and the value of $\pi$ would tend to decrease.

(c) Since the probability of Tyler missing $b$ consecutive free throws before making $a$ consecutive free throws is equal to

$$\theta(1-\pi, b, a) = \frac{(1-\pi)^{b-1}(1-\pi^{a})}{(1-\pi)^{b-1}+\pi^{a-1}-(1-\pi)^{b-1}\pi^{a-1}},$$

it follows directly that $\theta(\pi, a, b) + \theta(1-\pi, b, a) = 1$.

# 2

# *Univariate Distribution Theory*

## 2.1 Concepts and Notation

### 2.1.1 Discrete and Continuous Random Variables

A *discrete* random variable $X$ takes either a finite, or a countably infinite, number of values. A discrete random variable $X$ is characterized by its probability distribution $p_X(x) = \text{pr}(X = x)$, which is a formula giving the probability that $X$ takes the (permissible) value $x$. Hence, a valid discrete probability distribution $p_X(x)$ has the following two properties:

i. $0 \leq p_X(x) \leq 1$ for all (permissible) values of $x$ and
ii. $\sum_{\text{all } x} p_X(x) = 1$.

A *continuous* random variable $X$ can theoretically take all the real (and hence uncountably infinite) numerical values on a line segment of either finite or infinite length. A continuous random variable $X$ is characterized by its density function $f_X(x)$. A valid density function $f_X(x)$ has the following properties:

i. $0 \leq f_X(x) < +\infty$ for all (permissible) values of $x$;
ii. $\int_{\text{all } x} f_X(x)\,dx = 1$;
iii. For $-\infty < a < b < +\infty$, $\text{pr}(a < X < b) = \int_a^b f_X(x)\,dx$; and
iv. $\text{pr}(X = x) = 0$ for any particular value $x$, since $\int_x^x f_X(x)\,dx = 0$.

### 2.1.2 Cumulative Distribution Functions

In general, the cumulative distribution function (CDF) for a univariate random variable $X$ is the function $F_X(x) = \text{pr}(X \leq x), -\infty < x < +\infty$, which possesses the following properties:

i. $0 \leq F_X(x) \leq 1, -\infty < x < +\infty$;
ii. $F_X(x)$ is a monotonically nondecreasing function of $x$; and
iii. $\lim_{x \to -\infty} F_X(x) = 0$ and $\lim_{x \to +\infty} F_X(x) = 1$.

For an integer-valued discrete random variable $X$, it follows that

i. $F_X(x) = \sum_{\text{all } x^* \leq x} p_X(x^*)$;
ii. $p_X(x) = \text{pr}(X = x) = F_X(x) - F_X(x-1)$; and
iii. $[dF_X(x)]/dx \neq p_X(x)$ since $F_X(x)$ is a discontinuous function of $x$.

For a continuous random variable $X$, it follows that

i. $F_X(x) = \int_{\text{all} x^* \leq x} f_X(x^*)\, dx^*$;
ii. For $-\infty < a < x < b < +\infty, \text{pr}(a < X < b) = F_X(b) - F_X(a)$; and
iii. $[dF_X(x)]/dx = f_X(x)$ since $F_X(x)$ is an absolutely continuous function of $x$.

### 2.1.3 Median and Mode

For any discrete distribution $p_X(x)$ or density function $f_X(x)$, the population *median* $\xi$ satisfies the two inequalities

$$\text{pr}(X \leq \xi) \geq \tfrac{1}{2} \quad \text{and} \quad \text{pr}(X \geq \xi) \geq \tfrac{1}{2}.$$

For a density function $f_X(x)$, $\xi$ is that value of $X$ such that

$$\int_{-\infty}^{\xi} f_X(x)\, dx = \frac{1}{2}.$$

The population *mode* for either a discrete probability distribution $p_X(x)$ or a density function $f_X(x)$ is a value of $x$ that maximizes $p_X(x)$ or $f_X(x)$. The population mode is not necessarily unique, since $p_X(x)$ or $f_X(x)$ may achieve its maximum for several different values of $x$; in this situation, all these local maxima are called modes.

### 2.1.4 Expectation Theory

Let $g(X)$ be any scalar function of a univariate random variable $X$. Then, the *expected value* $E[g(X)]$ of $g(X)$ is defined to be

$$E[g(X)] = \sum_{\text{all } x} g(x) p_X(x) \text{ when } X \text{ is a discrete random variable,}$$

and is defined to be

$$E[g(X)] = \int_{\text{all } x} g(x) f_X(x)\, dx \text{ when } X \text{ is a continuous random variable.}$$

Note that $E[g(X)]$ is said to exist if $|E[g(X)]| < +\infty$; otherwise, $E[g(X)]$ is said not to exist.

Some general rules for computing expectations are:

i. If C is a constant independent of $X$, then $E(C) = C$;
ii. $E[Cg(X)] = CE[g(X)]$;
iii. If $C_1, C_2, \ldots, C_k$ are $k$ constants all independent of $X$, and if $g_1(X), g_2(X), \ldots, g_k(X)$ are $k$ scalar functions of $X$, then

$$E\left[\sum_{i=1}^{k} C_i g_i(X)\right] = \sum_{i=1}^{k} C_i E[g_i(X)];$$

iv. If $k \to \infty$, then

$$E\left[\sum_{i=1}^{\infty} C_i g_i(X)\right] = \sum_{i=1}^{\infty} C_i E[g_i(X)]$$

when $|\sum_{i=1}^{\infty} C_i E[g_i(X)]| < +\infty$.

### 2.1.5 Some Important Expectations

#### 2.1.5.1 Mean

$\mu = E(X)$ is the *mean* of X.

#### 2.1.5.2 Variance

$\sigma^2 = V(X) = E\{[X - E(X)]^2\}$ is the *variance* of X, and $\sigma = +\sqrt{\sigma^2}$ is the *standard deviation* of X.

#### 2.1.5.3 Moments

More generally, if $r$ is a positive integer, a binomial expansion of $[X - E(X)]^r$ gives

$$E\{[X - E(X)]^r\} = E\left\{\sum_{j=0}^{r} C_j^r X^j [-E(X)]^{r-j}\right\} = \sum_{j=0}^{r} C_j^r (-1)^{r-j} E(X^j)[E(X)]^{r-j},$$

where $E\{[X - E(X)]^r\}$ is the *rth moment about the mean.*

For example, for $r = 2$, we obtain

$$E\{[X - E(X)]^2\} = V(X) = E(X^2) - [E(X)]^2;$$

and, for $r = 3$, we obtain

$$E\{[X - E(X)]^3\} = E(X^3) - 3E(X^2)E(X) + 2[E(X)]^3,$$

which is a measure of the *skewness* of the distribution of X.

#### 2.1.5.4 Moment Generating Function

$M_X(t) = E(e^{tX})$ is called the *moment generating function* for the random variable X, provided that $M_X(t) < +\infty$ for $t$ in some neighborhood of 0 [i.e., for all $t \in (-\epsilon, \epsilon), \epsilon > 0$]. For $r$ a positive integer, and with $E(X^r)$ defined as the *rth moment about the origin* (i.e., about 0) for the random variable X, then $M_X(t)$ can be used to generate moments about the origin via the algorithm

$$\frac{d^r M_X(t)}{dt^r}_{|t=0} = E(X^r).$$

More generally, for $r$ a positive integer, the function

$$M_X^*(t) = E\left\{e^{t[X-E(X)]}\right\} = e^{-tE(X)}M_X(t)$$

can be used to generate moments about the mean via the algorithm

$$\frac{d^r M_X^*(t)}{dt^r}_{|t=0} = E\{[X - E(X)]^r\}.$$

#### 2.1.5.5 Probability Generating Function

If we let $e^t$ equal $s$ in $M_X(t) = E(e^{tX})$, we obtain the *probability generating function* $P_X(s) = E(s^X)$. Then, for $r$ a positive integer, and with

$$E\left[\frac{X!}{(X-r)!}\right] = E[X(X-1)(X-2)\cdots(X-r+1)]$$

defined as the *rth factorial moment* for the random variable X, then $P_X(s)$ can be used to generate factorial moments via the algorithm

$$\frac{d^r P_X(s)}{ds^r}_{|s=1} = E\left[\frac{X!}{(X-r)!}\right].$$

As an example, the probability generating function $P_X(s)$ can be used to find the variance of X when V(X) is written in the form

$$V(X) = E[X(X-1)] + E(X) - [E(X)]^2.$$

### 2.1.6 Inequalities Involving Expectations

#### 2.1.6.1 Markov's Inequality

If $X$ is a nonnegative random variable [i.e., $\text{pr}(X \geq 0) = 1$], then $\text{pr}(X > k) \leq \text{E}(X)/k$ for any constant $k > 0$. As a special case, for $r > 0$, if $X = |Y - \text{E}(Y)|^r$ when $Y$ is any random variable, then, with $\nu_r = \text{E}\left[|Y - \text{E}(Y)|^r\right]$, we have

$$\text{pr}\left[|Y - \text{E}(Y)|^r > k\right] \leq \frac{\nu_r}{k},$$

or equivalently with $k = t^r \nu_r$,

$$\text{pr}\left[|Y - \text{E}(Y)| > t\nu_r^{1/r}\right] \leq t^{-r}, \quad t > 0.$$

For $r = 2$, we obtain *Tchebyshev's Inequality*, namely,

$$\text{pr}\left[|Y - \text{E}(Y)| > t\sqrt{\text{V}(Y)}\right] \leq t^{-2}, \quad t > 0.$$

#### 2.1.6.2 Jensen's Inequality

Let $X$ be a random variable with $|\text{E}(X)| < \infty$. If $\text{g}(X)$ is a *convex* function of $X$, then $\text{E}[\text{g}(X)] \geq \text{g}[\text{E}(X)]$, provided that $|\text{E}[\text{g}(X)]| < \infty$. If $\text{g}(X)$ is a *concave* function of $X$, then the inequality is reversed, namely, $\text{E}[\text{g}(X)] \leq \text{g}[\text{E}(X)]$.

#### 2.1.6.3 Hölder's Inequality

Let $X$ and $Y$ be random variables, and let $p, 1 < p < \infty$, and $q, 1 < q < \infty$, satisfy the restriction $1/p + 1/q = 1$. Then,

$$\text{E}(|XY|) \leq \left[\text{E}(|X|^p)\right]^{1/p} \left[\text{E}(|Y|^q)\right]^{1/q}.$$

As a special case, when $p = q = 2$, we obtain the *Cauchy–Schwartz Inequality*, namely,

$$\text{E}(|XY|) \leq \sqrt{\text{E}(X^2)\text{E}(Y^2)}.$$

### 2.1.7 Some Important Probability Distributions for Discrete Random Variables

#### 2.1.7.1 Binomial Distribution

If $X$ is the number of successes in $n$ trials, where the trials are conducted independently with the probability $\pi$ of success remaining the same from trial to trial, then

$$\text{p}_X(x) = \text{C}_x^n \pi^x (1 - \pi)^{n-x}, \quad x = 0, 1, \ldots, n \quad \text{and} \quad 0 < \pi < 1.$$

When $X \sim \text{BIN}(n, \pi)$, then $\text{E}(X) = n\pi, V(X) = n\pi(1-\pi)$, and $\text{M}_X(t) = [\pi e^t + (1-\pi)]^n$.

When $n = 1$, X has the *Bernoulli* distribution.

#### 2.1.7.2 Negative Binomial Distribution

If $Y$ is the number of trials required to obtain exactly $k$ successes, where $k$ is a specified positive integer, and where the trials are conducted independently with the probability $\pi$ of success remaining the same from trial to trial, then

$$\text{p}_Y(y) = \text{C}_{k-1}^{y-1}\pi^k(1-\pi)^{y-k}, \quad y = k, k+1, \ldots, \infty \quad \text{and} \quad 0 < \pi < 1.$$

When $Y \sim \text{NEGBIN}(k, \pi)$, then $\text{E}(Y) = k/\pi, \text{V}(Y) = k(1-\pi)/\pi^2$, and

$$\text{M}_Y(t) = \left[\frac{\pi e^t}{1-(1-\pi)e^t}\right]^k.$$

In the special case when $k = 1$, then $Y$ has a *geometric* distribution, namely,

$$\text{p}_Y(y) = \pi(1-\pi)^{y-1}, \quad y = 1, 2, \ldots, \infty \quad \text{and} \quad 0 < \pi < 1.$$

When $Y \sim \text{GEOM}(\pi)$, then $\text{E}(Y) = 1/\pi, \text{V}(Y) = (1-\pi)/\pi^2$, and $\text{M}_Y(t) = \pi e^t/[1-(1-\pi)e^t]$.

When $X \sim \text{BIN}(n, \pi)$ and when $Y \sim \text{NEGBIN}(k, \pi)$, then $\text{pr}(X < k) = \text{pr}(Y > n)$.

#### 2.1.7.3 Poisson Distribution

As a model for rare events, the Poisson distribution can be derived as a limiting case of the binomial distribution as $n \to \infty$ and $\pi \to 0$ with $\lambda = n\pi$ held constant; this limit is

$$\text{p}_X(x) = \frac{\lambda^x e^{-\lambda}}{x!}, \quad x = 0, 1, \ldots, \infty \quad \text{and} \quad \lambda > 0.$$

When $X \sim \text{POI}(\lambda)$, then $\text{E}(X) = \text{V}(X) = \lambda$ and $\text{M}_X(t) = e^{\lambda(e^t-1)}$.

#### 2.1.7.4 Hypergeometric Distribution

Suppose that a finite-sized population of size $N(< +\infty)$ contains $a$ items of Type A and $b$ items of Type B, with $(a+b) = N$. If a sample of $n(< N)$ items is randomly selected *without replacement* from this population of $N$ items, then the number $X$ of items of Type A contained in this sample of $n$ items has the hypergeometric distribution, namely,

$$\text{p}_X(x) = \frac{\text{C}_x^a \text{C}_{n-x}^b}{\text{C}_n^{a+b}} = \frac{\text{C}_x^a \text{C}_{n-x}^{N-a}}{\text{C}_n^N}, \quad \max(0, n-b) \leq X \leq \min(n, a).$$

When $X \sim \text{HG}(a, N-a, n)$, then

$$\text{E}(X) = n\left(\frac{a}{N}\right) \quad \text{and} \quad \text{V}(X) = n\left(\frac{a}{N}\right)\left(\frac{N-a}{N}\right)\left(\frac{N-n}{N-1}\right).$$

### 2.1.8 Some Important Distributions (i.e., Density Functions) for Continuous Random Variables

#### 2.1.8.1 *Normal Distribution*

The normal distribution density function is

$$f_X(x) = \frac{1}{\sqrt{2\pi}\sigma}e^{-(x-\mu)^2/2\sigma^2}, \qquad -\infty < x < \infty, \quad -\infty < \mu < \infty, \quad 0 < \sigma^2 < \infty.$$

When $X \sim \text{N}(\mu, \sigma^2)$, then $\text{E}(X) = \mu$, $\text{V}(X) = \sigma^2$, and $\text{M}_X(t) = e^{\mu t + \sigma^2 t^2/2}$. Also, when $X \sim \text{N}(\mu, \sigma^2)$, then the standardized variable $Z = (X - \mu)/\sigma \sim \text{N}(0, 1)$, with density function

$$f_Z(z) = \frac{1}{\sqrt{2\pi}}e^{-z^2/2}, \quad -\infty < z < \infty.$$

#### 2.1.8.2 *Lognormal Distribution*

When $X \sim \text{N}(\mu, \sigma^2)$, then the random variable $Y = e^X$ has a *lognormal distribution*, with density function

$$f_Y(y) = \frac{1}{\sqrt{2\pi}\sigma y}e^{-[\ln(y)-\mu]^2/2\sigma^2}, \quad 0 < y < \infty, \quad -\infty < \mu < \infty, \quad 0 < \sigma^2 < \infty.$$

When $Y \sim \text{LN}(\mu, \sigma^2)$, then $\text{E}(Y) = e^{\mu + (\sigma^2/2)}$ and $\text{V}(Y) = [\text{E}(Y)]^2(e^{\sigma^2} - 1)$.

#### 2.1.8.3 *Gamma Distribution*

The gamma distribution density function is

$$f_X(x) = \frac{x^{\beta-1}e^{-x/\alpha}}{\Gamma(\beta)\alpha^\beta}, \qquad 0 < x < \infty, \quad 0 < \alpha < \infty, \quad 0 < \beta < \infty.$$

When $X \sim \text{GAMMA}(\alpha, \beta)$, then $\text{E}(X) = \alpha\beta$, $\text{V}(X) = \alpha^2\beta$, and $\text{M}_X(t) = (1 - \alpha t)^{-\beta}$. The Gamma distribution has two important special cases:

i. When $\alpha = 2$ and $\beta = \nu/2$, then $X \sim \chi^2_\nu$ (i.e., $X$ has a *chi-squared distribution* with $\nu$ degrees of freedom). When $X \sim \chi^2_\nu$, then

$$f_X(x) = \frac{x^{\frac{\nu}{2}-1}e^{-x/2}}{\Gamma\left(\frac{\nu}{2}\right)2^{\nu/2}}, \qquad 0 < x < \infty \quad \text{and} \quad \nu \text{ a positive integer;}$$

also, $\mathrm{E}(X) = \nu, \mathrm{V}(X) = 2\nu$, and $\mathrm{M}_X(t) = (1-2t)^{-\nu/2}$. And, if $Z \sim \mathrm{N}(0,1)$, then $Z^2 \sim \chi_1^2$.

ii. When $\beta = 1$, then $X$ has a *negative exponential* distribution with density function

$$\mathrm{f}_X(x) = \frac{1}{\alpha}\mathrm{e}^{-x/\alpha}, \quad 0 < x < \infty, \quad 0 < \alpha < \infty.$$

When $X \sim \mathrm{NEGEXP}(\alpha)$, then $\mathrm{E}(X) = \alpha, \mathrm{V}(X) = \alpha^2$, and $\mathrm{M}_X(t) = (1-\alpha t)^{-1}$.

#### 2.1.8.4 Beta Distribution

The Beta distribution density function is

$$\mathrm{f}_X(x) = \frac{\Gamma(\alpha+\beta)}{\Gamma(\alpha)\Gamma(\beta)} x^{\alpha-1}(1-x)^{\beta-1}, \qquad 0 < x < 1, \quad 0 < \alpha < \infty, \quad 0 < \beta < \infty.$$

When $X \sim \mathrm{BETA}(\alpha, \beta)$, then $\mathrm{E}(X) = \frac{\alpha}{\alpha+\beta}$ and $\mathrm{V}(X) = \frac{\alpha\beta}{(\alpha+\beta)^2(\alpha+\beta+1)}$.

#### 2.1.8.5 Uniform Distribution

The Uniform distribution density function is

$$\mathrm{f}_X(x) = \frac{1}{(\theta_2 - \theta_1)}, \qquad -\infty < \theta_1 < x < \theta_2 < \infty.$$

When $X \sim \mathrm{UNIF}(\theta_1, \theta_2)$, then $\mathrm{E}(X) = \frac{(\theta_1+\theta_2)}{2}$, $\mathrm{V}(X) = \frac{(\theta_2-\theta_1)^2}{12}$ and $\mathrm{M}_X(t) = \frac{(\mathrm{e}^{t\theta_2} - \mathrm{e}^{t\theta_1})}{t(\theta_2-\theta_1)}$.

## EXERCISES

### Exercise 2.1

(a) In a certain small group of seven people, suppose that exactly four of these people have a certain rare blood disorder. If individuals are selected at random one-at-a-time *without replacement* from this group of seven people, find the numerical value of the *expected number* of individuals that have to be selected in order to obtain one individual with this rare blood disorder and one individual without this rare blood disorder.

(b) Now, consider a finite-sized population of size $N\,(< +\infty)$ in which there are exactly $M\,(2 \leq M < N)$ individuals with this rare blood disorder. Suppose that individuals are selected from this population at random one-at-a-time *without replacement*. Let the random variable $X$ denote the number of individuals selected until exactly $k\,(1 \leq k \leq M < N)$ individuals are selected who have this rare blood disorder. *Derive* an *explicit expression* for the probability distribution of the random variable $X$.

(c) Given the conditions described in part (b), *derive* an *explicit expression* for the probability that the third individual selected has this rare blood disorder.

**Exercise 2.2.** Suppose that the positive integers $1, 2, \ldots, k$, $k \geq 3$, are arranged randomly in a horizontal line, thus occupying $k$ slots. Assume that all arrangements of these $k$ integers are equally likely.

(a) Derive the probability distribution $\mathrm{p}_X(x)$ of the discrete random variable $X$, where $X$ is the number of integers between the integers 1 and $k$. Also, show directly that $\mathrm{p}_X(x)$ is a valid discrete probability distribution.

(b) Develop an explicit expression for $\mathrm{E}(X)$.

**Exercise 2.3.** Consider an urn that contains four white balls and two black balls.

(a) Suppose that *pairs* of balls are selected from this urn *without replacement*; in particular, the first two balls selected (each ball selected without replacement) constitute the first pair, the next two balls selected constitute the second pair, and so on. Find numerical values for $\mathrm{E}(Y)$ and $\mathrm{V}(Y)$, where $Y$ is the number of black balls remaining in the urn after the first pair of white balls is selected.

(b) Now, suppose that *pairs* of balls are selected from this urn *with replacement* in the following manner: the first ball in a pair is randomly selected, its color is recorded, and then it is returned to the urn; then, the second ball making up this particular pair is randomly selected, its color is recorded, and then it is returned to the urn. Provide an explicit expression for the probability distribution of the random variable $X$, the number of pairs of balls that have to be selected in this manner until exactly two pairs of white balls are obtained (i.e., both balls in each of these two pairs are white)?

**Exercise 2.4.** To estimate the unknown size $N(< +\infty)$ of a population (e.g., the number of bass in a particular lake, the number of whales in a particular ocean, the number of birds of a specific species in a particular forest, etc.), a sampling procedure known as *capture–recapture* is often employed. This capture–recapture sampling method works as follows. For the first stage of sampling, $m$ animals are randomly chosen (i.e., captured) from the population of animals under study and are then individually marked to permit future identification. Then, these $m$ marked animals are released back into the population of animals under study. At the second stage of sampling, which occurs at some later time, $n(<m)$ animals are then randomly chosen (i.e., captured) from a population (of unknown size $N$) that now contains both $m$ marked animals and an unknown number of unmarked animals.

(a) Assuming (for now) that the size $N$ of the population under study is known, provide an explicit expression for the probability distribution of $X$, the number of marked animals in the set of $n(<m)$ randomly chosen animals obtained at the second stage of sampling.

(b) Again, assume (for now) that the size $N$ of the population under study is known. At the first stage of sampling, if the marks on the $m$ randomly chosen animals consist of the positive integers $1, 2, \ldots, m$, derive an explicit expression for the probability $\pi$ that a set of $n(4 < n < m)$ animals randomly chosen at the second stage of sampling contains at least two animals that were marked with any of the positive integers 1, 2, 3, and 4.

(c) Since the value of $N$ is actually unknown, the purpose of the capture–recapture sampling method is to provide an estimate of $N$ using the observed value $x$ of $X$ and the known sample sizes $m$ and $n$. Using logical arguments, suggest a formula for an estimate $\hat{N}$ of $N$ that is a function of $x$, $m$, and $n$. If $x = 22$, $m = 600$, and $n = 300$, compute the numerical value of $\hat{N}$. Do you notice any obvious problems associated with the use of the formula for $\hat{N}$ that you have developed?

**Exercise 2.5.** A researcher at the National Center for Health Statistics (NCHS) is interested in obtaining in-depth interviews from people in each of $k(\geq 2)$ health status categories. In what follows, assume that:

(i) this researcher interviews *exactly one randomly chosen person every day*;

(ii) each person randomly chosen to be interviewed is *equally likely* to be in any one of the $k$ health status categories.

This NCHS researcher is concerned that it will take her a considerable amount of time to interview at least one person in each of the $k$ health status categories, and so she asks the following design-related question: "*Given that* I have interviewed people in exactly $c$ different health status categories by the end of today, where $0 \leq c \leq (k-1)$, what is the (conditional) probability that I will encounter a person in a *new* health status category exactly $x(\geq 1)$ days from today?"

Develop an *explicit expression* for this conditional probability. Then, use this result to derive an expression for the *expected total number of days* required for this researcher to encounter at least one person in every health status category; also, find the numerical value of this expected value expression when $k = 4$.

**Exercise 2.6.** In a certain state lottery, suppose that the probability of buying a jackpot-winning ticket for a particular game is $\pi = 0.0005$.

(a) Suppose that a person wishes to buy $n$ tickets for this particular game. What is the smallest value, say $n^*$, of the number of tickets $n$ that this person needs to buy to have a probability of at least 0.90 of purchasing at least one jackpot-winning ticket? Use both the binomial and Poisson distributions to determine the value of $n^*$, and then comment on the numerical results.

(b) If each of the $n^*$ tickets purchased by this person costs \$1.00, what should be the smallest dollar amount A of the jackpot so that this person's expected net profit $\mathrm{E}(P)$ after purchasing $n^*$ tickets is nonnegative?

(c) If, in fact, a total of $N$ tickets are purchased for this game by lottery participants in this state, and if $K(0 < K < N)$ of these tickets are actually jackpot-winning tickets, develop an expression for the probability that at least $k(1 \leq k \leq n^*)$ of the $n^*(1 \leq n^* < N)$ tickets purchased by this person are jackpot-winning tickets.

**Exercise 2.7.** The Rhine Research Center, which studies parapsychology and related phenomena, is located near Duke University in Durham, North Carolina. It has been suggested by a certain parapsychologist employed by the Rhine Research Center that there could be extra-sensory perception (ESP) between monozygotic twins. To test this theory, this parapsychologist designs the following simple experiment. Each twin thinks of a particular whole number between 1 and $k$ inclusive (namely, each twin picks one of the $k$ numbers $1, 2, \ldots, k$), and then writes that number on a piece of paper. The two numbers that are written down are then compared to see whether or not they are the same. Let the random variable $Y$ take the value 1 if the two numbers are the same, and let $Y$ take the value 0 otherwise.

(a) Under the assumption that there is no ESP between a pair of monozygotic twins (i.e., each twin is picking his or her number totally at random), what is the exact probability distribution of the dichotomous random variable $Y$?

(b) Suppose that this parapsychologist is willing to declare that any pair of monozygotic twins possesses ESP if those twins choose numbers that are the same in one repetition of the experiment. However, this parapsychologist realizes that $k$ must be large enough to make such a declaration appear statistically credible. Help this parapsychologist out by determining the smallest value of $k$ required such that the probability of monozygotic twins with no ESP choosing matching numbers in one repetition of the experiment is no larger than 0.01?

(c) Using the value of $k$ determined in part (b), suppose that this parapsychologist runs this experiment independently on $n = 100$ different sets of monozygotic twins. If none of these sets of monozygotic twins actually has ESP, how likely is it that this parapsychologist will incorrectly declare that at least one set of monozygotic twins actually has ESP? Comment on this finding.

(d) For a particular set of monozygotic twins, suppose that this experiment is *independently repeated* 10 times using the value of $k$ determined in part (b). If these twins choose the same number in exactly 2 of the 10 independent repetitions of this parapsychological experiment, do you think that these data provide evidence of ESP or not?

(e) Suppose that only one repetition of the experiment is carried out using $k = 4$. Define the random variable $S$ to be the *sum* of the two numbers chosen by the monozygotic twins under study. *Given that* the two numbers chosen by the twins are *not* the same and *given that* the twins are choosing their numbers totally at random, derive the exact probability distribution of $S$ and find $\mathrm{E}(S)$ given the stated conditions.

**Exercise 2.8.** In order to have clinical expression of a mutagenic disease, it has been argued that two distinct steps have to occur. First, a mutagenic process starts with genetic damage. A mutagen (e.g., an agent like ionizing radiation) causes defects (or "breakpoints") in the DNA of human genetic material that produces the initial mutant cell. However, for a mutagenic process to be clinically expressed as a mutagenic disease, a second step is necessary, namely, the damaged (or mutant) cell (i.e., a cell with at least one breakpoint) must be able to clone (i.e., to reproduce its damaged self) effectively. A damaged cell that retains its ability to clone is said to be *viable*. In particular, the clinical expression of genetic damage (say, as a detectable cancer) cannot occur until the cell population cloned from the viable damaged cell is very large.

Suppose that we want to develop a statistical model for the above two-step mutagenic process involving a single cell exposed to ionizing radiation. To start, assume that the number $Y$ of breakpoints in the initial damaged (or mutant) cell has the truncated Poisson distribution

$$p_Y(y) = \frac{\lambda^y}{y!(e^{\lambda} - 1)}, \quad y = 1, 2, \ldots, \infty \quad \text{and} \quad \lambda > 0.$$

(a) Find an explicit expression for $\text{pr}(Y \leq 3 | Y \geq 2)$.

(b) For $r = 1, 2, \ldots$, derive an explicit expression for

$$\text{E}\left[\frac{Y!}{(Y-r)!}\right] = \text{E}[Y(Y-1)\cdots(Y-r+1)],$$

and then use this expression to find $\text{E}(Y)$ and $\text{V}(Y)$.

(c) Now, as a simple model, let the probability that there is no loss of viability (i.e., no serious inhibition of the damaged cell's reproductive capability) due to any one breakpoint be equal to $\pi, 0 < \pi < 1$. Then, if V is the event that a damaged cell is viable, assume that $\text{pr}(\text{V}|Y = y) = \pi^y$, where $\text{pr}(\text{V}|Y = y)$ is the probability that a cell is viable given that it has $y$ breakpoints, $y = 1, 2, \ldots, \infty$. This assumption is meant to reflect the fact that the viability of a damaged cell will decrease as the number of breakpoints increases. Develop an explicit expression for the probability $\theta$ that a damaged cell is viable.

**Exercise 2.9.** A certain production process is designed to make electric light bulbs, with each light bulb intended to have an exact wattage value of 30 watts. However, because of problems with the production process, the actual wattage of a light bulb made by this production process can be considered to be a continuous random variable $W$ that can be accurately modeled by the equation

$$W = 31 + (0.50)U,$$

where $U \sim \text{N}(0, 4)$. Find the exact numerical value of the probability that an electric light bulb made by this production process will have a wattage that does not deviate from the desired value of 30 watts by more than 0.50 watts.

**Exercise 2.10.** A certain company employs two manufacturing processes, Process 1 and Process 2, for producing very small square-shaped computer chips to be used in human hearing aids. Suppose that X, the *diagonal* of a computer chip in centimeters, is a continuous random variable with process-specific density functions defined as follows:

$$\text{Process 1:} \quad f_X(x) = 3.144e^{-x}, \quad 1.0 < x < 3.0;$$

$$\text{Process 2:} \quad f_X(x) = 2.574e^{-x}, \quad 0.8 < x < 2.8.$$

Only computer chips with diagonals between 1.0 and 2.0 centimeters are usable.

(a) Suppose that Process 1 produces 1000 computer chips per day, and that Process 2 produces 2000 computer chips per day. Further, at the end of each day, suppose that all 3000 computer chips are put into a large container and mixed together, thus making it impossible to tell which manufacturing process produced any particular computer chip. Suppose that two computer chips are selected randomly *with replacement* from this large container, and further suppose that one of these two computer chips is found to be usable and the other computer chip is found to be unusable. Determine the numerical value of the probability that both of these computer chips were produced by Process 1.

(b) If computer chips are selected randomly one-at-a-time *with replacement* from the container described in part (a), provide an *explicit expression* for the probability distribution $p_Y(y)$ of the discrete random variable $Y$, where $Y$ is the number of computer chips that have to be selected until at least two usable computer chips *and* at least one unusable computer chip are obtained. Also, prove directly that $p_Y(y)$ is a valid discrete probability distribution.

**Exercise 2.11.** Racing car windshields made of a new impact-resistant glass are tested for breaking strength by striking them repeatedly with a mechanical device that simulates the stresses caused by high-speed crashes in automobile races. A statistician claims that it is obviously unrealistic to assume that the probability of a windshield breaking on a given strike is independent of the number of strikes previously survived. More specifically, since any windshield would be expected to become progressively more prone to breaking as the number of strikes increases, this statistician suggests using the following probability model: Let $A_x$ be the event that a windshield survives the $x$th strike; then, for $0 < \theta < 1$,

$$\theta = \text{pr}(A_1) \quad \text{and} \quad \theta^x = \text{pr}(A_x | \cap_{i=1}^{x-1} A_i), \quad x = 2, 3, \ldots, \infty.$$

(a) Given this probability model, let the random variable $X$ denote the number of strikes required to break a windshield made of this new impact-resistant glass. Derive, using precise arguments, a general formula for $p_X(x)$, the probability distribution of $X$, and carefully prove that $p_X(x)$ satisfies all the requirements to be a valid discrete probability distribution.

(b) If terms of the form $\theta^j$ for $j > 3$ can be neglected, develop a reasonable approximation for $E(X)$.

**Exercise 2.12.** Suppose that the continuous random variable $X$ has the uniform distribution

$$f_X(x) = 1, 0 < x < 1.$$

Suppose that the continuous random variable $Y$ is related to $X$ via the equation $Y = [-\ln(1 - X)]^{1/3}$. By relating $F_Y(y)$ to $F_X(x)$, develop explicit expressions for $f_Y(y)$ and $E(Y^r)$ for $r \geq 0$.

**Exercise 2.13.** For a certain psychological test designed to measure work-related stress level, a score of zero is considered to reflect a normal level of work-related stress. Based on previous data, it is reasonable to assume that the score $X$ on this psychological test can be accurately modeled as a continuous random variable with density function

$$f_X(x) = \frac{1}{288}(36 - x^2), \quad -6 < x < 6,$$

where negative scores indicate lower-than-normal work-related stress levels and positive scores indicate higher-than-normal work-related stress levels.

(a) Find the numerical value of the probability that a randomly chosen person taking this psychological test makes a test score within two units of a test score of zero.

(b) Develop an explicit expression for $F_X(x)$, the cumulative distribution function (CDF) for X, and then use this result to compute the exact numerical value of the probability that a randomly chosen person makes a test score greater than three in value given that this person's test score suggests a higher-than-normal work-related stress level.

(c) Find the numerical value of the probability (say, $\pi$) that, on any particular day, the sixth person taking this psychological test is at least the third person to make a test score greater than one in value.

(d) Use Tchebyshev's Inequality to find numbers L and U such that

$$\text{pr}(L < X < U) \geq \tfrac{8}{9}.$$

Comment on your findings.

**Exercise 2.14.** Suppose that the continuous random variable X has the *mixture* distribution

$$f_X(x) = \pi f_1(x) + (1-\pi)f_2(x), \quad -\infty < x < +\infty,$$

where $f_1(x)$ is a normal density with mean $\mu_1$ and variance $\sigma_1^2$, where $f_2(x)$ is a normal density with mean $\mu_2$ and variance $\sigma_2^2$, where $\pi$ is the probability that X has distribution $f_1(x)$, and where $(1-\pi)$ is the probability that X has distribution $f_2(x)$.

(a) Develop an *explicit expression* for $P_X(s)$, the probability generating function of the random variable X, and then use this result directly to find E(X).

(b) Let $\pi = 0.60, \mu_1 = 1.00, \sigma_1^2 = 0.50, \mu_2 = 1.20,$ and $\sigma_2^2 = 0.40$. Suppose that one value of X is observed, and that value of X exceeds 1.10 in value. Find the numerical value of the probability that this observed value of X was obtained from $f_1(x)$.

(c) Now, suppose that $\pi = 1, \mu_1 = 0,$ and $\sigma_1^2 = 1$. Find the numerical value of $E(X|X > 1.00)$.

**Exercise 2.15.** If the random variable $Y \sim N(0,1)$, develop an explicit expression for $E\left(|Y^r|\right)$ when $r$ is an odd positive integer.

**Exercise 2.16.** Suppose that the discrete random variable Y has the *negative binomial* distribution

$$p_Y(y) = C_{k-1}^{y+k-1}\pi^k(1-\pi)^y, \quad y = 0,1,\ldots,\infty, \ 0 < \pi < 1,$$

with $k$ a known positive integer. Derive an explicit expression for $E[Y!/(Y-r)!]$ where $r$ is a nonnegative integer. Then, use this result to find E(X) and V(X) when $X = (Y+k)$.

**Exercise 2.17.** Suppose that $X$ is the concentration (in parts per million) of a certain airborne pollutant, and suppose that the random variable $Y = \ln(X)$ has a distribution that can be adequately modeled by the *double exponential* density function

$$f_Y(y) = (2\alpha)^{-1}e^{-|y-\beta|/\alpha}, \quad -\infty < y < \infty, \; -\infty < \beta < \infty, \; 0 < \alpha < \infty.$$

(a) Find an explicit expression for $F_Y(y)$, the cumulative distribution function (CDF) associated with the density function $f_Y(y)$. If $\alpha = 1$ and $\beta = 2$, use this CDF to find the numerical value of $\text{pr}(X > 4|X > 2)$.

(b) For the density function $f_Y(y)$ given above, derive an explicit expression for a generating function $\phi_Y(t)$ that can be used to generate the absolute-value moments $\nu_r = E\{|Y - E(Y)|^r\}$ for $r$ a nonnegative integer, and then use $\phi_Y(t)$ directly to find $\nu_1$ and $\nu_2 = V(Y)$.

**Exercise 2.18.** A certain statistical model describing the probability (or risk) $Y$ of an adult developing leukemia as a function of lifetime cumulative exposure $X$ to radiation (in microsieverts) is given by the equation

$$Y = g(X) = 1 - \alpha e^{-\beta X^2}, \qquad 0 < X < +\infty, \quad 0 < \alpha < 1, \quad 0 < \beta < +\infty,$$

where the continuous random variable $X$ has the distribution

$$f_X(x) = \left(\frac{2}{\pi\theta}\right)^{1/2} e^{-x^2/2\theta}, \quad 0 < x < +\infty, \quad 0 < \theta < +\infty.$$

Find an explicit expression relating average risk $E(Y)$ to average cumulative exposure $E(X)$. Comment on how the average risk varies as a function of $\alpha, \beta$, and $E(X)$.

**Exercise 2.19.** A conceptually infinitely large population consists of a proportion $\pi_0$ of nonsmokers, a proportion $\pi_l$ of *light* smokers (no more than one pack per day), and a proportion $\pi_h$ of *heavy* smokers (more than one pack per day), where $(\pi_0 + \pi_l + \pi_h) = 1$. Consider the following three random variables based on three different sampling schemes:

1. $X_1$ is the number of subjects that have to be randomly selected sequentially from this population until exactly two heavy smokers are obtained.
2. $X_2$ is the number of subjects that have to be randomly selected sequentially from this population until at least one light smoker and at least one heavy smoker are obtained.
3. $X_3$ is the number of subjects that have to be randomly selected sequentially from this population until at least one subject from each of the three smoking categories (i.e., nonsmokers, light smokers, and heavy smokers) is obtained.

(a) Develop an explicit expression for the probability distribution $p_{X_1}(x_1)$ of $X_1$.

(b) Develop an explicit expression for the probability distribution $p_{X_2}(x_2)$ of $X_2$.

(c) Develop an explicit expression for the probability distribution $p_{X_3}(x_3)$ of $X_3$.

**Exercise 2.20.** If $Y$ is a normally distributed random variable with mean $\mu$ and variance $\sigma^2$, then the random variable $X = e^Y$ is said to have a *lognormal* distribution. The lognormal distribution has been used in many important practical applications, one such important application being to model the distributions of chemical concentration levels to which workers are exposed in occupational settings.

(a) Using the fact that $Y \sim N(\mu, \sigma^2)$ and that $X = e^Y$, derive explicit expressions for $E(X)$ and $V(X)$.

(b) If the lognormal random variable $X = e^Y$ defined in part (a) represents the average concentration (in parts per million, or ppm) of a certain toxic chemical to which a typical worker in a certain chemical manufacturing industry is exposed over an 8-hour workday, and if $E(X) = V(X) = 1$, find the exact numerical value of $\text{pr}(X > 1)$, namely, the probability that such a typical worker will be exposed over an 8-hour workday to an average chemical concentration level greater than 1 ppm.

(c) To protect the health of workers in this chemical manufacturing industry, it is desirable to be highly confident that a typical worker will not be exposed to an average chemical concentration greater than $c$ ppm over an 8-hour workday, where $c$ is a known positive constant specified by federal guidelines.

Prove that

$$\text{pr}(X \le c) \ge (1 - \alpha), \quad 0 < \alpha < 0.50,$$

if

$$E(X) \le c e^{-0.50 z_{1-\alpha}^2},$$

where $\text{pr}(Z \le z_{1-\alpha}) = (1 - \alpha)$ when $Z \sim N(0, 1)$. The implication of this result is that it is possible to meaningfully reduce the chance that a worker will be exposed over an 8-hour workday to a high average concentration of a potentially harmful chemical by sufficiently lowering the mean concentration level $E(X)$, given the assumption that $Y = \ln(X) \sim N(\mu, \sigma^2)$.

**Exercise 2.21.** Let $X$ be a discrete random variable such that

$$\theta_x = \text{pr}(X = x) = \alpha\pi^x, \qquad x = 1, 2, \ldots, +\infty, \quad 0 < \pi < 1,$$

and let

$$\theta_0 = \text{pr}(X = 0) = 1 - \sum_{x=1}^{\infty} \alpha\pi^x.$$

Here, $\alpha$ is an appropriately chosen positive constant.

(a) Develop an explicit expression for $M_X(t) = E(e^{tX})$, and then use this expression to find $E(X)$. Be sure to specify appropriate ranges for $\alpha$ and $t$.

(b) Verify your answer for $E(X)$ in part (a) by computing $E(X)$ directly.

**Exercise 2.22.** A popular dimensionless measure of the skewness (or "asymmetry") of a density function $f_X(x)$ is the quantity

$$\alpha_3 = \frac{\mu_3}{\mu_2^{3/2}} = \frac{E\{[X - E(X)]^3\}}{[V(X)]^{3/2}}.$$

As a possible competitor to $\alpha_3$, a new dimensionless measure of asymmetry, denoted $\alpha_3^*$, is proposed, where

$$\alpha_3^* = \frac{E(X) - \theta}{\sqrt{V(X)}};$$

here, $\theta$ is defined as the *mode* of the density function $f_X(x)$, namely, that unique value of $x$ (if it exists) that maximizes $f_X(x)$.

For the gamma density function

$$f_X(x) = \frac{x^{\beta-1}e^{-x/\alpha}}{\Gamma(\beta)\alpha^\beta}, \qquad 0 < x < \infty, \quad \alpha > 0, \quad \beta > 0,$$

develop explicit expressions for $\alpha_3$ and $\alpha_3^*$, and comment on the findings.

**Exercise 2.23*.** Environmental scientists typically use personal exposure monitors to measure the average daily concentrations of chemicals to which workers are exposed during 8-h work shifts. In certain situations, some average concentration levels are very low and so fall below a known detection limit $L(>0)$ defined by the type of personal monitor being used; such unobservable average concentration levels are said to be *left-censored*.

To deal with this missing data problem, one suggested *ad hoc* approach is to replace such left-censored average concentration levels with some numerical function $g(L)(>0)$ of $L$, say, $L/\sqrt{2}$, $L/2$, or even $L$ itself. To study the statistical ramifications of such an *ad hoc* approach, let $X(\geq 0)$ be a continuous random variable representing the average concentration level for a randomly chosen worker in a certain industrial setting; further, assume that $X$ has the distribution $f_X(x)$ with mean $E(X)$ and variance $V(X)$. Then, define the random variable

$$U = X \quad \text{if } X \geq L \quad \text{and} \quad U = g(L) \quad \text{if } X < L.$$

(a) If $\pi = \text{pr}(X \geq L) = \int_L^\infty f_X(x)\,dx$, show that

$$E(U) = (1 - \pi)g(L) + \pi E(X|X \geq L)$$

and that

$$V(U) = \pi\left\{V(X|X \geq L) + (1 - \pi)\left[g(L) - E(X|X \geq L)\right]^2\right\}.$$

(b) Find an explicit expression for the optimal choice for $g(L)$ such that $E(U) = E(X)$, which is a very desirable equality when using $U$ as a *surrogate* for $X$. If $f_X(x) = e^{-x}, x \geq 0$, and $L = 0.05$, find the exact numerical value of this optimal choice for $g(L)$.

**Exercise 2.24***. Suppose that $X \sim N(\mu, \sigma^2)$. Develop an explicit expression for E(Y) when

$$Y = 1 - \alpha e^{-\beta X^2}, \qquad 0 < \alpha < 1, \quad 0 < \beta < +\infty.$$

**Exercise 2.25***. The *cumulant generating function* for a random variable X is defined as

$$\psi_X(t) = \ln[M_X(t)],$$

where $M_X(t) = E(e^{tX})$ is the moment generating function of X; and, the rth cumulant $\kappa_r$ is the coefficient of $t^r/r!$ in the series expansion

$$\psi_X(t) = \ln[M_X(t)] = \sum_{r=1}^{\infty} \kappa_r \frac{t^r}{r!}.$$

(a) If $Y = (X - c)$, where c is a constant independent of X, what is the relationship between the cumulants of Y and the cumulants of X?

(b) Find the cumulants of X when X is distributed as:

(i) $N(\mu, \sigma^2)$;

(ii) $POI(\lambda)$;

(iii) $GAMMA(\alpha, \beta)$.

(c) In general, show that $\kappa_1 = E(X)$, that $\kappa_2 = V(X)$, and that $\kappa_3 = E\{[X - E(X)]^3\}$.

**Exercise 2.26***. In the branch of statistics known as "survival analysis," interest concerns a continuous random variable T $(0 < T < \infty)$, the time until an event (such as death) occurs. For example, in a clinical trial evaluating the effectiveness of a new remission induction chemotherapy treatment for leukemia, investigators may wish to model the time (in months) in remission (or, equivalently, the time to the reappearance of leukemia) for patients who have received this chemotherapy treatment and who have gone into remission. In such settings, rather than modeling T directly, investigators will often model the *hazard function*, h(t), defined as

$$h(t) = \lim_{\Delta t \to 0} \frac{pr(t \leq T \leq t + \Delta t | T \geq t)}{\Delta t}, \qquad t > 0.$$

The hazard function, or "instantaneous failure rate," is the limiting value (as $\Delta t \to 0$) of the probability per unit of time of the occurrence of the event of interest during a small time interval $[t, t + \Delta t]$ of length $\Delta t$, given that the event has not occurred prior to time t.

(a) If $f_T(t) \equiv f(t)$ is the density function of T and if $F_T(t) \equiv F(t)$ is the corresponding CDF, show that

$$h(t) = \frac{f(t)}{S(t)},$$

where $S(t) = [1 - F(t)]$ is called the *survival function* and is the probability that the event of interest does not occur prior to time $t$.

(b) Using the result in part (a), show that

$$S(t) = e^{-H(t)},$$

where $H(t) = \int_0^t h(u)\,du$ is the *cumulative hazard function*.

(c) Prove that $E(T) = \int_0^{\infty} S(t)\,dt$.

(d) Due to funding restrictions, the chemotherapy clinical trial described above is to be terminated after a fixed period of time $c$ (in months). Suppose that patients remain in the trial until either their leukemia reappears or the clinical trial ends (i.e., assume that there is no loss to follow-up, so that all patients either come out of remission or remain in remission until the trial ends). The observed time on study for each patient is therefore $X = \min(T, c)$, where $T$ denotes the time in remission. Show that

$$E[H(X)] = F_T(c),$$

where $H(\cdot)$ is the cumulative hazard function for $T$.

For further details about survival analysis, see Hosmer, Lemeshow, and May (2008) and Kleinbaum and Klein (2005).

**Exercise 2.27***. A certain drug company produces and sells a popular insulin for the treatment of diabetes. At the beginning of each calendar year, the company produces a very large number of units of the insulin (where a unit is a dosage amount equivalent to one injection of the insulin), the production goal being to closely meet patient demand for the insulin during that year. The company makes a net *gain* of G dollars for each unit sold during the year, and the company suffers a net *loss* of $L$ dollars for each unit left unsold during the year. Further, suppose that the total number $X$ of units of insulin (if available) that patients would purchase during the year can be modeled approximately as a continuous random variable with probability density function $f_X(x),\ x > 0$.

(a) If $N$ is the total number of units of the insulin that should be produced at the beginning of the year to maximize the expected value of the profit $P$ of the company for the entire year, show that $N$ satisfies the equation

$$F_X(N) = \frac{G}{(G + L)},$$

where $F_X(x) = \text{pr}(X \leq x)$ is the CDF of the random variable $X$.

(b) Compute the value of $N$ if $G = 4, L = 1$, and

$$f_X(x) = (2 \times 10^{-10})xe^{-(10^{-10})x^2}, \quad x > 0.$$

**Exercise 2.28***. Suppose that a particular automobile insurance company adopts the following strategy with regard to setting the value of yearly premiums for coverage. Any policy holder must pay a premium of $P_1$ dollars for the first year of coverage. If a policy holder has a perfect driving record during this first year of coverage (i.e., this policy holder is not responsible for any traffic accidents or for any traffic violations during this first year of coverage), then the premium for the second year of coverage will be reduced to $\alpha P_1$, where $0 < \alpha < 1$. However, if this policy holder does not have a perfect driving record during the first year of coverage, then the premium for the second year of coverage will be increased to $\beta P_1$, where $1 < \beta < +\infty$.

More generally, let $\pi, 0 < \pi < 1$, be the probability that any policy holder has a perfect driving record during any particular year of coverage, and assume that any policy holder's driving record during any one particular year of coverage is independent of his or her driving record during any other year of coverage. Then, in general, for $k = 2, 3, \ldots, \infty$, let $P_{k-1}$ denote the premium for year $(k-1)$; thus, the premium $P_k$ for year $k$ will equal $\alpha P_{k-1}$ with probability $\pi$, and will equal $\beta P_{k-1}$ with probability $(1-\pi)$.

(a) For $k = 2, 3, \ldots, \infty$, develop an explicit expression for $\mathrm{E}(P_k)$, the average yearly premium for the $k$th year of coverage for any policy holder.

(b) This insurance company cannot afford to let the average yearly premium for any policy holder be smaller than a certain value, say, $P^*$. Find an expression (as a function of $P_1, P^*, \beta$, and $\pi$) for the smallest value of $\alpha$, say $\alpha^*$, such that the average yearly premium for year $k$ is not less than $P^*$. Then, consider the limiting value of $\alpha^*$ as $k \to \infty$; compute the numerical value of this limiting value of $\alpha^*$ when $\pi = 0.90$ and $\beta = 1.05$, and then comment on your findings.

**Exercise 2.29***. Suppose that the discrete random variable $X$ has the probability distribution

$$\mathrm{p}_X(x) = \mathrm{pr}(X = x) = \frac{1}{x!}\sum_{l=0}^{R-x}\frac{(-1)^l}{l!}, \quad x = 0, 1, \ldots, R,$$

where $R(>1)$ is a positive integer.

(a) Use an inductive argument to show that $\sum_{x=0}^{R} \mathrm{p}_X(x) = 1$.

(b) Find explicit expressions for $\mathrm{E}(X)$ and $\mathrm{V}(X)$. Also, find $\lim_{R\to\infty}\mathrm{p}_X(x)$. Comment on all these findings.

**Exercise 2.30***. Suppose that the number $X_T$ of incident (i.e., new) lung cancer cases developing in a certain disease-free population of size $N$ during a time interval of length $T$ (in years) has the Poisson distribution

$$\mathrm{p}_{X_T}(x) = \frac{(NT\lambda)^x \mathrm{e}^{-(NT\lambda)}}{x!}, \qquad x = 0, 1, \ldots, \infty; \quad N > 0, \quad \lambda > 0, \quad T > 0.$$

Here, $N$ and $T$ are known constants, and the parameter $\lambda$ is the unknown rate of lung cancer development per person-year (a quantity often referred to as the "incidence density" by epidemiologists).

(a) Starting at time zero, let the continuous random variable $W_n$ be the length of time in years that passes until exactly $n$ lung cancer cases have developed. $W_n$ is referred to as the "waiting time" until the $n$th lung cancer case has developed. By expressing the CDF $F_{W_n}(w_n)$ of the random variable $W_n$ in terms of a probability statement about the Poisson random variable $X_T$, develop an explicit expression for the density function of the random variable $W_n$.

(b) With $X_T \sim \text{POI}(NT\lambda)$, consider the standardized random variable $Z = [X_T - \text{E}(X_T)]/\sqrt{\text{V}(X_T)}$. Show that

$$\lim_{N\to\infty} \text{E}(e^{tZ}) = e^{t^2/2},$$

which is the moment generating function of a standard normal random variable. Then, if $N = 10^5$ and $\lambda = 10^{-4}$, use the above result to provide a reasonable value for the probability of observing no more than 90 new cases of lung cancer in any 10-year period of time.

**Exercise 2.31***. Important computational aids for the numerical evaluation of incomplete integrals of gamma and beta distributions involve expressing such integrals as sums of probabilities of particular Poisson and binomial distributions.

(a) Prove that

$$\int_c^\infty \frac{x^{\beta-1}e^{-x/\alpha}}{\Gamma(\beta)\alpha^\beta}\,dx = \sum_{j=0}^{\beta-1} e^{-c/\alpha}\frac{(c/\alpha)^j}{j!},$$

where $\alpha > 0$ and $c > 0$ and where $\beta$ is a positive integer.

(b) Prove that

$$\int_0^c \frac{\Gamma(\alpha+\beta)}{\Gamma(\alpha)\Gamma(\beta)}x^{\alpha-1}(1-x)^{\beta-1}\,dx = \sum_{i=\alpha}^{\alpha+\beta-1} C_i^{\alpha+\beta-1}c^i(1-c)^{\alpha+\beta-1-i},$$

where $\alpha$ and $\beta$ are positive integers and where $0 < c < 1$.

**Exercise 2.32***. Suppose that the probability that a sea turtle nest contains $n$ eggs is equal to $(1-\pi)\pi^{n-1}$, where $n = 1, 2, \ldots, \infty$ and $0 < \pi < 1$. Furthermore, each egg in any such nest has probability 0.30 of producing a live and healthy baby sea turtle, completely independent of what happens to any other egg in that same nest. Finally, because of predators (e.g., sea birds and other sea creatures) and other risk factors (e.g., shore erosion, harmful environmental conditions, etc.), each such live and healthy baby sea turtle then has probability 0.98 of NOT surviving to adulthood.

(a) Find the *exact numerical value* of the probability that any egg produces an adult sea turtle.

(b) Derive an *explicit expression* for the probability $\alpha$ that a randomly chosen sea turtle nest produces at least one adult sea turtle. Find the *exact numerical value* of $\alpha$ when $\pi = 0.20$.

(c) Suppose that a randomly chosen sea turtle nest is known to have produced exactly $k$ adult sea turtles, where $k \geq 0$. Derive an *explicit expression* for the probability $\beta_{nk}$ that this randomly chosen sea turtle nest originally contained exactly $n$ eggs, $n \geq 1$. Find the *exact numerical value* of $\beta_{nk}$ when $\pi = 0.20, k = 2$, and $n = 6$.

**Exercise 2.33***

(a) Prove Pascal's Identity, namely,

$$C_k^n = C_{k-1}^{n-1} + C_k^{n-1}$$

for any positive integers $n$ and $k$ such that $C_k^n \equiv 0$ if $k > n$.

(b) Prove Vandermonde's Identity, namely,

$$C_r^{m+n} = \sum_{k=0}^{r} C_{r-k}^m C_k^n,$$

where $m, n$, and $r$ are nonnegative integers satisfying $r \leq \min\{m, n\}$.

(c) For $y = 1, 2, \ldots, \min\{s, t\}$, suppose that the discrete random variable $X$ takes the value $x = 2y$ with probability

$$\pi_{2y} = \frac{2C_{y-1}^{s-1}C_{y-1}^{t-1}}{C_s^{s+t}},$$

and takes the value

$$\pi_{2y+1} = \frac{C_y^{s-1}C_{y-1}^{t-1} + C_{y-1}^{s-1}C_y^{t-1}}{C_s^{s+t}},$$

where $C_y^{s-1} \equiv 0$ when $y = s$ and $C_y^{t-1} \equiv 0$ when $y = t$.

Use Pascal's Identity and Vandermonde's Identity to show that $X$ has a *valid* discrete probability distribution.

## SOLUTIONS

**Solution 2.1**

(a) Let the random variable $Y$ denote the number of individuals that must be selected until one individual with the rare blood disorder *and* one individual without the rare blood disorder are selected. (Note that $Y$ can take the values 2, 3, 4, and 5.) If $D_i$ is the event that the $i$th individual selected has the rare blood disorder, then

$$\begin{aligned}
\text{pr}(Y=2) &= \text{pr}(\text{D}_1 \cap \bar{\text{D}}_2) + \text{pr}(\bar{\text{D}}_1 \cap \text{D}_2) = \text{pr}(\text{D}_1)\text{pr}(\bar{\text{D}}_2|\text{D}_1) \\
&\quad + \text{pr}(\bar{\text{D}}_1)\text{pr}(\text{D}_2|\bar{\text{D}}_1) \\
&= \left(\frac{4}{7}\right)\left(\frac{3}{6}\right) + \left(\frac{3}{7}\right)\left(\frac{4}{6}\right) = \frac{4}{7}; \\
\text{pr}(Y=3) &= \text{pr}(\text{D}_1 \cap \text{D}_2 \cap \bar{\text{D}}_3) + \text{pr}(\bar{\text{D}}_1 \cap \bar{\text{D}}_2 \cap \text{D}_3) \\
&= \text{pr}(\text{D}_1)\text{pr}(\text{D}_2|\text{D}_1)\text{pr}(\bar{\text{D}}_3|\text{D}_1 \cap \text{D}_2) \\
&\quad + \text{pr}(\bar{\text{D}}_1)\text{pr}(\bar{\text{D}}_2|\bar{\text{D}}_1)\text{pr}(\text{D}_3|\bar{\text{D}}_1 \cap \bar{\text{D}}_2) \\
&= \left(\frac{4}{7}\right)\left(\frac{3}{6}\right)\left(\frac{3}{5}\right) + \left(\frac{3}{7}\right)\left(\frac{2}{6}\right)\left(\frac{4}{5}\right) = \frac{10}{35}.
\end{aligned}$$

Similarly,

$$\begin{aligned}
\text{pr}(Y=4) &= \left(\frac{4}{7}\right)\left(\frac{3}{6}\right)\left(\frac{2}{5}\right)\left(\frac{3}{4}\right) + \left(\frac{3}{7}\right)\left(\frac{2}{6}\right)\left(\frac{1}{5}\right)\left(\frac{4}{4}\right) = \frac{4}{35}; \text{ and,} \\
\text{pr}(Y=5) &= \left(\frac{4}{7}\right)\left(\frac{3}{6}\right)\left(\frac{2}{5}\right)\left(\frac{1}{4}\right)\left(\frac{3}{3}\right) = \frac{1}{35} \\
&= 1 - \sum_{y=2}^{4} \text{pr}(Y=y) = 1 - \frac{34}{35}.
\end{aligned}$$

Finally, $\text{E}(Y) = 2\left(\frac{4}{7}\right) + 3\left(\frac{10}{35}\right) + 4\left(\frac{4}{35}\right) + 5\left(\frac{1}{35}\right) = 2.60$.

(b) Let A denote the event that "$(k-1)$ individuals have the rare blood disorder among the first $(x-1)$ individuals selected," and let B denote the event that "the $x$th individual selected has the rare blood disorder." Then,

$$\begin{aligned}
\text{p}_X(x) &= \text{pr}(X=x) = \text{pr}(\text{A} \cap \text{B}) = \text{pr}(\text{A})\text{pr}(\text{B}|\text{A}) \\
&= \frac{\text{C}^{M}_{k-1}\text{C}^{N-M}_{(x-1)-(k-1)}}{\text{C}^{N}_{x-1}} \cdot \frac{[M-(k-1)]}{[N-(x-1)]} \\
&= \frac{\text{C}^{M}_{k-1}\text{C}^{N-M}_{x-k}}{\text{C}^{N}_{x-1}}\left(\frac{M-k+1}{N-x+1}\right) \\
&= \frac{\text{C}^{x-1}_{k-1}\text{C}^{N-x}_{M-k}}{\text{C}^{N}_{M}}, \quad 1 \le k \le x \le (N-M+k).
\end{aligned}$$

(c) pr(third individual selected has the rare blood disorder)

$$= \text{pr}(D_1 \cap D_2 \cap D_3) + \text{pr}(D_1 \cap \bar{D}_2 \cap D_3) + \text{pr}(\bar{D}_1 \cap \bar{D}_2 \cap D_3) + \text{pr}(\bar{D}_1 \cap D_2 \cap D_3)$$

$$= \left(\frac{M}{N}\right)\left(\frac{M-1}{N-1}\right)\left(\frac{M-2}{N-2}\right) + \left(\frac{M}{N}\right)\left(\frac{N-M}{N-1}\right)\left(\frac{M-1}{N-2}\right)$$

$$+ \left(\frac{N-M}{N}\right)\left(\frac{N-M-1}{N-1}\right)\left(\frac{M}{N-2}\right) + \left(\frac{N-M}{N}\right)\left(\frac{M}{N-1}\right)\left(\frac{M-1}{N-2}\right) = \frac{M}{N}.$$

**Solution 2.2.** For $x = 0, 1, \ldots, (k-2)$, there are exactly $(k-x-1)$ pairs of slots for which the integer 1 precedes the integer $k$ and for which there are exactly $x$ integers between the integers 1 and $k$. Also, the integer $k$ can precede the integer 1, and the other $(k-2)$ integers can be arranged in the remaining $(k-2)$ slots in $(k-2)!$ ways. So,

$$\text{p}_X(x) = \frac{2(k-x-1)[(k-2)!]}{k!} = \frac{2(k-x-1)}{k(k-1)}, \quad x = 0, 1, \ldots, (k-2).$$

Clearly, $\text{p}_X(x) \geq 0, x = 0, 1, \ldots, (k-2)$, and

$$\sum_{x=0}^{k-2} \text{p}_X(x) = \sum_{x=0}^{k-2} \frac{2(k-x-1)}{k(k-1)} = \frac{2}{k(k-1)} \sum_{x=0}^{k-2} [(k-1) - x]$$

$$= \frac{2}{k(k-1)}\left[(k-1)^2 - \frac{(k-2)(k-1)}{2}\right]$$

$$= \frac{2}{k}\left[(k-1) - \frac{(k-2)}{2}\right] = 1.$$

So, $\text{p}_X(x)$ is a valid discrete probability distribution.

(b) Now,

$$\text{E}(X) = \sum_{x=0}^{k-2} x\text{p}_X(x) = \frac{2}{k(k-1)} \sum_{x=0}^{k-2} x[(k-1) - x]$$

$$= \frac{2}{k(k-1)}\left\{(k-1)\left[\frac{(k-2)(k-1)}{2}\right] - \frac{(k-2)(k-1)[2(k-2)+1]}{6}\right\}$$

$$= \frac{2}{k}\left[\frac{(k-2)(k-1)}{2} - \frac{(k-2)(2k-3)}{6}\right]$$

$$= \frac{(k-2)}{k}\left[\frac{3(k-1) - (2k-3)}{3}\right]$$

$$= \frac{(k-2)}{3}, \quad k \geq 3.$$

**Solution 2.3**

(a)

$$\begin{aligned}
\text{pr}(Y=2) &= \text{pr}(W_1 \cap W_2) = \left(\tfrac{4}{6}\right)\left(\tfrac{3}{5}\right) = \tfrac{2}{5}, \\
\text{pr}(Y=1) &= \text{pr}(B_1 \cap W_2 \cap W_3 \cap W_4) + \text{pr}(W_1 \cap B_2 \cap W_3 \cap W_4) \\
&= \left(\frac{2}{6}\right)\left(\frac{4}{5}\right)\left(\frac{3}{4}\right)\left(\frac{2}{3}\right) + \left(\frac{4}{6}\right)\left(\frac{2}{5}\right)\left(\frac{3}{4}\right)\left(\frac{2}{3}\right) = \frac{4}{15}, \\
\text{pr}(Y=0) &= \text{pr}(B_1 \cap W_2 \cap W_3 \cap B_4) + \text{pr}(B_1 \cap W_2 \cap B_3 \cap W_4) \\
&\quad + \text{pr}(W_1 \cap B_2 \cap B_3 \cap W_4) + \text{pr}(W_1 \cap B_2 \cap W_3 \cap B_4) \\
&\quad + \text{pr}(B_1 \cap B_2) \\
&= 4\left(\frac{2}{6}\right)\left(\frac{4}{5}\right)\left(\frac{3}{4}\right)\left(\frac{1}{3}\right) + \left(\frac{2}{6}\right)\left(\frac{1}{5}\right) = \frac{1}{3}.
\end{aligned}$$

Or, $\text{pr}(Y=0) = 1 - \text{pr}(Y=1) - \text{pr}(Y=2) = 1 - \frac{4}{15} - 2/5 = 1/3$.
Thus,

$$\text{E}(Y) = 0\left(\frac{1}{3}\right) + 1\left(\frac{4}{15}\right) + 2\left(\frac{2}{5}\right) = \frac{16}{15} = 1.0667.$$

Since

$$\text{E}(Y^2) = (0)^2\left(\frac{1}{3}\right) + (1)^2\left(\frac{4}{15}\right) + (2)^2\left(\frac{2}{5}\right) = \frac{28}{15},$$

$$\text{V}(Y) = \frac{28}{15} - \left(\frac{16}{15}\right)^2 = \frac{164}{225} = 0.7289.$$

(b) Clearly, pr(white ball) $= 2/3$, and this probability stays the same for each ball selected. So, pr(a pair contains 2 white balls) $= (2/3)^2 = 4/9$. Now, let $X =$ number of pairs that have to be selected to obtain exactly two pairs of white balls. Since $X \sim \text{NEGBIN}(k=2, \pi = 4/9)$, it follows that

$$\begin{aligned}
p_X(x) &= C_{2-1}^{x-1}\left(\frac{4}{9}\right)^2\left(\frac{5}{9}\right)^{x-2} \\
&= (x-1)\left(\frac{4}{9}\right)^2\left(\frac{5}{9}\right)^{x-2}, \quad x = 2, 3, \ldots, \infty.
\end{aligned}$$

**Solution 2.4**

(a) At the second stage of sampling, we are sampling without replacement from a finite population of $N$ animals, of which $m$ are marked and $(N - m)$ are unmarked. So, the hypergeometric distribution applies. In particular, the exact distribution of $X$ is

$$p_X(x) = \frac{C_x^m C_{n-x}^{N-m}}{C_n^N}, \quad \max[0, n - (N - m)] \le x \le n.$$

(b)

$$\pi = \sum_{j=2}^{4} \frac{C_j^4 C_{n-j}^{N-4}}{C_n^N}.$$

(c) Since $X$ has the hypergeometric distribution given in part (a), it follows directly that $E(X) = n(m/N)$. Since $x$, the observed value of $X$, is our best guess for $E(X)$, it is logical to equate $x$ to $E(X)$, obtaining $x = n(m/N)$. This leads to the expression $\hat{N} = mn/x$. When $x = 22$, $m = 600$, and $n = 300$, the computed value of $\hat{N}$ is 8181.82. Two obvious problems with the estimate $\hat{N}$ are that it does not necessarily take positive integer values, and it is not defined when $x = 0$.

**Solution 2.5.** Since each health status category has probability $1/k$ of being encountered, pr(encountering a new health status category|$c$ different health status categories have already been encountered) $= (1 - c/k)$. Also, the daily outcomes are mutually independent of one another, and the probability $(1 - c/k)$ remains the same from day to day.

So, pr(it takes exactly $x$ days to encounter a new health status category|$c$ different health status categories have already been encountered) = pr[*not* a new category in the first $(x - 1)$ days] × pr[new category on the $x$th day]

$$= \left(\frac{c}{k}\right)^{x-1} \cdot \left(1 - \frac{c}{k}\right) = (k - c)k^{-x}c^{x-1}, \quad 0 \le c \le (k - 1).$$

In other words, if $X$ is the random variable denoting the number of days required to encounter a new health status category, then $X$ has a *geometric distribution*, namely,

$$p_X(x) = (k - c)k^{-x}c^{x-1}, \quad x = 1, 2, \ldots, \infty.$$

(b) For $0 \le c \le (k - 1)$ and with $q = c/k$, we have E($X$|$c$ different health status categories have already been encountered)

$$= \sum_{x=1}^{\infty} x \left(\frac{c}{k}\right)^{x-1} \left(1 - \frac{c}{k}\right) = \left(1 - \frac{c}{k}\right) \sum_{x=1}^{\infty} x q^{x-1}$$

$$= \left(1 - \frac{c}{k}\right) \sum_{x=1}^{\infty} \frac{d(q^x)}{dq}$$

$$= \left(1 - \frac{c}{k}\right) \frac{d}{dq} \left\{ \sum_{x=1}^{\infty} q^x \right\} = \left(1 - \frac{c}{k}\right) \frac{d}{dq} \left[ \frac{q}{1 - q} \right]$$

$$= \left(1 - \frac{c}{k}\right) \left\{ \frac{(1)(1 - q) - q(-1)}{(1 - q)^2} \right\} = \frac{\left(1 - \frac{c}{k}\right)}{\left(1 - \frac{c}{k}\right)^2} = \frac{k}{(k - c)},$$

which follows directly since $X \sim \text{GEOM}\left(1 - \frac{c}{k}\right)$.

So, the expected total number of days $= k \sum_{c=0}^{k-1} (k - c)^{-1}$.

When $k = 4$, we get $4\sum_{c=0}^{3}(4-c)^{-1} = 8.33$; in other words, it will take, on average, nine days to encounter people in all $k = 4$ health status categories.

**Solution 2.6**

(a) If $X \sim \text{BIN}(n, \pi = 0.0005)$, then

$$\text{pr}(X \geq 1) = 1 - \text{pr}(X = 0) = 1 - (0.9995)^n \geq 0.90;$$

thus, we obtain

$$n \ln(0.9995) \leq \ln(0.10), \quad \text{or} \quad n \geq 4605.17, \quad \text{or} \quad n^* = 4606.$$

And, if $Y \sim \text{POI}(n\pi)$, then

$$\text{pr}(Y \geq 1) = 1 - \text{pr}(Y = 0) = 1 - e^{-n\pi} = 1 - e^{-0.0005n} \geq 0.90;$$

thus, we obtain

$$e^{-0.0005n} \leq 0.10, \quad \text{or} \quad -0.0005n \leq \ln(0.10), \quad \text{or} \quad n \geq 4605.17,$$

which again gives $n^* = 4606$. These numerical answers are the same because $\pi$ is very close to zero in value.

(b) With $X \sim \text{BIN}(n^*, \pi)$, then $P = (\text{A}X - n^*)$. Thus, requiring $\text{E}(P) = (\text{A}n^*\pi - n^*) \geq 0$ gives

$$\text{A}\pi - 1 \geq 0, \quad \text{or} \quad \text{A} \geq \pi^{-1}, \quad \text{or} \quad \text{A} \geq (0.0005)^{-1} = \$2000.00.$$

(c) Let $U$ be the discrete random variable denoting the number of the $n^*$ tickets purchased by this person that are jackpot-winning tickets. Then, $U \sim \text{HG}(N, K, n^*)$. So,

$$\text{pr}(k \leq U \leq 4{,}606) = \sum_{u=k}^{4606} \frac{C_u^K C_{4606-u}^{N-K}}{C_{4606}^N}.$$

**Solution 2.7**

(a)

$$\begin{aligned}
\text{pr}(Y = 1) &= \sum_{j=1}^{k} \text{pr(both twins choose the number } j) \\
&= \sum_{j=1}^{k} \text{pr(one twin chooses } j)\text{pr(other twin chooses } j) \\
&= \sum_{j=1}^{k} \left(\frac{1}{k}\right)\left(\frac{1}{k}\right) = \frac{1}{k}.
\end{aligned}$$

So,

$$p_Y(y) = \left(\frac{1}{k}\right)^y \left(\frac{k-1}{k}\right)^{1-y}, \quad y = 0, 1.$$

(b) We wish to choose the smallest value of $k$ such that $\frac{1}{k} \leq 0.01$, which requires $k = 100$.

(c) Let A be the event that "at least one set out of 100 sets of monozygotic twins chooses matching numbers" and let B be the event that "no set of monozygotic twins actually has ESP." Then,

$$\text{pr}(\text{A}|\text{B}) = 1 - \text{pr}(\bar{\text{A}}|\text{B}) = 1 - (0.99)^{100} = 1 - 0.366 = 0.634.$$

Thus, if this parapsychologist conducts this experiment on a reasonably large number of monozygotic twins, there is a very high probability of concluding incorrectly that one or more sets of monozygotic twins has ESP. Clearly, the chance of making this mistake increases as the number of sets of monozygotic twins studied increases.

(d) With $X$ defined as the number of matches in $n = 10$ independent repetitions of the experiment, then $X \sim \text{BIN}(n = 10, \pi = 0.01)$. So,

$$\begin{aligned}
\text{pr}(X \geq 2) &= 1 - \text{pr}(X \leq 1) \\
&= 1 - \sum_{x=0}^{1} C_x^{10} (0.01)^x (0.99)^{10-x} \\
&= 1 - (0.99)^{10} - (10)(0.01)(0.99)^9 \\
&= 1 - 0.9044 - 0.0914 \\
&= 0.0042.
\end{aligned}$$

So, for this particular set of monozygotic twins, there is some statistical evidence for the presence of ESP. Perhaps further study about this pair of twins is warranted, hopefully using other more sophisticated ESP detection experiments.

(e) Let D be the event that "the two *randomly chosen* numbers are *not* the same." Then,

$$\begin{aligned}
\text{pr}(S = 3|\text{D}) &= \frac{\text{pr}[(S = 3) \cap \text{D}]}{\text{pr}(\text{D})} \\
&= \frac{\text{pr}(1, 2) + \text{pr}(2, 1)}{\left(1 - \frac{1}{4}\right)} \\
&= \frac{\text{pr}(1)\text{pr}(2) + \text{pr}(2)\text{pr}(1)}{\left(\frac{3}{4}\right)}
\end{aligned}$$

$$= \frac{\left(\frac{1}{4}\right)\left(\frac{1}{4}\right)+\left(\frac{1}{4}\right)\left(\frac{1}{4}\right)}{\left(\frac{3}{4}\right)}$$

$$= \frac{1}{6};$$

$$\text{pr}(S=4|\text{D}) = \frac{\text{pr}(1,3)+\text{pr}(3,1)}{\left(\frac{3}{4}\right)} = \frac{\left(\frac{1}{16}+\frac{1}{16}\right)}{\left(\frac{3}{4}\right)} = \frac{1}{6};$$

$$\text{pr}(S=5|\text{D}) = \frac{\text{pr}(2,3)+\text{pr}(3,2)+\text{pr}(4,1)+\text{pr}(1,4)}{\left(\frac{3}{4}\right)}$$

$$= \frac{\left(\frac{4}{16}\right)}{\left(\frac{3}{4}\right)} = \frac{1}{3};$$

$$\text{pr}(S=6|\text{D}) = \frac{\text{pr}(2,4)+\text{pr}(4,2)}{\left(\frac{3}{4}\right)} = \frac{\left(\frac{2}{16}\right)}{\left(\frac{3}{4}\right)} = \frac{1}{6}$$

$$\text{pr}(S=7|\text{D}) = \frac{\text{pr}(3,4)+\text{pr}(4,3)}{\left(\frac{3}{4}\right)} = \frac{\left(\frac{2}{16}\right)}{\left(\frac{3}{4}\right)} = \frac{1}{6}.$$

Hence, the probability distribution is

$$\text{p}_S(s|\text{D}) = \begin{cases} \frac{1}{6} & \text{if } s = 3, 4, 6, \text{ or } 7; \\ \frac{1}{3} & \text{if } s = 5; \\ 0 & \text{otherwise.} \end{cases}$$

Note that this is a type of "truncated" distribution. The expected value is

$$\text{E}(S|\text{D}) = 3\left(\frac{1}{6}\right)+4\left(\frac{1}{6}\right)+5\left(\frac{1}{3}\right)+6\left(\frac{1}{6}\right)+7\left(\frac{1}{6}\right) = 5.$$

**Solution 2.8**

(a)

$$\text{pr}(Y \leq 3 | Y \geq 2) = \frac{\text{pr}[(Y \geq 2) \cap (Y \leq 3)]}{\text{pr}(Y \geq 2)} = \frac{\text{pr}(Y = 2) + \text{pr}(Y = 3)}{1 - \text{pr}(Y = 1)}$$

$$= \frac{\left[\frac{\lambda^2}{2!(e^{\lambda}-1)}\right] + \left[\frac{\lambda^3}{3!(e^{\lambda}-1)}\right]}{\left[1 - \frac{\lambda}{e^{\lambda}-1}\right]} = \frac{\lambda^2(\lambda+3)}{6(e^{\lambda} - \lambda - 1)}.$$

(b) For $r = 1, 2, \ldots,$

$$E\left[\frac{Y!}{(Y-r)!}\right] = \sum_{y=1}^{\infty} \left[\frac{y!}{(y-r)!}\right] \frac{\lambda^y}{y!(e^{\lambda} - 1)} = \sum_{y=r}^{\infty} \frac{\lambda^y}{(y-r)!(e^{\lambda} - 1)}$$

$$= \sum_{u=0}^{\infty} \frac{\lambda^{u+r}}{u!(e^{\lambda} - 1)} = \frac{\lambda^r e^{\lambda}}{(e^{\lambda} - 1)}.$$

So, for $r = 1$,

$$E(Y) = \frac{\lambda e^{\lambda}}{(e^{\lambda} - 1)}.$$

And, for $r = 2$,

$$E[Y(Y-1)] = \frac{\lambda^2 e^{\lambda}}{(e^{\lambda} - 1)},$$

so that

$$V(Y) = E[Y(Y-1)] + E(Y) - [E(Y)]^2 = \frac{\lambda e^{\lambda}(e^{\lambda} - \lambda - 1)}{(e^{\lambda} - 1)^2}.$$

(c)

$$\theta = \text{pr}(V) = \text{pr}[V \cap (Y \geq 1)] = \text{pr}\left\{V \cap [\cup_{y=1}^{\infty}(Y = y)]\right\}$$

$$= \text{pr}\left\{\cup_{y=1}^{\infty}[V \cap (Y = y)]\right\} = \sum_{y=1}^{\infty} \text{pr}[V \cap (Y = y)]$$

$$= \sum_{y=1}^{\infty} \text{pr}(V | Y = y)\text{pr}(Y = y) = \sum_{y=1}^{\infty} (\pi^y) \frac{\lambda^y}{y!(e^{\lambda} - 1)}$$

$$= (e^{\lambda} - 1)^{-1} \sum_{y=1}^{\infty} \frac{(\pi\lambda)^y}{y!} = (e^{\lambda} - 1)^{-1} \left[\sum_{y=0}^{\infty} \frac{(\pi\lambda)^y}{y!} - 1\right]$$

$$= \frac{(e^{\pi\lambda} - 1)}{(e^{\lambda} - 1)}.$$

**Solution 2.9.** First, note that $U/2 = Z \sim \mathrm{N}(0,1)$. Making use of this result, we then have

$$\begin{aligned}
\mathrm{pr}[|W-30|<0.50] &= \mathrm{pr}[-0.50<(W-30)<0.50] \\
&= \mathrm{pr}[29.5<W<30.5] \\
&= \mathrm{pr}[29.5<31+(0.50)U<30.5] \\
&= \mathrm{pr}\left[(29.5-31)<Z<(30.5-31)\right] \\
&= \mathrm{pr}(-1.50<Z<-0.50) \\
&= F_Z(-0.50)-F_Z(-1.50) \\
&= 0.3085-0.0668 \\
&= 0.2417.
\end{aligned}$$

**Solution 2.10**

(a) For Process 1,

$$\int_{1.0}^{2.0} 3.144e^{-x}\,dx = 0.7313.$$

For Process 2,

$$\int_{1.0}^{2.0} 2.574e^{-x}\,dx = 0.5987.$$

Let A be the event that "both computer chips were produced by Process 1," let B be the event that "one of the computer chips is acceptable and the other computer chip is unacceptable," and let C be the event that "any computer chip is acceptable." Clearly, $\mathrm{pr}(\mathrm{A}) = \left(\frac{1}{3}\right)^2 = 1/9 = 0.1111$. And, $\mathrm{pr}(\mathrm{B}|\mathrm{A}) = 2(0.7313)(0.2687) = 0.3930$. Also, since

$$\mathrm{pr}(\mathrm{C}) = (0.7313)\left(\frac{1}{3}\right) + (0.5987)\left(\frac{2}{3}\right) = 0.6429,$$

it follows that

$$\mathrm{pr}(\mathrm{B}) = C_1^2(0.6429)(0.3571) = 0.4592.$$

Finally,

$$\begin{aligned}
\mathrm{pr}(\mathrm{A}|\mathrm{B}) &= \frac{\mathrm{pr}(\mathrm{A}\cap\mathrm{B})}{\mathrm{pr}(\mathrm{B})} \\
&= \frac{\mathrm{pr}(\mathrm{B}|\mathrm{A})\mathrm{pr}(\mathrm{A})}{\mathrm{pr}(\mathrm{B})} = \frac{(0.3930)(0.1111)}{0.4592} = 0.0951.
\end{aligned}$$

(b) For $y = 3, 4, \ldots, \infty$, the event $Y = y$ can occur in one of two mutually exclusive ways: (i) The first $(y-1)$ chips selected are all acceptable, *and* then the $y$th chip selected is unacceptable; or (ii) The first $(y-1)$ chips selected include one acceptable chip and $(y-2)$ unacceptable chips, *and* then the $y$th chip selected is acceptable. So, if C is the event that a computer chip is acceptable, and if $P_1$ is the event that a computer chip is produced by Process 1, and if $\theta$ denotes the probability of selecting an acceptable chip, then

$$\theta = \text{pr}(C|P_1)\text{pr}(P_1) + \text{pr}(C|\bar{P}_1)\text{pr}(\bar{P}_1)$$
$$= (0.7313)\left(\frac{1}{3}\right) + (0.5987)\left(\frac{2}{3}\right) = 0.6429.$$

Thus, with $\theta = 0.6429$, we have

$$p_Y(y) = \theta^{y-1}(1-\theta) + (y-1)(1-\theta)^{y-2}\theta^2, \quad y = 3, 4, \ldots, \infty.$$

Now, $0 \leq p_Y(y), y = 3, 4, \ldots, \infty$, and

$$\sum_{y=3}^{\infty} p_Y(y) = \sum_{y=3}^{\infty} \left[\theta^{y-1}(1-\theta) + (y-1)(1-\theta)^{y-2}\theta^2\right]$$
$$= (1-\theta)\sum_{y=3}^{\infty}\theta^{y-1} + \theta\sum_{u=2}^{\infty} u(1-\theta)^{u-1}\theta$$
$$= (1-\theta)\left(\frac{\theta^2}{1-\theta}\right) + \theta\left[\sum_{u=1}^{\infty} u(1-\theta)^{u-1}\theta - \theta\right]$$
$$= \theta^2 + \theta\left(\frac{1}{\theta} - \theta\right) = 1,$$

so that $p_Y(y)$ is a valid discrete probability distribution.

**Solution 2.11**

(a)

$$\text{pr}(X > 1) = \text{pr}(A_1) = \theta;$$
$$\text{pr}(X > 2) = \text{pr}(A_1)\text{pr}(A_2|A_1) = \theta(\theta^2) = \theta^3;$$
$$\text{pr}(X > 3) = \text{pr}(A_1)\text{pr}(A_2|A_1)\text{pr}(A_3|A_1 \cap A_2) = \theta(\theta^2)(\theta^3) = \theta^6;$$

and, in general,

$$\text{pr}(X > x) = \prod_{i=1}^{x} \theta^i, x = 1, 2, \ldots, \infty.$$

So,

$$\begin{aligned}
\mathrm{p}_X(x) &= \mathrm{pr}(X = x) = \mathrm{pr}(X > x-1) - \mathrm{pr}(X > x) \\
&= \prod_{i=1}^{x-1} \theta^i - \prod_{i=1}^{x} \theta^i \\
&= \left(\prod_{i=1}^{x-1} \theta^i\right)(1-\theta^x) \\
&= \left(\theta^{\sum_{i=1}^{x-1} i}\right)(1-\theta^x) \\
&= \theta^{\frac{x(x-1)}{2}}(1-\theta^x), \quad x = 1, 2, \ldots, \infty.
\end{aligned}$$

Since $0 < \theta < 1$, clearly $0 \leq \mathrm{p}_X(x) \leq 1$ for $x = 1, 2, \ldots, \infty$. And,

$$\begin{aligned}
\sum_{x=1}^{\infty} \mathrm{p}_X(x) &= \sum_{x=1}^{\infty} \theta^{\frac{x(x-1)}{2}}(1-\theta^x) \\
&= \sum_{x=1}^{\infty} \theta^{\frac{x(x-1)}{2}} - \sum_{x=1}^{\infty} \theta^{\frac{x(x+1)}{2}} \\
&= \sum_{y=0}^{\infty} \theta^{\frac{y(y+1)}{2}} - \sum_{x=1}^{\infty} \theta^{\frac{x(x+1)}{2}} \\
&= 1 + \sum_{y=1}^{\infty} \theta^{\frac{y(y+1)}{2}} - \sum_{x=1}^{\infty} \theta^{\frac{x(x+1)}{2}} = 1.
\end{aligned}$$

(b) We have

$$\begin{aligned}
\mathrm{E}(X) &= \sum_{x=1}^{\infty} x\theta^{\frac{x(x-1)}{2}}(1-\theta^x) \\
&= \sum_{x=1}^{\infty} x\theta^{\frac{x(x-1)}{2}} - \sum_{x=1}^{\infty} x\theta^{\frac{x(x+1)}{2}} \\
&= \sum_{y=0}^{\infty} (y+1)\theta^{\frac{y(y+1)}{2}} - \sum_{x=1}^{\infty} x\theta^{\frac{x(x+1)}{2}} \\
&= \sum_{y=0}^{\infty} \theta^{\frac{y(y+1)}{2}} + \sum_{y=0}^{\infty} y\theta^{\frac{y(y+1)}{2}} - \sum_{x=1}^{\infty} x\theta^{\frac{x(x+1)}{2}}
\end{aligned}$$

$$= \sum_{y=0}^{\infty} \theta^{\frac{y(y+1)}{2}}$$

$$= 1 + \theta + \theta^3 + \theta^6 + \theta^{10} + \cdots \approx (1 + \theta + \theta^3),$$

assuming terms of the form $\theta^j$ for $j > 3$ can be neglected.

**Solution 2.12.** For $y \geq 0$,

$$\begin{aligned} F_Y(y) &= \text{pr}[Y \leq y] = \text{pr}\{[-\ln(1-X)]^{1/3} \leq y\} = \text{pr}[-\ln(1-X) \leq y^3] \\ &= \text{pr}[\ln(1-X) \geq -y^3] = \text{pr}[(1-X) \geq e^{-y^3}] \\ &= \text{pr}[X \leq (1-e^{-y^3})] = F_X(1-e^{-y^3}) = 1 - e^{-y^3}, \end{aligned}$$

since $F_X(x) = x, 0 < x < 1$.

So,

$$f_Y(y) = \frac{dF_Y(y)}{dy} = 3y^2 e^{-y^3}, \quad 0 < y < \infty.$$

So, for $r \geq 0$, and with $u = y^3$,

$$\begin{aligned} E(Y^r) &= \int_0^{\infty} (y^r) 3y^2 e^{-y^3} \, dy = \int_0^{\infty} u^{r/3} e^{-u} \, du \\ &= \Gamma\left(\frac{r}{3} + 1\right) \int_0^{\infty} \frac{u^{\left(\frac{r}{3}+1\right)-1} e^{-u}}{\Gamma\left(\frac{r}{3}+1\right)} \, du \\ &= \Gamma\left(\frac{r}{3} + 1\right), \quad r \geq 0. \end{aligned}$$

**Solution 2.13**

(a) $\text{pr}(-2 < X < 2) = \int_{-2}^{2} \frac{1}{288}(36 - x^2)\, dx = 0.4815.$

(b) $F_X(x) = \int_{-6}^{x} \frac{1}{288}(36 - t^2)\, dt = \frac{1}{288}(144 + 36x - \frac{x^3}{3}), -6 < x < 6.$ So,

$$\begin{aligned} \text{pr}(X > 3 | X > 0) &= \frac{\text{pr}[(X > 3) \cap (X > 0)]}{\text{pr}(X > 0)} = \frac{\text{pr}(X > 3)}{\text{pr}(X > 0)} \\ &= \frac{1 - F_X(3)}{1 - F_X(0)} = 0.3124. \end{aligned}$$

(c) Now, $\text{pr}(X > 1) = 1 - F_X(1) = 0.3762$. So, using the *negative binomial* distribution,

$$\pi = \sum_{k=3}^{6} C_{k-1}^{6-1} (0.3762)^k (0.6238)^{6-k} = 0.2332.$$

(d) Using Tchebyshev's Inequality, we know that $L = E(X) - 3\sqrt{V(X)}$ and $U = E(X) + 3\sqrt{V(X)}$. Since $f_X(x)$ is symmetric about zero, we know that $E(X) = 0$. So, $V(X) = E(X^2) = \int_{-6}^{6}(x^2)\frac{1}{288}(36 - x^2)\,dx = 7.20$, so that $L = -8.05$ and $U = 8.05$. These findings clearly illustrate the very conservative nature of Tchebyshev's Theorem, since $\text{pr}(-8.05 < X < 8.05) = 1$.

**Solution 2.14**

(a) Although it is possible to find $P_X(s)$ directly, it is easier to make use of the connection between the moment generating function of X and the probability generating function of X. In particular,

$$\begin{aligned} M_X(t) = E(e^{tX}) &= \int_{-\infty}^{\infty} e^{tx}[\pi f_1(x) + (1-\pi)f_2(x)]\,dx \\ &= \pi\int_{-\infty}^{\infty} e^{tx}f_1(x)\,dx + (1-\pi)\int_{-\infty}^{\infty} e^{tx}f_2(x)\,dx \\ &= \pi e^{(\mu_1 t + \sigma_1^2 t^2/2)} + (1-\pi)e^{(\mu_2 t + \sigma_2^2 t^2/2)}. \end{aligned}$$

So, using the fact that $s = e^t$ and $\ln(s) = t$, it follows directly that

$$P_X(s) = \pi s^{\mu_1} e^{\frac{\sigma_1^2[\ln(s)]^2}{2}} + (1-\pi)s^{\mu_2} e^{\frac{\sigma_2^2[\ln(s)]^2}{2}}.$$

So,

$$\begin{aligned} \frac{dP_X(s)}{ds} &= \pi\left[\mu_1 s^{\mu_1 - 1} e^{(\sigma_1^2/2)[\ln(s)]^2} + s^{\mu_1} e^{(\sigma_1^2/2)[\ln(s)]^2}\sigma_1^2 s^{-1}\ln(s)\right] \\ &\quad + (1-\pi)\left[\mu_2 s^{\mu_2 - 1} e^{(\sigma_2^2/2)[\ln(s)]^2} + s^{\mu_2} e^{(\sigma_2^2/2)[\ln(s)]^2} \right. \\ &\quad \left. \times\ \sigma_2^2 s^{-1}\ln(s)\right]. \end{aligned}$$

Finally,

$$E(X) = \frac{dP_X(s)}{ds}_{|s=1} = \pi\mu_1 + (1-\pi)\mu_2.$$

(b) Let A be the event that "X is from $f_1(x)$," so that $\bar{A}$ is the event that "X is from $f_2(x)$"; and, let B be the event that "$X > 1.10$." Then, as a direct application of Bayes' Theorem, we have

$$\begin{aligned} \text{pr}(A|B) &= \frac{\text{pr}(B|A)\text{pr}(A)}{\text{pr}(B|A)\text{pr}(A) + \text{pr}(B|\bar{A})\text{pr}(\bar{A})} \\ &= \frac{\pi\text{pr}(B|A)}{\pi\text{pr}(B|A) + (1-\pi)\text{pr}(B|\bar{A})}. \end{aligned}$$

Now, with $Z \sim N(0,1)$, we have

$$\text{pr}(B|A) = \text{pr}\left(\frac{X - 1.00}{\sqrt{0.50}} > \frac{1.10 - 1.00}{\sqrt{0.50}}\right) = \text{pr}(Z > 0.1414) = 0.44$$

and

$$\text{pr}(B|\bar{A}) = \text{pr}\left(\frac{X - 1.20}{\sqrt{0.40}} > \frac{1.10 - 1.20}{\sqrt{0.40}}\right) = \text{pr}(Z > -0.1581) = 0.56.$$

Thus,

$$\text{pr}(A|B) = \frac{(0.60)(0.44)}{(0.60)(0.44) + (0.40)(0.56)} = 0.54.$$

(c) Since $\text{pr}(X > 1) = 0.16$, it follows that the appropriate truncated density function for $X$ is

$$f_X(x|X > 1) = (0.16)^{-1}\frac{1}{\sqrt{2\pi}}e^{-x^2/2}, \quad 1 < x < \infty.$$

So,

$$E(X|X > 1) = (0.16)^{-1}\int_1^{\infty} x\frac{1}{\sqrt{2\pi}}e^{-x^2/2}\,dx.$$

Letting $y = x^2/2$, so that $dy = x\,dx$, we have

$$E(X|X > 1) = [\sqrt{2\pi}(0.16)]^{-1}\int_{1/2}^{\infty} e^{-y}\,dy$$

$$= 2.4934\left[-e^{-y}\right]_{1/2}^{\infty} = 2.4934\left(e^{-1/2}\right) = 1.5123.$$

**Solution 2.15.** For $r$ an odd positive integer,

$$E\left(|Y^r|\right) = \int_{-\infty}^{\infty} |y^r|\frac{1}{\sqrt{2\pi}}e^{-y^2/2}\,dy$$

$$= \int_{-\infty}^{0} (-y^r)\frac{1}{\sqrt{2\pi}}e^{-y^2/2}\,dy + \int_0^{\infty} y^r\frac{1}{\sqrt{2\pi}}e^{-y^2/2}\,dy$$

$$= 2\int_0^{\infty} y^r\frac{1}{\sqrt{2\pi}}e^{-y^2/2}\,dy$$

$$= \int_0^{\infty} \frac{1}{\sqrt{2\pi}}\left(u^{1/2}\right)^{r-1} e^{-u/2}\,du$$

$$= \int_0^{\infty} \frac{1}{\sqrt{2\pi}}u^{\left(\frac{r+1}{2}\right)-1}e^{-u/2}\,du$$

$$= \frac{1}{\sqrt{2\pi}}\Gamma\left(\frac{r+1}{2}\right)2^{\left(\frac{r+1}{2}\right)}\int_0^{\infty}\frac{u^{\left(\frac{r+1}{2}\right)-1}e^{-u/2}}{\Gamma\left(\frac{r+1}{2}\right)2^{\left(\frac{r+1}{2}\right)}}\,du$$

$$= \frac{1}{\sqrt{2\pi}}\Gamma\left(\frac{r+1}{2}\right)2^{\left(\frac{r+1}{2}\right)}.$$

**Solution 2.16**

$$
\begin{aligned}
\mathrm{E}\left[\frac{Y!}{(Y-r)!}\right] &= \sum_{y=0}^{\infty} \frac{y!}{(y-r)!} \mathrm{C}_{k-1}^{y+k-1} \pi^{k}(1-\pi)^{y} \\
&= \sum_{y=r}^{\infty} \frac{(y+k-1)!}{(k-1)!(y-r)!} \pi^{k}(1-\pi)^{y} \\
&= \sum_{u=0}^{\infty} \frac{(u+r+k-1)!}{(k-1)!u!} \pi^{k}(1-\pi)^{u+r} \\
&= \frac{\pi^{-r}(1-\pi)^{r}(k+r-1)!}{(k-1)!} \sum_{u=0}^{\infty} \mathrm{C}_{(k+r)-1}^{u+(k+r)-1} \pi^{(k+r)}(1-\pi)^{u} \\
&= \frac{\pi^{-r}(1-\pi)^{r}(k+r-1)!}{(k-1)!} \\
&= \frac{(k+r-1)!}{(k-1)!}\left(\frac{1-\pi}{\pi}\right)^{r}, \quad r = 0, 1, \ldots, \infty.
\end{aligned}
$$

So, $r = 1$ gives

$$
\mathrm{E}(Y) = k\left(\frac{1-\pi}{\pi}\right).
$$

And, $r = 2$ gives

$$
\mathrm{E}[Y(Y-1)] = k(k+1)\left(\frac{1-\pi}{\pi}\right)^{2},
$$

so that

$$
\begin{aligned}
\mathrm{V}(Y) &= k(k+1)\left(\frac{1-\pi}{\pi}\right)^{2} + k\left(\frac{1-\pi}{\pi}\right) - \left[k\left(\frac{1-\pi}{\pi}\right)\right]^{2} \\
&= \frac{k(1-\pi)}{\pi^{2}}.
\end{aligned}
$$

Since $X = (Y + k)$,

$$
\mathrm{E}(X) = \mathrm{E}(Y) + k = k\left(\frac{1-\pi}{\pi}\right) + k = \frac{k}{\pi}
$$

and

$$
\mathrm{V}(X) = \mathrm{V}(Y) = k(1-\pi)/\pi^{2}.
$$

These are expected answers, since $X \sim \mathrm{NEGBIN}(k, \pi)$.

**Solution 2.17**

(a) For $-\infty < y \leq \beta$,

$$F_Y(y) = \int_{-\infty}^{y} (2\alpha)^{-1} e^{-(\beta - t)/\alpha}\, dt$$
$$= \frac{e^{-\beta/\alpha}}{2} \int_{-\infty}^{y} \frac{1}{\alpha} e^{t/\alpha}\, dt$$
$$= \frac{e^{-\beta/\alpha}}{2} \left(e^{y/\alpha}\right)$$
$$= \frac{1}{2} e^{(y-\beta)/\alpha}.$$

Note that $F_Y(\beta) = \frac{1}{2}$; this is an expected result because the density function $f_Y(y)$ is symmetric around $\beta$.

For $\beta < y < +\infty$,

$$F_Y(y) = F_Y(\beta) + \int_{\beta}^{y} \frac{1}{2\alpha} e^{-(t-\beta)/\alpha}\, dt$$
$$= \frac{1}{2} + \frac{e^{\beta/\alpha}}{2} \int_{\beta}^{y} \frac{1}{\alpha} e^{-t/\alpha}\, dt$$
$$= \frac{1}{2} + \frac{e^{\beta/\alpha}}{2} [e^{-\beta/\alpha} - e^{-y/\alpha}]$$
$$= 1 - \frac{1}{2} e^{-(y-\beta)/\alpha}.$$

Thus,

$$F_Y(y) = \begin{cases} \frac{1}{2} e^{(y-\beta)/\alpha}, & -\infty < y \leq \beta; \\ 1 - \frac{1}{2} e^{-(y-\beta)/\alpha}, & \beta < y < +\infty. \end{cases}$$

Now, if $\alpha = 1$ and $\beta = 2$,

$$\text{pr}(X > 4|X > 2) = \text{pr}(Y > \ln 4|Y > \ln 2)$$
$$= \frac{\text{pr}\left[(Y > \ln 4) \cap (Y > \ln 2)\right]}{\text{pr}(Y > \ln 2)}$$
$$= \frac{\text{pr}(Y > \ln 4)}{\text{pr}(Y > \ln 2)}$$
$$= \frac{1 - F_Y(1.3863)}{1 - F_Y(0.6931)}$$

$$= \frac{1 - \frac{1}{2}e^{(1.3863-2)}}{1 - \frac{1}{2}e^{(0.6931-2)}}$$

$$= 0.8434.$$

(b) Now,

$$\begin{aligned}
\phi_Y(t) &= \mathrm{E}\left\{e^{t|Y-\mathrm{E}(Y)|}\right\} \\
&= \mathrm{E}\left\{e^{t|Y-\beta|}\right\} \\
&= \int_{-\infty}^{\infty} e^{t|y-\beta|}(2\alpha)^{-1}e^{-|y-\beta|/\alpha}\,dy \\
&= \int_{-\infty}^{\infty} \frac{1}{2\alpha}e^{-|y-\beta|\left(\frac{1}{\alpha}-t\right)}dy \\
&= \int_{-\infty}^{\infty} \frac{1}{2\alpha}e^{-|y-\beta|/\left[\frac{\alpha}{(1-\alpha t)}\right]}dy \\
&= \frac{[\alpha/(1-\alpha t)]}{\alpha} = (1-\alpha t)^{-1}, \quad \alpha t < 1.
\end{aligned}$$

So,

$$\begin{aligned}
\left[\frac{d\phi_Y(t)}{dt}\right]_{|t=0} &= \nu_1 = \mathrm{E}\{|Y - \mathrm{E}(Y)|\} \\
&= [-(1-\alpha t)^{-2}(-\alpha)]_{|t=0} = \alpha.
\end{aligned}$$

And,

$$\begin{aligned}
\left[\frac{d^2\phi_Y(t)}{dt^2}\right]_{|t=0} &= \nu_2 = \mathrm{E}\{|Y - \mathrm{E}(Y)|^2\} = \mathrm{V}(Y) \\
&= [\alpha(-2)(1-\alpha t)^{-3}(-\alpha)]_{|t=0} = 2\alpha^2.
\end{aligned}$$

**Solution 2.18.** First, with $y = x^2/2$ so that $dy = x\,dx$, we have

$$\begin{aligned}
\mathrm{E}(X) &= \int_0^{\infty} x\left(\frac{2}{\pi\theta}\right)^{1/2} e^{-x^2/2\theta}\,dx \\
&= \int_0^{\infty} \left(\frac{2}{\pi\theta}\right)^{1/2} e^{-y/\theta}\,dy \\
&= \left(\frac{2}{\pi\theta}\right)^{1/2}(\theta)\int_0^{\infty}\frac{1}{\theta}e^{-y/\theta}\,dy \\
&= \left(\frac{2\theta}{\pi}\right)^{1/2}.
\end{aligned}$$

Now, we have

$$
\begin{aligned}
E(Y) = E[g(X)] &= 1 - \int_0^{\infty} \left(\alpha e^{-\beta x^2}\right)\left(\frac{2}{\pi\theta}\right)^{1/2} e^{-x^2/2\theta}\,dx \\
&= 1 - \alpha\left(\frac{2}{\pi\theta}\right)^{1/2}\int_0^{\infty} e^{-\left(\beta+\frac{1}{2\theta}\right)x^2}\,dx \\
&= 1 - \alpha\left(\frac{2}{\pi\theta}\right)^{1/2}\left(\frac{1}{2}\right)\int_{-\infty}^{\infty} e^{-x^2\Big/\left[2\left(\frac{\theta}{2\theta\beta+1}\right)\right]}dx \\
&= 1 - \frac{\alpha}{\sqrt{\theta}}\int_{-\infty}^{\infty}\frac{1}{\sqrt{2\pi}}e^{-x^2\Big/\left[2\left(\frac{\theta}{2\theta\beta+1}\right)\right]}dx \\
&= 1 - \frac{\alpha}{\sqrt{\theta}}\sqrt{\frac{\theta}{(2\theta\beta+1)}} \\
&= 1 - \frac{\alpha}{\sqrt{2\theta\beta+1}} \\
&= 1 - \frac{\alpha}{\sqrt{\pi\beta[E(X)]^2+1}}.
\end{aligned}
$$

Note that the average risk increases as both $\beta$ and $E(X)$ increase, but the average risk decreases as $\alpha$ increases.

**Solution 2.19**

(a) Clearly, the distribution of $X_1$ is negative binomial, namely,

$$p_{X_1}(x_1) = C_{(2-1)}^{(x_1-1)}\pi_h^2(1-\pi_h)^{(x_1-2)}, \quad x_1 = 2, 3, \ldots, \infty.$$

(b) $p_{X_2}(x_2) = \text{pr}(X_2 = x_2) = \text{pr}\left[\cup_{j=1}^{x_2-1}(A_j \cap B)\right] + \text{pr}\left[\cup_{j=1}^{x_2-1}(C_j \cap D)\right]$, where $A_j$ is the event that "the first $(x_2 - 1)$ subjects selected consist of $j$ heavy smokers and $(x_2 - 1 - j)$ nonsmokers," B is the event that "the $x_2$th subject selected is a light smoker," $C_j$ is the event that "the first $(x_2 - 1)$ subjects selected consist of $j$ light smokers and $(x_2 - 1 - j)$ nonsmokers," and D is the event that "the $x_2$th subject selected is a heavy smoker."

So,

$$
\begin{aligned}
p_{X_2}(x_2) &= \left[\sum_{j=1}^{(x_2-1)} C_j^{(x_2-1)}\pi_h^j(\pi_0)^{(x_2-1-j)}\right]\pi_l \\
&+ \left[\sum_{j=1}^{(x_2-1)} C_j^{(x_2-1)}\pi_l^j(\pi_0)^{(x_2-1-j)}\right]\pi_h
\end{aligned}
$$

$$= [(\pi_h + \pi_0)^{(x_2-1)} - \pi_0^{(x_2-1)}]\pi_l + [(\pi_l + \pi_0)^{(x_2-1)} - \pi_0^{(x_2-1)}]\pi_h$$

$$= \pi_l(1-\pi_l)^{(x_2-1)} + \pi_h(1-\pi_h)^{(x_2-1)} - (1-\pi_0)\pi_0^{(x_2-1)},$$

$x_2 = 2, 3, \ldots, \infty.$

(c) Via a direct extension of the reasoning used in part (b), we obtain the following:

$$\begin{aligned} p_{X_3}(x_3) &= \left[\sum_{j=1}^{(x_3-2)} C_j^{(x_3-1)} \pi_l^j \pi_h^{(x_3-1-j)}\right]\pi_0 \\ &\quad + \left[\sum_{j=1}^{(x_3-2)} C_j^{(x_3-1)} \pi_0^j \pi_h^{(x_3-1-j)}\right]\pi_l \\ &\quad + \left[\sum_{j=1}^{(x_3-2)} C_j^{(x_3-1)} \pi_0^j \pi_l^{(x_3-1-j)}\right]\pi_h \\ &= \pi_0\left[(\pi_l+\pi_h)^{(x_3-1)} - \pi_l^{(x_3-1)} - \pi_h^{(x_3-1)}\right] \\ &\quad + \pi_l\left[(\pi_0+\pi_h)^{(x_3-1)} - \pi_0^{(x_3-1)} - \pi_h^{(x_3-1)}\right] \\ &\quad + \pi_h\left[(\pi_0+\pi_l)^{(x_3-1)} - \pi_0^{(x_3-1)} - \pi_l^{(x_3-1)}\right] \\ &= \pi_0(1-\pi_0)^{(x_3-1)} + \pi_l(1-\pi_l)^{(x_3-1)} + \pi_h(1-\pi_h)^{(x_3-1)} \\ &\quad - (1-\pi_0)\pi_0^{(x_3-1)} - (1-\pi_l)\pi_l^{(x_3-1)} - (1-\pi_h)\pi_h^{(x_3-1)}, \end{aligned}$$

$x_3 = 3, 4, \ldots, \infty.$

**Solution 2.20**

(a) Since $Y \sim \mathrm{N}(\mu, \sigma^2)$, it follows that the moment generating function for $Y = e^X$ is

$$E(e^{tY}) = E(X^t) = e^{\left(\mu t + \frac{\sigma^2 t^2}{2}\right)}, \quad -\infty < t < +\infty.$$

So, for $t = 1$,

$$E(X) = e^{(\mu + 0.50\sigma^2)}.$$

And, for $t = 2$,

$$V(X) = E(X^2) - [E(X)]^2 = e^{(2\mu + 2\sigma^2)} - \left[e^{(\mu+0.50\sigma^2)}\right]^2$$

$$= e^{(2\mu+\sigma^2)}(e^{\sigma^2} - 1).$$

(b) Since $E(X) = V(X) = 1$, we have

$$\frac{V(X)}{[E(X)]^2} = \left(e^{\sigma^2} - 1\right) = 1,$$

which gives $\sigma = 0.8326$.

And, the equation

$$[E(X)]^2 = e^{(2\mu+\sigma^2)} = e^{[2\mu+(0.8326)^2]} = 1 \text{ gives } \mu = -0.3466.$$

So,

$$\text{pr}(X > 1) = \text{pr}(Y > 0) = \text{pr}\left[\frac{Y - (-0.3466)}{0.8326} > \frac{0 - (-0.3466)}{0.8326}\right]$$
$$= \text{pr}(Z > 0.4163) = 0.339, \quad \text{since } Z \sim N(0, 1).$$

(c) Now,

$$\text{pr}(X \leq c) = \text{pr}[Y \leq \ln(c)] = \text{pr}\left[Z \leq \frac{\ln(c) - \mu}{\sigma}\right],$$

$$\text{where } Z = \frac{Y - \mu}{\sigma} \sim N(0, 1).$$

Thus, to satisfy $\text{pr}(X \leq c) \geq (1 - \alpha)$ requires $\{[\ln(c) - \mu]/\sigma\} \geq z_{1-\alpha}$.

And, since $E(X) = e^{(\mu+0.50\sigma^2)}$, so that $\mu = \ln[E(X)] - 0.50\sigma^2$, the inequality $\{[\ln(c) - \mu]/\sigma\} \geq z_{1-\alpha}$ is equivalent to the inequality

$$\frac{\ln(c) - [\ln E(X) - 0.50\sigma^2]}{\sigma} \geq z_{1-\alpha},$$

which, in turn, can be written in the form

$$\ln\left[\frac{c}{E(X)}\right] \geq \sigma z_{1-\alpha} - 0.50\sigma^2 = 0.50z^2_{1-\alpha} - 0.50(z_{1-\alpha} - \sigma)^2.$$

So, if $\ln[c/E(X)] \geq 0.50z^2_{1-\alpha}$, then the above inequality will be satisfied. Equivalently, we need to pick $E(X)$ small enough so that

$$E(X) \leq ce^{-0.50z^2_{1-\alpha}}.$$

**Solution 2.21**

(a) Since $\theta_0 = 1 - \alpha[\pi/(1 - \pi)]$ and $\theta_x = \alpha\theta^x$ for $x \geq 1$, we require that $0 < \alpha < [(1 - \pi)/\pi]$ so that $0 < \pi_x < 1, x = 0, 1, 2, \ldots, +\infty$. Now,

$$E(e^{tX}) = \sum_{x=0}^{\infty} e^{tx}\pi_x$$

$$= \left[1 - \alpha\left(\frac{\pi}{1-\pi}\right)\right] + \sum_{x=1}^{\infty} e^{tx}(\alpha\pi^x)$$

$$= \left[1 - \alpha\left(\frac{\pi}{1-\pi}\right)\right] + \alpha\sum_{x=1}^{\infty} (\pi e^t)^x$$

$$= \left[1 - \alpha\left(\frac{\pi}{1-\pi}\right)\right] + \alpha\left[\frac{\pi e^t}{1-\pi e^t}\right]$$

provided that $0 < \pi e^t < 1$, or that $-\infty < t < -\ln \pi$. So,

$$\mathrm{M}_X(t) = \left[1 - \alpha\left(\frac{\pi}{1-\pi}\right)\right] + \alpha\left[\frac{\pi e^t}{1-\pi e^t}\right],$$

$0 < \alpha < [(1-\pi)/\pi], -\infty < t < -\ln \pi$.
So,

$$\mathrm{E}(X) = \left.\frac{d\mathrm{M}_X(t)}{dt}\right|_{t=0} = \left\{\alpha\pi\left[\frac{e^t(1-\pi e^t) - e^t(-\pi e^t)}{(1-\pi e^t)^2}\right]\right\}_{|t=0}$$

$$= \alpha\pi\left\{\left[\frac{e^t}{(1-\pi e^t)^2}\right]\right\}_{|t=0}$$

$$= \frac{\alpha\pi}{(1-\pi)^2}.$$

(b)

$$\mathrm{E}(X) = \sum_{x=0}^{\infty} x\theta_x = \sum_{x=1}^{\infty} x\alpha\pi^x$$

$$= \alpha\pi\sum_{x=1}^{\infty} x\pi^{x-1} = \alpha\pi\sum_{x=1}^{\infty}\frac{d}{d\pi}(\pi^x)$$

$$= \alpha\pi\frac{d}{d\pi}\sum_{x=1}^{\infty}\pi^x = \alpha\pi\frac{d}{d\pi}\left(\frac{\pi}{1-\pi}\right)$$

$$= \frac{\alpha\pi}{(1-\pi)^2}.$$

**Solution 2.22.** For the gamma distribution,

$$\mathrm{E}(X^r) = \int_0^\infty x^r \frac{x^{\beta-1}e^{-x/\alpha}}{\Gamma(\beta)\alpha^\beta}\,dx = \frac{\Gamma(\beta+r)}{\Gamma(\beta)}\alpha^r, \quad (\beta+r) > 0.$$

So,

$$\begin{aligned}\mu_3 &= \mathrm{E}\{[X-\mathrm{E}(X)]^3\} = \mathrm{E}(X^3) - 3\mathrm{E}(X^2)\mathrm{E}(X) + 2[\mathrm{E}(X)]^3 \\ &= \beta(\beta+1)(\beta+2)\alpha^3 - 3[\beta(\beta+1)\alpha^2](\alpha\beta) + 2\alpha^3\beta^3 \\ &= 2\alpha^3\beta.\end{aligned}$$

Thus,

$$\alpha_3 = \frac{2\alpha^3\beta}{(\alpha^2\beta)^{3/2}} = \frac{2}{\sqrt{\beta}}.$$

Now, to find the mode of the gamma distribution, we need to find that value of $x$, say $\theta$, which maximizes $f_X(x)$, or equivalently, which maximizes the function

$$h(x) = \ln\left(x^{\beta-1}e^{-x/\alpha}\right) = (\beta-1)\ln(x) - \frac{x}{\alpha}.$$

So,

$$\frac{dh(x)}{dx} = \frac{(\beta-1)}{x} - \frac{1}{\alpha} = 0$$

gives $\theta = \alpha(\beta-1)$, which, for $\beta > 1$, maximizes $f_X(x)$; in particular, note that $[d^2h(x)]/dx^2 = (1-\beta)/x^2$, when evaluated at $x = \theta = \alpha(\beta-1)$, is negative for $\beta > 1$.

Finally, we have

$$\alpha_3^* = \frac{\alpha\beta - \alpha(\beta-1)}{\sqrt{\alpha^2\beta}} = \frac{1}{\sqrt{\beta}}.$$

Thus, we have $\alpha_3 = 2\alpha_3^*$, so that the two measures are essentially equivalent with regard to quantifying the degree of asymmetry for the gamma distribution.

NOTE: For the beta distribution,

$$f_X(x) = \frac{\Gamma(\alpha+\beta)}{\Gamma(\alpha)\Gamma(\beta)}x^{\alpha-1}(1-x)^{\beta-1}, \quad 0 < x < 1,\ \alpha > 0,\ \beta > 0,$$

the interested reader can verify that the mode of the beta distribution is

$$\theta = \frac{(\alpha-1)}{(\alpha+\beta-2)}, \quad \alpha > 1,\ \beta > 1,$$

and that

$$\alpha_3 = \frac{2(\beta-\alpha)}{(\alpha+\beta+2)}\sqrt{\frac{(\alpha+\beta+1)}{\alpha\beta}} = \frac{2(\alpha+\beta-2)}{(\alpha+\beta+2)}\alpha_3^*.$$

**Solution 2.23***

(a) We have

$$\begin{aligned}
E(U) &= \int_0^L g(L)f_X(x)dx + \int_L^\infty xf_X(x)\,dx \\
&= g(L)\int_0^L f_X(x)dx + \pi\int_L^\infty x\left[\frac{f_X(x)}{\pi}\right]dx \\
&= (1-\pi)g(L) + \pi\int_L^\infty xf_X(x|X \geq L)\,dx \\
&= (1-\pi)g(L) + \pi E(X|X \geq L).
\end{aligned}$$

And, using a similar development, we have

$$\begin{aligned}
E(U^2) &= \int_0^L [g(L)]^2 f_X(x)\,dx + \int_L^\infty x^2 f_X(x)\,dx \\
&= (1-\pi)\left[g(L)\right]^2 + \pi\int_L^\infty x^2 f_X(x|X \geq L)\,dx \\
&= (1-\pi)\left[g(L)\right]^2 + \pi E(X^2|X \geq L).
\end{aligned}$$

Thus,

$$\begin{aligned}
V(U) &= E(U^2) - [E(U)]^2 \\
&= (1-\pi)\left[g(L)\right]^2 + \pi E(X^2|X \geq L) - [(1-\pi)g(L) \\
&\quad + \pi E(X|X \geq L)]^2 \\
&= (1-\pi)\left[g(L)\right]^2 + \pi E(X^2|X \geq L) - (1-\pi)^2\left[g(L)\right]^2 \\
&\quad - 2\pi(1-\pi)g(L)E(X|X \geq L) - \pi^2\,[E(X|X \geq L)]^2 \\
&= \left[(1-\pi) - (1-\pi)^2\right]\left[g(L)\right]^2 \\
&\quad + \pi E(X^2|X \geq L) - 2\pi(1-\pi)g(L)E(X|X \geq L) \\
&\quad - \pi^2\,[E(X|X \geq L)]^2 + \pi\,[E(X|X \geq L)]^2 - \pi\,[E(X|X \geq L)]^2 \\
&= \pi(1-\pi)\left[g(L)\right]^2 + \pi\left\{E(X^2|X \geq L) - [E(X|X \geq L)]^2\right\} \\
&\quad - 2\pi(1-\pi)g(L)E(X|X \geq L) + \pi(1-\pi)\,[E(X|X \geq L)]^2 \\
&= \pi V(X|X \geq L) + \pi(1-\pi)\left[g(L) - E(X|X \geq L)\right]^2 \\
&= \pi\left\{V(X|X \geq L) + (1-\pi)\left[g(L) - E(X|X \geq L)\right]^2\right\}.
\end{aligned}$$

(b) Since

$$\begin{aligned}
\mathrm{E}(X) &= \int_0^{\infty} x\mathrm{f}_X(x)\,\mathrm{d}x = \int_0^{L} x\mathrm{f}_X(x)\,\mathrm{d}x + \int_L^{\infty} x\mathrm{f}_X(x)\,\mathrm{d}x \\
&= (1-\pi)\int_0^{L} x\left[\frac{\mathrm{f}_X(x)}{(1-\pi)}\right]\mathrm{d}x + \pi\int_L^{\infty} x\left[\frac{\mathrm{f}_X(x)}{\pi}\right]\mathrm{d}x \\
&= (1-\pi)\mathrm{E}(X|X<L) + \pi\mathrm{E}(X|X\ge L),
\end{aligned}$$

it follows directly that choosing $\mathrm{g}(L)$ to be equal to $\mathrm{E}(X|X<L)$ will insure that $\mathrm{E}(U)=\mathrm{E}(X)$.

When $\mathrm{f}_X(x)=\mathrm{e}^{-x},\ x\ge 0$, and $L=0.05$, then

$$(1-\pi) = \int_0^{0.05} \mathrm{e}^{-x}\,\mathrm{d}x = \left[-\mathrm{e}^{-x}\right]_0^{0.05} = 0.0488.$$

Thus, using integration by parts with $u=x$ and $\mathrm{d}v=\mathrm{e}^{-x}\,\mathrm{d}x$, we find that the optimal choice for $\mathrm{g}(L)$ has the numerical value

$$\begin{aligned}
\mathrm{E}(X|X<L) &= \int_0^{L} x\mathrm{f}_X(x|X<L)\,\mathrm{d}x = \int_0^{L} x\left[\frac{\mathrm{f}_X(x)}{(1-\pi)}\right]\mathrm{d}x \\
&= \int_0^{0.05} x\left[\frac{\mathrm{e}^{-x}}{0.0488}\right]\mathrm{d}x = (0.0488)^{-1}\int_0^{0.05} x\mathrm{e}^{-x}\,\mathrm{d}x \\
&= (20.4918)\left\{\left[-x\mathrm{e}^{-x}\right]_0^{0.05} + \int_0^{0.05}\mathrm{e}^{-x}\,\mathrm{d}x\right\} \\
&= (20.4918)\left(-0.05\mathrm{e}^{-0.05}+0.0488\right) = 0.0254.
\end{aligned}$$

For information about a more rigorous statistical approach for dealing with this left-censoring issue, see Taylor et al. (2001).

**Solution 2.24***. Now,

$$\begin{aligned}
\mathrm{E}(Y) &= \int_{-\infty}^{\infty} (1-\alpha\mathrm{e}^{-\beta x^2})\frac{1}{\sqrt{2\pi}\sigma}\mathrm{e}^{-(x-\mu)^2/2\sigma^2}\,\mathrm{d}x \\
&= 1 - \frac{\alpha}{\sqrt{2\pi}\sigma}\int_{-\infty}^{\infty} \mathrm{e}^{-\left[\beta x^2+\frac{(x-\mu)^2}{2\sigma^2}\right]}\mathrm{d}x.
\end{aligned}$$

And,

$$\beta x^2 + \frac{(x-\mu)^2}{2\sigma^2} = \left(\beta + \frac{1}{2\sigma^2}\right)x^2 - \left(\frac{\mu}{\sigma^2}\right)x + \frac{\mu^2}{2\sigma^2}$$

$$= \left[x\sqrt{\beta + \frac{1}{2\sigma^2}} - \frac{\mu}{2\sigma^2\sqrt{\beta + \frac{1}{2\sigma^2}}}\right]^2$$

$$- \frac{\mu^2}{4\sigma^4\left(\beta + \frac{1}{2\sigma^2}\right)} + \frac{\mu^2}{2\sigma^2}$$

$$= \left(\frac{2\beta\sigma^2 + 1}{2\sigma^2}\right)\left[x - \frac{\mu}{(2\beta\sigma^2+1)}\right]^2 + \frac{\beta\mu^2}{(2\beta\sigma^2+1)}.$$

Finally,

$$E(Y) = 1 - \frac{\alpha}{\sqrt{2\pi}\sigma}\int_{-\infty}^{\infty} e^{-\left\{\frac{\left[x - \frac{\mu}{(2\beta\sigma^2+1)}\right]^2}{2\left(\frac{\sigma^2}{2\beta\sigma^2+1}\right)}\right\}} e^{-\frac{\beta\mu^2}{(2\beta\sigma^2+1)}}\,dx$$

$$= 1 - \left(\frac{\alpha}{\sigma}\right)e^{-\frac{\beta\mu^2}{(2\beta\sigma^2+1)}}\sqrt{\frac{\sigma^2}{(2\beta\sigma^2+1)}}$$

$$= 1 - \frac{\alpha}{\sqrt{2\beta\sigma^2+1}}e^{-\frac{\beta\mu^2}{(2\beta\sigma^2+1)}}.$$

**Solution 2.25***

(a) Now,

$$\psi_Y(t) = \ln\left[E(e^{tY})\right] = \ln\left[E\left(e^{t(X-c)}\right)\right]$$

$$= \ln\left[e^{-tc}E\left(e^{tX}\right)\right] = -tc + \ln\left[E\left(e^{tX}\right)\right] = -tc + \sum_{r=1}^{\infty}\kappa_r\frac{t^r}{r!}$$

$$= (\kappa_1 - c)t + \sum_{r=2}^{\infty}\kappa_r\frac{t^r}{r!}.$$

Hence, the cumulants of $Y$ are identical to those for $X$, except for the first cumulant. In particular, if $Y = (X - c)$, then the first cumulant of $Y$ is $(\kappa_1 - c)$, where $\kappa_1$ is the first cumulant of $X$.

(b)

(i) If $X \sim N(\mu, \sigma^2)$, then the moment generating function of $X$ is $M_X(t) = e^{\mu t + \sigma^2 t^2/2}$. So,

$$\psi_X(t) = \mu t + \frac{\sigma^2 t^2}{2}.$$

Hence, $\kappa_1 = \mu, \kappa_2 = \sigma^2$, and $\kappa_r = 0$ for $r = 3, 4, \ldots, \infty$.

(ii) If X ~POI($\lambda$), then $\mathrm{M_X}(t) = \mathrm{e}^{\lambda(\mathrm{e}^t - 1)}$. So,

$$\psi_X(t) = \lambda(\mathrm{e}^t - 1) = \lambda \sum_{r=1}^{\infty} \frac{t^r}{r!} = \sum_{r=1}^{\infty} (\lambda) \frac{t^r}{r!}.$$

Thus, $\kappa_r = \lambda$ for $r = 1, 2, \ldots, \infty$.

(iii) If X ~GAMMA($\alpha, \beta$), then $\mathrm{M_X}(t) = (1 - \alpha t)^{-\beta}$.
So, $\psi_X(t) = -\beta \ln(1 - \alpha t)$. Now,

$$\ln(1 + y) = \sum_{r=1}^{\infty} (-1)^{r+1} \frac{y^r}{r}, \quad -1 < y < +1.$$

If $y = -\alpha t$, and $t$ is chosen so that $|\alpha t| < 1$, then

$$\ln(1 - \alpha t) = \sum_{r=1}^{\infty} (-1)^{r+1} \frac{(-\alpha t)^r}{r}$$

$$= \sum_{r=1}^{\infty} (-1)^{2r+1} \alpha^r (r-1)! \frac{t^r}{r!} = -\sum_{r=1}^{\infty} [(r-1)! \alpha^r] \frac{t^r}{r!}.$$

So,

$$\psi_X(t) = -\beta \ln(1 - \alpha t) = \sum_{r=1}^{\infty} [(r-1)! \alpha^r \beta] \frac{t^r}{r!}, \quad |\alpha t| < 1;$$

thus, $\kappa_r = (r-1)! \alpha^r \beta$ for $r = 1, 2, \ldots, \infty$.

(c) First, for $r = 1, 2, \ldots, \infty$, since $\psi_X(t) = \sum_{r=1}^{\infty} \kappa_r \frac{t^r}{r!}$, it follows directly that

$$\kappa_r = \frac{\mathrm{d}^r \psi_X(t)}{\mathrm{d}t^r}_{|t=0}.$$

So, since $\psi_X(t) = \ln[\mathrm{M_X}(t)]$ and since $[\mathrm{d}^r \mathrm{M_X}(t)]/\mathrm{d}t^r{}_{|t=0} = \mathrm{E}(X^r)$ for $r = 1, 2, \ldots, \infty$, we have

$$\kappa_1 = \frac{\mathrm{d}\psi_X(t)}{\mathrm{d}t}_{|t=0} = \left\{ [\mathrm{M_X}(t)]^{-1} \frac{\mathrm{dM_X}(t)}{\mathrm{d}t} \right\}_{|t=0}$$

$$= (1)^{-1} \mathrm{E}(X) = \mathrm{E}(X).$$

Next, since

$$\frac{\mathrm{d}^2 \psi_X(t)}{\mathrm{d}t^2} = -[\mathrm{M_X}(t)]^{-2} \left[ \frac{\mathrm{dM_X}(t)}{\mathrm{d}t} \right]^2 + [\mathrm{M_X}(t)]^{-1} \frac{\mathrm{d}^2 \mathrm{M_X}(t)}{\mathrm{d}t^2},$$

it follows that

$$\kappa_2 = \frac{d^2\psi_X(t)}{dt^2}_{|t=0} = -(1)^{-2}[E(X)]^2 + (1)^{-1}E(X^2)$$
$$= E(X^2) - [E(X)]^2 = V(X).$$

Finally, since

$$\frac{d^3\psi_X(t)}{dt^3} = 2[M_X(t)]^{-3}\left[\frac{dM_X(t)}{dt}\right]^3 - 2[M_X(t)]^{-2}\left[\frac{dM_X(t)}{dt}\right]$$
$$\times\left[\frac{d^2M_X(t)}{dt^2}\right] - [M_X(t)]^{-2}\left[\frac{dM_X(t)}{dt}\right]\left[\frac{d^2M_X(t)}{dt^2}\right]$$
$$+ [M_X(t)]^{-1}\left[\frac{d^3M_X(t)}{dt^3}\right],$$

we have

$$\kappa_3 = \frac{d^3\psi_X(t)}{dt^3}_{|t=0}$$
$$= 2(1)^{-3}[E(X)]^3 - 2(1)^{-2}[E(X)][E(X^2)] - (1)^{-2}[E(X)]$$
$$\times [E(X^2)] + (1)^{-1}E(X^3)$$
$$= E(X^3) - 3E(X)E(X^2) + 2[E(X)]^3 = E\{[X - E(X)]^3\}.$$

**Solution 2.26***

(a)

$$h(t) = \lim_{\Delta t\to 0}\frac{pr(t \le T \le t + \Delta t | T \ge t)}{\Delta t}$$
$$= \frac{\lim_{\Delta t\to 0} pr(t \le T \le t + \Delta t)/\Delta t}{pr(T \ge t)}$$
$$= \frac{dF(t)/dt}{1 - F(t)} = \frac{f(t)}{S(t)}.$$

(b) From part (a), $H(t) = \int_0^t h(u)\,du = \int_0^t \frac{f(u)}{S(u)}\,du$. Since

$$dS(u) = d[1 - F(u)]\,du = -f(u)\,du,$$

we have

$$H(t) = -\int_0^t \frac{1}{S(u)}\,dS(u) = -\ln[S(t)] + \ln[S(0)] = -\ln[S(t)] + \ln(1)$$
$$= -\ln[S(t)], \quad \text{or} \quad S(t) = e^{-H(t)}.$$

(c) Now,

$$\mathrm{E}(T) = \int_0^\infty t\mathrm{f}(t)\,dt = \int_0^\infty \left[\int_0^t du\right] \mathrm{f}(t)\,dt = \int_0^\infty \left[\int_0^\infty \mathrm{I}(t > u)du\right] \mathrm{f}(t)\,dt,$$

where I(A) is an indicator function taking the value 1 if event A holds and taking the value 0 otherwise. Hence,

$$\begin{aligned}
\mathrm{E}(T) &= \int_0^\infty \left[\int_0^\infty \mathrm{I}(t > u)\,du\right] \mathrm{f}(t)\,dt \\
&= \int_0^\infty \left[\int_0^\infty \mathrm{I}(t > u)\mathrm{f}(t)\,dt\right] du \\
&= \int_0^\infty \left[\int_u^\infty \mathrm{f}(t)\,dt\right] du \\
&= \int_0^\infty \mathrm{S}(u)\,du.
\end{aligned}$$

(d) Note that

$$X = \begin{cases} T & \text{if } T < c; \\ c & \text{if } T \geq c \end{cases}$$

So, $\mathrm{f}_X(x) = \mathrm{f}_T(x)$ if $T < c$ (so that $0 < x < c$) and $\mathrm{f}_X(x) = c$ if $T \geq c$, an event which occurs with probability $[1 - \mathrm{F}_T(c)]$. Thus,

$$\mathrm{f}_X(x) = \mathrm{f}_T(x)\mathrm{I}(x < c) + [1 - \mathrm{F}_T(c)]\mathrm{I}(x = c), \quad 0 < x \leq c,$$

where, as in part (c), $\mathrm{I}(\cdot)$ denotes the indicator function. In other words, $\mathrm{f}_X(x)$ is a mixture of a continuous density [namely, $\mathrm{f}_T(x)$] for $x < c$ and a point mass at $c$ occurring with probability $[1 - \mathrm{F}_T(c)]$.

So,

$$\begin{aligned}
\mathrm{E}[\mathrm{H}(X)] &= \mathrm{E}\left[\mathrm{H}(X)\mathrm{I}(X < c) + \mathrm{H}(c)\mathrm{I}(X = c)\right] \\
&= \mathrm{E}[\mathrm{H}(X)\mathrm{I}(X < c)] + \mathrm{H}(c)\mathrm{E}[\mathrm{I}(X = c)] \\
&= \int_0^c \mathrm{H}(x)\mathrm{f}_T(x)\,dx + \mathrm{H}(c)\mathrm{pr}(X = c) \\
&= \int_0^c \mathrm{H}(x)\mathrm{f}_T(x)\,dx + \mathrm{H}(c)[1 - \mathrm{F}_T(c)].
\end{aligned}$$

Using integration by parts with $u = \mathrm{H}(x)$ and $dv = \mathrm{f}_T(x)\,dx$, we have

$$\begin{aligned}
\mathrm{E}[\mathrm{H}(X)] &= \mathrm{H}(x)\mathrm{F}_T(x)\big|_0^c - \int_0^c \mathrm{h}(x)\mathrm{F}_T(x)\,dx + \mathrm{H}(c) - \mathrm{H}(c)\mathrm{F}_T(c) \\
&= \mathrm{H}(c)\mathrm{F}_T(c) - 0 - \int_0^c \mathrm{h}(x)\mathrm{F}_T(x)\,dx + \mathrm{H}(c) - \mathrm{H}(c)\mathrm{F}_T(c) \\
&= -\int_0^c \mathrm{h}(x)\mathrm{F}_T(x)\,dx + \mathrm{H}(c) = -\int_0^c \mathrm{h}(x)[1-\mathrm{S}(x)]\,dx + \mathrm{H}(c) \\
&= -\int_0^c \mathrm{h}(x)\left[1 - \frac{\mathrm{f}_T(x)}{\mathrm{h}(x)}\right]dx + \mathrm{H}(c) \\
&= -\int_0^c \mathrm{h}(x)\,dx + \int_0^c \mathrm{f}_T(x)\,dx + \mathrm{H}(c) \\
&= -[\mathrm{H}(c) - \mathrm{H}(0)] + [\mathrm{F}_T(c) - 0] + \mathrm{H}(c) \\
&= \mathrm{H}(0) + \mathrm{F}_T(c) = -\ln[\mathrm{S}(0)] + \mathrm{F}_T(c) = -\ln(1) + \mathrm{F}_T(c) \\
&= \mathrm{F}_T(c).
\end{aligned}$$

**Solution 2.27***

(a) If $N$ units are produced, then it follows that $P = NG$ if $X \geq N$, and $P = [XG - (N - X)L] = [(G+L)X - NL]$ if $X < N$. Hence,

$$\begin{aligned}
\mathrm{E}(P) &= \int_0^N [(G+L)x - NL]\mathrm{f}_X(x)\,dx + \int_N^\infty (NG)\mathrm{f}_X(x)\,dx \\
&= (G+L)\int_0^N x\mathrm{f}_X(x)\,dx - NL\mathrm{F}_X(N) + NG[1 - \mathrm{F}_X(N)] \\
&= (G+L)\int_0^N x\mathrm{f}_X(x)\,dx + NG - N(G+L)\mathrm{F}_X(N).
\end{aligned}$$

Now, via integration by parts,

$$\int_0^N x\mathrm{f}_X(x)\,dx = [x\mathrm{F}_X(x)]_0^N - \int_0^N \mathrm{F}_X(x)\,dx = N\mathrm{F}_X(N) - \int_0^N \mathrm{F}_X(x)\,dx,$$

so that we finally obtain

$$\mathrm{E}(P) = NG - (G+L)\int_0^N \mathrm{F}_X(x)\,dx.$$

So,

$$\frac{d\mathrm{E}(P)}{dN} = G - (G+L)[\mathrm{F}_X(N)] = G - (G+L)\mathrm{F}_X(N) = 0,$$

which gives

$$\mathrm{F}_X(N) = \frac{G}{G+L};$$

since

$$\frac{d^2E(P)}{dN^2} = -(G+L)f_X(N) < 0,$$

this choice for $N$ maximizes $E(P)$.

(b) Since $f_X(x) = 2kxe^{-kx^2}$, with $k = 10^{-10}$, $F_X(x) = 1 - e^{-kx^2}$. So, solving the equation

$$F_X(N) = 1 - e^{-kN^2} = \frac{G}{(G+L)}$$

gives

$$N = \left[\frac{\ln\left(\frac{L}{G+L}\right)}{-k}\right]^{1/2}$$

So, using the values $G = 4, L = 1$, and $k = 10^{-10}$, we obtain $N = 126,860$ units.

**Solution 2.28***

(a) For $k = 2$, $\text{pr}(P_2 = \alpha P_1) = \pi$ and $\text{pr}(P_2 = \beta P_1) = (1-\pi)$, so that $E(P_2) = P_1[\alpha\pi + \beta(1-\pi)]$.

For $k = 3$, $\text{pr}(P_3 = \alpha^2 P_1) = \pi^2$, $\text{pr}(P_3 = \alpha\beta P_1) = 2\pi(1-\pi)$, and $\text{pr}(P_3 = \beta^2 P_1) = (1-\pi)^2$, so that

$$\begin{aligned} E(P_3) &= P_1\left[\alpha^2\pi^2 + 2\alpha\beta\pi(1-\pi) + \beta^2(1-\pi)^2\right] \\ &= P_1[\alpha\pi + \beta(1-\pi)]^2. \end{aligned}$$

In general,

$$\text{pr}\left[P_k = \alpha^j\beta^{(k-1)-j}P_1\right] = C_j^{k-1}\pi^j(1-\pi)^{(k-1)-j},$$
$$j = 0, 1, \ldots, (k-1),$$

so that

$$\begin{aligned} E(P_k) &= \sum_{j=0}^{k-1}\left[\alpha^j\beta^{(k-1)-j}P_1\right]C_j^{k-1}\pi^j(1-\pi)^{(k-1)-j} \\ &= P_1\sum_{j=0}^{k-1}C_j^{k-1}(\alpha\pi)^j[\beta(1-\pi)]^{(k-1)-j} \\ &= P_1[\alpha\pi + \beta(1-\pi)]^{k-1}, \quad k = 2, 3, \ldots, \infty. \end{aligned}$$

(b) For $k = 2, 3, \ldots, \infty$, we consider the inequality

$$\mathrm{E}(P_k) = P_1\,[\alpha\pi + \beta(1-\pi)]^{k-1} \geq P^*,$$

or equivalently

$$\alpha \geq \frac{1}{\pi}\left[\left(\frac{P^*}{P_1}\right)^{[1/(k-1)]} - \beta(1-\pi)\right],$$

which gives

$$\alpha^* = \frac{1}{\pi}\left[\left(\frac{P^*}{P_1}\right)^{[1/(k-1)]} - \beta(1-\pi)\right].$$

Now,

$$\lim_{k\to\infty}\alpha^* = \frac{1}{\pi}\,[1 - \beta(1-\pi)] = \beta - \frac{(\beta-1)}{\pi}.$$

Since $1 < \beta < +\infty$, this limiting value of $\alpha^*$ varies directly with $\pi$ (i.e., the larger is $\pi$, the larger is this limiting value of $\alpha^*$). In particular, when $\pi = 1$, so that every policy holder has a perfect driving record every year, then this insurance company should never reduce the yearly premium from its first-year value of $P_1$.

If $\beta = 1.05$ and $\pi = 0.90$, then this limiting value equals 0.9944. So, for these particular values of $\beta$ and $\pi$, this insurance company should never allow the yearly premium to be below $0.9944P_1$ in value.

**Solution 2.29***

(a) For $R = 2$, we have

$$\sum_{x=0}^{2}\left(\frac{1}{x!}\right)\sum_{l=0}^{2-x}\frac{(-1)^l}{l!} = \left(\frac{1}{0!}\right)\left(1 - 1 + \frac{1}{2!}\right) + \left(\frac{1}{1!}\right)(1-1) + \left(\frac{1}{2!}\right)(1) = 1.$$

Then, assuming that the result holds for the value $R$, we obtain

$$\begin{aligned}
&\sum_{x=0}^{R+1}\left(\frac{1}{x!}\right)\sum_{l=0}^{(R+1)-x}\frac{(-1)^l}{l!} \\
&= \sum_{x=0}^{R}\left(\frac{1}{x!}\right)\sum_{l=0}^{(R-x)+1}\frac{(-1)^l}{l!} + \frac{1}{(R+1)!} \\
&= \sum_{x=0}^{R}\left(\frac{1}{x!}\right)\sum_{l=0}^{R-x}\frac{(-1)^l}{l!} + \sum_{x=0}^{R}\left(\frac{1}{x!}\right)\frac{(-1)^{(R+1)-x}}{[(R+1)-x]!} + \frac{1}{(R+1)!} \\
&= \sum_{x=0}^{R}\mathrm{p}_X(x) + \frac{1}{(R+1)!}\sum_{x=0}^{R}C_x^{R+1}(-1)^{(R+1)-x} + \frac{1}{(R+1)!}
\end{aligned}$$

$$= \sum_{x=0}^{R} p_X(x) + \frac{1}{(R+1)!} \sum_{x=0}^{R+1} C_x^{R+1} (1)^x (-1)^{(R+1)-x}$$
$$- \frac{1}{(R+1)!} + \frac{1}{(R+1)!}$$
$$= \sum_{x=0}^{R} p_X(x) + [1 + (-1)]^{R+1} = \sum_{x=0}^{R} p_X(x) = 1,$$

which completes the proof by induction.

(b) We have

$$\text{E}(X) = \sum_{x=0}^{R} x \left(\frac{1}{x!}\right) \sum_{l=0}^{R-x} \frac{(-1)^l}{l!}$$
$$= \sum_{x=1}^{R} \left[\frac{1}{(x-1)!}\right] \sum_{l=0}^{R-x} \frac{(-1)^l}{l!}$$
$$= \sum_{y=0}^{R-1} \left(\frac{1}{y!}\right) \sum_{l=0}^{(R-1)-y} \frac{(-1)^l}{l!} = 1.$$

And,

$$\text{E}[X(X-1)] = \sum_{x=0}^{R} x(x-1) \left(\frac{1}{x!}\right) \sum_{x=0}^{R-x} \frac{(-1)^l}{l!}$$
$$= \sum_{x=2}^{R} \left[\frac{1}{(x-2)!}\right] \sum_{l=0}^{R-x} \frac{(-1)^l}{l!}$$
$$= \sum_{y=0}^{R-2} \left(\frac{1}{y!}\right) \sum_{l=0}^{(R-2)-y} \frac{(-1)^l}{l!} = 1,$$

so that $\text{V}(X) = \text{E}[X(X-1)] + \text{E}(X) - [\text{E}(X)]^2 = 1 + 1 - (1)^2 = 1$.

It seems counterintuitive that neither $\text{E}(X)$ nor $\text{V}(X)$ depends on the value of $R$. Also,

$$\lim_{R\to\infty} p_X(x) = \lim_{R\to\infty} \left[\frac{1}{x!} \sum_{l=0}^{R-x} \frac{(-1)^l}{l!}\right]$$
$$= \frac{1}{x!} \sum_{l=0}^{\infty} \frac{(-1)^l}{l!} = \frac{e^{-1}}{x!}$$
$$= \frac{(1)^x e^{-1}}{x!}, \quad x = 0, 1, \ldots, \infty.$$

So, as $R \to \infty$, the distribution of $X$ becomes Poisson with $\text{E}(X) = \text{V}(X) = 1$.

**Solution 2.30***

(a) First,

$$\begin{aligned} 1 - \text{F}_{W_n}(w_n) &= \text{pr}(W_n > w_n) \\ &= \text{pr}[X_{w_n} \leq (n-1) \text{ in the time interval}(0, w_n)] \\ &= \sum_{x=0}^{n-1} \frac{(Nw_n\lambda)^x \text{e}^{-(Nw_n\lambda)}}{x!}, \end{aligned}$$

so that

$$\text{F}_{W_n}(w_n) = 1 - \sum_{x=0}^{n-1} \frac{(Nw_n\lambda)^x \text{e}^{-(Nw_n\lambda)}}{x!}.$$

So,

$$\begin{aligned} \text{f}_{W_n}(w_n) &= \frac{\text{dF}_{W_n}(w_n)}{\text{d}w_n} \\ &= -\text{e}^{-Nw_n\lambda} \sum_{x=0}^{n-1} \frac{1}{x!} \left[ xN\lambda(Nw_n\lambda)^{x-1} - N\lambda(Nw_n\lambda)^x \right] \\ &= -N\lambda \text{e}^{-Nw_n\lambda} \left[ \sum_{x=1}^{n-1} \frac{(Nw_n\lambda)^{x-1}}{(x-1)!} - \sum_{x=0}^{n-1} \frac{(Nw_n\lambda)^x}{x!} \right] \\ &= -N\lambda \text{e}^{-Nw_n\lambda} \left[ -\frac{(Nw_n\lambda)^{n-1}}{(n-1)!} \right] \\ &= \frac{w_n^{n-1} \text{e}^{-N\lambda w_n}}{\Gamma(n)(N\lambda)^{-n}}, \quad w_n > 0. \end{aligned}$$

So, $W_n \sim \text{GAMMA}\left[\alpha = (N\lambda)^{-1}, \beta = n\right]$.

(b) Note that $\text{E}(X_T) = \text{V}(X_T) = NT\lambda$. So,

$$\begin{aligned} \text{E}(\text{e}^{tZ}) &= E\left[ \text{e}^{t\left(\frac{X_T - NT\lambda}{\sqrt{NT\lambda}}\right)} \right] \\ &= \text{e}^{-t\sqrt{NT\lambda}} \text{E}\left[ \text{e}^{\frac{t}{\sqrt{NT\lambda}} X_T} \right] \\ &= \text{e}^{-t\sqrt{NT\lambda}} \text{e}^{NT\lambda\left(\text{e}^{t/\sqrt{NT\lambda}} - 1\right)}. \end{aligned}$$

Now,

$$-t\sqrt{NT\lambda} + NT\lambda\left[\sum_{j=0}^{\infty}\frac{\left(t/\sqrt{NT\lambda}\right)^j}{j!} - 1\right]$$

$$= -t\sqrt{NT\lambda} + t\sqrt{NT\lambda} + t^2/2 + \sum_{j=3}^{\infty}\frac{\left(t/\sqrt{NT\lambda}\right)^j}{j!}(NT\lambda),$$

which converges to $t^2/2$ as $N \rightarrow \infty$. Thus,

$$\lim_{N\rightarrow\infty} \mathrm{E}(\mathrm{e}^{tZ}) = \mathrm{e}^{t^2/2},$$

so that, for large $N$,

$$Z = \frac{X_T - NT\lambda}{\sqrt{NT\lambda}} \mathrel{\dot\sim} \mathrm{N}(0, 1).$$

Then, if $N = 10^5$, $\lambda = 10^{-4}$, and $T = 10$, so that $NT\lambda = 100$, then

$$\begin{aligned}\mathrm{pr}(X_T \leq 90|NT\lambda = 100) &= \mathrm{pr}\left[\frac{X_T - 100}{\sqrt{100}} \leq \frac{90 - 100}{\sqrt{100}}\right] \\ &\doteq \mathrm{pr}(Z \leq -1.00) \\ &\doteq 0.16,\end{aligned}$$

since $Z = (X_T - 100)/\sqrt{100} \mathrel{\dot\sim} \mathrm{N}(0, 1)$ for large $N$.

**Solution 2.31***

(a) With $y = (x - c)$, we have

$$\begin{aligned}\int_c^{\infty}\frac{x^{\beta-1}\mathrm{e}^{-x/\alpha}}{\Gamma(\beta)\alpha^{\beta}}\,dx &= \int_0^{\infty}\frac{(y+c)^{\beta-1}\mathrm{e}^{-(y+c)/\alpha}}{\Gamma(\beta)\alpha^{\beta}}\,dy \\ &= \frac{\mathrm{e}^{-c/\alpha}}{\Gamma(\beta)\alpha^{\beta}}\int_0^{\infty}(y+c)^{\beta-1}\mathrm{e}^{-y/\alpha}\,dy \\ &= \frac{\mathrm{e}^{-c/\alpha}}{\Gamma(\beta)\alpha^{\beta}}\int_0^{\infty}\left[\sum_{j=0}^{\beta-1}\mathrm{C}_j^{\beta-1}c^j y^{\beta-1-j}\right]\mathrm{e}^{-y/\alpha}\,dy \\ &= \frac{\mathrm{e}^{-c/\alpha}}{\Gamma(\beta)\alpha^{\beta}}\sum_{j=0}^{\beta-1}\mathrm{C}_j^{\beta-1}c^j\int_0^{\infty}y^{(\beta-j)-1}\mathrm{e}^{-y/\alpha}\,dy\end{aligned}$$

$$= \frac{e^{-c/\alpha}}{(\beta-1)!\alpha^{\beta}} \sum_{j=0}^{\beta-1} \frac{(\beta-1)!}{(\beta-j-1)!j!} c^j \left[\Gamma(\beta-j)\alpha^{\beta-j}\right]$$

$$= \sum_{j=0}^{\beta-1} \frac{e^{-c/\alpha}(c/\alpha)^j}{j!},$$

which is $\text{pr}[X \leq (\beta-1)]$ when $X \sim \text{POI}(c/\alpha)$.

(b) With $x = c(1-y)$, we have

$$\int_0^c \frac{\Gamma(\alpha+\beta)}{\Gamma(\alpha)\Gamma(\beta)} x^{\alpha-1}(1-x)^{\beta-1}\,dx$$

$$= \int_0^1 \frac{\Gamma(\alpha+\beta)}{\Gamma(\alpha)\Gamma(\beta)} [c(1-y)]^{\alpha-1}[1-c(1-y)]^{\beta-1}(c)\,dy$$

$$= \frac{\Gamma(\alpha+\beta)}{\Gamma(\alpha)\Gamma(\beta)} c^{\alpha} \int_0^1 (1-y)^{\alpha-1} \left[\sum_{j=0}^{\beta-1} C_j^{\beta-1}(cy)^j(1-c)^{\beta-1-j}\right] dy$$

$$= \frac{\Gamma(\alpha+\beta)}{\Gamma(\alpha)\Gamma(\beta)} c^{\alpha} \sum_{j=0}^{\beta-1} C_j^{\beta-1} c^j (1-c)^{\beta-1-j} \int_0^1 y^j(1-y)^{\alpha-1}\,dy.$$

Thus, since

$$\int_0^1 y^j(1-y)^{\alpha-1}\,dy = \frac{\Gamma(j+1)\Gamma(\alpha)}{\Gamma(\alpha+j+1)},$$

we have

$$\frac{(\alpha+\beta-1)!}{(\alpha-1)!(\beta-1)!} \sum_{j=0}^{\beta-1} \frac{(\beta-1)!}{j!(\beta-1-j)!} c^{\alpha+j}(1-c)^{\beta-1-j} \frac{\Gamma(j+1)\Gamma(\alpha)}{\Gamma(\alpha+j+1)}$$

$$= \sum_{j=0}^{\beta-1} \frac{(\alpha+\beta-1)!}{(\alpha+j)!(\beta-1-j)!} c^{\alpha+j}(1-c)^{\beta-1-j}$$

$$= \sum_{i=\alpha}^{\alpha+\beta-1} C_i^{\alpha+\beta-1} c^i (1-c)^{\alpha+\beta-1-i},$$

which is $\text{pr}(X \geq \alpha)$ when $X \sim \text{BIN}(\alpha+\beta-1, c)$.

**Solution 2.32***

(a) Let A be the event that "any egg produces a live and healthy baby sea turtle," let B be the event that "a live and healthy baby sea turtle grows to adulthood," and let C be the event that "any egg produces an adult sea turtle." Then,

$$\begin{aligned}\text{pr}(C) &= \text{pr}(A \cap B) = \text{pr}(A)\text{pr}(B|A) = (0.30)(1 - 0.98)\\ &= (0.30)(0.02) = 0.006.\end{aligned}$$

(b) Let $T_0$ be the event that "any randomly chosen sea turtle nest produces no adult sea turtles" and let $E_n$ be the event that "any randomly chosen sea turtle nest contains exactly $n$ eggs." Then,

$$\begin{aligned}\alpha = \text{pr}(\bar{T}_0) &= \text{pr}[\cup_{n=1}^{\infty}(\bar{T}_0 \cap E_n] = \sum_{n=1}^{\infty} \text{pr}(\bar{T}_0|E_n)\text{pr}(E_n)\\ &= 1 - \text{pr}(T_0) = 1 - \sum_{n=1}^{\infty} \text{pr}(T_0|E_n)\text{pr}(E_n)\\ &= 1 - \sum_{n=1}^{\infty}[(0.994)^n](1-\pi)\pi^{n-1}\\ &= 1 - 0.994(1-\pi)\sum_{n=1}^{\infty}(0.994\pi)^{n-1}\\ &= 1 - 0.994(1-\pi)\left[\frac{1}{1-0.994\pi}\right]\\ &= 1 - \frac{0.994(1-\pi)}{1-0.994\pi} = \frac{0.006}{1-0.994\pi}.\end{aligned}$$

When $\pi = 0.20$, then $\alpha = 0.0075$.

(c) Let $T_k$ be the event that "a randomly chosen sea turtle nest produces exactly $k$ adult sea turtles." Then, based on the stated assumptions, it follows that

$$\text{pr}(T_k|E_n) = C_k^n(0.006)^k(0.994)^{n-k}, \quad k = 0, 1, \ldots, n.$$

Then,

$$\begin{aligned}\text{pr}(E_n|T_k) &= \frac{\text{pr}(E_n \cap T_k)}{\text{pr}(T_k)} = \frac{\text{pr}(T_k|E_n)\text{pr}(E_n)}{\text{pr}(T_k)}\\ &= \frac{[C_k^n(0.006)^k(0.994)^{n-k}][\pi(1-\pi)^{n-1}]}{\text{pr}(T_k)}.\end{aligned}$$

Now, for $n \geq k \geq 1$, we have

$$\begin{aligned}\text{pr}(T_k) &= \sum_{n=k}^{\infty} \text{pr}(T_k \cap E_n) = \sum_{n=k}^{\infty} \text{pr}(T_k|E_n)\text{pr}(E_n)\\ &= \sum_{n=k}^{\infty} C_k^n(0.006)^k(0.994)^{n-k}(1-\pi)\pi^{n-1}\end{aligned}$$

$$= \left(\frac{0.006}{0.994}\right)^k \left(\frac{1-\pi}{\pi}\right) \sum_{n=k}^{\infty} C_k^n (0.994\pi)^n$$

$$= \left(\frac{0.006}{0.994}\right)^k \left(\frac{1-\pi}{\pi}\right) \sum_{m=0}^{\infty} C_k^{m+k} (0.994\pi)^{m+k}$$

$$= (0.006)^k (1-\pi)\pi^{k-1} \sum_{m=0}^{\infty} C_k^{m+k} (0.994\pi)^m$$

$$= (0.006)^k (1-\pi)\pi^{k-1} (1-0.994\pi)^{-(k+1)}$$

So,

$$\beta_{nk} = \text{pr}(\text{E}_n|\text{T}_k) = \frac{[C_k^n (0.006)^k (0.994)^{n-k}][(1-\pi)\pi^{n-1}]}{(0.006)^k (1-\pi)\pi^{k-1}(1-0.994\pi)^{-(k+1)}}$$

$$= C_k^n (1-0.994\pi)^{k+1} (0.994\pi)^{n-k}, \quad 1 \le k \le n < \infty.$$

When $k = 0$,

$$\beta_{n0} = \frac{(0.994)^n[(1-\pi)\pi^{n-1}]}{\left[\frac{0.994(1-\pi)}{1-0.994\pi}\right]}$$

$$= (0.994\pi)^{n-1}(1-0.994\pi), \quad n = 1, 2, \ldots, \infty.$$

For any fixed $k \ge 0$, note that, as required, $\sum_{n=k}^{\infty} \text{pr}(\text{E}_n|\text{T}_k) = 1$. Finally, when $\pi = 0.20, k = 2$, and $n = 6$, $\beta_{62} = \text{pr}(\text{E}_6|\text{T}_2) = 0.0123$.

**Solution 2.33***

(a) If $k > n$, the result is obvious since $0 = (0 + 0)$; so, we only need to consider the case when $k \le n$. Now,

$$C_{k-1}^{n-1} + C_k^{n-1} = \frac{(n-1)!}{(k-1)!(n-k)!} + \frac{(n-1)!}{k!(n-k-1)!}$$

$$= (n-1)!\left[\frac{k}{k!(n-k)!} + \frac{(n-k)}{k!(n-k)!}\right]$$

$$= (n-1)!\left[\frac{n}{k!(n-k)!}\right] = \frac{n!}{k!(n-k)!} = C_k^n,$$

which completes the proof.

(b) The left-hand side of Vandermonde's Identity is the number of ways of choosing $r$ objects from a total of $(m + n)$ objects. For $k = 0, 1, \ldots, r$, this can be accomplished by choosing $k$ objects from the set of $n$ objects (which can be done in $C_k^n$ ways) and by choosing $(r - k)$ objects from the set of $m$ objects (which can be done in $C_{r-k}^m$ ways), giving the product $C_k^n C_{r-k}^m$ as the total number of ways of choosing

$r$ objects from a total of $(m+n)$ objects *given that exactly* $k$ objects must be chosen from the set of $n$ objects. Vandermonde's Identity follows directly by summing this product over all the values of $k$.

(c) Without loss of generality, assume that $s \leq t$. Then, we wish to show that

$$\sum_{y=1}^{s}(\pi_{2y}+\pi_{2y+1}) = \frac{\sum_{y=1}^{s}\left[2C_{y-1}^{s-1}C_{y-1}^{t-1}+C_{y}^{s-1}C_{y-1}^{t-1}+C_{y-1}^{s-1}C_{y}^{t-1}\right]}{C_{s}^{s+t}} = 1,$$

or, equivalently, that the numerator $N$ in the above ratio expression is equal to $C_s^{s+t}$. Now, using Pascal's Identity, we have

$$\begin{aligned}
N &= \sum_{y=1}^{s}\left[2C_{y-1}^{s-1}C_{y-1}^{t-1}+C_{y}^{s-1}C_{y-1}^{t-1}+C_{y-1}^{s-1}C_{y}^{t-1}\right] \\
&= \sum_{y=1}^{s}\left\{C_{y-1}^{s-1}\left[C_{y-1}^{t-1}+C_{y}^{t-1}\right]+C_{y-1}^{t-1}\left[C_{y-1}^{s-1}+C_{y}^{s-1}\right]\right\} \\
&= \sum_{y=1}^{s}\left[C_{y-1}^{s-1}C_{y}^{t}+C_{y-1}^{t-1}C_{y}^{s}\right] \\
&= \sum_{y=1}^{s}\left[\frac{(s-1)!t!}{(y-1)!(s-y)!y!(t-y)!}+\frac{(t-1)!s!}{(y-1)!(t-y)!y!(s-y)!}\right] \\
&= (s+t)\sum_{y=1}^{s}\left[\frac{(s-1)!(t-1)!}{(y-1)!(s-y)!y!(t-y)!}\right] \\
&= \frac{(s+t)}{s}\sum_{y=1}^{s}C_{y}^{s}C_{y-1}^{t-1} \\
&= \frac{(s+t)}{s}\sum_{k=0}^{s-1}C_{k+1}^{s}C_{k}^{t-1} \\
&= \frac{(s+t)}{s}\sum_{k=0}^{s-1}C_{(s-1)-k}^{s}C_{k}^{t-1}.
\end{aligned}$$

Then, in the above summation, if we let $r=(s-1), m=s$, and $n=(t-1)$, in which case $(s-1)\leq \min\{s,(t-1)\}$ since $s \leq t$, then Vandermonde's Identity gives

$$\begin{aligned}
\sum_{k=0}^{s-1}C_{(s-1)-k}^{s}C_{k}^{t-1} &= \sum_{k=0}^{r}C_{r-k}^{m}C_{k}^{n} \\
&= C_{r}^{m+n} \\
&= C_{s-1}^{s+t-1}.
\end{aligned}$$

Finally,

$$N = \frac{(s+t)}{s} C_{s-1}^{s+t-1} = \frac{(s+t)}{s} \left[ \frac{(s+t-1)!}{(s-1)!t!} \right]$$
$$= \frac{(s+t)!}{s!t!} = C_s^{s+t}.$$

This completes the proof since it then follows that

$$0 \leq \pi_{2y} \leq 1 \quad \text{and} \quad 0 \leq \pi_{2y+1} \leq 1, \quad y = 1, 2, \ldots, \min\{s, t\}.$$

# 3

# *Multivariate Distribution Theory*

## 3.1 Concepts and Notation

### 3.1.1 Discrete and Continuous Multivariate Distributions

A *discrete multivariate* probability distribution for $k$ discrete random variables $X_1, X_2, \ldots, X_k$ is denoted

$$\mathrm{p}_{X_1,X_2,\ldots,X_k}(x_1, x_2, \ldots, x_k) = \mathrm{pr}\left[\cap_{i=1}^{k}(X_i = x_i)\right] \equiv \mathrm{p}_{\mathbf{X}}(\boldsymbol{x}) = \mathrm{pr}(\mathbf{X} = \boldsymbol{x}), \quad \boldsymbol{x} \in \mathcal{D},$$

where the row vector $\mathbf{X} = (X_1, X_2, \ldots, X_k)$, the row vector $\boldsymbol{x} = (x_1, x_2, \ldots, x_k)$, and $\mathcal{D}$ is the *domain* (i.e., the set of all permissible values) of the discrete random vector $\mathbf{X}$. A valid multivariate discrete probability distribution has the following properties:

(i) $0 \leq \mathrm{p}_{\mathbf{X}}(\boldsymbol{x}) \leq 1$ for all $\boldsymbol{x} \in \mathcal{D}$;

(ii) $\sum\sum_{\mathcal{D}} \cdots \sum \mathrm{p}_{\mathbf{X}}(\boldsymbol{x}) = 1$;

(iii) If $\mathcal{D}_1$ is a subset of $\mathcal{D}$, then

$$\mathrm{pr}[\mathbf{X} \in \mathcal{D}_1] = \sum\sum_{\mathcal{D}_1} \cdots \sum \mathrm{p}_{\mathbf{X}}(\boldsymbol{x}).$$

A *continuous multivariate* probability distribution (i.e., a multivariate density function) for $k$ continuous random variables $X_1, X_2, \ldots, X_k$ is denoted

$$\mathrm{f}_{X_1,X_2,\ldots,X_k}(x_1, x_2, \ldots, x_k) \equiv \mathrm{f}_{\mathbf{X}}(\boldsymbol{x}), \quad \boldsymbol{x} \in \mathcal{D},$$

where $\mathcal{D}$ is the domain of the continuous random vector $\mathbf{X}$. A valid multivariate density function has the following properties:

(i) $0 \leq \mathrm{f}_{\mathbf{X}}(\boldsymbol{x}) < +\infty$ for all $\boldsymbol{x} \in \mathcal{D}$;

(ii) $\int\int_{\mathcal{D}} \cdots \int \mathrm{f}_{\mathbf{X}}(\boldsymbol{x})\, \mathbf{d}\boldsymbol{x} = 1$, where $\mathbf{d}\boldsymbol{x} = \mathrm{d}x_1 \mathrm{d}x_2 \ldots \mathrm{d}x_k$;

(iii) If $\mathcal{D}_1$ is a subset of $\mathcal{D}$, then

$$\text{pr}[\boldsymbol{X} \in \mathcal{D}_1] = \int\int_{\mathcal{D}_1}\cdots\int \text{f}_{\boldsymbol{X}}(\boldsymbol{x})\,d\boldsymbol{x}.$$

### 3.1.2 Multivariate Cumulative Distribution Functions

In general, the multivariate CDF for a random vector $\boldsymbol{X}$ is the scalar function

$$\text{F}_{\boldsymbol{X}}(\boldsymbol{x}) = \text{pr}(\boldsymbol{X} \leq \boldsymbol{x}) = \text{pr}\left[\cap_{i=1}^{k}(X_i \leq x_i)\right].$$

For a discrete random vector, $\text{F}_{\boldsymbol{X}}(\boldsymbol{x})$ is a discontinuous function of $\boldsymbol{x}$. For a continuous random vector, $\text{F}_{\boldsymbol{X}}(\boldsymbol{x})$ is an absolutely continuous function of $\boldsymbol{x}$, so that

$$\frac{\partial^k \text{F}_{\boldsymbol{X}}(\boldsymbol{x})}{\partial x_1 \partial x_2 \cdots \partial x_k} = \text{f}_{\boldsymbol{X}}(\boldsymbol{x}).$$

### 3.1.3 Expectation Theory

Let $\text{g}(\boldsymbol{X})$ be a scalar function of $\boldsymbol{X}$. If $\boldsymbol{X}$ is a discrete random vector with probability distribution $\text{p}_{\boldsymbol{X}}(\boldsymbol{x})$, then

$$\text{E}[\text{g}(\boldsymbol{X})] = \sum\sum_{\mathcal{D}}\cdots\sum \text{g}(\boldsymbol{x})\text{p}_{\boldsymbol{X}}(\boldsymbol{x}).$$

And, if $\boldsymbol{X}$ is a continuous random vector with density function $\text{f}_{\boldsymbol{X}}(\boldsymbol{x})$, then

$$\text{E}[\text{g}(\boldsymbol{X})] = \int\int_{\mathcal{D}}\cdots\int \text{g}(\boldsymbol{x})\text{f}_{\boldsymbol{X}}(\boldsymbol{x})\,d\boldsymbol{x}.$$

Some important expectations of interest in the multivariate setting are:

#### *3.1.3.1 Covariance*

For $i \neq j$, the *covariance* between the two random variables $X_i$ and $X_j$ is defined as

$$\begin{aligned}\text{cov}(X_i, X_j) &= \text{E}\{[X_i - \text{E}(X_i)][X_j - \text{E}(X_j)]\}\\ &= \text{E}(X_iX_j) - \text{E}(X_i)\text{E}(X_j), \quad -\infty < \text{cov}(X_i, X_j) < +\infty.\end{aligned}$$

#### 3.1.3.2 Correlation

For $i \neq j$, the *correlation* between the two random variables $X_i$ and $X_j$ is defined as

$$\mathrm{corr}(X_i, X_j) = \frac{\mathrm{cov}(X_i, X_j)}{\sqrt{\mathrm{V}(X_i)\mathrm{V}(X_j)}}, \quad -1 \leq \mathrm{corr}(X_i, X_j) \leq +1.$$

#### 3.1.3.3 Moment Generating Function

With the row vector $\boldsymbol{t} = (t_1, t_2, \ldots, t_k)$,

$$\mathrm{M}_{\mathbf{X}}(\boldsymbol{t}) = \mathrm{E}\left(\mathrm{e}^{\boldsymbol{t}\mathbf{X}'}\right) = \mathrm{E}\left(\mathrm{e}^{\sum_{i=1}^{k} t_i X_i}\right)$$

is called the *multivariate moment generating function* for the random vector $\mathbf{X}$. In particular, with $r_1, r_2, \ldots, r_k$ being nonnegative integers satisfying the restriction $\sum_{i=1}^{k} r_i = r$, we have

$$\mathrm{E}[\mathrm{X}_1^{r_1} \mathrm{X}_2^{r_2} \cdots \mathrm{X}_k^{r_k}] = \left. \frac{\partial^r \mathrm{M}_{\mathbf{X}}(\boldsymbol{t})}{\partial t_1^{r_1} \partial t_2^{r_2} \cdots \partial t_k^{r_k}} \right|_{\boldsymbol{t}=\mathbf{0}},$$

where the notation $\boldsymbol{t} = \mathbf{0}$ means that $t_i = 0, i = 1, 2, \ldots, k$.

### 3.1.4 Marginal Distributions

When $\mathbf{X}$ is a discrete random vector, the *marginal distribution* of any proper subset of the $k$ random variables $X_1, X_2, \ldots, X_k$ can be found by summing over all the random variables *not* in the subset of interest. In particular, for $1 \leq j < k$, the marginal distribution of the random variables $X_1, X_2, \ldots, X_j$ is equal to

$$\mathrm{p}_{X_1, X_2, \ldots, X_j}(x_1, x_2, \ldots, x_j) = \sum_{\text{all } x_{j+1}} \sum_{\text{all } x_{j+2}} \cdots \sum_{\text{all } x_{k-1}} \sum_{\text{all } x_k} \mathrm{p}_{\mathbf{X}}(\boldsymbol{x}).$$

When $\mathbf{X}$ is a continuous random vector, the *marginal distribution* of any proper subset of the $k$ random variables $X_1, X_2, \ldots, X_k$ can be found by integrating over all the random variables *not* in the subset of interest. In particular, for $1 \leq j < k$, the marginal distribution of the random variables $X_1, X_2, \ldots, X_j$ is equal to

$$\mathrm{f}_{X_1, X_2, \ldots, X_j}(x_1, x_2, \ldots, x_j)$$
$$= \int_{\text{all } x_{j+1}} \int_{\text{all } x_{j+2}} \cdots \int_{\text{all } x_{k-1}} \int_{\text{all } x_k} \mathrm{f}_{\mathbf{X}}(\boldsymbol{x}) \mathrm{d}x_k \, \mathrm{d}x_{k-1} \cdots \mathrm{d}x_{j+2} \, \mathrm{d}x_{j+1}.$$

### 3.1.5 Conditional Distributions and Expectations

For $\boldsymbol{X}$ a discrete random vector, let $\boldsymbol{X}_1$ denote a proper subset of the $k$ discrete random variables $X_1, X_2, \ldots, X_k$, let $\boldsymbol{X}_2$ denote another proper subset of $X_1, X_2, \ldots, X_k$, and assume that the subsets $\boldsymbol{X}_1$ and $\boldsymbol{X}_2$ have no elements in common. Then, the *conditional distribution* of $\boldsymbol{X}_2$ given that $\boldsymbol{X}_1 = \boldsymbol{x}_1$ is defined as the joint distribution of $\boldsymbol{X}_1$ and $\boldsymbol{X}_2$ divided by the marginal distribution of $\boldsymbol{X}_1$, namely,

$$\begin{aligned} \mathrm{p}_{\boldsymbol{X}_2}(\boldsymbol{x}_2|\boldsymbol{X}_1 = \boldsymbol{x}_1) &= \frac{\mathrm{p}_{\boldsymbol{X}_1,\boldsymbol{X}_2}(\boldsymbol{x}_1, \boldsymbol{x}_2)}{\mathrm{p}_{\boldsymbol{X}_1}(\boldsymbol{x}_1)} \\ &= \frac{\mathrm{pr}[(\boldsymbol{X}_1 = \boldsymbol{x}_1) \cap (\boldsymbol{X}_2 = \boldsymbol{x}_2)]}{\mathrm{pr}(\boldsymbol{X}_1 = \boldsymbol{x}_1)}, \quad \mathrm{pr}(\boldsymbol{X}_1 = \boldsymbol{x}_1) > 0. \end{aligned}$$

Then, if $\mathrm{g}(\boldsymbol{X}_2)$ is a scalar function of $\boldsymbol{X}_2$, it follows that

$$\mathrm{E}[\mathrm{g}(\boldsymbol{X}_2)|\boldsymbol{X}_1 = \boldsymbol{x}_1] = \underset{\text{all } \boldsymbol{x}_2}{\sum\sum\cdots\sum} \mathrm{g}(\boldsymbol{x}_2)\mathrm{p}_{\boldsymbol{X}_2}(\boldsymbol{x}_2|\boldsymbol{X}_1 = \boldsymbol{x}_1).$$

For $\boldsymbol{X}$ a continuous random vector, let $\boldsymbol{X}_1$ denote a proper subset of the $k$ continuous random variables $X_1, X_2, \ldots, X_k$, let $\boldsymbol{X}_2$ denote another proper subset of $X_1, X_2, \ldots, X_k$, and assume that the subsets $\boldsymbol{X}_1$ and $\boldsymbol{X}_2$ have no elements in common. Then, the *conditional density function* of $\boldsymbol{X}_2$ given that $\boldsymbol{X}_1 = \boldsymbol{x}_1$ is defined as the joint density function of $\boldsymbol{X}_1$ and $\boldsymbol{X}_2$ divided by the marginal density function of $\boldsymbol{X}_1$, namely,

$$\mathrm{f}_{\boldsymbol{X}_2}(\boldsymbol{x}_2|\boldsymbol{X}_1 = \boldsymbol{x}_1) = \frac{\mathrm{f}_{\boldsymbol{X}_1,\boldsymbol{X}_2}(\boldsymbol{x}_1, \boldsymbol{x}_2)}{\mathrm{f}_{\boldsymbol{X}_1}(\boldsymbol{x}_1)}, \quad \mathrm{f}_{\boldsymbol{X}_1}(\boldsymbol{x}_1) > 0.$$

Then, if $\mathrm{g}(\boldsymbol{X}_2)$ is a scalar function of $\boldsymbol{X}_2$, it follows that

$$\mathrm{E}[\mathrm{g}(\boldsymbol{X}_2)|\boldsymbol{X}_1 = \boldsymbol{x}_1] = \underset{\text{all } \boldsymbol{x}_2}{\int\int\cdots\int} \mathrm{g}(\boldsymbol{x}_2)\mathrm{f}_{\boldsymbol{X}_2}(\boldsymbol{x}_2|\boldsymbol{X}_1 = \boldsymbol{x}_1)\, \mathbf{d}\boldsymbol{x}_2.$$

More generally, if $\mathrm{g}(\boldsymbol{X}_1, \boldsymbol{X}_2)$ is a scalar function of $\boldsymbol{X}_1$ and $\boldsymbol{X}_2$, then useful *iterated expectation* formulas are:

$$\mathrm{E}[\mathrm{g}(\boldsymbol{X}_1, \boldsymbol{X}_2)] = \mathrm{E}_{\boldsymbol{x}_1}\{\mathrm{E}[\mathrm{g}(\boldsymbol{X}_1, \boldsymbol{X}_2)|\boldsymbol{X}_1 = \boldsymbol{x}_1]\} = \mathrm{E}_{\boldsymbol{x}_2}\{\mathrm{E}[\mathrm{g}(\boldsymbol{X}_1, \boldsymbol{X}_2)|\boldsymbol{X}_2 = \boldsymbol{x}_2]\}$$

and

$$\begin{aligned} \mathrm{V}[\mathrm{g}(\boldsymbol{X}_1, \boldsymbol{X}_2)] &= \mathrm{E}_{\boldsymbol{x}_1}\{\mathrm{V}[\mathrm{g}(\boldsymbol{X}_1, \boldsymbol{X}_2)|\boldsymbol{X}_1 = \boldsymbol{x}_1]\} + \mathrm{V}_{\boldsymbol{x}_1}\{\mathrm{E}[\mathrm{g}(\boldsymbol{X}_1, \boldsymbol{X}_2)|\boldsymbol{X}_1 = \boldsymbol{x}_1]\} \\ &= \mathrm{E}_{\boldsymbol{x}_2}\{\mathrm{V}[\mathrm{g}(\boldsymbol{X}_1, \boldsymbol{X}_2)|\boldsymbol{X}_2 = \boldsymbol{x}_2]\} + \mathrm{V}_{\boldsymbol{x}_2}\{\mathrm{E}[\mathrm{g}(\boldsymbol{X}_1, \boldsymbol{X}_2)|\boldsymbol{X}_2 = \boldsymbol{x}_2]\}. \end{aligned}$$

Also,

$$\mathrm{p}_{\boldsymbol{X}}(\boldsymbol{x}) \equiv \mathrm{p}_{X_1,X_2,\ldots,X_k}(x_1, x_2, \ldots, x_k) = \mathrm{p}_{X_1}(x_1) \prod_{i=2}^{k} \mathrm{p}_{X_i}\left[x_i \middle| \cap_{j=1}^{i-1} (X_j = x_j)\right]$$

and

$$\mathrm{f}_{\boldsymbol{X}}(\boldsymbol{x}) \equiv \mathrm{f}_{X_1,X_2,\ldots,X_k}(x_1, x_2, \ldots, x_k) = \mathrm{f}_{X_1}(x_1) \prod_{i=2}^{k} \mathrm{f}_{X_i}\left[x_i \middle| \cap_{j=1}^{i-1} (X_j = x_j)\right].$$

Note that there are $k!$ ways of writing each of the above two expressions.

### 3.1.6 Mutual Independence among a Set of Random Variables

The random vector $\boldsymbol{X}$ is said to consist of a set of $k$ mutually independent random variables if and only if

$$\mathrm{F}_{\boldsymbol{X}}(\boldsymbol{x}) = \prod_{i=1}^{k} \mathrm{F}_{X_i}(x_i) = \prod_{i=1}^{k} \mathrm{pr}(X_i \leq x_i)$$

for all possible choices of $x_1, x_2, \ldots, x_k$.

Given mutual independence, then

$$\mathrm{p}_{\boldsymbol{X}}(\boldsymbol{x}) \equiv \mathrm{p}_{X_1,X_2,\ldots,X_k}(x_1, x_2, \ldots, x_k) = \prod_{i=1}^{k} \mathrm{p}_{X_i}(x_i)$$

when $\boldsymbol{X}$ is a discrete random vector, and

$$\mathrm{f}_{\boldsymbol{X}}(\boldsymbol{x}) \equiv \mathrm{f}_{X_1,X_2,\ldots,X_k}(x_1, x_2, \ldots, x_k) = \prod_{i=1}^{k} \mathrm{f}_{X_i}(x_i)$$

when $\boldsymbol{X}$ is a continuous random vector.

Also, for $i = 1, 2, \ldots, k$, let $\mathrm{g}_i(X_i)$ be a scalar function of $X_i$. Then, if $X_1, X_2, \ldots, X_k$ constitute a set of $k$ mutually independent random variables, it follows that

$$\mathrm{E}\left[\prod_{i=1}^{k} \mathrm{g}_i(X_i)\right] = \prod_{i=1}^{k} \mathrm{E}[\mathrm{g}_i(X_i)].$$

And, if $X_1, X_2, \ldots, X_k$ are mutually independent random variables, then any subset of these $k$ random variables also constitutes a group of mutually independent random variables. Also, for $i \neq j$, if $X_i$ and $X_j$ are independent random variables, then $\mathrm{corr}(X_i, X_j) = 0$; however, if $\mathrm{corr}(X_i, X_j) = 0$, it does *not* necessarily follow that $X_i$ and $X_j$ are independent random variables.

### 3.1.7 Random Sample

Using the notation $\mathbf{X}_i = (X_{i1}, X_{i2}, \ldots, X_{ik})$, the random vectors $\mathbf{X}_1, \mathbf{X}_2, \ldots, \mathbf{X}_n$ are said to constitute a random sample of size $n$ from the discrete parent population $\mathrm{p}_{\mathbf{X}}(\mathbf{x})$ if the following two conditions hold:

(i) $\mathbf{X}_1, \mathbf{X}_2, \ldots, \mathbf{X}_n$ constitute a set of mutually independent random vectors;

(ii) For $i = 1, 2, \ldots, n, \mathrm{p}_{\mathbf{X}_i}(\mathbf{x}_i) = \mathrm{p}_{\mathbf{X}}(\mathbf{x}_i)$; in other words, $\mathbf{X}_i$ follows the discrete parent population distribution $\mathrm{p}_{\mathbf{X}}(\mathbf{x})$.

A completely analogous definition holds for a random sample from a continuous parent population $\mathrm{f}_{\mathbf{X}}(\mathbf{x})$.

Standard statistical terminology describes a random sample $\mathbf{X}_1, \mathbf{X}_2, \ldots, \mathbf{X}_n$ of size $n$ as consisting of a set of *independent and identically distributed* (i.i.d.) random vectors. In this regard, it is important to note that the mutual independence property pertains to the relationship among the *random vectors*, *not* to the relationship among the $k$ (*possibly mutually dependent*) scalar random variables within a random vector.

### 3.1.8 Some Important Multivariate Discrete and Continuous Probability Distributions

#### *3.1.8.1 Multinomial*

The *multinomial* distribution is often used as a statistical model for the analysis of categorical data. In particular, for $i = 1, 2, \ldots, k$, suppose that $\pi_i$ is the probability that an observation falls into the $i$th of $k$ distinct categories, where $0 < \pi_i < 1$ and where $\sum_{i=1}^{k} \pi_i = 1$. If the discrete random variable $X_i$ is the number of observations out of $n$ that fall into the $i$th category, then the $k$ random variables $X_1, X_2, \ldots, X_k$ jointly follow a $k$-variate multinomial distribution, namely,

$$\mathrm{p}_{\mathbf{X}}(\mathbf{x}) \equiv \mathrm{p}_{X_1, X_2, \ldots, X_k}(x_1, x_2, \ldots, x_k) = \frac{n!}{x_1! x_2! \cdots x_k!} \pi_1^{x_1} \pi_2^{x_2} \cdots \pi_k^{x_k}, \quad \mathbf{x} \in \mathcal{D},$$

where $\mathcal{D} = \{\mathbf{x} : 0 \leq x_i \leq n, i = 1, 2, \ldots, k, \text{ and } \sum_{i=1}^{k} x_i = n\}$.

When $(X_1, X_2, \ldots, X_k) \sim \mathrm{MULT}(n; \pi_1, \pi_2, \ldots, \pi_k)$, then $X_i \sim \mathrm{BIN}(n, \pi_i)$ for $i = 1, 2, \ldots, k$, and $\mathrm{cov}(X_i, X_j) = -n\pi_i\pi_j$ for $i \neq j$.

#### *3.1.8.2 Multivariate Normal*

The *multivariate normal* distribution is often used to model the joint behavior of $k$ possibly mutually correlated continuous random variables. The multivariate normal density function for $k$ continuous random variables $X_1, X_2, \ldots, X_k$

is defined as

$$f_{\boldsymbol{X}}(\boldsymbol{x}) \equiv f_{X_1,X_2,\ldots,X_k}(x_1,x_2,\ldots,x_k) = \frac{1}{(2\pi)^{k/2}|\boldsymbol{\Sigma}|^{1/2}} e^{-(1/2)(\boldsymbol{x}-\boldsymbol{\mu})\boldsymbol{\Sigma}^{-1}(\boldsymbol{x}-\boldsymbol{\mu})'},$$

where $-\infty < x_i < \infty$ for $i = 1, 2, \ldots, k$, where $\boldsymbol{\mu} = (\mu_1, \mu_2, \ldots, \mu_k) = [\mathrm{E}(X_1), \mathrm{E}(X_2), \ldots, \mathrm{E}(X_k)]$, and where $\boldsymbol{\Sigma}$ is the $(k \times k)$ covariance matrix of $\boldsymbol{X}$ with $i$th diagonal element equal to $\sigma_i^2 = \mathrm{V}(X_i)$ and with $(i, j)$th element $\sigma_{ij}$ equal to $\mathrm{cov}(X_i, X_j)$ for $i \neq j$.

Also, when $\boldsymbol{X} \sim \mathrm{MVN}_k(\boldsymbol{\mu}, \boldsymbol{\Sigma})$, then the moment generating function for $\boldsymbol{X}$ is

$$\mathrm{M}_{\boldsymbol{X}}(\boldsymbol{t}) = e^{\boldsymbol{t}\boldsymbol{\mu}' + (1/2)\boldsymbol{t}\boldsymbol{\Sigma}\boldsymbol{t}'}.$$

And, for $i = 1, 2, \ldots, k$, the marginal distribution of $X_i$ is normal with mean $\mu_i$ and variance $\sigma_i^2$.

As an important special case, when $k = 2$, we obtain the *bivariate normal* distribution, namely,

$$f_{X_1,X_2}(x_1, x_2) = \frac{1}{2\pi\sigma_1\sigma_2\sqrt{(1-\rho^2)}} e^{-\frac{1}{2(1-\rho^2)}\left[\left(\frac{x_1-\mu_1}{\sigma_1}\right)^2 - 2\rho\left(\frac{x_1-\mu_1}{\sigma_1}\right)\left(\frac{x_2-\mu_2}{\sigma_2}\right) + \left(\frac{x_2-\mu_2}{\sigma_2}\right)^2\right]},$$

where $-\infty < x_1 < \infty$ and $-\infty < x_2 < \infty$, and where $\rho = \mathrm{corr}(X_1, X_2)$.

When $(X_1, X_2) \sim \mathrm{BVN}(\mu_1, \mu_2; \sigma_1^2, \sigma_2^2; \rho)$, then the moment generating function for $X_1$ and $X_2$ is

$$\mathrm{M}_{X_1,X_2}(t_1, t_2) = e^{t_1\mu_1 + t_2\mu_2 + (1/2)(t_1^2\sigma_1^2 + 2t_1t_2\rho\sigma_1\sigma_2 + t_2^2\sigma_2^2)}.$$

The conditional distribution of $X_2$ given $X_1 = x_1$ is normal with

$$\mathrm{E}(X_2|X_1 = x_1) = \mu_2 + \rho\frac{\sigma_2}{\sigma_1}(x_1 - \mu_1) \quad \text{and} \quad \mathrm{V}(X_2|X_1 = x_1) = \sigma_2^2(1 - \rho^2).$$

And, the conditional distribution of $X_1$ given $X_2 = x_2$ is normal with

$$\mathrm{E}(X_1|X_2 = x_2) = \mu_1 + \rho\frac{\sigma_1}{\sigma_2}(x_2 - \mu_2) \quad \text{and} \quad \mathrm{V}(X_1|X_2 = x_2) = \sigma_1^2(1 - \rho^2).$$

These conditional expectation expressions for the bivariate normal distribution are special cases of a more general result. More generally, for a pair of either discrete or continuous random variables $X_1$ and $X_2$, if the conditional expectation of $X_2$ given $X_1 = x_1$ is a *linear* (or straightline) function of $x_1$, namely $\mathrm{E}(X_2|X_1 = x_1) = \alpha_1 + \beta_1 x_1, -\infty < \alpha_1 < +\infty, -\infty < \beta_1 < +\infty$, then $\mathrm{corr}(X_1, X_2) = \rho = \beta_1\sqrt{[\mathrm{V}(X_1)]/[\mathrm{V}(X_2)]}$. Analogously, if $\mathrm{E}(X_1|X_2 = x_2) = \alpha_2 + \beta_2 x_2, -\infty < \alpha_2 < +\infty, -\infty < \beta_2 < +\infty$, then $\rho = \beta_2\sqrt{[\mathrm{V}(X_2)]/[\mathrm{V}(X_1)]}$.

### 3.1.9 Special Topics of Interest

#### 3.1.9.1 *Mean and Variance of a Linear Function of Random Variables*

For $i = 1, 2, \ldots, k$, let $g_i(X_i)$ be a scalar function of the random variable $X_i$. Then, if $a_1, a_2, \ldots, a_k$ are known constants, and if $L = \sum_{i=1}^{k} a_i g_i(X_i)$, we have

$$\mathrm{E}(L) = \sum_{i=1}^{k} a_i \mathrm{E}[g_i(X_i)],$$

and

$$\mathrm{V}(L) = \sum_{i=1}^{k} a_i^2 \mathrm{V}[g_i(X_i)] + 2\sum_{i=1}^{k-1}\sum_{j=i+1}^{k} a_i a_j \mathrm{cov}[g_i(X_i), g_j(X_j)].$$

In the special case when the random variables $X_i$ and $X_j$ are uncorrelated for all $i \neq j$, then

$$\mathrm{V}(L) = \sum_{i=1}^{k} a_i^2 \mathrm{V}[g_i(X_i)].$$

#### 3.1.9.2 *Convergence in Distribution*

A sequence of random variables $U_1, U_2, \ldots, U_n, \ldots$ *converges in distribution* to a random variable $U$ if

$$\lim_{n\to\infty} \mathrm{F}_{U_n}(u) = \mathrm{F}_U(u)$$

for all values of $u$ where $\mathrm{F}_U(u)$ is continuous. Notationally, we write $U_n \xrightarrow{\mathrm{D}} U$.

As an important example, suppose that $X_1, X_2, \ldots, X_n$ constitute a random sample of size $n$ from either a univariate discrete probability distribution $p_X(x)$ or a univariate density function $f_X(x)$, where $\mathrm{E}(X) = \mu(-\infty < \mu < +\infty)$ and $\mathrm{V}(X) = \sigma^2(0 < \sigma^2 < +\infty)$. With $\bar{X} = n^{-1}\sum_{i=1}^{n} X_i$, consider the standardized random variable

$$U_n = \frac{\bar{X} - \mu}{\sigma/\sqrt{n}} = \frac{\sum_{i=1}^{n} X_i - n\mu}{\sqrt{n}\sigma}.$$

Then, it can be shown that $\lim_{n\to\infty} \mathrm{M}_{U_n}(t) = \mathrm{e}^{t^2/2}$, leading to the conclusion that $U_n \xrightarrow{\mathrm{D}} Z$, where $Z \sim$N(0,1). This is the well-known *Central Limit Theorem*.

#### 3.1.9.3 *Order Statistics*

Let $X_1, X_2, \ldots, X_n$ constitute a random sample of size $n$ from a univariate density function $f_X(x), -\infty < x < +\infty$, with corresponding cumulative distribution function $F_X(x) = \int_{-\infty}^{x} f_X(t)\,dt$. Then, the $n$ *order statistics* $X_{(1)}, X_{(2)}, \ldots, X_{(n)}$ satisfy the relationship

$$-\infty < X_{(1)} < X_{(2)} < \cdots < X_{(n-1)} < X_{(n)} < +\infty.$$

For $r = 1, 2, \ldots, n$, the random variable $X_{(r)}$ is called the $r$th order statistic. In particular, $X_{(1)} = \min\{X_1, X_2, \ldots, X_n\}$, $X_{(n)} = \max\{X_1, X_2, \ldots, X_n\}$, and $X_{((n+1)/2)} = \text{median}\{X_1, X_2, \ldots, X_n\}$ when $n$ is an odd positive integer.

For $r = 1, 2, \ldots, n$, the distribution of $X_{(r)}$ is

$$f_{X_{(r)}}(x_{(r)}) = nC_{r-1}^{n-1}[F_X(x_{(r)})]^{r-1}[1 - F_X(x_{(r)})]^{n-r} f_X(x_{(r)}), -\infty < x_{(r)} < +\infty.$$

For $1 \le r < s \le n$, the joint distribution of $X_{(r)}$ and $X_{(s)}$ is equal to

$$\begin{aligned} f_{X_{(r)},X_{(s)}}(x_{(r)}, x_{(s)}) &= \frac{n!}{(r-1)!(s-r-1)!(n-s)!}[F_X(x_{(r)})]^{r-1} \\ &\quad \times [F_X(x_{(s)}) - F_X(x_{(r)})]^{s-r-1} \\ &\quad \times [1 - F_X(x_{(s)})]^{n-s} f_X(x_{(r)}) f_X(x_{(s)}), \end{aligned}$$

$$-\infty < x_{(r)} < x_{(s)} < +\infty.$$

And, the joint distribution of $X_{(1)}, X_{(2)}, \ldots, X_{(n)}$ is

$$f_{X_{(1)},X_{(2)},\ldots,X_{(n)}}(x_{(1)}, x_{(2)}, \ldots, x_{(n)}) = n! \prod_{i=1}^{n} f_X(x_{(i)}),$$

$$-\infty < x_{(1)} < x_{(2)} < \cdots < x_{(n-1)} < x_{(n)} < +\infty.$$

#### 3.1.9.4 Method of Transformations

With $k = 2$, let $X_1$ and $X_2$ be two continuous random variables with joint density function $f_{X_1,X_2}(x_1, x_2), (x_1, x_2) \in \mathcal{D}$. Let $Y_1 = g_1(X_1, X_2)$ and $Y_2 = g_2(X_1, X_2)$ be random variables, where the functions $y_1 = g_1(x_1, x_2)$ and $y_2 = g_2(x_1, x_2)$ define a one-to-one transformation from the domain $\mathcal{D}$ in the $(x_1, x_2)$-plane to the domain $\mathcal{D}^*$ in the $(y_1, y_2)$-plane. Further, let $x_1 = h_1(y_1, y_2)$ and $x_2 = h_2(y_1, y_2)$ be the inverse functions expressing $x_1$ and $x_2$ as functions of $y_1$ and $y_2$. Then, the joint density function of the random variables $Y_1$ and $Y_2$ is

$$f_{Y_1,Y_2}(y_1, y_2) = f_{X_1,X_2}[h_1(y_1, y_2), h_2(y_1, y_2)]|J|, \quad (y_1, y_2) \in \mathcal{D}^*,$$

where the *Jacobian* $J, J \neq 0$, of the transformation is the second-order determinant

$$J = \begin{vmatrix} \dfrac{\partial h_1(y_1, y_2)}{\partial y_1} & \dfrac{\partial h_1(y_1, y_2)}{\partial y_2} \\ \dfrac{\partial h_2(y_1, y_2)}{\partial y_1} & \dfrac{\partial h_2(y_1, y_2)}{\partial y_2} \end{vmatrix}.$$

For the special case $k = 1$ when $Y_1 = g_1(X_1)$ and $X_1 = h_1(Y_1)$, it follows that

$$f_{Y_1}(y_1) = f_{X_1}[h_1(y_1)]\left|\frac{dh_1(y_1)}{dy_1}\right|, \quad y_1 \in \mathcal{D}^*.$$

It is a straightforward generalization to the situation when $Y_i = g_i(X_1, X_2, \ldots, X_k), i = 1, 2, \ldots, k$, with the Jacobian $J$ being the determinant of a $(k \times k)$ matrix.

## EXERCISES

**Exercise 3.1.** Two balls are selected sequentially at random *without replacement* from an urn containing $N$ (>1) balls numbered individually from 1 to $N$. Let the discrete random variable $X$ be the number on the first ball selected, and let the discrete random variable $Y$ be the number on the second ball selected.

(a) Provide an explicit expression for the joint distribution of the random variables X and $Y$, and also provide explicit expressions for the marginal distributions of X and $Y$.

(b) Provide an explicit expression for $\text{pr}[X \geq (N-1)|Y = y]$, where $y$ is a fixed positive integer satisfying the inequality $1 \leq y \leq N$.

(c) Derive an explicit expression for $\text{corr}(X, Y)$, the correlation between $X$ and $Y$. Find the limiting value of $\text{corr}(X, Y)$ as $N \to \infty$, and then comment on your finding.

**Exercise 3.2.** Consider an experiment consisting of $n$ mutually independent Bernoulli trials, where each trial results in either a success (denoted by the letter S) or a failure (denoted by the letter F). For any trial, the probability of a success is equal to $\pi, 0 < \pi < 1$, and so the probability of a failure is equal to $(1 - \pi)$. For any set of $n$ trials with outcomes arranged in a linear sequence, a *run* is a *subsequence* of outcomes of the same type which is both preceded and succeeded by outcomes of the opposite type or by the beginning or by the end of the complete sequence. The number of successes in a success (or S) run is referred to as its *length*.

For any such sequence of $n$ Bernoulli trial outcomes, let the discrete random variable $M_n$ denote the length of the *shortest* S run in the sequence, and let the discrete random variable $L_n$ denote the length of the *longest* S run in the sequence. For example, for the sequence of $n = 12$ outcomes given by

FFSFSSSFFFSS,

the observed value of $M_{12}$ is $m_{12} = 1$ and the observed value of $L_{12}$ is $l_{12} = 3$.

(a) If $n = 5$, find the joint distribution of the random variables $M_5$ and $L_5$.

(b) Find the marginal distribution of the random variable $L_5$, and then find the numerical value of $E(L_5)$ when $\pi = 0.90$.

**Exercise 3.3.** Suppose that $p_U(u) = \text{pr}(U = u) = n^{-1}, u = 1, 2, \ldots, n$. Further, suppose that, given (or conditional on) $U = u$, X and $Y$ are independent geometric random variables, with

$$p_X(x|U = u) = u^{-1}(1 - u^{-1})^{x-1}, \quad x = 1, 2, \ldots, \infty$$

and

$$p_Y(y|U=u) = u^{-1}(1-u^{-1})^{y-1}, \quad y = 1, 2, \ldots, \infty.$$

(a) Derive an explicit expression for $\text{corr}(X, Y)$, the correlation between the random variables $X$ and $Y$.

(b) Develop an expression for $\text{pr}(X \neq Y)$. What is the numerical value of this probability when $n = 4$?

**Exercise 3.4.** Suppose that $Z \sim N(0, 1)$, that $U \sim \chi^2_\nu$, and that $Z$ and $U$ are independent random variables. Then, the random variable

$$T_\nu = \frac{Z}{\sqrt{U/\nu}}$$

has a (central) $t$-distribution with $\nu$ degrees of freedom.

(a) By considering the conditional density function of $T_\nu$ given $U = u$, develop an explicit expression for the density function of $T_\nu$.

(b) Find $E(T_\nu)$ and $V(T_\nu)$.

**Exercise 3.5**

(a) Suppose that $Y$ is a random variable with conditional mean $E(Y|X = x) = \beta_0 + \beta_1 x$ and that $X$ is a random variable with mean $E(X)$ and variance $V(X)$. Use conditional expectation theory to show that

$$\text{corr}(X, Y) = \beta_1 \sqrt{\frac{V(X)}{V(Y)}},$$

and then comment on this finding.

(b) Now, given the above assumptions, suppose also that $E(X|Y = y) = \alpha_0 + \alpha_1 y$. Develop an explicit expression relating $\text{corr}(X, Y)$ to $\alpha_1$ and $\beta_1$, and then comment on this finding.

(c) Now, suppose that $E(Y|X = x) = \beta_0 + \beta_1 x + \beta_2 x^2$. Derive an explicit expression for $\text{corr}(X, Y)$, and then comment on how the addition of the quadratic term $\beta_2 x^2$ affects the relationship between $\text{corr}(X, Y)$ and $\beta_1$ given in part (a).

**Exercise 3.6.** Suppose that the amounts $X$ and $Y$ (in milligrams) of two toxic chemicals in a liter of water selected at random from a river near a certain manufacturing plant can be modeled by the bivariate density function

$$f_{X,Y}(x, y) = 6\theta^{-3}(x - y), \quad 0 < y < x < \theta.$$

(a) Derive an explicit expression for $\text{corr}(X, Y)$, the correlation between the two continuous random variables $X$ and $Y$.

(b) Set up appropriate integrals that are needed to find

$$\mathrm{pr}\left[(X+Y)<\theta|(X+2Y)>\frac{\theta}{4}\right].$$

Note that the appropriate integrals do not have to be evaluated, but the integrands and the limits of integration must be correctly specified for all integrals that are used.

(c) Let $(X_1,Y_1),(X_2,Y_2),\ldots,(X_n,Y_n)$ constitute a random sample of size $n$ from $f_{X,Y}(x,y)$, and let $\bar{X}=n^{-1}\sum_{i=1}^{n}X_i$ and $\bar{Y}=n^{-1}\sum_{i=1}^{n}Y_i$. Develop explicit expressions for $E(L)$ and $V(L)$ when $L=(3\bar{X}-2\bar{Y})$.

**Exercise 3.7.** For a certain type of chemical reaction involving two chemicals A and B, let $X$ denote the proportion of the initial amount (in grams) of chemical A that remains unreacted at equilibrium, and let $Y$ denote the corresponding proportion of the initial amount (in grams) of chemical B that remains unreacted at equilibrium. The bivariate density function for the continuous random variables $X$ and $Y$ is assumed to be of the form

$$f_{X,Y}(x,y)=\frac{\Gamma(\alpha+\beta+3)}{\Gamma(\alpha+1)\Gamma(\beta+1)}(1-x)^{\alpha}y^{\beta},\quad 0<y<x,\quad 0<x<1,$$

$$\alpha>-1,\quad \beta>-1.$$

(a) Derive explicit expressions for $f_X(x)$ and $f_Y(y)$, the marginal distributions of the random variables $X$ and $Y$, and for $f_Y(y|X=x)$, the conditional density function of $Y$ given $X=x$.

(b) Use the results obtained in part (a) to develop an expression for $\rho_{X,Y}=\mathrm{corr}(X,Y)$, the correlation between the random variables $X$ and $Y$. What is the numerical value of this correlation coefficient when $\alpha=2$ and $\beta=3$?

(c) Let $(X_1,Y_1),(X_2,Y_2),\ldots,(X_n,Y_n)$ constitute a random sample of size $n$ from $f_{X,Y}(x,y)$. If $\bar{X}=n^{-1}\sum_{i=1}^{n}X_i$ and $\bar{Y}=n^{-1}\sum_{i=1}^{n}Y_i$, find the expected value and variance of the random variable $L=(3\bar{X}-5\bar{Y})$ when $n=10,\alpha=2$, and $\beta=3$.

**Exercise 3.8.** A certain simple biological system involves exactly two independently functioning components. If one of these two components fails, then the entire system fails. For $i=1,2$, let $Y_i$ be the random variable representing the time (in weeks) to failure of the $i$th component, with the distribution of $Y_i$ being negative exponential, namely,

$$f_{Y_i}(y_i)=\theta_i e^{-\theta_i y_i},\quad 0<y_i<\infty,\quad \theta_i>0.$$

Further, assume that $Y_1$ and $Y_2$ are independent random variables. Clearly, if component 1 fails first, then $Y_1$ is observable, but $Y_2$ is not observable (i.e., $Y_2$ is then said to be *censored*); conversely, if component 2 fails first, then $Y_2$ is observable, but $Y_1$ is not observable (i.e., $Y_1$ is censored). Thus, if this biological system fails, then only two random variables, call them $U$ and $W$, are observable, where $U=\min(Y_1,Y_2)$ and where $W=1$ if $Y_1<Y_2$ and $W=0$ if $Y_2<Y_1$.

(a) Develop an explicit expression for the joint distribution $f_{U,W}(u,w)$ of the random variables $U$ and $W$.

(b) Find the marginal distribution $p_W(w)$ of the random variable $W$.

(c) Find the marginal distribution $f_U(u)$ of the random variable $U$.

(d) Are $U$ and $W$ independent random variables?

**Exercise 3.9.** It has been documented via numerous research studies that the eldest child in a family with multiple children generally has a higher IQ than his or her siblings. In a certain large population of U.S. families with two children, suppose that the random variable $Y_1$ denotes the IQ of the older child and that the random variable $Y_2$ denotes the IQ of the younger child. Assume that $Y_1$ and $Y_2$ have a joint bivariate normal distribution with parameter values $E(Y_1) = 110$, $E(Y_2) = 100$, $V(Y_1) = V(Y_2) = 225$, and $\rho = \text{corr}(Y_1, Y_2) = 0.80$.

(a) Suppose that three families are randomly chosen from this large population of U.S. families with two children. What is the probability that the older child has an IQ at least 15 points higher than the younger child for at least two of these three families?

(b) For a family randomly chosen from this population, if the older child is known to have an IQ of 120, what is the probability that the younger child has an IQ greater than 120?

**Exercise 3.10.** Discrete choice statistical models are useful in many situations, including transportation research. For example, transportation researchers may want to know why certain individuals choose to use public bus transportation instead of a car. As a starting point, the investigators typically assume that each mode of transportation carries with it a certain value, or "utility," that makes it more or less desirable to consumers. For instance, cars may be more convenient, but a bus may be more environmentally friendly. According to the "maximum utility principle," consumers select the alternative that has the greatest desirability or utility.

As a simple illustration of a discrete choice statistical model, suppose that there are only two possible discrete choices, A and B. Let the random variable Y take the value 1 if choice A is made, and let Y take the value 0 if choice B is made. Furthermore, let $U$ and $V$ be the utilities associated with the choices A and B, respectively, and assume that $U$ and $V$ are independent random variables, each having the same standard *Gumbel* (Type-I Extreme-Value) distribution. In particular, both $U$ and $V$ are assumed to have CDFs of the general form

$$F_X(x) = \text{pr}(X \leq x) = e^{-e^{-x}}, \quad -\infty < x < \infty.$$

According to the maximum utility principle, $Y = 1$ if and only if $U > V$, or equivalently, if $W = (U - V) > 0$.

(a) Show that $W$ follows a *logistic* distribution, with CDF $F_W(w) = 1/(1 + e^{-w}), -\infty < w < \infty$.

(b) Suppose that $U = \alpha + E_1$ and $V = E_2$, where $E_1$ and $E_2$ are independent error terms, each following the standard Gumbel CDF of the general form $F_X(x)$ given above. Here, $\alpha$ represents the average population difference between the two utilities. (More generally, $U$ and $V$ can be modeled as functions of covariates, although this extension is not considered here.) Again, assume that we observe $Y = 1$ if choice A is made and $Y = 0$ if choice B is made. Find an explicit expression as a function of $\alpha$ for $\text{pr}(Y = 1)$ under the maximum utility principle.

**Exercise 3.11.** Let $X_1, X_2, \ldots, X_n$ constitute a random sample of size $n(n \geq 3)$ from the parent population

$$f_X(x) = \lambda e^{-\lambda x}, \qquad 0 < x < +\infty, \quad 0 < \lambda < +\infty.$$

(a) Find the conditional density function of $X_1, X_2, \ldots, X_n$ given that $S = \sum_{i=1}^{n} X_i = s$.

(b) Consider the $(n - 1)$ random variables

$$Y_1 = \frac{X_1}{S}, \quad Y_2 = \frac{(X_1 + X_2)}{S}, \ldots, \quad Y_{n-1} = \frac{(X_1 + X_2 + \cdots + X_{n-1})}{S}.$$

Find the joint distribution of $Y_1, Y_2, \ldots, Y_{n-1}$ given that $S = s$.

(c) When $n = 3$ and when $n = 4$, find the marginal distribution of $Y_1$ given that $S = s$, and then use these results to infer the structure of the marginal distribution of $Y_1$ given that $S = s$ for any $n \geq 3$.

**Exercise 3.12.** Let $X_1, X_2, \ldots, X_n$ constitute a random sample of size $n$ from a $N(\mu, \sigma^2)$ population. Then, consider the $n$ random variables $Y_1, Y_2, \ldots, Y_n$, where $Y_i = e^{X_i}$, $i = 1, 2, \ldots, n$. Finally, consider the following two random variables:

(i) The arithmetic mean $\bar{Y}_a = n^{-1} \sum_{i=1}^{n} Y_i$;

(ii) The geometric mean $\bar{Y}_g = \left( \prod_{i=1}^{n} Y_i \right)^{1/n}$.

Develop an explicit expression for $\text{corr}(\bar{Y}_a, \bar{Y}_g)$, the correlation between the two random variables $\bar{Y}_a$ and $\bar{Y}_g$. Then, find the limiting value of this correlation as $n \to \infty$, and comment on your finding.

**Exercise 3.13.** For a certain public health research study, an epidemiologist is interested in determining via blood tests which particular subjects in a random sample of $N(= Gn)$ human subjects possess a certain antibody; here, $G$ and $n$ are positive integers. For the population from which the random sample of $N$ subjects is selected, the proportion of subjects in that population possessing the antibody is equal to $\pi(0 < \pi < 1)$, a known quantity. The epidemiologist is considering two possible blood testing plans:

*Plan* #1: Perform the blood test separately on each of the $N$ subjects in the random sample;

*Plan* #2: Divide the $N$ subjects in the random sample into $G$ groups of $n$ subjects each; then, for each group of size $n$, take a blood sample from each of the $n$ subjects in that group, mix the $n$ blood samples together, and do one blood test on the mixture; if the blood test on the mixture is negative (indicating that the antibody is not present in that mixture), then none of those $n$ subjects possesses the antibody; however, if the blood test on the mixture is positive (indicating that the antibody is present), then the blood test will have to be performed on each of the $n$ subjects in that group.

(a) Let $T_2$ be the random variable denoting the number of blood tests required for Plan #2. Develop an explicit expression for $E(T_2)$.

(b) Clearly, the larger the value of $\pi$, the more likely it is that the blood test on a mixture of $n$ blood samples will be positive, necessitating a blood test on every one of those $n$ subjects. Determine the optimal value of $n$ (say, $n^*$) and the associated desired largest value of $\pi$ (say, $\pi^*$) for which $E(T_2) < N$ (i.e., for which Plan #2 is preferred to Plan #1).

**Exercise 3.14.** For the state of North Carolina (NC), suppose that the number $Y$ of female residents who are homicide victims in any particular calendar year follows a Poisson distribution with mean $E(Y) = L\lambda$, where $L$ is the total number of person-months at risk for homicide for all female NC residents during that year and where $\lambda$ is the rate of female homicides per person-month. Let $\pi$ be the proportion of all homicide victims who were pregnant at the time of the homicide; more specifically, $\pi$ =pr(woman was pregnant at the time of the homicide | woman was a homicide victim). It can be assumed that women in NC function independently of one another with regard to homicide-related and pregnancy-related issues.

Domestic violence researchers are interested in making statistical inferences about the true average (or expected value) of the number $Y_p$ of homicide victims who were pregnant at the time of the homicide and about the true average (or expected value) of the number $Y_{\bar{p}} = (Y - Y_p)$ of homicide victims who were not pregnant at the time of the homicide.

Find the conditional joint moment generating function

$$M_{Y_p, Y_{\bar{p}}}(s, t|Y = y) = M_{Y_p, (Y - Y_p)}(s, t|Y = y)$$

of $Y_p$ and $Y_{\bar{p}} = (Y - Y_p)$ given $Y = y$, and then unconditionalize to determine the distributions of $Y_p$ and $Y_{\bar{p}}$. Are $Y_p$ and $Y_{\bar{p}}$ independent random variables? If $L$ has a known value, and if estimates $\hat{\lambda}$ and $\hat{\pi}$ of $\lambda$ and $\pi$ are available, provide reasonable estimates of $E(Y_p)$ and $E(Y_{\bar{p}})$.

**Exercise 3.15.** A chemical test for the presence of a fairly common protein in human blood produces a continuous measurement $X$. Let the random variable $D$ take the value 1 if a person's blood contains the protein in question, and let $D$ take the value 0 if a person's blood does not contain the protein in question. Among all those people carrying the protein, $X$ has a *lognormal* distribution with mean $E(X|D = 1) = 2.00$ and variance $V(X|D = 1) = 2.60$. Among all those people not carrying the protein, $X$ has a *lognormal* distribution with mean $E(X|D = 0) = 1.50$ and variance $V(X|D = 0) = 3.00$. In addition, it is known that 60% of all human beings actually carry this particular protein in their blood.

(a) If a person is randomly chosen and is given the chemical test, what is the numerical value of the probability that this person's blood truly contains the protein in question given that the event "$1.60 < X < 1.80$" has occurred (i.e., it is known that the observed value of $X$ for this person lies between 1.60 and 1.80)?

(b) Let the random variable $X$ be the value of the chemical test for a person chosen completely randomly. Provide numerical values for $E(X)$ and $V(X)$.

(c) Suppose that the following diagnostic rule is proposed: "classify a randomly chosen person as carrying the protein if $X > c$, and classify that person as not carrying the protein if $X \leq c$, where $0 < c < \infty$." Thus, a carrier for which $X \leq c$ is misclassified, as is a noncarrier for which $X > c$. For this diagnostic rule, develop an expression (as a function of $c$) for the *probability of misclassification* $\theta$ of a randomly chosen human being, and then find the numerical value $c^*$ of $c$ that minimizes $\theta$. Comment on your finding.

**Exercise 3.16.** Let $(X_1, Y_1), (X_2, Y_2), \ldots, (X_n, Y_n)$ constitute a random sample of size $n$ from a bivariate population involving two random variables $X$ and $Y$, where $E(X) = \mu_x, E(Y) = \mu_y, V(X) = \sigma_x^2, V(Y) = \sigma_y^2$, and $\rho = \text{corr}(X, Y)$. Show that the random variable

$$U = (n-1)^{-1} \sum_{i=1}^{n} (X_i - \bar{X})(Y_i - \bar{Y})$$

has an expected value equal to the parametric function $\text{cov}(X, Y) = \rho\sigma_x\sigma_y$.

**Exercise 3.17.** A certain large community in the United States receives its drinking water supply from a nearby lake, which itself is located in close proximity to a plant that uses benzene, among other chemicals, to manufacture styrene. Because this community has recently experienced elevated rates of leukemia, a blood cancer that has been associated with benzene exposure, the EPA decides to send a team to sample the drinking water used by this community and to determine whether or not this drinking water contains a benzene level exceeding the EPA standard of 5 parts of benzene per billion parts of water (i.e., a standard of 5 ppb). Suppose that the continuous random variable $X$ represents the measured benzene concentration in ppb in the drinking water used by this community, and assume that $X$ has a lognormal distribution. More specifically, assume that $Y = \ln(X)$ has a normal distribution with unknown mean $\mu$ and variance $\sigma^2 = 2$. The EPA decides to take $n = 10$ independently chosen drinking water samples and to measure the benzene concentration in each of these 10 drinking water samples. Based on the results of these 10 benzene concentration measurements (denoted $X_1, X_2, \ldots, X_{10}$), the EPA team has to decide whether the true mean benzene concentration in this community's drinking water is in violation of the EPA standard (i.e., exceeds 5 ppb). Three decision rules are proposed:

*Decision Rule* #1: Decide that the drinking water is in violation of the EPA standard if at least 3 of the 10 benzene concentration measurements exceed 5 ppb.

*Decision Rule* #2: Decide that the drinking water is in violation of the EPA standard if the geometric mean of the 10 benzene concentration measurements exceeds 5 ppb, where

$$\bar{X}_g = \left( \prod_{i=1}^{10} X_i \right)^{1/10}.$$

*Decision Rule* #3: Decide that the drinking water is in violation of the EPA standard if the maximum of the 10 benzene concentration measurements, denoted $X_{(10)}$, exceeds 5 ppb.

(a) For *each* of these three different decision rules, develop, as a function of the unknown parameter $\mu$, a *general expression* for the probability of deciding that the drinking water is in violation of the EPA standard. Also, if $\mathrm{E}(X) = 7$, find the numerical value of each of these three probabilities.

(b) For Decision Rule #2, examine expressions for $\mathrm{pr}(\bar{X}_g > 5)$ and $\mathrm{E}(\bar{X}_g)$ to provide analytical arguments as to why Decision Rule #2 performs so poorly.

**Exercise 3.18.** Let $X_1, X_2, \ldots, X_n$ constitute a random sample of size $n$ from a normal distribution with mean $\mu = 0$ and variance $\sigma^2 = 2$. Determine the smallest value of $n$, say $n^*$, such that $\mathrm{pr}[\min\{X_1^2, X_2^2, \ldots, X_n^2\} \leq 0.002] \geq 0.80$. [HINT: If $Z \sim \mathrm{N}(0,1)$, then $Z^2 \sim \chi_1^2$.]

**Exercise 3.19.** A large hospital wishes to determine the appropriate number of coronary bypass grafts that it can perform during the upcoming calendar year based both on the size of its coronary bypass surgery staff (e.g., surgeons, nurses, anesthesiologists, technicians, etc.) and on other logistical and space considerations. National data suggest that a typical coronary bypass surgery patient would require exactly one (vessel) graft with probability $\pi_1 = 0.54$, would require exactly two grafts with probability $\pi_2 = 0.22$, would require exactly three grafts with probability $\pi_3 = 0.15$, and would require exactly four grafts with probability $\pi_4 = 0.09$. Further, suppose that it is known that this hospital cannot feasibly perform more than about 900 coronary bypass grafts in any calendar year.

(a) An administrator for this hospital suggests that it might be reasonable to perform coronary bypass surgery on $n = 500$ different patients during the upcoming calendar year and still have a reasonably high probability (say, $\geq 0.95$) of not exceeding the yearly upper limit of 900 coronary bypass grafts. Use the *Central Limit Theorem* to assess the reasonableness of this administrator's suggestion.

(b) Provide a reasonable value for the largest number $n^*$ of patients that can undergo coronary bypass surgery at this hospital during the upcoming year so that, with probability at least equal to 0.95, no more than 900 grafts will need to be performed.

**Exercise 3.20.** For the $i$th of $k$ drug treatment centers ($i = 1, 2, \ldots, k$) in a certain large U.S. city, suppose that the distribution of the number $X_i$ of adult male drug users that have to be tested until *exactly one* such adult drug user tests positively for HIV is assumed to be *geometric*, namely

$$\mathrm{p}_{X_i}(x_i) = \pi(1-\pi)^{x_i - 1}, \qquad x_i = 1, 2, \ldots, +\infty; \quad 0 < \pi < 1.$$

In all that follows, assume that $X_1, X_2, \ldots, X_k$ constitute a set of mutually independent random variables.

(a) If $\pi = 0.05$ and if $S = \sum_{i=1}^{k} X_i$, provide a reasonable numerical value for $\text{pr}(S > 1,100)$ if $k = 50$.

(b) Use moment generating function (MGF) theory to show that the distribution of the random variable $U = 2\pi S$ is, *for small* $\pi$, approximately chi-squared with $2k$ degrees of freedom.

(c) Use the result in part (b) to compute a numerical value for $\text{pr}(S > 1,100)$ when $\pi = 0.05$ and $k = 50$, and then compare your answer to the one found in part (a).

**Exercise 3.21.** Let $Y_1, Y_2, \ldots, Y_n$ constitute a random sample of size $n$ (>1) from the *Pareto* density function

$$f_Y(y;\theta) = \theta c^{\theta} y^{-(\theta+1)} \qquad 0 < c < y < +\infty \quad \text{and} \quad \theta > 0,$$

where $c$ is a known positive constant and where $\theta$ is an unknown parameter. The Pareto density function has been used to model the distribution of family incomes in certain populations.

Consider the random variable

$$U_n = \theta n[Y_{(1)} - c]/c,$$

where $Y_{(1)} = \min\{Y_1, Y_2, \ldots, Y_n\}$. Directly evaluate $\lim_{n\to\infty} F_{U_n}(u)$, where $F_{U_n}(u)$ is the CDF of $U_n$, to find the asymptotic distribution of $U_n$. In other words, derive an explicit expression for the CDF of $U$ when $U_n$ converges in distribution to $U$.

**Exercise 3.22.** For a certain laboratory experiment involving mice, suppose that the random variable $X, 0 < X < 1$, represents the *proportion* of a fixed time period (in minutes) that it takes a mouse to locate food at the end of a maze, and further suppose that $X$ follows a uniform distribution on the interval $(0, 1)$, namely,

$$f_X(x) = 1, \quad 0 < x < 1.$$

Suppose that the experiment involves $n$ randomly chosen mice. Further, suppose that $x_1, x_2, \ldots, x_n$ are the $n$ realized values (i.e., the $n$ observed proportions) of the $n$ mutually independent random variables $X_1, X_2, \ldots, X_n$, which themselves can be considered to constitute a random sample of size $n$ from $f_X(x)$. Let the random variable $U$ be the smallest proportion based on the shortest time required for a mouse to locate the food, and let the random variable $V$ be the proportion of the fixed time period still remaining based on the longest time required for a mouse to locate the food.

(a) Find an explicit expression for the joint distribution of the random variables $U$ and $V$.

(b) Let $R = nU$ and let $S = nV$. Find the asymptotic joint distribution of $R$ and $S$. $\left[\text{HINT: Evaluate } \lim_{n\to\infty} \left\{\text{pr}[(R > r) \cap (S > s)]\right\}\right]$.

**Exercise 3.23.** Suppose that the total cost C (in millions of dollars) for repairs due to floods occurring in the United States in any particular year can be modeled by defining the random variable $C$ as follows:

$$C = 0 \text{ if } X = 0 \quad \text{and} \quad C = \sum_{j=1}^{X} C_j \text{ if } X > 0;$$

here, the number of floods $X$ in any particular year in the United States is assumed to have a Poisson distribution with mean $\text{E}(X) = \lambda$, and $C_j$ is the cost (in millions of dollars) for repairs due to the $j$th flood in that particular year. Also, it is assumed that $C_1, C_2, \ldots$ are i.i.d. random variables, each with the same expected value $\mu$, the same variance $\sigma^2$, and the same moment generating function $\text{M}(t) = \text{E}(e^{tC_j})$. Note that the actual distribution of the random variables $C_1, C_2, \ldots$ has not been specified.

(a) Develop an explicit expression for $\text{corr}(X, C)$, the correlation between the random variables $X$ and $C$, and then comment on the structure of the expression that you obtained.

(b) Develop an explicit expression for $\text{M}_C(t) = \text{E}(e^{tC})$, the moment generating function of the random variable $C$, and then use this result to find $\text{E}(C)$.

**Exercise 3.24.** To evaluate the performance of a new cholesterol-lowering drug, a large drug company plans to enlist a randomly chosen set of $k$ private medical practices to help conduct a clinical trial. Under the protocol proposed by the drug company, each private medical practice is to enroll into the clinical trial a set of $n$ randomly chosen subjects with high cholesterol. The cholestorol level (in mg/dL) of each subject is to be measured both before taking the new drug and after taking the new drug on a daily basis for 6 months. The continuous response variable of interest is $Y$, the change in a subject's cholesterol level over the 6-month period. The following statistical model will be used:

$$Y_{ij} = \mu + \beta_i + \epsilon_{ij}, \qquad i = 1, 2, \ldots, k \quad \text{and} \quad j = 1, 2, \ldots, n.$$

Here, $\mu$ is the average change in cholesterol level for a typical subject with high cholesterol who takes this new cholesterol-lowering drug on a daily basis for a 6-month period, $\beta_i$ is the *random effect* associated with the $i$th private medical practice, and $\epsilon_{ij}$ is the random effect associated with the $j$th subject in the $i$th private medical practice. Here, it is assumed that $\beta_i \sim \text{N}(0, \sigma_\beta^2)$, that $\epsilon_{ij} \sim \text{N}(0, \sigma_\epsilon^2)$, and that the sets $\{\beta_i\}$ and $\{\epsilon_{ij}\}$ constitute a group of $(k + kn)$ mutually independent random variables. Finally, let

$$\bar{Y} = (kn)^{-1} \sum_{i=1}^{k} \sum_{j=1}^{n} Y_{ij}$$

be the overall sample mean.

(a) Develop explicit expressions for $\text{E}(\bar{Y})$ and $\text{V}(\bar{Y})$.

(b) Suppose that it will cost $D_c$ dollars for each clinic to enroll and monitor $n$ subjects over the duration of the proposed clinical trial, and further suppose that each subject is to be paid $D_p$ dollars for participating in the clinical trial. Thus, the total cost of the clinical trial is equal to $C = (kD_c + knD_p)$. Suppose that this drug company can only afford to spend $C^*$ dollars to conduct the proposed clinical trial. Find specific expressions for $n^*$ and $k^*$, the specific values of $n$ and $k$ that minimize the variance of $\bar{Y}$ subject to the condition that $C = (kD_c + knD_p) = C^*$.

(c) If $C^* = 100,000, D_c = 10,000, D_p = 100, \sigma_\beta^2 = 4,$ and $\sigma_\epsilon^2 = 9,$ find appropriate numerical values for $n^*$ and $k^*$.

**Exercise 3.25.** For $i = 1, 2$, suppose that the conditional distribution of $Y_i$ given that $Y_3 = y_3$ is

$$p_{Y_i}(y_i|Y_3 = y_3) = y_3^{y_i} e^{-y_3}/y_i!, \quad y_i = 0, 1, \ldots, \infty.$$

Further, assume that the random variable $Y_3$ has the truncated Poisson distribution

$$p_{Y_3}(y_3) = \frac{\lambda_3^{y_3}}{y_3!(e^{\lambda_3} - 1)}, \qquad y_3 = 1, 2, \ldots, \infty \quad \text{and} \quad \lambda_3 > 0;$$

and, also assume that the random variables $Y_1$ and $Y_2$ are *conditionally independent* given that $Y_3 = y_3$.

Then, consider the random variables

$$R = (Y_1 + Y_3) \quad \text{and} \quad S = (Y_2 + Y_3).$$

Derive an explicit expression for the moment generating function $M_U(t)$ of the random variable $U = (R + S)$, and then use $M_U(t)$ directly to find an explicit expression for $E(U)$. Verify that your expression for $E(U)$ is correct by finding $E(U)$ directly.

**Exercise 3.26.** Suppose that $n(>1)$ balls are randomly tossed into $C(>1)$ cells, so that the probability is $1/C$ of any ball ending up in the $i$th cell, $i = 1, 2, \ldots, C$.

Find the expected value and the variance of the number $X$ of cells that will end up being empty (i.e., that will contain no balls). For the special case when $C = 6$ and $n = 5$, find the numerical values of $E(X)$ and $V(X)$.

**Exercise 3.27.** A researcher at the Federal Highway Administration (FHWA) proposes the following statistical model for traffic fatalities. Let the random variable $N$ be the number of automobile accidents occurring on a given stretch of heavily traveled interstate highway over a specified time period. For $i = 1, 2, \ldots, N$, let the random variable $Y_i$ take the value 1 if the $i$th automobile accident involved at least one fatality, and let $Y_i$ take the value 0 otherwise. Let $\text{pr}(Y_i = 1) = \pi, 0 < \pi < 1$, and further assume that the $\{Y_i\}$ are mutually independent dichotomous random variables. Also, let the random variable $N$ have the geometric distribution

$$p_N(n) = \theta(1 - \theta)^{n-1}, \quad n = 1, \ldots, \infty; \quad 0 < \theta < 1.$$

This researcher is interested in the random variable

$$T = Y_1 + Y_2 + \cdots + Y_N,$$

the total number of automobile accidents involving fatalities on that stretch of interstate highway during the specified time period.

(a) Find explicit expressions for $E(T)$, $V(T)$, and $\text{corr}(N, T)$.

(b) Find an explicit expression for $\text{pr}(T = 0)$.

**Exercise 3.28.** Let $X_1, X_2, \ldots, X_m$ constitute a random sample of size $m$ from a POI($\lambda_1$) population, and let $Y_1, Y_2, \ldots, Y_n$ constitute a random sample of size $n$ from a POI($\lambda_2$) population. Consider the random variable

$$U = (\bar{X} - \bar{Y}) = m^{-1}\sum_{i=1}^{m} X_i - n^{-1}\sum_{i=1}^{n} Y_i.$$

(a) Find explicit expressions for $E(U)$ and $V(U)$.

(b) Use the Lagrange multiplier method to find expressions (which are functions of $N, \lambda_1$, and $\lambda_2$) for $m$ and $n$ that minimize $V(U)$ subject to the restriction $(m + n) = N$, where $N$ is the total sample size that can be selected from these two Poisson populations due to cost considerations. Provide an interpretation for your findings. If $N = 60$, $\lambda_1 = 2$, and $\lambda_2 = 8$, use these expressions to find numerical values for $m$ and $n$.

**Exercise 3.29*.** Let the discrete random variables $X$ and $Y$ denote the numbers of AIDS cases that will be detected yearly in two different NC counties, one in the eastern part of the state and the other in the western part of the state. Further, assume that $X$ and $Y$ are *independent* random variables, and that they have the respective distributions

$$p_X(x) = (1 - \pi_x)\pi_x^x, \qquad x = 0, 1, \ldots, \infty, \quad 0 < \pi_x < 1$$

and

$$p_Y(y) = (1 - \pi_y)\pi_y^y, \qquad y = 0, 1, \ldots, \infty, \quad 0 < \pi_y < 1.$$

(a) Derive an explicit expression for $\theta = \text{pr}(X = Y)$.

(b) The *absolute difference* in the numbers of AIDS cases that will be detected yearly in both counties is the random variable $U = |X - Y|$. Derive an explicit expression for $p_U(u)$, the probability distribution of the random variable $U$.

(c) For a particular year, suppose that the observed values of $X$ and $Y$ are $x = 9$ and $y = 7$. Provide a quantitative answer regarding the question of whether or not these observed values of $X$ and $Y$ provide statistical evidence that $\pi_x \neq \pi_y$. For your calculations, you may assume that $\pi_x \leq 0.10$ and that $\pi_y \leq 0.10$.

**Exercise 3.30*.** Let the random variable $Y$ denote the number of Lyme disease cases that develop in the state of NC during any one calendar year. The event $Y = 0$ is *not* observable since the observational apparatus (i.e., diagnosis) is activated only when $Y > 0$. Since Lyme disease is a rare disease, it seems appropriate to model the distribution of $Y$ by the zero-truncated Poisson distribution (ZTPD)

$$p_Y(y) = \frac{\left(e^{\theta} - 1\right)^{-1} \theta^y}{y!}, \quad y = 1, 2, \ldots, \infty,$$

where $\theta(>0)$ is called the "incidence parameter."

(a) Find an explicit expression for

$$\psi(t) = \mathrm{E}\left[(t+1)^Y\right].$$

(b) Use $\psi(t)$ to show that

$$\mathrm{E}(Y) = \frac{\theta e^{\theta}}{(e^{\theta}-1)} \quad \text{and} \quad \mathrm{V}(Y) = \frac{\theta e^{\theta}(e^{\theta}-\theta-1)}{(e^{\theta}-1)^2}.$$

(c) To lower the incidence of Lyme disease in NC, the state health department mounts a vigorous media campaign to educate NC residents about all aspects of Lyme disease (including information about preventing and dealing with tick bites, using protective measures such as clothing and insect repellents, recognizing symptoms of Lyme disease, treating Lyme disease, etc.) Assume that this media campaign has the desired effect of lowering $\theta$ to $\pi\theta$, where $0 < \pi < 1$. Let $Z$ be the number of Lyme disease cases occurring during a 1-year period *after* the media campaign is over. Assume that

$$p_Z(z) = \frac{(\pi\theta)^z e^{-\pi\theta}}{z!}, \quad z = 0, 1, \ldots, \infty,$$

and that $Y$ and $Z$ are independent random variables.

There is interest in the random variable $X = (Y + Z)$, the *total* number of Lyme disease cases that occur altogether (namely, 1 year before and 1 year after the media campaign). Find an explicit expression for

$$p_X(x) = \mathrm{pr}(X = x) = \mathrm{pr}[(Y + Z) = x], \quad x = 1, 2, \ldots, \infty.$$

(d) Find $\mathrm{E}(X)$ and $\mathrm{V}(X)$.

**Exercise 3.31***. For patients receiving a double kidney transplant, let $X_i$ be the lifetime (in months) of the $i$th kidney, $i = 1, 2$. Also, assume that the density function of $X_i$ is negative exponential with mean $\alpha^{-1}$, namely,

$$f_{X_i}(x_i) = \alpha e^{-\alpha x_i}, \quad x_i > 0,\ \alpha > 0,\ i = 1, 2,$$

and further assume that $X_1$ and $X_2$ are independent random variables. As soon as one of the two kidneys fails, the lifetime $Y$ (in months) of the remaining functional kidney follows the conditional density function

$$f_Y(y|U = u) = \beta e^{-\beta(y-u)}, \quad 0 < u < y < \infty,\ \beta > 2\alpha,$$

where $U = \min(X_1, X_2)$.

(a) Show that the probability that both organs are still functioning at time $t$ is equal to

$$\pi_2(t) = e^{-2\alpha t}, \quad t \geq 0.$$

(b) Show that the probability that exactly one organ is still functioning at time $t$ is equal to

$$\pi_1(t) = \frac{2\alpha}{(\beta - 2\alpha)}\left(e^{-2\alpha t} - e^{-\beta t}\right), \quad t \geq 0.$$

(c) Using the results in parts (a) and (b), develop an explicit expression for $f_T(t)$, the density function of the length of life $T$ (in months) of the two-kidney system [i.e., $T$ is the length of time (in months) until both kidneys have failed].

(d) Develop an explicit expression for the marginal distribution $f_Y(y)$ of the random variable $Y$. How are the random variables $T$ and $Y$ related? Also, find explicit expressions for the expected value and variance of the length of life of the two-kidney system.

**Exercise 3.32*.** Let $X_1, X_2, \ldots, X_n$ constitute a random sample of size $n(>3)$ from a $N(\mu, \sigma^2)$ parent population. Further, define

$$\bar{X} = n^{-1}\sum_{i=1}^{n} X_i, \quad S^2 = (n-1)^{-1}\sum_{i=1}^{n}(X_i - \bar{X})^2 \quad \text{and} \quad T_{(n-1)} = \frac{\bar{X} - \mu}{S/\sqrt{n}}.$$

(a) Develop an explicit expression for $\text{corr}[\bar{X}, T_{(n-1)}]$. Find the numerical value of this correlation when $n = 4$ and when $n = 6$.

(b) Using the fact that $\Gamma(x) \approx \sqrt{2\pi} e^{-x} x^{(x-1/2)}$ for large $x$, find the limiting value of $\text{corr}[\bar{X}, T_{(n-1)}]$ as $n \to \infty$, and then interpret this limit in a meaningful way.

**Exercise 3.33*.** Suppose that there are three identical looking die. Two of these three die are perfectly balanced, so that the probability is $\frac{1}{6}$ of obtaining any one of the six numbers 1, 2, 3, 4, 5, and 6. The third die is an unbalanced die. For this unbalanced die, the probability of obtaining a 1 is equal to $\left(\frac{1}{6} - \epsilon\right)$ and the probability of obtaining a 6 is equal to $\left(\frac{1}{6} + \epsilon\right)$, where $\epsilon, 0 < \epsilon < \frac{1}{6}$, has a known value; for this unbalance die, the probability is $\frac{1}{6}$ of obtaining any of the remaining numbers 2, 3, 4, and 5.

In a simple attempt to identify which of these die is the unbalanced one, it is decided that each of the three die will be tossed $n$ times, and then that die producing the smallest number of ones in $n$ tosses will be identified as the unbalanced die. Develop an expression (which may involve summation signs) that can be used to find the minimum value of $n$ (say, $n^*$) required so that the probability of correctly identifying the unbalanced die will be at least 0.99.

**Exercise 3.34***

(a) If $X_1$ and $X_2$ are i.i.d. random variables, each with the same CDF

$$F_{X_i}(x_i) = \exp(-e^{-x_i}), \qquad -\infty < x_i < +\infty, \quad i = 1, 2,$$

prove that the random variable $Y = (X_1 - X_2)$ has CDF

$$F_Y(y) = (1 + e^{-y})^{-1}, \quad -\infty < y < +\infty.$$

(b) In extreme value theory, under certain validating conditions, the largest observation $X_{(n)}$ in a random sample $X_1, X_2, \ldots, X_n$ of size $n$ has a CDF which can be approximated for large $n$ by the expression

$$F_{X_{(n)}}(x_{(n)}) = \exp\{-\exp[-n\theta(x_{(n)} - \beta)]\}.$$

The parameters $\theta$ ($\theta > 0$) and $\beta(-\infty < \beta < +\infty)$ depend on the structure of the population being sampled. Using this large-sample approximation and the result from part (a), find an explicit expression for a random variable $U = g[X_{1(m)}, X_{2(m)}]$ such that $\text{pr}(\theta \leq U) \doteq (1 - \alpha), 0 < \alpha < 1$. Assume that there is a random sample of size $2m$ ($m$ large) available that has been selected from a population of unspecified structure satisfying the validating conditions, and consider the random variable

$$m\theta\left[X_{1(m)} - \beta\right] - m\theta\left[X_{2(m)} - \beta\right],$$

where $X_{1(m)}$ is the largest observation in the first set of $m$ observations and where $X_{2(m)}$ is the largest observation in the second set of $m$ observations.

**Exercise 3.35*.** Let $X_1, X_2, \ldots, X_n$ constitute a random sample of size $n$ from the density function $f_X(x), -\infty < x < +\infty$.

(a) For $i = 1, 2, \ldots, n$, let $U_i = F_X(X_i)$, where $F_X(x) = \int_{-\infty}^{x} f_X(t)\, dt$. Find the distribution of the random variable $U_i$. The transformation $U_i = F_X(X_i)$ is called the *Probability Integral Transformation.*

(b) Let $U_{(1)}, U_{(2)}, \ldots, U_{(n)}$ be the $n$ order statistics corresponding to the i.i.d. random variables $U_1, U_2, \ldots, U_n$. For $1 \leq r < s \leq n$, prove that the random variable

$$V_{rs} = [U_{(s)} - U_{(r)}] \sim \text{BETA}(\alpha = s - r, \beta = n - s + r + 1).$$

(c) For $0 < \theta < 1$ and $0 < \text{p} < 1$, consider the probability statement

$$\begin{aligned}\theta &= \text{pr}(V_{rs} \geq \text{p}) = \text{pr}\{[U_{(s)} - U_{(r)}] \geq \text{p}\} \\ &= \text{pr}\left[F_X(X_{(s)}) - F_X(X_{(r)}) \geq \text{p}\right].\end{aligned}$$

The random interval $[X_{(r)}, X_{(s)}]$ is referred to as a $100\theta$ percent *tolerance interval* for the density function $f_X(x)$. More specifically, this random interval has probability $\theta$ of containing *at least* a proportion p of the total area (equal to 1) under $f_X(x)$, regardless of the particular structure of $f_X(x)$. As an example, find the numerical value of $\theta$ when $n = 10$, $r = 1$, $s = 10$, and $\text{p} = 0.80$.

**Exercise 3.36*.** Clinical studies where several clinics participate, using a standardized protocol, in the evaluation of new drug therapies have become quite common. In what follows, assume that a statistical design is being used for which patients who meet protocol requirements are each randomly assigned to one of $t$ new drug therapies and to one of $c$ clinics, where the $c$ clinics participating in a particular study can be considered to represent a random sample from a conceptually very large population of clinics that might use the new drug therapies.

For $i = 1, 2, \ldots, t$, $j = 1, 2, \ldots, c$, and $k = 1, 2, \ldots, n_{ij}$, consider the linear model

$$Y_{ijk} = \mu_i + \beta_j + \gamma_{ij} + \epsilon_{ijk},$$

where $Y_{ijk}$ is a continuous random variable representing the response to the $i$th drug therapy of the $k$th patient at the $j$th clinic, $\mu_i$ is the *fixed* average effect of the $i$th drug therapy, $\beta_j$ is a random variable representing the *random* effect of the $j$th clinic, $\gamma_{ij}$ is a random variable representing the *random* effect due to the interaction between the $i$th drug therapy and the $j$th clinic, and $\epsilon_{ijk}$ is a random variable representing the *random* effect of the $k$th patient receiving the $i$th drug therapy at the $j$th clinic. The random variables $\beta_j$, $\gamma_{ij}$, and $\epsilon_{ijk}$ are assumed to be mutually independent random variables for all $i, j$, and $k$, each with an expected value equal to 0 and with respective variances equal to $\sigma^2_\beta, \sigma^2_\gamma$, and $\sigma^2_\epsilon$.

(a) Develop an explicit expression for $\text{V}(Y_{ijk})$, the variance of $Y_{ijk}$.

(b) Develop an explicit expression for the covariance between the responses of two different patients receiving the same drug therapy at the same clinic.

(c) Develop an explicit expression for the covariance between the responses of two different patients receiving different drug therapies at the same clinic.

(d) For $i = 1, 2, \ldots, t$, let $\bar{Y}_{ij} = n_{ij}^{-1} \sum_{k=1}^{n_{ij}} Y_{ijk}$ be the mean of the $n_{ij}$ responses for patients receiving drug therapy $i$ at clinic $j$. Develop explicit expressions for $\text{E}(\bar{Y}_{ij})$, for $\text{V}(\bar{Y}_{ij})$, and for $\text{cov}\left(\bar{Y}_{ij}, \bar{Y}_{i'j}\right)$ when $i \neq i'$.

(e) Let

$$L = \sum_{i=1}^{t} a_i \bar{Y}_i,$$

where the $\{a_i\}_{i=1}^{t}$ are a set of known constants satisfying the constraint $\sum_{i=1}^{t} a_i = 0$ and where $\bar{Y}_i = c^{-1} \sum_{j=1}^{c} \bar{Y}_{ij}$. Develop explicit general expressions for $\text{E}(L)$ and for $\text{V}(L)$. For the special case when $a_1 = +1, a_2 = -1, a_3 = a_4 = \cdots = a_t = 0$, how do the general expressions for $\text{E}(L)$ and $\text{V}(L)$ simplify? More generally, comment on why $L$ can be considered to be an important random variable when analyzing data from multicenter clinical studies that simultaneously evaluate several drug therapies.

**Exercise 3.37*.** Let $X_1, X_2, \ldots, X_n$ constitute a random sample of size $n(>1)$ from a parent population of *unspecfied structure*, where $\text{E}(X_i) = \mu$, $\text{V}(X_i) = \sigma^2$, and $\text{E}[(X_i - \mu)^4] = \mu_4, i = 1, 2, \ldots, n$. Define the sample mean and the sample variance, respectively, as

$$\bar{X} = n^{-1} \sum_{i=1}^{n} X_i \quad \text{and} \quad S^2 = (n-1)^{-1} \sum_{i=1}^{n} (X_i - \bar{X})^2.$$

(a) Prove that

$$\text{V}(S^2) = \frac{1}{n}\left[\mu_4 - \left(\frac{n-3}{n-1}\right)\sigma^4\right].$$

(b) How does the general expression in part (a) simplify if the parent population is $POI(\lambda)$ and if the parent population is $N(\mu, \sigma^2)$?

**Exercise 3.38***. Let $X_1, X_2, \ldots, X_n$ constitute a random sample of size $n$ from a parent population of *unspecified structure*, where $E(X_i) = \mu, V(X_i) = \sigma^2$, and $E[(X_i - \mu)^3] = \mu_3, i = 1, 2, \ldots, n$. Define the sample mean and the sample variance, respectively, as

$$\bar{X} = n^{-1} \sum_{i=1}^{n} X_i \quad \text{and} \quad S^2 = (n-1)^{-1} \sum_{i=1}^{n} (X_i - \bar{X})^2.$$

(a) Show that $\text{cov}(\bar{X}, S^2)$ can be written as an explicit function of $n$ and $\mu_3$.

(b) Suppose that $X_1$ and $X_2$ constitute a random sample of size $n = 2$ from the parent population

$$p_X(x) = \left(\frac{1}{4}\right)^{|x|} \left(\frac{1}{2}\right)^{1-|x|}, \quad x = -1, 0, 1.$$

Show directly that $\text{cov}(\bar{X}, S^2) = 0$, but that $\bar{X}$ and $S^2$ are *dependent* random variables. Comment on this finding relative to the general result developed in part (a).

## SOLUTIONS

### Solution 3.1

(a) The joint distribution of X and Y is

$$p_{X,Y}(x, y) = \text{pr}(X = x)\text{pr}(Y = y | X = x) = \left(\frac{1}{N}\right)\left(\frac{1}{N-1}\right),$$
$$x = 1, 2, \ldots, N \quad \text{and} \quad y = 1, 2, \ldots, N \text{ with } x \neq y.$$

Hence, the marginal distribution of X is

$$p_X(x) = \sum_{\text{all } y, y \neq x} p_{X,Y}(x, y) = \frac{(N-1)}{N(N-1)} = \frac{1}{N}, \quad x = 1, 2, \ldots, N.$$

Analogously, the marginal distribution of Y is

$$p_Y(y) = \frac{1}{N}, \quad y = 1, 2, \ldots, N.$$

(b)

$$
\begin{aligned}
&\text{pr}[X \geq (N-1)|Y = y] \\
&= \frac{\text{pr}\{[X \geq (N-1)] \cap (Y = y)\}}{\text{pr}(Y = y)} \\
&= \frac{\text{pr}[(X = N-1) \cap (Y = y)] + \text{pr}[(X = N) \cap (Y = y)]}{1/N} \\
&= \begin{cases} \frac{2}{N-1}, & y = 1, 2, \ldots, (N-2); \\ \frac{1}{N-1}, & y = (N-1), N. \end{cases}
\end{aligned}
$$

(c)

$$
\text{E}(X) = \text{E}(Y) = \frac{1}{N}\sum_{i=1}^{N} i = \frac{N(N+1)/2}{N} = \frac{(N+1)}{2}.
$$

$$
\begin{aligned}
\text{V}(X) = \text{V}(Y) &= \frac{1}{N}\sum_{i=1}^{N} i^2 - \left[\frac{(N+1)}{2}\right]^2 \\
&= \frac{N(N+1)(2N+1)/6}{N} - \frac{(N+1)^2}{4} = \frac{(N^2-1)}{12}.
\end{aligned}
$$

Now, $\text{p}_Y(y|X = x) = 1/(N-1)$, $y = 1, 2, \ldots, N$ with $y \neq x$.
So,

$$
\text{E}(Y|X = x) = \left[\sum_{y=1}^{N} \frac{y}{(N-1)}\right] - \frac{x}{(N-1)} = \frac{N(N+1)}{2(N-1)} - \frac{x}{(N-1)}.
$$

So,

$$
\begin{aligned}
\text{E}(XY) &= \text{E}_x\{\text{E}(XY|X = x)\} = \text{E}_x\{x\text{E}(Y|X = x)\} \\
&= \text{E}_x\left\{\frac{N(N+1)}{2(N-1)}x - \frac{x^2}{(N-1)}\right\} \\
&= \frac{N(N+1)}{2(N-1)}\text{E}(X) - \frac{\text{E}(X^2)}{(N-1)} \\
&= \frac{N(N+1)^2}{4(N-1)} - \frac{\left[(N^2-1)/12 + (N+1)^2/4\right]}{(N-1)} \\
&= \frac{(N+1)(3N+2)}{12}.
\end{aligned}
$$

So,

$$
\text{cov}(X, Y) = \frac{(N+1)(3N+2)}{12} - \frac{(N+1)^2}{4} = \frac{-(N+1)}{12}.
$$

Hence,

$$\begin{aligned}\text{corr}(X,Y) &= \frac{\text{cov}(X,Y)}{\sqrt{\text{V}(X)\text{V}(Y)}}\\ &= \frac{-(N+1)/12}{\sqrt{[(N^2-1)/12][(N^2-1)/12]}}\\ &= \frac{-1}{(N-1)}.\end{aligned}$$

As $N \to \infty$, $\text{corr}(X,Y) \to 0$ as expected, since the population of balls is becoming infinitely large.

**Solution 3.2**

(a) The most direct approach is to list all the $2^5 = 32$ possible sequences and their associated individual probabilities of occurring. If we let

$$\pi_{ml} = \text{pr}[(M_5 = m) \cap (L_5 = l)], \qquad m = 0,1,\ldots,5 \quad \text{and} \quad l = 0,1,\ldots,5,$$

it then follows directly that

$$\pi_{00} = (1-\pi)^5, \pi_{11} = [\pi^3(1-\pi)^2 + 6\pi^2(1-\pi)^3 + 5\pi(1-\pi)^4],$$

$$\pi_{22} = [\pi^4(1-\pi) + 4\pi^2(1-\pi)^3], \pi_{33} = 3\pi^3(1-\pi)^2,$$

$$\pi_{44} = 2\pi^4(1-\pi), \pi_{55} = \pi^5, \pi_{12} = 6\pi^3(1-\pi)^2, \pi_{13} = 2\pi^4(1-\pi),$$

and $\pi_{ml} = 0$ otherwise.

(b) With $\text{p}_{L_5}(l) = \text{pr}(L_5 = l) = \pi_l, l = 0,1,\ldots,5$, then

$$\pi_0 = (1-\pi)^5, \pi_1 = [\pi^3(1-\pi)^2 + 6\pi^2(1-\pi)^3 + 5\pi(1-\pi)^4],$$

$$\pi_2 = [\pi^4(1-\pi) + 4\pi^2(1-\pi)^3 + 6\pi^3(1-\pi)^2,$$

$$\pi_3 = [3\pi^3(1-\pi)^2 + 2\pi^4(1-\pi)],$$

$$\pi_4 = 2\pi^4(1-\pi), \quad \text{and} \quad \pi_5 = \pi^5.$$

Thus,

$$\text{E}(L_5) = \sum_{l=0}^{5} l\pi_l = 5\pi(1-\pi)^4 + 14\pi^2(1-\pi)^3 + 22\pi^3(1-\pi)^2 + 16\pi^4(1-\pi) + 5\pi^5.$$

When $\pi = 0.90$, then $\text{E}(L_5) = 4.1745$. For further details, see Makri, Philippou, and Psillakis (2007).

**Solution 3.3**

(a) Since $E(U) = (n+1)/2$, it follows from conditional expectation theory that $E(X) = E_u[E(X|U=u)] = E_u(u) = E(U) = (n+1)/2$. Completely analogously, $E(Y) = (n+1)/2$. Also, $V(X|U=u) = (1-u^{-1})/(u^{-1})^2 = u(u-1)$, so that $V(X) = V_u[E(X|U=u)] + E_u[V(X|U=u)] = V_u(u) + E_u[u(u-1)] = V(U) + E(U^2) - E(U) = 2V(U) + [E(U)]^2 - E(U) = \frac{5}{12}(n^2-1)$. Completely analogously, $V(Y) = \frac{5}{12}(n^2-1)$. Also, $E(XY) = E_u[E(XY|U=u)] = E_u[E(X|U=u)E(Y|U=u)] = E_u(u^2) = V(U) + [E(U)]^2 = (n+1)(2n+1)/6$.

Thus, based on the above results,

$$\text{corr}(X,Y) = \frac{E(XY) - E(X)E(Y)}{\sqrt{V(X)V(Y)}} = \frac{1}{5}.$$

(b) Now,

$$\begin{aligned}
\text{pr}(X \neq Y) &= 1 - \text{pr}(X = Y) \\
&= 1 - \sum_{u=1}^{n} \text{pr}(X = Y|U = u)\text{pr}(U = u) \\
&= 1 - \frac{1}{n}\sum_{u=1}^{n} \text{pr}(X = Y|U = u).
\end{aligned}$$

And,

$$\begin{aligned}
\text{pr}(X = Y|U = u) &= \sum_{k=1}^{\infty} \text{pr}(X = k|U = u)\text{pr}(Y = k|U = u) \\
&= \sum_{k=1}^{\infty} u^{-1}(1-u^{-1})^{k-1}u^{-1}(1-u^{-1})^{k-1} \\
&= u^{-2}\sum_{k=1}^{\infty} [(1-u^{-1})^2]^{k-1} \\
&= u^{-2}\left[\frac{1}{1-(1-u^{-1})^2}\right] = \frac{1}{2u-1}.
\end{aligned}$$

So,

$$\text{pr}(X \neq Y) = 1 - \frac{1}{n}\sum_{u=1}^{n} \frac{1}{2u-1}.$$

And, when $n = 4$, $\text{pr}(X \neq Y) = 0.581$.

**Solution 3.4**

(a) First, the conditional density function of $T_\nu$ given $U = u$ is $\mathrm{N}(0, \nu/u)$. Since $U \sim \chi^2_\nu = \mathrm{GAMMA}(\alpha = 2, \beta = \frac{\nu}{2})$, we have

$$\begin{aligned}
f_{T_\nu}(t_\nu) &= \int_0^\infty f_{T_\nu,U}(t_\nu, u)\, du = \int_0^\infty f_{T_\nu}(t_\nu|U = u) f_U(u)\, du \\
&= \int_0^\infty \frac{u^{1/2}}{\sqrt{2\pi\nu}} e^{-ut_\nu^2/2\nu} \cdot \frac{u^{\frac{\nu}{2}-1} e^{-u/2}}{\Gamma\left(\frac{\nu}{2}\right) 2^{\nu/2}} du \\
&= \frac{1}{\sqrt{2\pi\nu}\Gamma\left(\frac{\nu}{2}\right) 2^{\nu/2}} \int_0^\infty u^{\left(\frac{\nu+1}{2}\right)-1} e^{-(t_\nu^2/2\nu + 1/2)u}\, du \\
&= \frac{\Gamma[(\nu+1)/2] \left(\frac{t_\nu^2}{2\nu} + \frac{1}{2}\right)^{-[(\nu+1)/2]}}{\sqrt{2\pi\nu}\Gamma\left(\frac{\nu}{2}\right) 2^{\nu/2}} \\
&= \frac{\Gamma[(\nu+1)/2]}{\sqrt{\pi\nu}\Gamma\left(\frac{\nu}{2}\right)} \left(1 + \frac{t_\nu^2}{\nu}\right)^{-\left(\frac{\nu+1}{2}\right)}, \quad -\infty < t_\nu < \infty.
\end{aligned}$$

(b) Since $Z$ and $U$ are independent random variables, we have

$$\mathrm{E}(T_\nu) = \sqrt{\nu}\mathrm{E}\left(ZU^{-1/2}\right) = \sqrt{\nu}\mathrm{E}(Z)\mathrm{E}\left(U^{-1/2}\right) = 0 \quad \text{since } \mathrm{E}(Z) = 0.$$

And,

$$\begin{aligned}
\mathrm{V}(T_\nu) = \mathrm{E}(T_\nu^2) &= \nu\mathrm{E}\left(Z^2U^{-1}\right) = \nu\mathrm{E}(Z^2)\mathrm{E}(U^{-1}) \\
&= \nu(1)\mathrm{E}(U^{-1}) = \nu\mathrm{E}(U^{-1}).
\end{aligned}$$

Since $U \sim \mathrm{GAMMA}(\alpha = 2, \beta = \nu/2)$, we know that

$$\mathrm{E}(U^r) = \frac{\Gamma(\beta + r)}{\Gamma(\beta)}\alpha^r = \frac{\left(\frac{\nu}{2} + r\right)}{\Gamma\left(\frac{\nu}{2}\right)} 2^r, \left(\frac{\nu}{2} + r\right) > 0.$$

Finally, with $r = -1$,

$$\mathrm{V}(T_\nu) = \nu\frac{\Gamma\left(\frac{\nu}{2} - 1\right)}{\Gamma\left(\frac{\nu}{2}\right)} 2^{-1} = \nu\frac{\Gamma\left(\frac{\nu}{2} - 1\right)}{\left(\frac{\nu}{2} - 1\right)\Gamma\left(\frac{\nu}{2} - 1\right)} 2^{-1} = \frac{\nu}{(\nu - 2)}, \nu > 2.$$

**Solution 3.5**

(a) First, using conditional expectation theory, we have

$$\mathrm{E}(Y) = \mathrm{E}_x[\mathrm{E}(Y|X = x)] = \mathrm{E}_x[\beta_0 + \beta_1 x] = \beta_0 + \beta_1\mathrm{E}(X).$$

And, since

$$E(XY) = E_x[E(XY|X = x)] = E_x[xE(Y|X = x)]$$
$$= E_x[x(\beta_0 + \beta_1 x)] = \beta_0 E(X) + \beta_1 E(X^2),$$

we have

$$\text{cov}(X, Y) = \beta_0 E(X) + \beta_1 E(X^2) - E(X)[\beta_0 + \beta_1 E(X)] = \beta_1 V(X).$$

Thus,

$$\text{corr}(X, Y) = \frac{\beta_1 V(X)}{\sqrt{V(X)V(Y)}} = \beta_1 \sqrt{\frac{V(X)}{V(Y)}}.$$

Thus, $\text{corr}(X, Y) = k\beta_1$, where $k = \sqrt{V(X)/V(Y)} > 0$. In particular, when $\beta_1 = 0$, indicating no linear relationship between $X$ and $Y$ in the sense that $E(Y|X = x) = \beta_0$ does not depend on $x$, then $\text{corr}(X, Y) = 0$. When $\beta_1 < 0$, then $\text{corr}(X, Y) < 0$. And, when $\beta_1 > 0$, then $\text{corr}(X, Y) > 0$. In general, $\text{corr}(X, Y)$ is reflecting the strength of the *linear*, or straight-line, relationship between $X$ and $Y$.

(b) When $E(X|Y = y) = \alpha_0 + \alpha_1 y$, it follows, using arguments identical to those used in part (a), that

$$\text{corr}(X, Y) = \frac{\alpha_1 V(Y)}{\sqrt{V(X)V(Y)}} = \alpha_1 \sqrt{\frac{V(Y)}{V(X)}}.$$

Thus, it follows directly that $\alpha_1$ and $\beta_1$ have the same sign and that $[\text{corr}(X, Y)]^2 = \alpha_1\beta_1$. In particular, when both $\alpha_1$ and $\beta_1$ are negative, then $\text{corr}(X, Y) = -\sqrt{\alpha_1\beta_1}$; and, when both $\alpha_1$ and $\beta_1$ are positive, then $\text{corr}(X, Y) = +\sqrt{\alpha_1\beta_1}$.

(c) If $E(Y|X = x) = \beta_0 + \beta_1 x + \beta_2 x^2$, then

$$E(Y) = E_x[E(Y|X = x)] = E_x[\beta_0 + \beta_1 x + \beta_2 x^2]$$
$$= \beta_0 + \beta_1 E(X) + \beta_2 E(X^2).$$

And, since

$$E(XY) = E_x[E(XY|X = x)] = E_x[xE(Y|X = x)]$$
$$= E_x[x(\beta_0 + \beta_1 x + \beta_2 x^2)] = \beta_0 E(X) + \beta_1 E(X^2) + \beta_2 E(X^3),$$

we have

$$\text{cov}(X, Y) = \beta_0 E(X) + \beta_1 E(X^2) + \beta_2 E(X^3)$$
$$- E(X)[\beta_0 + \beta_1 E(X) + \beta_2 E(X^2)]$$
$$= \beta_1 V(X) + \beta_2[E(X^3) - E(X)E(X^2)].$$

Finally,

$$\text{corr}(X, Y) = \beta_1\sqrt{\frac{\text{V}(X)}{\text{V}(Y)}} + \beta_2\left[\frac{\text{E}(X^3) - \text{E}(X)\text{E}(X^2)}{\sqrt{\text{V}(X)\text{V}(Y)}}\right].$$

Thus, unless $\beta_2 = 0$ or unless $\text{E}(X^3) = \text{E}(X)\text{E}(X^2)$, the direct connection between $\text{corr}(X, Y)$ and $\beta_1$ is lost.

**Solution 3.6**

(a)

$$\text{f}_X(x) = 6\theta^{-3}\int_0^x (x - y)\,dy = 6\theta^{-3}\left(x^2 - \frac{x^2}{2}\right) = \frac{3x^2}{\theta^3}, \quad 0 < x < \theta.$$

So,

$$\text{E}(X^r) = \int_0^\theta x^r \cdot 3\theta^{-3}x^2\,dx = \frac{3\theta^r}{(r+3)}, \quad r = 1, 2, \ldots;$$

thus,

$$\text{E}(X) = \frac{3\theta}{4}, \ \text{E}(X^2) = \frac{3\theta^2}{5}, \quad \text{and} \quad \text{V}(X) = \frac{3\theta^2}{5} - \left(\frac{3\theta}{4}\right)^2 = \frac{3\theta^2}{80}.$$

Now,

$$\text{f}_Y(y|X = x) = \frac{\text{f}_{X,Y}(x, y)}{\text{f}_X(x)} = \frac{2(x - y)}{x^2}, \quad 0 < y < x.$$

So,

$$\text{E}(Y^r|X = x) = \int_0^x y^r\left(\frac{2}{x} - \frac{2y}{x^2}\right)dy = \frac{2x^r}{(r+1)(r+2)}, \quad r = 1, 2, \ldots$$

So,

$$\text{E}(Y|X = x) = \frac{x}{3} \text{ (which is a linear function of } x\text{)}.$$

Also,

$$\text{E}(Y^2|X = x) = \frac{x^2}{6}, \text{ so that } \text{V}(Y|X = x) = \frac{x^2}{6} - \left(\frac{x}{3}\right)^2 = \frac{x^2}{18}.$$

Since $\text{corr}(X, Y) = \left(\frac{1}{3}\right)\sqrt{\text{V}(X)/\text{V}(Y)}$, we need $\text{V}(Y)$.

Now,

$$\begin{aligned}\text{V}(Y) &= \text{V}_x[\text{E}(Y|X = x)] + \text{E}_x[\text{V}(Y|X = x)]\\ &= \text{V}\left(\frac{X}{3}\right) + \text{E}\left(\frac{X^2}{18}\right)\\ &= \frac{3\theta^2}{80} = \text{V}(X), \text{ so that } \text{corr}(X, Y) = \frac{1}{3}.\end{aligned}$$

Equivalently,

$$\mathrm{E}(Y) = \mathrm{E}_x[\mathrm{E}(Y|X=x)] = \mathrm{E}\left(\frac{X}{3}\right) = \frac{\theta}{4}$$

and

$$\mathrm{E}(XY) = \mathrm{E}_x[x\mathrm{E}(Y|X=x)] = \mathrm{E}\left(\frac{X^2}{3}\right) = \frac{\theta^2}{5},$$

so that

$$\begin{aligned}\mathrm{cov}(X,Y) &= \mathrm{E}(XY) - \mathrm{E}(X)\mathrm{E}(Y)\\ &= \frac{\theta^2}{5} - \left(\frac{3\theta}{4}\right)\left(\frac{\theta}{4}\right) = \frac{\theta^2}{80}.\end{aligned}$$

Hence,

$$\mathrm{corr}(X,Y) = \frac{\mathrm{cov}(X,Y)}{\sqrt{\mathrm{V}(X)\mathrm{V}(Y)}} = \frac{\theta^2/80}{\sqrt{(3\theta^2/80)(3\theta^2/80)}} = \frac{1}{3} \text{ as before.}$$

(b)

$$\mathrm{pr}\left[(X+Y) < \theta \middle| (X+2Y) > \frac{\theta}{4}\right] = \frac{\mathrm{pr}\left\{[(X+Y) < \theta] \cap \left[(X+2Y) > \frac{\theta}{4}\right]\right\}}{\mathrm{pr}\left[(X+2Y) > \frac{\theta}{4}\right]}.$$

So,

$$\begin{aligned}&\mathrm{pr}\left\{[(X+Y) < \theta] \cap \left[(X+2Y) > \frac{\theta}{4}\right]\right\}\\ &= \int_{\frac{\theta}{12}}^{\frac{\theta}{4}} \int_{\left(\frac{\theta-4x}{8}\right)}^{x} \mathrm{f}_{X,Y}(x,y)\,dy\,dx + \int_{\frac{\theta}{4}}^{\frac{\theta}{2}} \int_0^x \mathrm{f}_{X,Y}(x,y)\,dy\,dx\\ &\quad + \int_{\frac{\theta}{2}}^{\theta} \int_0^{(\theta-x)} \mathrm{f}_{X,Y}(x,y)\,dy\,dx\\ &= \int_0^{\frac{\theta}{12}} \int_{\left(\frac{\theta-8y}{4}\right)}^{(\theta-y)} \mathrm{f}_{X,Y}(x,y)\,dx\,dy + \int_{\frac{\theta}{12}}^{\frac{\theta}{2}} \int_y^{(\theta-y)} \mathrm{f}_{X,Y}(x,y)\,dx\,dy.\end{aligned}$$

And,

$$\begin{aligned}\mathrm{pr}\left[(X+2Y) > \frac{\theta}{4}\right] &= \int_{\frac{\theta}{12}}^{\frac{\theta}{4}} \int_{\left(\frac{\theta-4x}{8}\right)}^{x} \mathrm{f}_{X,Y}(x,y)\,dy\,dx + \int_{\frac{\theta}{4}}^{\theta} \int_0^x \mathrm{f}_{X,Y}(x,y)\,dy\,dx\\ &= \int_0^{\frac{\theta}{12}} \int_{\left(\frac{\theta-8y}{4}\right)}^{\theta} \mathrm{f}_{X,Y}(x,y)\,dx\,dy + \int_{\frac{\theta}{12}}^{\theta} \int_y^{\theta} \mathrm{f}_{X,Y}(x,y)\,dx\,dy,\end{aligned}$$

where $\mathrm{f}_{X,Y}(x,y) = 6\theta^{-3}(x-y), 0 < y < x < \theta$.

(c) From part (a), we know that $\mathrm{E}(X_i) = \frac{3\theta}{4}, \mathrm{V}(X_i) = \frac{3\theta^2}{80}, \mathrm{E}(Y_i) = \frac{\theta}{4}, \mathrm{V}(Y_i) = \frac{3\theta^2}{80}$, and $\mathrm{cov}(X_i, Y_i) = \frac{\theta^2}{80}$.

So,

$\mathrm{E}(L) = \mathrm{E}(3\bar{X} - 2\bar{Y}) = 3\mathrm{E}(\bar{X}) - 2\mathrm{E}(\bar{Y}) = \frac{7\theta}{4}$, since $\mathrm{E}(\bar{X}) = \frac{1}{n}\sum_{i=1}^{n}\mathrm{E}(X_i)$ and $\mathrm{E}(\bar{Y}) = \frac{1}{n}\sum_{i=1}^{n}\mathrm{E}(Y_i)$.

And,

$\mathrm{V}(L) = \mathrm{V}(3\bar{X} - 2\bar{Y}) = \mathrm{V}\left[(3)\frac{1}{n}\sum_{i=1}^{n} X_i - (2)\frac{1}{n}\sum_{i=1}^{n} Y_i\right] = \frac{1}{n^2}\sum_{i=1}^{n}\mathrm{V}(3X_i - 2Y_i)$, since the pairs are mutually independent.

Now,

$$\begin{aligned}\mathrm{V}(3X_i - 2Y_i) &= 9\mathrm{V}(X_i) + 4\mathrm{V}(Y_i) + 2(3)(-2)\mathrm{cov}(X_i, Y_i)\\ &= 9\left(\frac{3\theta^2}{80}\right) + 4\left(\frac{3\theta^2}{80}\right) - 12\left(\frac{\theta^2}{80}\right) = \frac{27\theta^2}{80}.\end{aligned}$$

Thus,

$$\mathrm{V}(L) = \frac{1}{n^2}\sum_{i=1}^{n}\left(\frac{27\theta^2}{80}\right) = \frac{27\theta^2}{80n}.$$

**Solution 3.7**

(a) Since

$$\begin{aligned}\mathrm{f}_X(x) &= \frac{\Gamma(\alpha+\beta+3)}{\Gamma(\alpha+1)\Gamma(\beta+1)}(1-x)^{\alpha}\int_0^x y^{\beta}\,dy\\ &= \frac{\Gamma(\alpha+\beta+3)}{\Gamma(\alpha+1)\Gamma(\beta+2)}x^{\beta+1}(1-x)^{\alpha}, \quad 0 < x < 1,\end{aligned}$$

it follows that

$$\mathrm{f}_Y(y|X = x) = \frac{\mathrm{f}_{X,Y}(x,y)}{\mathrm{f}_X(x)} = (\beta+1)y^{\beta}x^{-\beta-1}, \quad 0 < y < x < 1.$$

Also,

$$\begin{aligned}\mathrm{f}_Y(y) &= \frac{\Gamma(\alpha+\beta+3)}{\Gamma(\alpha+1)\Gamma(\beta+1)}y^{\beta}\int_y^1 (1-x)^{\alpha}\,dx\\ &= \frac{\Gamma(\alpha+\beta+3)}{\Gamma(\alpha+2)\Gamma(\beta+1)}y^{\beta}(1-y)^{\alpha+1}, \quad 0 < y < 1.\end{aligned}$$

(b) It is clear that $\mathrm{f}_X(x)$ and $\mathrm{f}_Y(y)$ are beta distributions, with variances

$$\mathrm{V}(X) = \frac{(\beta+2)(\alpha+1)}{(\alpha+\beta+3)^2(\alpha+\beta+4)} \quad \text{and} \quad \mathrm{V}(Y) = \frac{(\beta+1)(\alpha+2)}{(\alpha+\beta+3)^2(\alpha+\beta+4)}.$$

And,

$$E(Y|X=x) = \int_0^x y f_Y(y|X=x)\,dy = \frac{(\beta+1)}{x^{\beta+1}}\int_0^x y^{\beta+1}\,dy = \left(\frac{\beta+1}{\beta+2}\right)x.$$

Thus, appealing to the mathematical relationship between the correlation coefficient and the slope for a simple linear (i.e., straightline) regression model, we have

$$\rho_{X,Y} = \left(\frac{\beta+1}{\beta+2}\right)\sqrt{\frac{V(X)}{V(Y)}} = \left[\frac{(\alpha+1)(\beta+1)}{(\alpha+2)(\beta+2)}\right]^{1/2}.$$

When $\alpha = 2$ and $\beta = 3$, $\rho_{X,Y} = 0.7746$.
Alternatively, $\rho_{X,Y}$ can be computed using the formula

$$\rho_{X,Y} = \frac{E(XY) - E(X)E(Y)}{\sqrt{V(X)V(Y)}},$$

where, for example,

$$E(XY) = E_x[E(XY|X=x)] = E_x[xE(Y|X=x)] = \left(\frac{\beta+1}{\beta+2}\right)E(X^2),$$

and

$$E(X^2) = V(X) + [E(X)]^2 = \frac{(\beta+2)(\alpha+1)}{(\alpha+\beta+3)^2(\alpha+\beta+4)} + \left[\frac{\beta+2}{\alpha+\beta+3}\right]^2.$$

(c) Since

$$E(X_i) = E(\bar{X}) = \frac{(\beta+2)}{(\alpha+\beta+3)} = \frac{5}{8} \quad \text{and} \quad E(Y_i) = E(\bar{Y}) = \frac{(\beta+1)}{(\alpha+\beta+3)} = \frac{1}{2},$$

it follows that $E(L) = 3(\frac{5}{8}) - 5(\frac{1}{2}) = -\frac{5}{8}$.
And,

$$V(L) = V(3\bar{X} - 5\bar{Y}) = V\left[\frac{3}{n}\sum_{i=1}^{n} X_i - \frac{5}{n}\sum_{i=1}^{n} Y_i\right] = \frac{1}{n^2}V\left[\sum_{i=1}^{n}(3X_i - 5Y_i)\right]$$

$$= \frac{1}{n}V(3X_i - 5Y_i) = \frac{1}{n}\left[9V(X_i) + 25V(Y_i) - 2(3)(5)\rho_{X,Y}\sqrt{V(X_i)V(Y_i)}\right].$$

When $n = 10, \alpha = 2$, and $\beta = 3$, we then find that

$$V(L) = \frac{1}{10}\left[9\left(\frac{5}{192}\right) + 25\left(\frac{1}{36}\right) - 30(0.7746)\sqrt{\left(\frac{5}{192}\right)\left(\frac{1}{36}\right)}\right] = 0.0304.$$

**Solution 3.8**

(a) Now,

$$\text{pr}[(U \le u) \cap (W=0)] = \text{pr}[(Y_2 \le u) \cap (Y_2 < Y_1)]$$

$$= \int_0^u \int_{y_2}^{\infty} \left(\theta_1 e^{-\theta_1 y_1}\right)\left(\theta_2 e^{-\theta_2 y_2}\right) dy_1\, dy_2$$

$$= \frac{\theta_1}{(\theta_1+\theta_2)}\left[1 - e^{-(\theta_1+\theta_2)u}\right], \quad 0 < u < \infty,\ w = 0.$$

So,

$$f_{U,W}(u,0) = \theta_1 e^{-(\theta_1+\theta_2)u}, \quad 0 < u < \infty,\ w = 0.$$

And,

$$\text{pr}[(U \le u) \cap (W = 1)] = \text{pr}[(Y_1 \le u) \cap (Y_1 < Y_2)]$$

$$= \int_0^u \int_{y_1}^{\infty} \left(\theta_1 e^{-\theta_1 y_1}\right)\left(\theta_2 e^{-\theta_2 y_2}\right) dy_2\, dy_1$$

$$= \frac{\theta_2}{(\theta_1+\theta_2)}\left[1 - e^{-(\theta_1+\theta_2)u}\right],$$

$$0 < u < \infty,\ w = 1.$$

So,

$$f_{U,W}(u,1) = \theta_2 e^{-(\theta_1+\theta_2)u}, \quad 0 < u < \infty,\ w = 1.$$

So, we can compactly combine the above two results notationally as follows:

$$f_{U,W}(u,w) = \theta_1^{(1-w)}\theta_2^{w} e^{-(\theta_1+\theta_2)u}, \quad 0 < u < \infty,\ w = 0, 1.$$

(b) We have

$$p_W(w) = \int_0^{\infty} f_{U,W}(u,w)\, du = \theta_1^{(1-w)}\theta_2^{w} \int_0^{\infty} e^{-(\theta_1+\theta_2)u}\, du$$

$$= \theta_1^{(1-w)}\theta_2^{w}(\theta_1+\theta_2)^{-1} = \left(\frac{\theta_1}{\theta_1+\theta_2}\right)^{(1-w)}\left(\frac{\theta_2}{\theta_1+\theta_2}\right)^{w}, \quad w = 0, 1.$$

(c) We have

$$f_U(u) = \sum_{w=0}^{1} f_{U,W}(u,w) = e^{-(\theta_1+\theta_2)u} \sum_{w=0}^{1} \theta_1^{(1-w)}\theta_2^{w}$$

$$= (\theta_1+\theta_2)e^{-(\theta_1+\theta_2)u}, \quad 0 < u < \infty.$$

(d) Since $f_{U,W}(u,w) = f_U(u)p_W(w), 0 < u < \infty, w = 0, 1$, it follows that $U$ and $W$ are independent random variables.

**Solution 3.9**

(a) First, since $Y_1$ and $Y_2$ have a joint bivariate normal distribution, it follows that the random variable $(Y_1 - Y_2)$ is normally distributed.

Also, $E(Y_1 - Y_2) = E(Y_1) - E(Y_2) = 110 - 100 = 10$. And,

$$V(Y_1 - Y_2) = V(Y_1) + V(Y_2) - 2\rho\sqrt{V(Y_1)V(Y_2)}$$
$$= 225 + 225 - 2(0.80)(15)(15) = 90.$$

Thus, we have $(Y_1 - Y_2) \sim N(10, 90)$.
Hence,

$$\text{pr}(Y_1 - Y_2 > 15) = \text{pr}\left[\frac{(Y_1 - Y_2) - 10}{\sqrt{90}} > \frac{15 - 10}{\sqrt{90}}\right] = \text{pr}(Z > 0.527),$$

where $Z \sim N(0, 1)$, so that $\text{pr}(Y_1 - Y_2 > 15) \approx 0.30$.

So, using the $\text{BIN}(n = 3, \pi = 0.30)$ distribution, the probability that the older child has an IQ at least 15 points higher than the younger child for at least two of three randomly chosen families is equal to

$$C_2^3(0.30)^2(0.70)^1 + C_3^3(0.30)^3(0.70)^0 = 0.216.$$

(b) From general properties of the bivariate normal distribution, we have

$$E(Y_2|Y_1 = y_1) = E(Y_2) + \rho\sqrt{\frac{V(Y_2)}{V(Y_1)}}\left[y_1 - E(Y_1)\right],$$

and

$$V(Y_2|Y_1 = y_1) = V(Y_2)(1 - \rho^2).$$

Also, $Y_2$ given $Y_1 = y_1$ is normally distributed. In our particular situation,

$$E(Y_2|Y_1 = 120) = 100 + (0.80)\sqrt{\frac{225}{225}}(120 - 110) = 108,$$

and

$$V(Y_2|Y_1 = 120) = 225[1 - (0.80)^2] = 81.$$

So,

$$\text{pr}(Y_2 > 120|Y_1 = 120) = \text{pr}\left[\frac{Y_2 - 108}{\sqrt{81}} > \frac{120 - 108}{\sqrt{81}}\right]$$
$$= \text{pr}(Z > 1.333)$$

where $Z \sim N(0, 1)$, so that $\text{pr}(Y_2 > 120|Y_1 = 120) \approx 0.09$.

**Solution 3.10**

(a) Now,

$$F_W(w) = \text{pr}(W \leq w) = \text{pr}[(U - V) \leq w) = \text{E}_v[\text{pr}(U - v \leq w|V = v)]$$
$$= \text{E}_v[\text{pr}(U \leq w + v|V = v)] = \int_{-\infty}^{\infty} F_U(w + v|V = v)\, f_V(v)\, dv,$$

where $f_V(v)$ is the density for $V$. Thus, we obtain

$$F_W(w) = \int_{-\infty}^{\infty} e^{-e^{-(w+v)}} e^{-v} e^{-e^{-v}}\, dv$$
$$= \int_{-\infty}^{\infty} e^{-\left(e^{-w}e^{-v}\right)} e^{-v} e^{-e^{-v}}\, dv$$
$$= \int_{-\infty}^{\infty} e^{-v} e^{-\left[(1+e^{-w})e^{-v}\right]}\, dv.$$

Letting $z = 1 + e^{-w}$, we obtain

$$F_W(w) = \int_{-\infty}^{\infty} e^{-v} e^{-ze^{-v}}\, dv$$
$$= \frac{1}{z} \int_{-\infty}^{\infty} ze^{-v} e^{-ze^{-v}}\, dv = z^{-1}.$$

Thus, $F_W(w) = 1/(1 + e^{-w})$, $-\infty < w < \infty$, and hence $W$ has a logistic distribution.

(b) Now, $\text{pr}(Y=1) = \text{pr}(U > V) = \text{pr}(\alpha + E_1 > E_2) = \text{pr}(E_1 - E_2 > -\alpha) = 1 - F_W(-\alpha)$ $= F_W(\alpha) = 1/(1 + e^{-\alpha})$. This expression is exactly $\text{pr}(Y = 1)$ for an ordinary logistic regression model with a single intercept term $\alpha$. Thus, logistic regression, and, more generally, multinomial logistic regression, can be motivated via a random utility framework, where the utilities involve i.i.d. standard Gumbel error terms. Likewise, probit regression can be motivated by assuming i.i.d. standard normal error terms for the utilities.

**Solution 3.11**

(a) Since $X_1, X_2, \ldots, X_n$ constitute a set of i.i.d. negative exponential random variables with $\text{E}(X_i) = \lambda^{-1}, i = 1, 2, \ldots, n$, it follows directly that $S \sim \text{GAMMA}(\alpha = \lambda^{-1}, \beta = n)$.

Hence, with $s = \sum_{i=1}^{n} x_i$, we have

$$f_{X_1,X_2,\ldots,X_n}(x_1,x_2,\ldots,x_n|S=s) = \frac{f_{X_1,X_2,\ldots,X_n,S}(x_1,x_2,\ldots,x_n,s)}{f_S(s)}$$
$$= \frac{f_{X_1,X_2,\ldots,X_n}(x_1,x_2,\ldots,x_n)}{f_S(s)}$$
$$= \frac{\prod_{i=1}^{n} \lambda e^{-\lambda x_i}}{\lambda^n s^{n-1} e^{-\lambda s}/(n-1)!}$$
$$= \frac{(n-1)!}{s^{n-1}}, \quad x_i > 0, \quad i = 1,2,\ldots,n,$$
$$\text{and } \sum_{i=1}^{n} x_i = s.$$

(b) The inverse functions for this transformation are

$$X_1 = SY_1, X_2 = S(Y_2 - Y_1), \ldots, X_{n-1} = S(Y_{n-1} - Y_{n-2});$$

hence, it follows that the Jacobian is equal to $S^{n-1}$, since it is the determinant of the $(n-1) \times (n-1)$ matrix with $(i,j)$th element equal to $\partial X_i / \partial Y_j, i = 1,2,\ldots,(n-1)$ and $j = 1,2,\ldots,(n-1)$. Thus, using the result from part (a), we have

$$f_{Y_1,Y_2,\ldots,Y_{n-1}}(y_1,y_2,\ldots,y_{n-1}|S=s)$$
$$= \frac{(n-1)!}{s^{n-1}} \left| s^{n-1} \right| = (n-1)!, \quad 0 < y_1 < y_2 < \cdots < y_{n-1} < 1.$$

(c) When $n = 3$,

$$f_{Y_1}(y_1) = \int_{y_1}^{1} (2!)\, dy_2 = 2(1 - y_1), \quad 0 < y_1 < 1.$$

When $n = 4$,

$$f_{Y_1}(y_1) = \int_{y_1}^{1} \int_{y_2}^{1} (3!)\, dy_3\, dy_2 = 3(1 - y_1)^2, \quad 0 < y_1 < 1.$$

In general,

$$f_{Y_1}(y_1) = (n-1)(1 - y_1)^{n-2}, \quad 0 < y_1 < 1.$$

**Solution 3.12** From moment generating function theory, $\mathrm{E}\left(Y_i^r\right) = \mathrm{E}\left(e^{rX_i}\right) = e^{r\mu + r^2\sigma^2/2}$, $-\infty < r < \infty$, since $X_i \sim \mathrm{N}(\mu, \sigma^2)$. So, $\mathrm{E}(Y_i) = e^{\mu + \sigma^2/2}$ and $\mathrm{E}(Y_i^2) = e^{2\mu + 2\sigma^2}$, $i = 1,\ldots,n$; also, $Y_1, Y_2, \ldots, Y_n$ are mutually independent random variables.

So, $\mathrm{E}(\bar{Y}_a) = \mathrm{e}^{\mu+\sigma^2/2}$ and, by mutual independence,

$$\mathrm{V}(\bar{Y}_a) = \frac{1}{n^2}\sum_{i=1}^{n}\mathrm{V}(Y_i) = \frac{\mathrm{e}^{2\mu+2\sigma^2} - \left(\mathrm{e}^{\mu+\frac{\sigma^2}{2}}\right)^2}{n} = \frac{\mathrm{e}^{2\mu+\sigma^2}\left(\mathrm{e}^{\sigma^2}-1\right)}{n}.$$

Also, by mutual independence,

$$\mathrm{E}(\bar{Y}_g) = \prod_{i=1}^{n}\mathrm{E}\left(Y_i^{1/n}\right) = \mathrm{e}^{\mu+\frac{\sigma^2}{2n}} \text{ and } \mathrm{E}\left(\bar{Y}_g^2\right) = \prod_{i=1}^{n}\mathrm{E}\left(Y_i^{2/n}\right)$$
$$= \mathrm{e}^{2\mu+2\sigma^2/n},$$

so that

$$\mathrm{V}(\bar{Y}_g) = \mathrm{e}^{2\mu+2\sigma^2/n} - \left(\mathrm{e}^{\mu+\sigma^2/2n}\right)^2 = \mathrm{e}^{2\mu+\sigma^2/n}\left(\mathrm{e}^{\sigma^2/n}-1\right).$$

Finally,

$$\mathrm{E}(\bar{Y}_a\bar{Y}_g) = \mathrm{E}\left[\frac{1}{n}\sum_{i=1}^{n}Y_i\bar{Y}_g\right] = \frac{1}{n}\sum_{i=1}^{n}\mathrm{E}(Y_i\bar{Y}_g).$$

Now,

$$\mathrm{E}(Y_i\bar{Y}_g) = \mathrm{E}\left[Y_i\left(\prod_{i=1}^{n}Y_i\right)^{1/n}\right] = \mathrm{E}\left[Y_i^{(1+\frac{1}{n})}\cdot\prod_{\text{all} j\neq i}Y_j^{1/n}\right]$$
$$= \mathrm{E}\left(Y_i^{1+\frac{1}{n}}\right)\prod_{\text{all } j\neq i}\mathrm{E}\left(Y_j^{1/n}\right)$$
$$= \mathrm{e}^{\left(\frac{n+1}{n}\right)\mu+\frac{(n+1)^2\sigma^2}{2n^2}}\cdot\left[\mathrm{e}^{\frac{\mu}{n}+\frac{\sigma^2}{2n^2}}\right]^{(n-1)} = \mathrm{e}^{2\mu+\frac{(n+3)\sigma^2}{2n}}.$$

So,

$$\mathrm{corr}(\bar{Y}_a,\bar{Y}_g) = \frac{\mathrm{E}(\bar{Y}_a\bar{Y}_g) - \mathrm{E}(\bar{Y}_a)\mathrm{E}(\bar{Y}_g)}{\sqrt{\mathrm{V}(\bar{Y}_a)\mathrm{V}(\bar{Y}_g)}} = \frac{\mathrm{e}^{2\mu+\frac{(n+3)\sigma^2}{2n}} - \left(\mathrm{e}^{\mu+\frac{\sigma^2}{2}}\right)\left(\mathrm{e}^{\mu+\frac{\sigma^2}{2n}}\right)}{\sqrt{\left[\frac{\mathrm{e}^{2\mu+\sigma^2}\left(\mathrm{e}^{\sigma^2}-1\right)}{n}\right]\left[\mathrm{e}^{2\mu+\frac{\sigma^2}{n}}\left(\mathrm{e}^{\frac{\sigma^2}{n}}-1\right)\right]}}$$

$$= \frac{\mathrm{e}^{2\mu+\left(\frac{n+1}{2n}\right)\sigma^2}\left(n\mathrm{e}^{\frac{\sigma^2}{n}}-n\right)^{1/2}}{\sqrt{\mathrm{e}^{4\mu+\left(\frac{n+1}{n}\right)\sigma^2}\left(\mathrm{e}^{\sigma^2}-1\right)}} = \frac{\left(n\mathrm{e}^{\frac{\sigma^2}{n}}-n\right)^{1/2}}{\sqrt{\mathrm{e}^{\sigma^2}-1}},$$

which does *not* depend on $\mu$.

Since

$$ne^{\sigma^2/n} - n = n\sum_{j=0}^{\infty}\frac{\left(\sigma^2/n\right)^j}{j!} - n = n + \sigma^2 + n\sum_{j=2}^{\infty}\frac{\left(\sigma^2/n\right)^j}{j!} - n = \sigma^2 + \sum_{j=2}^{\infty}\frac{\sigma^{2j}n^{1-j}}{j!},$$

it follows that

$$\lim_{n\to\infty}\left[\text{corr}(\bar{Y}_a, \bar{Y}_g)\right] = \frac{\sigma}{\sqrt{e^{\sigma^2} - 1}},$$

which monotonically goes to 0 as $\sigma^2 \to \infty$. Hence, the larger is $\sigma^2$, the smaller is the correlation.

**Solution 3.13**

(a) For the $i$th group, $i = 1, 2, \ldots, G$, if $Y_i$ denotes the number of blood tests required for Plan #2, then

$$\begin{aligned}\text{E}(Y_i) &= (1)(1-\pi)^n + (n+1)[1-(1-\pi)^n] \\ &= (n+1) - n(1-\pi)^n.\end{aligned}$$

Then, since $T_2 = \sum_{i=1}^{G} Y_i$, it follows that

$$\begin{aligned}\text{E}(T_2) &= \sum_{i=1}^{G}\text{E}(Y_i) = G[(n+1) - n(1-\pi)^n] \\ &= N + G[1 - n(1-\pi)^n].\end{aligned}$$

(b) For $N - \text{E}(T_2) = G[n(1-\pi)^n - 1] > 0$, we require $n(1-\pi)^n > 1$, or equivalently,

$$\frac{\ln(n)}{n} > \ln\left(\frac{1}{1-\pi}\right).$$

Now, it is clear that we want to pick the largest value of $n$, say $n^*$, that maximizes the quantity $\ln(n)/n$, thus providing the desired largest value of $\pi$, say $\pi^*$, for which $\text{E}(T_2) < N$. It is straightforward to show that $n^* = 3$, which then gives $\pi^*$=0.3066. So, if we use groups of size three, then the expected number of blood tests required under Plan #2 will be smaller than the number $N$ of blood tests required under Plan #1 for all values of $\pi$ less than 0.3066.

**Solution 3.14.** Since the conditional distribution of $Y_{\text{p}}$, given $Y = y$, is BIN($y, \pi$), we have

$$\begin{aligned}\text{M}_{Y_{\text{p}},Y_{\bar{\text{p}}}}(s, t|Y = y) &= \text{E}\left[e^{sY_{\text{p}}+tY_{\bar{\text{p}}}}|Y = y\right] = \text{E}\left[e^{sY_{\text{p}}+t(Y-Y_{\text{p}})}|Y = y\right] \\ &= e^{ty}\text{E}\left[e^{(s-t)Y_{\text{p}}}|Y = y\right] = e^{ty}\left[\pi e^{(s-t)} + (1-\pi)\right]^y \\ &= \left[\pi e^s + (1-\pi)e^t\right]^y.\end{aligned}$$

Hence, letting $\theta = \left[\pi e^s + (1-\pi)e^t\right]$ and recalling that $Y \sim \text{POI}(L\lambda)$, we have

$$\begin{aligned}
M_{Y_p,Y_{\bar{p}}}(s,t) &= E_y\left[M_{Y_p,Y_{\bar{p}}}(s,t|Y=y)\right] = E(\theta^Y) \\
&= \sum_{y=0}^{\infty}(\theta^y)\frac{(L\lambda)^y e^{-L\lambda}}{y!} = e^{-L\lambda}\sum_{y=0}^{\infty}\frac{(L\lambda\theta)^y}{y!} \\
&= e^{L\lambda(\theta-1)} = e^{L\lambda[\pi e^s+(1-\pi)e^t-1]} \\
&= e^{L\lambda\pi(e^s-1)}e^{L\lambda(1-\pi)(e^t-1)} \\
&= M_{Y_p}(s)M_{Y_{\bar{p}}}(t).
\end{aligned}$$

Hence, we have shown that $Y_p \sim \text{POI}(L\lambda\pi)$, that $Y_{\bar{p}} \sim \text{POI}[L\lambda(1-\pi)]$, and that $Y_p$ and $Y_{\bar{p}}$ are independent random variables. Finally, reasonable estimates of $E(Y_p)$ and $E(Y_{\bar{p}})$ are $L\hat{\lambda}\hat{\pi}$ and $L\hat{\lambda}(1-\hat{\pi})$, respectively.

**Solution 3.15**

(a) In general, if $Y = \ln(X) \sim N(\mu, \sigma^2)$, then

$$E(X) = e^{\mu+\sigma^2/2} \quad \text{and} \quad V(X) = [E(X)]^2(e^{\sigma^2}-1),$$

so that

$$\sigma^2 = \ln\left[1 + \frac{V(X)}{[E(X)]^2}\right] \quad \text{and} \quad \mu = \ln[E(X)] - \frac{\sigma^2}{2}.$$

So, since $E(X|D=1) = 2.00$ and $V(X|D=1) = 2.60$, it follows that $E(Y|D=1) = 0.443$ and $V(Y|D=1) = 0.501$. Also, since $E(X|D=0) = 1.50$ and $V(X|D=0) = 3.00$, we have $E(Y|D=0) = -0.018$ and $V(Y|D=0) = 0.847$. Thus,

$$\begin{aligned}
&\text{pr}(D=1|1.60 < X < 1.80) \\
&\quad = \frac{\text{pr}\left[(D=1)\cap(1.60 < X < 1.80)\right]}{\text{pr}(1.60 < X < 1.80)} \\
&\quad = \frac{\text{pr}(1.60 < X < 1.80|D=1)\text{pr}(D=1)}{\text{pr}(1.60 < X < 1.80|D=1)\text{pr}(D=1) + \text{pr}(1.60 < X < 1.80|D=0)\text{pr}(D=0)}.
\end{aligned}$$

Now,

$$\begin{aligned}
\text{pr}(1.60 < X < 1.80|D=1) &= \text{pr}\left(\frac{0.470-0.443}{0.708} < Z < \frac{0.588-0.443}{0.708}\right) \\
&= \text{pr}(0.038 < Z < 0.205) \\
&= 0.070,
\end{aligned}$$

where $Z \sim N(0,1)$. And,

$$\begin{aligned}\text{pr}(1.60 < X < 1.80|D=0) &= \text{pr}\left(\frac{0.470-(-0.018)}{0.920} < Z < \frac{0.588-(-0.018)}{0.920}\right)\\ &= \text{pr}(0.530 < Z < 0.659)\\ &= 0.043.\end{aligned}$$

So,

$$\begin{aligned}\text{pr}(D=1|1.60 < X < 1.80) &= \frac{0.070(0.60)}{0.070(0.60)+0.043(0.40)}\\ &= \frac{0.042}{0.042+0.017} = \frac{0.042}{0.059} = 0.712.\end{aligned}$$

(b) Clearly, $X$ is a *mixture* of lognormal densities, namely,

$$f_X(x) = 0.60\left[\frac{1}{\sqrt{2\pi}(0.708)x}e^{\frac{-[\ln(x)-0.441]^2}{2(0.501)}}\right] + 0.40\left[\frac{1}{\sqrt{2\pi}(0.920)x}e^{\frac{-[\ln(x)-(-0.018)]^2}{2(0.847)}}\right],$$

$0 < x < +\infty$. So,

$$\begin{aligned}E(X) &= E(X|D=1)\text{pr}(D=1) + E(X|D=0)\text{pr}(D=0)\\ &= (2.00)(0.60) + (1.50)(0.40) = 1.20 + 0.60 = 1.80.\end{aligned}$$

And,

$$\begin{aligned}E(X^2) &= E(X^2|D=1)\text{pr}(D=1) + E(X^2|D=0)\text{pr}(D=0)\\ &= \left[(2.60)+(2.00)^2\right](0.60) + \left[(3.00)+(1.50)^2\right](0.40)\\ &= (6.60)(0.60) + (5.25)(0.40) = 3.96 + 2.10 = 6.06.\end{aligned}$$

So,

$$V(X) = 6.06 - (1.80)^2 = 6.06 - 3.24 = 2.82.$$

(c)

$$\begin{aligned}\theta &= \text{pr(misclassification)}\\ &= \text{pr}\left[(X \le c) \cap (D=1)\right] + \text{pr}\left[(X > c) \cap (D=0)\right]\\ &= \text{pr}(X \le c|D=1)\text{pr}(D=1) + \text{pr}(X > c|D=0)\text{pr}(D=0)\\ &= \text{pr}\left(Z \le \frac{\ln(c)-0.443}{0.708}\right)(0.60) + \text{pr}\left(Z > \frac{\ln(c)-(-0.018)}{0.920}\right)(0.40)\\ &= (0.60)F_Z\left(\frac{\ln(c)-0.443}{0.708}\right) + 0.40\left[1 - F_Z\left(\frac{\ln(c)+0.018}{0.920}\right)\right],\end{aligned}$$

where $F_Z(z) = \text{pr}(Z \le z)$ when $Z \sim N(0,1)$.

So, with $k = \ln(c)$, we have

$$\frac{d\theta}{dk} = \left(\frac{0.60}{0.708}\right)\frac{1}{\sqrt{2\pi}}e^{-\frac{1}{2}\left(\frac{k-0.443}{0.708}\right)^2} - \left(\frac{0.40}{0.920}\right)\frac{1}{\sqrt{2\pi}}e^{-\frac{1}{2}\left(\frac{k+0.018}{0.920}\right)^2} = 0$$

$$\Rightarrow \ln(0.848) - \frac{(k^2 - 0.886k + 0.196)}{1.003} - \ln(0.435) + \frac{(k^2 + 0.036k + 0.0003)}{1.693} = 0$$

$$\Rightarrow \left(\frac{1}{1.693} - \frac{1}{1.003}\right)k^2 + \left(\frac{0.886}{1.003} + \frac{0.036}{1.693}\right)k$$

$$+ \left[\frac{0.0003}{1.693} - 0.165 - \frac{0.196}{1.003} + 0.832\right] = 0$$

$$\Rightarrow 0.406k^2 - 0.905k - 0.472 = 0.$$

The two roots of this quadratic equation are:

$$\frac{0.905 \pm \sqrt{(-0.905)^2 - 4(0.406)(-0.472)}}{2(0.406)} = \frac{0.905 \pm 1.259}{0.812},$$

or $-0.436$ and $2.665$. The value $c^* = e^{-0.436} = 0.647$ minimizes $\theta$.
Note that

$$c^* < E(X|D = 0) < E(X|D = 1),$$

which appears to be a counterintuitive finding. However, note that the value of $c^*$ is inversely proportional to the value of the prevalence of the protein (i.e., the higher the prevalence of the protein, the lower the value of $c^*$). In the extreme, if the prevalence is 0%, then the value of $c^*$ is $+\infty$; and, if the prevalence is 100%, then the value of $c^*$ is 0. In our particular situation, the prevalence is 60% (a fairly high value), so that a "low" value of $c^*$ would be anticipated.

**Solution 3.16.** Now,

$$\sum_{i=1}^{n}(X_i - \bar{X})(Y_i - \bar{Y}) = \sum_{i=1}^{n}(X_iY_i - X_i\bar{Y} - \bar{X}Y_i + \bar{X}\bar{Y}) = \sum_{i=1}^{n}X_iY_i - n\bar{X}\bar{Y}$$

$$= \sum_{i=1}^{n}X_iY_i - n^{-1}\left[\left(\sum_{i=1}^{n}X_i\right)\left(\sum_{i=1}^{n}Y_i\right)\right]$$

$$= \sum_{i=1}^{n}X_iY_i - n^{-1}\left[\sum_{i=1}^{n}X_iY_i + \sum_{\text{all } i \neq j}X_iY_j\right]$$

$$= (1 - n^{-1})\sum_{i=1}^{n}X_iY_i - n^{-1}\sum_{\text{all } i \neq j}X_iY_j.$$

Since

$$\mathrm{E}\left(\sum_{i=1}^{n} X_i Y_i\right) = n\mathrm{E}(X_i Y_i) = n[\mathrm{cov}(X_i, Y_i) + \mu_x \mu_y]$$
$$= n(\rho\sigma_x\sigma_y + \mu_x\mu_y),$$

we have

$$\begin{aligned}\mathrm{E}(U) &= (n-1)^{-1}\{(1-n^{-1})n(\rho\sigma_x\sigma_y + \mu_x\mu_y) - n^{-1}[n(n-1)\mu_x\mu_y]\} \\ &= (n-1)^{-1}[(n-1)(\rho\sigma_x\sigma_y + \mu_x\mu_y) - (n-1)\mu_x\mu_y] \\ &= \rho\sigma_x\sigma_y.\end{aligned}$$

**Solution 3.17**

(a) Let $X_i$ denote the $i$th benzene concentration measurement, $i = 1, 2, \ldots, 10$. Then, we know that $Y_i = \ln X_i \sim \mathrm{N}(\mu, \sigma^2 = 2)$ and that the $\{Y_i\}$ are mutually independent.

*Decision Rule* #1:

$$\mathrm{pr}(X_i > 5) = \mathrm{pr}[\ln X_i > \ln 5] = \mathrm{pr}\left[\frac{Y_i - \mu}{\sqrt{2}} > \frac{1.6094 - \mu}{\sqrt{2}}\right]$$
$$= \mathrm{pr}\left[Z > \frac{1.6094 - \mu}{1.4142}\right] = 1 - \mathrm{F}_Z\left(\frac{1.6094 - \mu}{1.4142}\right), \quad Z \sim \mathrm{N}(0, 1).$$

So, if $\theta_1 = \mathrm{pr}$(Decision that drinking water violates EPA standard|Decision Rule #1), then

$$\begin{aligned}\theta_1 &= \sum_{j=3}^{10} \mathrm{C}_j^{10}\left[1 - \mathrm{F}_Z\left(\frac{1.6094 - \mu}{1.4142}\right)\right]^j \left[\mathrm{F}_Z\left(\frac{1.6094 - \mu}{1.4142}\right)\right]^{10-j} \\ &= 1 - \sum_{j=0}^{2} \mathrm{C}_j^{10}\left[1 - \mathrm{F}_Z\left(\frac{1.6094 - \mu}{1.4142}\right)\right]^j \left[\mathrm{F}_Z\left(\frac{1.6094 - \mu}{1.4142}\right)\right]^{10-j}.\end{aligned}$$

Now, with $\mathrm{E}(X) = e^{\mu + \frac{\sigma^2}{2}} = e^{\mu+1} = 7$, then $\mu = \ln(7) - 1 = 1.9459 - 1 = 0.9459$. Thus, with $\mu = 0.9459$, we have $\mathrm{F}_Z\left(\frac{1.6094 - 0.9459}{1.4142}\right) = \mathrm{F}_Z(0.4692) \approx 0.680$, so that

$$\theta_1 = 1 - \sum_{j=0}^{2} \mathrm{C}_j^{10}(0.320)^j (0.680)^{10-j} = 1 - 0.0211 - 0.0995 - 0.2107 = 0.6687.$$

*Decision Rule* #2:
Since $\bar{Y} = \ln\bar{X}_g = \frac{1}{10}\sum_{i=1}^{10} Y_i$, then $\bar{Y} \sim \text{N}(\mu, 2/10)$. So, with $\theta_2$ =pr(Decision that drinking water violates EPA standard|Decision Rule #2), then

$$\theta_2 = \text{pr}(\bar{X}_g > 5) = \text{pr}\left[\frac{\bar{Y}-\mu}{\sqrt{0.20}} > \frac{\ln 5 - \mu}{\sqrt{0.20}}\right]$$
$$= 1 - F_Z\left(\frac{1.6094-\mu}{0.4472}\right), \quad Z \sim \text{N}(0,1).$$

With $\mu = 0.9459$, we have

$$\theta_2 = 1 - F_Z\left(\frac{1.6094 - 0.9459}{0.4472}\right)$$
$$= 1 - F_Z(1.4837) \approx 1 - 0.931 = 0.069.$$

*Decision Rule* #3:
With $\theta_3$ = pr(Decision that drinking water violates EPA standard|Decision Rule #3), we have

$$\theta_3 = \text{pr}\left[X_{(10)} > 5\right] = 1 - \text{pr}[\cap_{i=1}^{10}(X_i \leq 5)]$$
$$= 1 - \{\text{pr}(Y_i \leq \ln 5)\}^{10} = 1 - \left\{\text{pr}\left(\frac{Y_i-\mu}{\sqrt{2}} \leq \frac{1.6094-\mu}{\sqrt{2}}\right)\right\}^{10}$$
$$= 1 - \left[\text{pr}\left(Z \leq \frac{1.6094-\mu}{1.4142}\right)\right]^{10}, \quad Z \sim \text{N}(0,1).$$

With $\mu = 0.9459$, we have

$$\theta_3 = 1 - \left[\text{pr}\left(Z \leq \frac{1.6094-0.9459}{1.4142}\right)\right]^{10}$$
$$= 1 - [\text{pr}\,(Z \leq 0.4692)]^{10} = 1 - [F_Z(0.4692)]^{10}$$
$$= 1 - (0.680)^{10} = 1 - 0.0211 = 0.9789.$$

(b) With $\text{E}(X) = 7$, so that $\mu = 0.9459$, we have

$$\text{pr}(\bar{X}_g > 5) = \text{pr}(\bar{Y} > \ln 5) = \text{pr}\left[\frac{\bar{Y}-0.9459}{\sqrt{2/n}} > \frac{\ln 5 - 0.9459}{\sqrt{2/n}}\right]$$
$$= \text{pr}(Z > 0.4692\sqrt{n}), \quad Z \sim \text{N}(0,1).$$

Thus, $\text{pr}(\bar{X}_g > 5)$ gets smaller as $n$ increases! The reason for this phenomenon can be determined by examining $\text{E}(\bar{X}_g)$. In general,

$$\begin{aligned}
\text{E}(\bar{X}_g) &= \text{E}\left[\left(\prod_{i=1}^{n} X_i\right)^{1/n}\right] = \prod_{i=1}^{n} \text{E}\left(X_i^{1/n}\right) \\
&= \left[\text{E}\left(e^{Y_i/n}\right)\right]^n = \left[e^{\frac{\mu}{n}+\frac{\sigma^2(1/n)^2}{2}}\right]^n \\
&= e^{\mu+\frac{\sigma^2}{2n}}.
\end{aligned}$$

Hence, for $n > 1$,

$$\text{E}(\bar{X}_g) < \text{E}(X) = e^{\mu+\sigma^2/2},$$

with the size of the bias increasing as $n$ increases. In particular,

$$\lim_{n\to+\infty} \text{E}(\bar{X}_g) = e^{\mu},$$

which is the *median*, not the mean, of the lognormal distribution of the random variable $X$.

**Solution 3.18**

$$\begin{aligned}
\text{pr}\left[\min\{X_1^2, X_2^2, \ldots, X_n^2\} \le 0.002\right] &= 1 - \text{pr}\left[\min\{X_1^2, X_2^2, \ldots, X_n^2\} > 0.002\right] \\
&= 1 - \text{pr}\left[\cap_{i=1}^{n}(X_i^2 > 0.002)\right] \\
&= 1 - \prod_{i=1}^{n} \text{pr}\left[\left(\frac{X_i}{\sqrt{2}}\right)^2 > \frac{0.002}{2}\right] \\
&= 1 - \text{pr}\left[(U_i > 0.001)\right]^n \\
&= 1 - (0.975)^n,
\end{aligned}$$

since $U_i \sim \chi_1^2$, $i = 1, 2, \ldots, n$.

So, $n^*$ is the smallest positive integer such that

$$1 - (0.975)^n \ge 0.80,$$

or

$$n \ge \frac{0.20}{-\ln(0.975)} = \frac{0.20}{0.0253} = 7.9051,$$

so that $n^* = 8$.

**Solution 3.19**

(a) Let the random variable $X_i$ denote the number of coronary bypass grafts needed by the $i$th patient, $i = 1, 2, \ldots, 500$. It is reasonable to assume that $X_1, X_2, \ldots, X_{500}$ constitute a set of 500 i.i.d random variables. Also, $E(X_i) = \sum_{j=1}^{4} j\pi_j = 1.79$ and $V(X_i) = \sum_{j=1}^{4} j^2\pi_j - (1.79)^2 = 1.0059$. Thus, with the random variable $T = \sum_{i=1}^{500} X_i$ denoting the *total* number of coronary bypass grafts to be performed during the upcoming year, it follows that $E(T) = 500(1.79) = 895.00$ and $V(T) = 500(1.0059) = 502.95$. Thus, by the Central Limit Theorem, the standardized random variable $Z = [T - E(T)]/\sqrt{V(T)} \dot{\sim} N(0, 1)$ for large $n$. Hence, we have

$$\begin{aligned} \mathrm{pr}(T \leq 900) &= \mathrm{pr}\left[\frac{T - E(T)}{\sqrt{V(T)}} \leq \frac{900 - E(T)}{\sqrt{V(T)}}\right] \\ &\approx \mathrm{pr}(Z \leq 0.223) \approx 0.59. \end{aligned}$$

Thus, the hospital administrator's suggestion is not reasonable.

(b) In general, if this hospital plans to perform coronary bypass surgery on $n$ patients during the upcoming year, then $E(T) = 1.79n$, $V(T) = 1.0059n$, and $\sqrt{V(T)} = 1.0029\sqrt{n}$. Again, by the Central Limit Theorem, the standardized random variable $Z = (T - 1.79n)/(1.0029\sqrt{n}) \dot{\sim} N(0, 1)$ for large $n$. Hence, we have

$$\begin{aligned} \mathrm{pr}(T \leq 900) &= \mathrm{pr}\left[\frac{T - 1.79n}{1.0029\sqrt{n}} \leq \frac{900 - 1.79n}{1.0029\sqrt{n}}\right] \\ &\approx \mathrm{pr}\left[Z \leq \frac{897.3975}{\sqrt{n}} - 1.7848\sqrt{n}\right]. \end{aligned}$$

Hence, for $\mathrm{pr}(T \leq 900) \geq 0.95$, $n^*$ is the largest value of $n$ satisfying the inequality

$$\frac{897.3975}{\sqrt{n^*}} - 1.7848\sqrt{n^*} \geq 1.645.$$

It is straightforward to show that $n^* = 482$.

**Solution 3.20**

(a) Since $S = \sum_{i=1}^{k} X_i$, where the $\{X_i\}$ are i.i.d. random variables, the Central Limit Theorem allows us to say that

$$\frac{S - E(S)}{\sqrt{V(S)}} \dot{\sim} N(0, 1)$$

for large $k$. So, for $\pi = 0.05$ and $k = 50$,

$$E(S) = \frac{k}{\pi} = \frac{50}{0.05} = 1000$$

and

$$\mathrm{V}(S) = \frac{k(1-\pi)}{\pi^2} = \frac{50(0.95)}{(0.05)^2} = 19{,}000;$$

thus,

$$\sqrt{\mathrm{V}(S)} = 137.84.$$

So, with $Z \mathrel{\dot\sim} \mathrm{N}(0,1)$ for large $k$, we have

$$\begin{aligned}
\mathrm{pr}[S > (1100)] &= \mathrm{pr}\left[\frac{S - \mathrm{E}(S)}{\sqrt{\mathrm{V}(S)}} > \frac{(1100) - \mathrm{E}(S)}{\sqrt{\mathrm{V}(S)}}\right] \\
&\doteq \mathrm{pr}\left[Z > \frac{(1100) - (1000)}{137.84}\right] \\
&= \mathrm{pr}(Z > 0.7255) \doteq 0.235.
\end{aligned}$$

(b)

$$\begin{aligned}
\mathrm{M}_U(t) &= \mathrm{E}(\mathrm{e}^{tU}) = \mathrm{E}\left[\mathrm{e}^{t(2\pi S)}\right] = \mathrm{E}\left[\mathrm{e}^{2\pi t \sum_{i=1}^{k} X_i}\right] \\
&= \mathrm{E}\left[\prod_{i=1}^{k} \mathrm{e}^{2\pi t X_i}\right] = \prod_{i=1}^{k} \mathrm{M}_{X_i}(2\pi t),
\end{aligned}$$

so that

$$\lim_{\pi \to 0} \mathrm{M}_U(t) = \prod_{i=1}^{k} \left[\lim_{\pi \to 0} \mathrm{M}_{X_i}(2\pi t)\right].$$

Now,

$$\lim_{\pi \to 0} \mathrm{M}_{X_i}(2\pi t) = \lim_{\pi \to 0} \left\{\frac{\pi \mathrm{e}^{2\pi t}}{1 - (1-\pi)\mathrm{e}^{2\pi t}}\right\} = \frac{0}{0},$$

so we can employ L'Hôpital's Rule.

So,

$$\frac{\partial(\pi \mathrm{e}^{2\pi t})}{\partial \pi} = \mathrm{e}^{2\pi t} + 2\pi t \mathrm{e}^{2\pi t}$$

and

$$\frac{\partial\left[1 - (1-\pi)\mathrm{e}^{2\pi t}\right]}{\partial \pi} = \mathrm{e}^{2\pi t} - (1-\pi)(2t)\mathrm{e}^{2\pi t}.$$

So,

$$\lim_{\pi\to 0} \mathrm{M}_{X_i}(2\pi t) = \lim_{\pi\to 0}\left\{\frac{e^{2\pi t}+2\pi t e^{2\pi t}}{e^{2\pi t}-(1-\pi)(2t)e^{2\pi t}}\right\}$$

$$= \lim_{\pi\to 0}\left\{\frac{1+2\pi t}{1-(1-\pi)2t}\right\} = (1-2t)^{-1},$$

so that

$$\lim_{\pi\to 0} \mathrm{M}_U(t) = \prod_{i=1}^{k}\left[(1-2t)^{-1}\right] = (1-2t)^{-k},$$

which is the MGF for a GAMMA[$\alpha = 2, \beta = k$], or $\chi^2_{2k}$, random variable. So, for small $\pi$, $U = 2\pi S \dot{\sim} \chi^2_{2k}$.

(c) For small $\pi$,

$$\mathrm{pr}[S > (1100)] = \mathrm{pr}[2\pi S > (2\pi)(1100)] \doteq \mathrm{pr}[U > (2\pi)(1100)],$$

where $U \dot{\sim} \chi^2_{2k}$.

When $\pi = 0.05$ and $k = 50$, we have

$$\mathrm{pr}[S > (1100)] \doteq \mathrm{pr}[U > (2)(0.05)(1100)] = \mathrm{pr}(U > 110) \approx 0.234,$$

since $U \dot{\sim} \chi^2_{2k} = \chi^2_{100}$. This number agrees quite well with the numerical answer computed in part (a).

**Solution 3.21**

$$\mathrm{F}_{U_n}(u) = \mathrm{pr}(U_n \leq u) = 1 - \mathrm{pr}(U_n > u).$$

Now,

$$\mathrm{pr}(U_n > u) = \mathrm{pr}\left[\frac{\theta n[Y_{(1)} - c]}{c} > u\right]$$

$$= \mathrm{pr}\left\{Y_{(1)} > \frac{uc}{\theta n} + c\right\}$$

$$= \mathrm{pr}\left\{\cap_{i=1}^{n}\left(Y_i > \frac{uc}{\theta n} + c\right)\right\}$$

$$= \prod_{i=1}^{n}\mathrm{pr}\left(Y_i > \frac{uc}{\theta n} + c\right)$$

$$= \left[1 - \mathrm{F}_Y\left(\frac{uc}{\theta n} + c; \theta\right)\right]^n,$$

where

$$\begin{aligned}
F_Y(y;\theta) &= \int_c^y \theta c^{\theta} t^{-(\theta+1)}\,dt \\
&= c^{\theta}\left[-t^{-\theta}\right]_c^y \\
&= c^{\theta}[c^{-\theta} - y^{-\theta}] \\
&= 1 - (y/c)^{-\theta}, \quad 0 < c < y < +\infty.
\end{aligned}$$

So, $F_{U_n}(u) = 1 - \text{pr}(U_n > u)$, where

$$\begin{aligned}
\text{pr}(U_n > u) &= \left\{1 - \left[1 - \left(\frac{\frac{uc}{\theta n} + c}{c}\right)^{-\theta}\right]\right\}^n \\
&= \left[\left(1 + \frac{u}{\theta n}\right)^{-\theta}\right]^n \\
&= \left[\left(1 + \frac{u}{\theta n}\right)^n\right]^{-\theta}.
\end{aligned}$$

So,

$$\lim_{n\to\infty} \text{pr}(U_n > u) = \lim_{n\to\infty}\left[\left(1 + \frac{u}{\theta n}\right)^n\right]^{-\theta} = (e^{u/\theta})^{-\theta} = e^{-u}.$$

So,

$$\lim_{n\to\infty} F_{U_n}(u) = 1 - e^{-u},$$

so that $f_U(u) = e^{-u}, 0 < u < +\infty$.

**Solution 3.22**

(a) First, note that $U = X_{(1)} = \min\{X_1, X_2, \ldots, X_n\}$ and that $V = (1 - X_{(n)})$, where $X_{(n)} = \max\{X_1, X_2, \ldots, X_n\}$. Since $F_X(x) = x, 0 < x < 1$, direct application of the general formula for the joint distribution of any two-order statistics based on a random sample of size $n$ from $f_X(x)$ (see the introductory material for this chapter) gives

$$f_{X_{(1)},X_{(n)}}(x_{(1)}, x_{(n)}) = n(n-1)\left(x_{(n)} - x_{(1)}\right)^{n-2}, 0 < x_{(1)} < x_{(n)} < 1.$$

For the transformation $U = X_{(1)}$ and $V = (1 - X_{(n)})$, with inverse functions $X_{(1)} = U$ and $X_{(n)} = (1 - V)$, the absolute value of the Jacobian is equal to 1; so, it follows directly that

$$f_{U,V}(u, v) = n(n-1)(1 - u - v)^{n-2}, 0 < u < 1, 0 < (u + v) < 1.$$

(b) Now,

$$\begin{aligned}\theta_n &= \text{pr}\left[(R > r) \cap (S > s)\right] = \text{pr}\left[(nU > r) \cap (nV > r)\right] \\ &= \text{pr}\left[\left(U > \frac{r}{n}\right) \cap \left(V > \frac{s}{n}\right)\right] \\ &= \int_{r/n}^{1-s/n} \int_{s/n}^{1-u} n(n-1)(1-u-v)^{n-2}\,dv\,du \\ &= \left(1 - \frac{r}{n} - \frac{s}{n}\right)^n.\end{aligned}$$

So, we have

$$\begin{aligned}\lim_{n\to\infty} \theta_n &= \lim_{n\to\infty} \left(1 - \frac{r}{n} - \frac{s}{n}\right)^n \\ &= \lim_{n\to\infty} \left\{1 + \frac{[-(r+s)]}{n}\right\}^n \\ &= e^{-(r+s)} = e^{-r}e^{-s}, \quad 0 < r < \infty, \quad 0 < s < \infty.\end{aligned}$$

So, asymptotically, $R$ and $S$ are independent random variables with exponential distributions, namely,

$$f_R(r) = e^{-r},\ 0 < r < \infty \quad \text{and} \quad f_S(s) = e^{-s},\ 0 < s < \infty.$$

**Solution 3.23**

(a)

$$\begin{aligned}E(C) &= E_x[E(C|X = x)] = \sum_{x=0}^{\infty} E(C|X = x)\text{pr}(X = x) \\ &= E(C|X = 0)\text{pr}(X = 0) + \sum_{x=1}^{\infty} E(C|X = x)\,\text{pr}(X = x) \\ &= 0 + \sum_{x=1}^{\infty} E\left[\sum_{j=1}^{X} C_j \Big| X = x\right] \text{pr}(X = x) \\ &= \sum_{x=1}^{\infty} (x\mu)\text{pr}(X = x) = \mu \sum_{x=0}^{\infty} x\,\text{pr}(X = x) = \mu E(X) = \mu\lambda;\end{aligned}$$

and,

$$\begin{aligned}E(C^2) &= \sum_{x=0}^{\infty} E(C^2|X = x)\text{pr}(X = x) \\ &= 0 + \sum_{x=1}^{\infty} E\left[\left(\sum_{j=1}^{X} C_j\right)^2 \Big| X = x\right] \text{pr}(X = x)\end{aligned}$$

$$= \sum_{x=1}^{\infty} \left\{ \mathrm{E}\left[ \sum_{j=1}^{x} C_j^2 + 2 \sum_{\text{all } j<k} C_j C_k \middle| X = x \right] \right\} \mathrm{pr}(X = x)$$

$$= \sum_{x=1}^{\infty} \left[ x(\sigma^2 + \mu^2) + x(x-1)\mu^2 \right] \mathrm{pr}(X = x)$$

$$= \sum_{x=0}^{\infty} (x\sigma^2 + x^2\mu^2) \mathrm{pr}(X = x)$$

$$= \sigma^2 \mathrm{E}(X) + \mu^2 \mathrm{E}(X^2) = \sigma^2\lambda + \mu^2(\lambda + \lambda^2).$$

Thus, $\mathrm{V}(C) = \sigma^2\lambda + \mu^2(\lambda + \lambda^2) - (\mu\lambda)^2 = \lambda(\sigma^2 + \mu^2)$. Alternatively, $\mathrm{V}(C) = \mathrm{V}_x\left[\mathrm{E}(C|X = x)\right] + \mathrm{E}_x\left[\mathrm{V}(C|X = x)\right] = \mathrm{V}(X\mu) + \mathrm{E}(X\sigma^2) = \lambda(\mu^2 + \sigma^2)$.
Now,

$$\mathrm{E}(XC) = \sum_{x=0}^{\infty} \mathrm{E}(XC|X = x)\mathrm{pr}(X = x) = \sum_{x=1}^{\infty} x\mathrm{E}(C|X = x)\mathrm{pr}(X = x)$$

$$= \sum_{x=1}^{\infty} x\mathrm{E}\left[ \sum_{j=1}^{X} C_j \middle| X = x \right] \mathrm{pr}(X = x) = \sum_{x=1}^{\infty} x(x\mu)\mathrm{pr}(X = x)$$

$$= \sum_{x=0}^{\infty} \mu x^2 \mathrm{pr}(X = x) = \mu \mathrm{E}(X^2) = \mu(\lambda + \lambda^2).$$

So,

$$\mathrm{corr}(X, C) = \frac{\mathrm{cov}(X, C)}{\sqrt{\mathrm{V}(X)\mathrm{V}(C)}} = \frac{\mathrm{E}(XC) - \mathrm{E}(X)\mathrm{E}(C)}{\sqrt{\mathrm{V}(X)\mathrm{V}(C)}}$$

$$= \frac{\mu(\lambda + \lambda^2) - (\lambda)(\mu\lambda)}{\sqrt{\lambda\left[\lambda(\sigma^2 + \mu^2)\right]}} = \frac{\mu}{(\sigma^2 + \mu^2)^{1/2}},$$

which does *not* depend on $\lambda$.

(b)

$$\mathrm{M}_C(t) = \mathrm{E}(\mathrm{e}^{tC}) = \mathrm{E}_x[\mathrm{E}(\mathrm{e}^{tC}|X = x)] = \sum_{x=0}^{\infty} \mathrm{E}(\mathrm{e}^{tC}|X = x)\mathrm{pr}(X = x)$$

$$= \mathrm{E}(\mathrm{e}^{tC}|X = 0)\mathrm{pr}(X = 0) + \sum_{x=1}^{\infty} \mathrm{E}(\mathrm{e}^{tC}|X = x)\mathrm{pr}(X = x)$$

$$= (1)(e^{-\lambda}) + \sum_{x=1}^{\infty} E\left[ e^{t\sum_{j=1}^{X} C_j} \middle| X = x \right] pr(X = x)$$

$$= e^{-\lambda} + \sum_{x=1}^{\infty} E\left[ \prod_{j=1}^{x} e^{tC_j} \right] pr(X = x)$$

$$= e^{-\lambda} + \sum_{x=1}^{\infty} [M(t)]^x \, pr(X = x)$$

$$= e^{-\lambda} + \sum_{x=0}^{\infty} [M(t)]^x \frac{\lambda^x e^{-\lambda}}{x!} - e^{-\lambda}$$

$$= e^{-\lambda} \sum_{x=0}^{\infty} \frac{[\lambda M(t)]^x}{x!} = e^{-\lambda} e^{\lambda M(t)} = e^{\lambda[M(t)-1]}.$$

So,

$$E(C) = \frac{dM_C(t)}{dt}\bigg|_{t=0} = \left\{ e^{\lambda[M(t)-1]} \cdot \lambda \frac{dM(t)}{dt} \right\}\bigg|_{t=0}$$

$$= e^{\lambda[M(0)-1]} \cdot \lambda \left[ \frac{dM(t)}{dt} \right]\bigg|_{t=0}$$

$$= e^{\lambda(1-1)} \cdot \lambda \cdot E(C_j) = \lambda\mu,$$

which agrees with the result derived in part (a).

**Solution 3.24**

(a) Since $E(Y_{ij}) = \mu$, it follows directly that $E(\bar{Y}) = \mu$. Now, $\bar{Y} = k^{-1} \sum_{i=1}^{k} \bar{Y}_i$, where

$$\bar{Y}_i = n^{-1} \sum_{j=1}^{n} Y_{ij} = n^{-1} \sum_{j=1}^{n} (\mu + \beta_i + \epsilon_{ij}) = \mu + \beta_i + \bar{\epsilon}_i,$$

where $\bar{\epsilon}_i = n^{-1} \sum_{j=1}^{n} \epsilon_{ij}$. Thus, $V(\bar{Y}_i) = \sigma_\beta^2 + \sigma_\epsilon^2/n$.
Since $\{\bar{Y}_1, \bar{Y}_2, \ldots, \bar{Y}_k\}$ constitute a set of $k$ mutually independent random variables, it follows that

$$V(\bar{Y}) = k^{-2} \sum_{i=1}^{k} V(\bar{Y}_i) = \frac{\sigma_\beta^2}{k} + \frac{\sigma_\epsilon^2}{kn}.$$

(b) We can employ the method of Lagrange multipliers to solve this problem. In particular, consider the function

$$Q = \frac{\sigma_\beta^2}{k} + \frac{\sigma_\epsilon^2}{kn} + \lambda\left[(kD_c + knD_p) - C^*\right],$$

where $\lambda$ is the Lagrange multiplier. Now, consider the following three equations:

$$\frac{\partial Q}{\partial k} = \frac{-\sigma_\beta^2}{k^2} - \frac{\sigma_\epsilon^2}{k^2 n} + \lambda(D_c + nD_p) = 0;$$

$$\frac{\partial Q}{\partial n} = \frac{-\sigma_\epsilon^2}{kn^2} + \lambda k D_p = 0;$$

$$\frac{\partial Q}{\partial \lambda} = kD_c + knD_p - C^* = 0.$$

Solving these three equations gives

$$n^* = \left(\frac{\sigma_\epsilon}{\sigma_\beta}\right)\left(\frac{D_c}{D_p}\right)^{1/2} \quad \text{and} \quad k^* = \frac{C^*}{(D_c + n^* D_p)}.$$

(c) Since

$$n^* = \frac{3}{2}\left(\frac{10,000}{100}\right)^{1/2} = 15 \quad \text{and} \quad k^* = \frac{100,000}{10,000 + 15(100)} = 8.70,$$

the clinical trial should involve 9 private medical practices, with each private medical practice being required to enroll 15 patients.

**Solution 3.25.** First, we have

$$\begin{aligned}
\mathrm{M}_{Y_3}(t) = \mathrm{E}(\mathrm{e}^{tY_3}) &= \sum_{y_3=1}^{\infty} \mathrm{e}^{ty_3} \frac{\lambda_3^{y_3}}{y_3!(\mathrm{e}^{\lambda_3} - 1)} \\
&= \frac{\mathrm{e}^{\lambda_3}}{(\mathrm{e}^{\lambda_3} - 1)} \sum_{y_3=0}^{\infty} \mathrm{e}^{ty_3} \frac{\lambda_3^{y_3}\mathrm{e}^{-\lambda_3}}{y_3!} - \frac{1}{(\mathrm{e}^{\lambda_3} - 1)} \\
&= \frac{\mathrm{e}^{\lambda_3}[\mathrm{e}^{\lambda_3(\mathrm{e}^t - 1)}]}{(\mathrm{e}^{\lambda_3} - 1)} - \frac{1}{(\mathrm{e}^{\lambda_3} - 1)} \\
&= \frac{\mathrm{e}^{\lambda_3 \mathrm{e}^t} - 1}{(\mathrm{e}^{\lambda_3} - 1)}.
\end{aligned}$$

So,

$$\begin{aligned}
\mathrm{M}_U(t) &= \mathrm{E}[\mathrm{e}^{t(R+S)}] = \mathrm{E}[\mathrm{e}^{t(Y_1+Y_2+2Y_3)}] \\
&= \mathrm{E}[\mathrm{e}^{tY_1}\mathrm{e}^{tY_2}\mathrm{e}^{2tY_3}] \\
&= \mathrm{E}_{y_3}\left\{\mathrm{E}(\mathrm{e}^{tY_1}\mathrm{e}^{tY_2}\mathrm{e}^{2tY_3}|Y_3 = y_3)\right\} \\
&= \mathrm{E}_{y_3}\left\{\mathrm{e}^{2ty_3}\mathrm{E}(\mathrm{e}^{tY_1}\mathrm{e}^{tY_2}|Y_3 = y_3)\right\}
\end{aligned}$$

$$= \mathrm{E}_{y_3}\left\{ e^{2ty_3}\mathrm{E}(e^{tY_1}|Y_3 = y_3)\mathrm{E}(e^{tY_2}|Y_3 = y_3) \right\}$$

$$= \mathrm{E}_{y_3}\left\{ e^{2ty_3}e^{y_3(e^t-1)}e^{y_3(e^t-1)} \right\}$$

$$= \mathrm{E}_{y_3}\left\{ e^{2(e^t+t-1)y_3} \right\}$$

$$= \frac{e^{\lambda_3[e^{2(e^t+t-1)}]} - 1}{(e^{\lambda_3} - 1)}.$$

So,

$$\mathrm{E}(U) = \frac{d\mathrm{M}_U(t)}{dt}_{|t=0}$$

$$= \left[ \frac{e^{\lambda_3[e^{2(e^t+t-1)}]} \cdot \lambda_3 e^{2(e^t+t-1)} \cdot 2(e^t + 1)}{(e^{\lambda_3} - 1)} \right]_{|t=0}$$

$$= \frac{4\lambda_3 e^{\lambda_3}}{(e^{\lambda_3} - 1)}.$$

And, since

$$\mathrm{E}(Y_3) = \frac{d\mathrm{M}_{Y_3}(t)}{dt}_{|t=0} = \frac{\lambda_3 e^{\lambda_3}}{(e^{\lambda_3} - 1)},$$

we have

$$\mathrm{E}(U) = \mathrm{E}(R + S) = \mathrm{E}(Y_1 + Y_2 + 2Y_3)$$

$$= \mathrm{E}_{y_3}[\mathrm{E}(Y_1|Y_3 = y_3)] + \mathrm{E}_{y_3}[\mathrm{E}(Y_2|Y_3 = y_3)] + 2\mathrm{E}(Y_3)$$

$$= \mathrm{E}(Y_3) + \mathrm{E}(Y_3) + 2\mathrm{E}(Y_3) = \frac{4\lambda_3 e^{\lambda_3}}{(e^{\lambda_3} - 1)}.$$

**Solution 3.26.** Consider the random variable $X = \sum_{i=1}^{C} X_i$, where the dichotomous random variable $X_i$ takes the value 1 if the $i$th cell is empty and takes the value 0 if the $i$th cell contains at least one ball.

Now,

$$\mathrm{pr}(X_i = 1) = \left(1 - \frac{1}{C}\right)^n = \pi, \text{ say,}$$

so that $\mathrm{E}(X_i) = \pi$ and $\mathrm{V}(X_i) = \pi(1 - \pi), i = 1, 2, \ldots, C$.

Hence,

$$\mathrm{E}(X) = \sum_{i=1}^{C} \mathrm{E}(X_i) = C\pi = C\left(\frac{C-1}{C}\right)^n.$$

Now,

$$\mathrm{V}(X) = \sum_{i=1}^{C} \mathrm{V}(X_i) + 2\sum_{i=1}^{C-1}\sum_{j=i+1}^{C} \mathrm{cov}(X_i, X_j).$$

And,

$$\text{cov}(X_i, X_j) = \text{E}(X_iX_j) - \text{E}(X_i)\text{E}(X_j),$$

with

$$\text{E}(X_iX_j) = \text{pr}[(X_i = 1) \cap (X_j = 1)] = \left(\frac{C-2}{C}\right)^n,$$

so that

$$\text{cov}(X_i, X_j) = \left(\frac{C-2}{C}\right)^n - \left(\frac{C-1}{C}\right)^{2n}.$$

Finally,

$$\text{V}(X) = C\left(\frac{C-1}{C}\right)^n\left[1 - \left(\frac{C-1}{C}\right)^n\right] + C(C-1)\left[\left(\frac{C-2}{C}\right)^n - \left(\frac{C-1}{C}\right)^{2n}\right].$$

When $C = 6$ and $n = 5$, it follows that $\text{E}(X) = 2.4113$ and $\text{V}(X) = 0.5483$.

**Solution 3.27**

(a) First,

$$\text{E}(Y_i) = \pi \quad \text{and} \quad \text{V}(Y_i) = \pi(1-\pi), \quad i = 1, 2, \ldots, N.$$

So,

$$\begin{aligned}
\text{E}(T) &= \text{E}_n\left[\text{E}(T|N = n)\right] \\
&= \text{E}_n\left[\text{E}\left(\sum_{i=1}^{N} Y_i|N = n\right)\right] \\
&= \text{E}_n[n\text{E}(Y_i)] = \text{E}_n(n\pi) = \pi\text{E}(N).
\end{aligned}$$

Since $\text{E}(N) = \theta^{-1}$ and $\text{V}(N) = (1-\theta)/\theta^2$, we have

$$\text{E}(T) = \pi/\theta.$$

Now,

$$\begin{aligned}
\text{V}(T) &= \text{V}_n[\text{E}(T|N = n)] + \text{E}_n[\text{V}(T|N = n)] \\
&= \text{V}_n(n\pi) + \text{E}_n[n\pi(1-\pi)] \\
&= \pi^2\text{V}(N) + \pi(1-\pi)\text{E}(N) \\
&= \pi^2\frac{(1-\theta)}{\theta^2} + \frac{\pi(1-\pi)}{\theta} \\
&= \frac{\pi(\pi + \theta - 2\pi\theta)}{\theta^2}.
\end{aligned}$$

Now,

$$\begin{aligned}
E(NT) &= E_n[E(NT|N=n)] = E_n[nE(T|N=n)] \\
&= E_n(n^2\pi) = \pi E(N^2) = \pi\left[\frac{(1-\theta)}{\theta^2} + \frac{1}{\theta^2}\right] \\
&= \pi(2-\theta)/\theta^2.
\end{aligned}$$

So,

$$\begin{aligned}
\text{corr}(N,T) &= \frac{\frac{\pi(2-\theta)}{\theta^2} - \left(\frac{\pi}{\theta}\right)\left(\frac{1}{\theta}\right)}{\sqrt{\frac{\pi(\pi+\theta-2\pi\theta)}{\theta^2} \cdot \frac{(1-\theta)}{\theta^2}}} \\
&= \sqrt{\frac{\pi(1-\theta)}{\pi(1-\theta)+\theta(1-\pi)}}.
\end{aligned}$$

(b)

$$\text{pr}(T=0) = \sum_{n=1}^{\infty} \text{pr}[(T=0)\cap(N=n)] = \sum_{n=1}^{\infty} \text{pr}(T=0|N=n)\text{pr}(N=n).$$

Now, since the $\{Y_i\}$ are i.i.d. Bernoulli random variables, we know that $T \sim \text{BIN}(n, \pi)$ given $N = n$. So,

$$p_T(t|N=n) = C_t^n \pi^t (1-\pi)^{n-t}, \quad t = 0, 1, 2, \ldots, n.$$

So,

$$\begin{aligned}
\text{pr}(T=0) &= \sum_{n=1}^{\infty} (1-\pi)^n \theta(1-\theta)^{n-1} \\
&= \frac{\theta}{(1-\theta)} \sum_{n=1}^{\infty} [(1-\pi)(1-\theta)]^n \\
&= \frac{\theta}{(1-\theta)} \left\{\frac{(1-\pi)(1-\theta)}{1-(1-\pi)(1-\theta)}\right\} \\
&= \frac{\theta(1-\pi)}{\pi+\theta(1-\pi)}.
\end{aligned}$$

**Solution 3.28**

(a) Now, since $E(X_i) = \lambda_1$ and $E(Y_i) = \lambda_2$, we have

$$\begin{aligned}
E(U) &= E(\bar{X}) - E(\bar{Y}) = m^{-1}\sum_{i=1}^{m} E(X_i) - n^{-1}\sum_{i=1}^{n} E(Y_i) \\
&= (\lambda_1 - \lambda_2).
\end{aligned}$$

And, since $\{X_1, X_2, \ldots, X_m; Y_1, Y_2, \ldots, Y_n\}$ constitute a set of $(m+n)$ mutually independent random variables with $V(X_i) = \lambda_1$ and $V(Y_i) = \lambda_2$, it follows that

$$V(U) = V(\bar{X} - \bar{Y}) = V(\bar{X}) + V(\bar{Y}) = \frac{\lambda_1}{m} + \frac{\lambda_2}{n}.$$

(b) We wish to minimize $V(U)$ subject to the restriction $(m+n) = N$. So, consider the function

$$H(m, n) = \frac{\lambda_1}{m} + \frac{\lambda_2}{n} + \gamma(m + n - N),$$

where $\gamma$ is a Lagrange multiplier.

So,

$$\frac{\partial H(m, n)}{\partial m} = \frac{-\lambda_1}{m^2} + \gamma = 0 \text{ gives } \lambda_1 = \gamma m^2,$$

and

$$\frac{\partial H(m, n)}{\partial n} = \frac{-\lambda_2}{n^2} + \gamma = 0 \text{ gives } \lambda_2 = \gamma n^2.$$

Thus,

$$\frac{\lambda_2}{\lambda_1} = \frac{\gamma n^2}{\gamma m^2} = \frac{n^2}{m^2} \text{ gives } \frac{n}{m} = \sqrt{\theta},$$

where $\theta = \lambda_2/\lambda_1$.

Hence, $n = m\sqrt{\theta}$ gives $N = (m+n) = m(1+\sqrt{\theta})$, so that

$$m = \frac{N}{(1+\sqrt{\theta})} \quad \text{and} \quad n = \frac{N\sqrt{\theta}}{(1+\sqrt{\theta})}.$$

Note that

$$\frac{n}{m} = \sqrt{\theta} = \sqrt{\frac{\lambda_2}{\lambda_1}} = \sqrt{\frac{V(Y_i)}{V(X_i)}},$$

indicating that a larger sample should be selected from the more variable Poisson population. In particular, $V(X_i) < V(Y_i)$ requires $n > m$, $V(X_i) > V(Y_i)$ requires $m > n$, and $V(X_i) = V(Y_i)$ requires $m = n$.

When $N = 60$, $\lambda_1 = 2$, and $\lambda_2 = 8$, so that $\sqrt{\theta} = 2$, we obtain $m = 20$ and $n = 40$.

**Solution 3.29***

(a)

$$\theta = \text{pr}(X = Y) = \sum_{s=0}^{\infty} \text{pr}[(X = s) \cap (Y = s)]$$

$$= \sum_{s=0}^{\infty} \text{pr}(X = s)\text{pr}(Y = s) = \sum_{s=0}^{\infty} (1 - \pi_x)\pi_x^s (1 - \pi_y)\pi_y^s$$

$$= (1-\pi_x)(1-\pi_y)\sum_{s=0}^{\infty}(\pi_x\pi_y)^s = (1-\pi_x)(1-\pi_y)\left[\frac{1}{(1-\pi_x\pi_y)}\right]$$

$$= \frac{(1-\pi_x)(1-\pi_y)}{(1-\pi_x\pi_y)}.$$

(b) First, $p_U(0) = \text{pr}(U = 0) = \text{pr}(X = Y) = \theta$. And, for $u = 1, 2, \ldots, \infty$,

$$p_U(u) = \text{pr}(|X - Y| = u) = \text{pr}[(X - Y) = u] + \text{pr}[(X - Y) = -u].$$

So, for $u = 1, 2, \ldots, \infty$, we have

$$\text{pr}(X - Y = u) = \sum_{k=0}^{\infty} \text{pr}[(X = k + u) \cap (Y = k)]$$

$$= \sum_{k=0}^{\infty} \text{pr}(X = k + u)\text{pr}(Y = k) = \sum_{k=0}^{\infty}(1-\pi_x)\pi_x^{k+u}(1-\pi_y)\pi_y^k$$

$$= (1-\pi_x)(1-\pi_y)\pi_x^u \sum_{k=0}^{\infty}(\pi_x\pi_y)^k = \frac{(1-\pi_x)(1-\pi_y)}{(1-\pi_x\pi_y)}\pi_x^u = \theta\pi_x^u.$$

And,

$$\text{pr}(X - Y = -u) = \sum_{k=0}^{\infty} \text{pr}(X = k)\text{pr}(Y = k + u)$$

$$= \sum_{k=0}^{\infty}(1-\pi_x)\pi_x^k(1-\pi_y)\pi_y^{k+u} = (1-\pi_x)(1-\pi_y)\pi_y^u \sum_{k=0}^{\infty}(\pi_x\pi_y)^k$$

$$= \frac{(1-\pi_x)(1-\pi_y)}{(1-\pi_x\pi_y)}\pi_y^u = \theta\pi_y^u.$$

Hence, we have

$$p_U(0) = \theta \text{ and } p_U(u) = \theta\left(\pi_x^u + \pi_y^u\right), u = 1, 2, \ldots, \infty.$$

It can be shown directly that $\sum_{u=0}^{\infty} p_U(u) = 1$.

(c) For the available data, the observed value of $U$ is $u = 2$. So, under the assumption that $\pi_x = \pi_y = \pi$, say, it follows that

$$\text{pr}(U \geq 2|\pi_x = \pi_y = \pi) = \sum_{u=2}^{\infty} 2\theta\pi^u = 2\left(\frac{1-\pi}{1+\pi}\right)\sum_{u=2}^{\infty}\pi^u$$

$$= 2\left(\frac{1-\pi}{1+\pi}\right)\left(\frac{\pi^2}{1-\pi}\right) = \frac{2\pi^2}{(1+\pi)}.$$

Given the restrictions $\pi_x \leq 0.10$ and $\pi_y \leq 0.10$, the largest possible value of $\text{pr}(U \geq 2|\pi_x = \pi_y = \pi) = 2\pi^2/(1+\pi)$ is $2(0.10)^2/(1+0.10) = 0.018$. So, these data provide fairly strong statistical evidence that $\pi_x \neq \pi_y$.

**Solution 3.30***

(a)

$$\psi(t) = \text{E}[(t+1)^Y] = \sum_{y=1}^{\infty}(t+1)^y \frac{\left(e^{\theta}-1\right)^{-1}\theta^y}{y!}$$

$$= (e^{\theta}-1)^{-1}\sum_{y=1}^{\infty}\frac{[\theta(t+1)]^y}{y!} = (e^{\theta}-1)^{-1}\left\{\sum_{y=0}^{\infty}\frac{[\theta(t+1)]^y}{y!} - 1\right\}$$

$$= \frac{e^{\theta(t+1)}-1}{(e^{\theta}-1)}.$$

(b) For $Y$ a positive integer,

$$\text{E}\left[(t+1)^Y\right] = \text{E}\left[\sum_{j=0}^{Y} C_j^Y t^j (1)^{Y-j}\right]$$

$$= \text{E}\left\{1 + tY + \frac{t^2}{2}Y(Y-1) + \cdots\right\}$$

$$= 1 + t\text{E}(Y) + \frac{t^2}{2}\text{E}[Y(Y-1)] + \cdots.$$

So,

$$\left.\frac{d\psi(t)}{dt}\right|_{t=0} = \text{E}(Y), \quad \left.\frac{d^2\psi(t)}{dt^2}\right|_{t=0} = \text{E}[Y(Y-1)].$$

Thus,

$$\frac{d}{dt}\left[\frac{e^{\theta(t+1)}-1}{(e^{\theta}-1)}\right]_{t=0} = \left[\frac{\theta e^{\theta(t+1)}}{(e^{\theta}-1)}\right]_{t=0} = \frac{\theta e^{\theta}}{(e^{\theta}-1)} = \text{E}(Y).$$

And,

$$\frac{d^2}{dt^2}\left[\frac{e^{\theta(t+1)}-1}{(e^{\theta}-1)}\right]_{t=0} = \frac{d}{dt}\left[\frac{\theta e^{\theta(t+1)}}{(e^{\theta}-1)}\right]_{t=0} = \left[\frac{\theta^2 e^{\theta(t+1)}}{(e^{\theta}-1)}\right]_{t=0}$$

$$= \frac{\theta^2 e^{\theta}}{(e^{\theta}-1)} = \text{E}[Y(Y-1)].$$

Finally,

$$\begin{aligned}
V(Y) &= E[Y(Y-1)] + E(Y) - [E(Y)]^2 \\
&= \frac{\theta^2 e^{\theta}}{(e^{\theta}-1)} + \frac{\theta e^{\theta}}{(e^{\theta}-1)} - \frac{\theta^2 e^{2\theta}}{(e^{\theta}-1)^2} \\
&= \frac{\theta e^{\theta}(e^{\theta}-\theta-1)}{(e^{\theta}-1)^2}.
\end{aligned}$$

(c)

$$\begin{aligned}
p_X(x) &= \text{pr}(X=x) = \text{pr}[(Y+Z)=x] \\
&= \sum_{l=0}^{x-1} \text{pr}(Y=x-l)\text{pr}(Z=l) \\
&= \sum_{l=0}^{x-1} \frac{(e^{\theta}-1)^{-1}\theta^{x-l}}{(x-l)!} \cdot \frac{(\pi\theta)^l e^{-\pi\theta}}{l!} \\
&= \left[e^{\pi\theta}(e^{\theta}-1)\right]^{-1} \theta^x \sum_{l=0}^{x-1} \frac{\pi^l}{l!(x-l)!} \\
&= \frac{\left[e^{\pi\theta}(e^{\theta}-1)\right]^{-1}\theta^x}{x!} \sum_{l=0}^{x-1} C_l^x \pi^l \\
&= \frac{\left[e^{\pi\theta}(e^{\theta}-1)\right]^{-1}\theta^x}{x!} \left[\sum_{l=0}^{x} C_l^x \pi^l (1)^{x-l} - \pi^x\right] \\
&= \left[e^{\pi\theta}(e^{\theta}-1)\right]^{-1} \frac{\theta^x}{x!}\left[(\pi+1)^x - \pi^x\right], \quad x = 1, 2, \ldots, \infty.
\end{aligned}$$

(d) Since $Z \sim \text{POI}(\pi\theta)$,

$$E(X) = E(Y+Z) = E(Y) + E(Z) = \frac{\theta e^{\theta}}{(e^{\theta}-1)} + \pi\theta = \theta\left[\frac{e^{\theta}}{(e^{\theta}-1)} + \pi\right].$$

And,

$$V(X) = V(Y) + V(Z) = \frac{\theta e^{\theta}(e^{\theta}-\theta-1)}{(e^{\theta}-1)^2} + \pi\theta = \theta\left[\frac{e^{\theta}(e^{\theta}-\theta-1)}{(e^{\theta}-1)^2} + \pi\right].$$

**Solution 3.31***

(a) pr(both kidneys are still functioning at time $t$) =

$$\text{pr}[(X_1 \geq t) \cap (X_2 \geq t)] = \text{pr}(X_1 \geq t)\text{pr}(X_2 \geq t) = \left[\int_t^{\infty} \alpha e^{-\alpha x}\,dx\right]^2 = e^{-2\alpha t}.$$

(b) pr(exactly one kidney is functioning at time $t$) =pr[$(U < t) \cap (Y \geq t)$]. Now, from part (a),

$$\mathrm{F}_U(u) = \mathrm{pr}(U \leq u) = 1 - \mathrm{e}^{-2\alpha u},$$

so that

$$\mathrm{f}_U(u) = 2\alpha \mathrm{e}^{-2\alpha u}, \quad u > 0.$$

Hence,

$$\mathrm{f}_{U,Y}(u,y) = \mathrm{f}_U(u)\mathrm{f}_Y(y|U=u) = \left(2\alpha\mathrm{e}^{-2\alpha u}\right)\left[\beta\mathrm{e}^{-\beta(y-u)}\right],$$
$$0 < u < y < \infty.$$

Thus,

$$\mathrm{pr}[(U < t) \cap (Y \geq t)] = \int_0^t \int_t^\infty (2\alpha\mathrm{e}^{-2\alpha u})[\beta\mathrm{e}^{-\beta(y-u)}]\,\mathrm{d}y\,\mathrm{d}u$$
$$= \frac{2\alpha}{(\beta - 2\alpha)}\left(\mathrm{e}^{-2\alpha t} - \mathrm{e}^{-\beta t}\right), \quad t \geq 0.$$

(c) $\mathrm{F}_T(t) = \mathrm{pr}(T \leq t) = 1 - \pi_0(t) - \pi_1(t)$, so that

$$\mathrm{f}_T(t) = \frac{\mathrm{d}}{\mathrm{d}t}[\mathrm{F}_T(t)] = \frac{2\alpha\beta}{(\beta - 2\alpha)}\left(\mathrm{e}^{-2\alpha t} - \mathrm{e}^{-\beta t}\right), \quad t \geq 0.$$

(d) The marginal density of $Y$ is given by the expression

$$\mathrm{f}_Y(y) = \int_0^y (2\alpha\mathrm{e}^{-2\alpha u})[\beta\mathrm{e}^{-\beta(y-u)}]\,\mathrm{d}u = \frac{2\alpha\beta}{(\beta - 2\alpha)}\left(\mathrm{e}^{-2\alpha y} - \mathrm{e}^{-\beta y}\right),$$
$$y \geq 0.$$

So, as expected, $T$ and $Y$ have exactly the same distribution (i.e., $T = Y$). Finally,

$$\mathrm{E}(T) = \mathrm{E}(Y) = \mathrm{E}_u[\mathrm{E}(Y|U=u)] = \mathrm{E}_u\left(u + \frac{1}{\beta}\right) = \frac{1}{2\alpha} + \frac{1}{\beta},$$

and

$$\mathrm{V}(T) = \mathrm{V}(Y) = \mathrm{V}_u[\mathrm{E}(Y|U=u)] + \mathrm{E}_u[\mathrm{V}(Y|U=u)]$$
$$= \mathrm{V}_u\left(u + \frac{1}{\beta}\right) + \mathrm{E}\left(\frac{1}{\beta^2}\right) = \frac{1}{4\alpha^2} + \frac{1}{\beta^2}.$$

**Solution 3.32***

(a) Since $\bar{X}$ and $S^2$ are independent random variables, and since $\bar{X} \sim \mathrm{N}(\mu, \sigma^2/n)$, it follows that $\mathrm{E}[T_{(n-1)}] = \sqrt{n}\mathrm{E}(\bar{X} - \mu)\mathrm{E}(S^{-1}) = 0$. Thus,

$$\begin{aligned}\mathrm{cov}\left[\bar{X}, \frac{\sqrt{n}(\bar{X} - \mu)}{S}\right] &= \mathrm{E}\left\{\bar{X}\left[\frac{\sqrt{n}(\bar{X} - \mu)}{S}\right]\right\} = \sqrt{n}\mathrm{E}[\bar{X}(\bar{X} - \mu)]\mathrm{E}(S^{-1}) \\ &= \sqrt{n}\mathrm{E}(\bar{X}^2 - \mu^2)\mathrm{E}(S^{-1}) = \sqrt{n}\mathrm{V}(\bar{X})\mathrm{E}(S^{-1}) \\ &= \frac{\sigma^2}{\sqrt{n}}\mathrm{E}(S^{-1}).\end{aligned}$$

Now, since

$$U = \frac{(n-1)S^2}{\sigma^2} \sim \chi^2_{(n-1)} = \mathrm{GAMMA}\left[\alpha = 2, \beta = \frac{(n-1)}{2}\right],$$

it follows that

$$\mathrm{E}(U^r) = \int_0^\infty u^r \frac{u^{\left(\frac{n-1}{2}\right)-1} e^{-u/2}}{\Gamma\left(\frac{n-1}{2}\right) 2^{\left(\frac{n-1}{2}\right)}} du = \frac{\Gamma\left(\frac{n-1}{2} + r\right)}{\Gamma\left(\frac{n-1}{2}\right)} 2^r, \left(\frac{n-1}{2}\right) + r > 0.$$

So,

$$\mathrm{E}(U^{-1/2}) = \mathrm{E}\left\{\left[\frac{(n-1)S^2}{\sigma^2}\right]^{-1/2}\right\} = \frac{\sigma}{\sqrt{n-1}}\mathrm{E}(S^{-1}) = \frac{\Gamma\left(\frac{n-1}{2} - \frac{1}{2}\right)}{\Gamma\left(\frac{n-1}{2}\right)} 2^{-1/2},$$

so that

$$\mathrm{E}(S^{-1}) = \frac{\Gamma\left(\frac{n-2}{2}\right)}{\Gamma\left(\frac{n-1}{2}\right)} \sqrt{\frac{(n-1)}{2\sigma^2}}, \quad n > 2.$$

Thus,

$$\mathrm{cov}[\bar{X}, T_{(n-1)}] = \sigma\sqrt{\frac{(n-1)}{2n}} \frac{\Gamma\left(\frac{n-2}{2}\right)}{\Gamma\left(\frac{n-1}{2}\right)}, \quad n > 2.$$

Now,

$$\mathrm{V}[T_{(n-1)}] = \mathrm{E}\left[T^2_{n-1}\right] = \mathrm{E}\left[\frac{n(\bar{X} - \mu)^2}{S^2}\right] = n\mathrm{E}[(\bar{X} - \mu)^2]\mathrm{E}(S^{-2}) = \sigma^2\mathrm{E}(S^{-2}).$$

And, since

$$\mathrm{E}(U^{-1}) = \frac{\sigma^2}{(n-1)}\mathrm{E}(S^{-2}) = \frac{\Gamma\left(\frac{n-1}{2}-1\right)}{\Gamma\left(\frac{n-1}{2}\right)}2^{-1} = (n-3)^{-1}, \quad n > 3,$$

it follows that

$$\mathrm{E}(S^{-2}) = \frac{(n-1)}{(n-3)\sigma^2}$$

and hence

$$\mathrm{V}[T_{(n-1)}] = \frac{(n-1)}{(n-3)}, \quad n > 3.$$

Finally,

$$\mathrm{corr}[\bar{X}, T_{(n-1)}] = \sqrt{\frac{(n-3)}{2}}\frac{\Gamma\left(\frac{n-2}{2}\right)}{\Gamma\left(\frac{n-1}{2}\right)}, \quad n > 3.$$

When $n = 4$, $\mathrm{corr}[\bar{X}, T_{(n-1)}] = \sqrt{2/\pi} = 0.798$; for $n = 6$, $\mathrm{corr}[\bar{X}, T_{(n-1)}] = 2\sqrt{2/3\pi} = 0.921$.

(b) Using the stated "large $x$" approximation for $\Gamma(x)$, we have

$$\mathrm{corr}[\bar{X}, T_{(n-1)}] \approx \sqrt{\frac{(n-3)}{2}}\frac{\sqrt{2\pi}\mathrm{e}^{-\left(\frac{n-2}{2}\right)}\left(\frac{n-2}{2}\right)^{\left[\left(\frac{n-2}{2}\right)-\frac{1}{2}\right]}}{\sqrt{2\pi}\mathrm{e}^{-\left(\frac{n-1}{2}\right)}\left(\frac{n-1}{2}\right)^{\left[\left(\frac{n-1}{2}\right)-\frac{1}{2}\right]}}$$

$$= \left[\frac{\mathrm{e}(n-3)(n-2)^{(n-3)}}{(n-1)^{(n-2)}}\right]^{1/2},$$

so that $\lim_{n\to\infty} \mathrm{corr}[\bar{X}, T_{(n-1)}] = 1$.

As $n \to \infty$, the distribution of $T_{(n-1)}$ becomes that of a standard normal random variable $Z = (\bar{X} - \mu)/(\sigma/\sqrt{n})$, and the random variable $Z = -\sqrt{n}\mu/\sigma + (\sqrt{n}/\sigma)\bar{X}$ is a straight line function of the random variable $\bar{X}$.

**Solution 3.33***. Let $X$ and $Y$ denote the numbers of ones obtained when the two balanced die are each tossed $n$ times, and let $Z$ be the number of ones obtained when the unbalanced die is tossed $n$ times. Further, let $U = \min(X, Y)$. Then, $n^*$ is the smallest value of $n$ such that $\mathrm{pr}(Z < U) \geq 0.99$.

Now, for $u = 0, 1, \ldots, n$,

$$\begin{aligned}
\text{pr}(U = u) &= \text{pr}[(X = u) \cap (Y > u)] + \text{pr}[(X > u) \cap (Y = u)] \\
&\quad + \text{pr}[(X = u) \cap (Y = u)] \\
&= \text{pr}(X = u)\text{pr}(Y > u) + \text{pr}(X > u)\text{pr}(Y = u) \\
&\quad + \text{pr}(X = u)\text{pr}(Y = u) \\
&= 2C_u^n \left(\frac{1}{6}\right)^u \left(\frac{5}{6}\right)^{n-u} \left[\sum_{j=u+1}^{n} C_j^n \left(\frac{1}{6}\right)^j \left(\frac{5}{6}\right)^{n-j}\right] \\
&\quad + \left[C_u^n \left(\frac{1}{6}\right)^u \left(\frac{5}{6}\right)^{n-u}\right]^2.
\end{aligned}$$

Finally, determine $n^*$ as the smallest value of $n$ such that

$$\begin{aligned}
\text{pr}(Z < U) &= \sum_{z=0}^{n-1} \sum_{u=z+1}^{n} \text{pr}(Z = z)\text{pr}(U = u) \\
&= \sum_{z=0}^{n-1} \sum_{u=z+1}^{n} C_z^n \left(\frac{1}{6} - \epsilon\right)^z \left(\frac{5}{6} + \epsilon\right)^{n-z} \text{pr}(U = u) \geq 0.99.
\end{aligned}$$

**Solution 3.34**

(a)

$$\begin{aligned}
F_Y(y) = \text{pr}(Y \leq y) &= \text{pr}[(X_1 - X_2) \leq y] = \text{pr}[X_1 \leq (X_2 + y)] \\
&= \int_{-\infty}^{\infty} \text{pr}[X_1 \leq (x_2 + y) | X_2 = x_2] f_{X_2}(x_2)\, dx_2 \\
&= \int_{-\infty}^{\infty} e^{-e^{-(x_2+y)}} e^{-e^{-x_2}} e^{-x_2}\, dx_2 \\
&= \int_{-\infty}^{\infty} e^{-e^{-x_2}(1+e^{-y})} e^{-x_2}\, dx_2.
\end{aligned}$$

Let $u = -e^{-x_2}(1 + e^{-y})$, so that $du = e^{-x_2}(1 + e^{-y})\, dx_2$. So,

$$\begin{aligned}
F_Y(y) &= \int_{-\infty}^{0} e^u (1 + e^{-y})^{-1}\, du \\
&= (1 + e^{-y})^{-1} \int_{-\infty}^{0} e^u\, du
\end{aligned}$$

$$= (1 + e^{-y})^{-1}[e^u]_{-\infty}^0$$

$$= (1 + e^{-y})^{-1}, \quad -\infty < y < +\infty.$$

(b) Let $X_{1(m)}$ denote the largest observation in the first $m$ observations, and let $X_{2(m)}$ denote the largest observation in the second $m$ observations. Then, from part (a), the variable

$$m\theta[X_{1(m)} - \beta] - m\theta[X_{2(m)} - \beta] = m\theta[X_{1(m)} - X_{2(m)}]$$

has the CDF

$$\left[1 + e^{-m\theta(X_{1(m)} - X_{2(m)})}\right]^{-1}.$$

So,

$$\text{pr}\left\{\left|m\theta(X_{1(m)} - X_{2(m)})\right| \leq k\right\} = \text{pr}\left\{\theta \leq \frac{k}{m\left|X_{1(m)} - X_{2(m)}\right|}\right\}$$

$$= \text{pr}\left\{-k \leq m\theta\left(X_{1(m)} - X_{2(m)}\right) \leq k\right\}$$

$$= \frac{1}{(1 + e^{-k})} - \frac{1}{(1 + e^{k})} = \frac{\left(e^k - 1\right)}{\left(e^k + 1\right)}, \quad k > 0.$$

So, if $k_{1-\alpha}$ is chosen so that

$$\frac{\left(e^{k_{1-\alpha}} - 1\right)}{\left(e^{k_{1-\alpha}} + 1\right)} = (1 - \alpha),$$

then

$$U = \frac{k_{1-\alpha}}{m\left|X_{1(m)} - X_{2(m)}\right|}.$$

**Solution 3.35***

(a) Clearly, $0 < U_i = F_X(X_i) < 1$. And,

$$F_{U_i}(u_i) = \text{pr}(U_i \leq u_i) = \text{pr}\left[F_X(X_i) \leq u_i\right]$$

$$= \text{pr}\left\{F_X^{-1}\left[F_X(X_i)\right] \leq F_X^{-1}(u_i)\right\}$$

$$= \text{pr}\left[X_i \leq F_X^{-1}(u_i)\right] = F_X\left[F_X^{-1}(u_i)\right] = u_i.$$

So, since $dF_{U_i}(u_i)/du_i = f_{U_i}(u_i) = 1, 0 < u_i < 1$, it follows that $U_i = F_X(X_i)$ has a uniform distribution on the interval $(0, 1)$.

(b) Given the result in part (a), it follows that $U_{(1)}, U_{(2)}, \ldots, U_{(n)}$ can be considered to be the order statistics based on a random sample $U_1, U_2, \ldots, U_n$ of size $n$ from a uniform distribution on the interval $(0, 1)$. Hence, from the theory of order statistics, it follows directly that

$$f_{U_{(r)},U_{(s)}}(u_{(r)}, u_{(s)}) = \frac{n!}{(r-1)!(s-r-1)!(n-s)!} u_{(r)}^{r-1} (u_{(s)} - u_{(r)})^{s-r-1} \times (1 - u_{(s)})^{n-s}, \quad 0 < u_{(r)} < u_{(s)} < 1.$$

Now, using the method of transformations, let $V_{rs} \equiv V = [U_{(s)} - U_{(r)}]$ and $W = U_{(r)}$, so that $U_{(s)} = (V + W)$ and $U_{(r)} = W$. Then, the Jacobian $J = 1$, and so

$$f_{V,W}(v, w) = \frac{n!}{(r-1)!(s-r-1)!(n-s)!} w^{r-1} v^{s-r-1} (1 - v - w)^{n-s},$$

$$0 < (v + w) < 1.$$

Then, using the relationship $y = w/(1 - v)$, so that $dy = dw/(1 - v)$, and making use of properties of the beta distribution, we have

$$\begin{aligned} f_V(v) &= \int_0^{1-v} \frac{n!}{(r-1)!(s-r-1)!(n-s)!} w^{r-1} v^{s-r-1} \\ &\quad \times (1 - v - w)^{n-s}\, dw \\ &= \int_0^1 \frac{n!}{(r-1)!(s-r-1)!(n-s)!} [(1-v)y]^{r-1} v^{s-r-1} \\ &\quad \times [(1-v) - (1-v)y]^{n-s} (1-v)\, dy \\ &= v^{s-r-1} (1-v)^{[(r-1)+(n-s)+1]} \int_0^1 \frac{n!}{(r-1)!(s-r-1)!(n-s)!} \\ &\quad \times y^{r-1} (1-y)^{n-s}\, dy \\ &= \frac{\Gamma(n+1)}{\Gamma(s-r)\Gamma(n-s+r+1)} v^{s-r-1} (1-v)^{n-s+r}, \quad 0 < v < 1, \end{aligned}$$

so that $V_{rs} \sim \text{BETA}(\alpha = s - r, \beta = n - s + r + 1)$.

(c) If $n = 10, r = 1, s = 10$, and $p = 0.80$, then $f_{V_{rs}}(v) = 90v^8(1 - v), 0 < v < 1$, so that

$$\theta = \text{pr}(V_{1n} \geq 0.80) = \int_{0.80}^1 90v^8(1 - v)\, dv = 0.6242.$$

**Solution 3.36***

(a) $V(Y_{ijk}) = V(\beta_j) + V[\gamma_{ij}] + V(\epsilon_{ijk}) = \sigma_\beta^2 + \sigma_{\alpha\beta}^2 + \sigma_\epsilon^2.$

(b) For fixed $i$ and $j$, and for $k \neq k'$, we have

$$\begin{aligned}\text{cov}(Y_{ijk}, Y_{ijk'}) &= \text{cov}[\mu_i + \beta_j + \gamma_{ij} + \epsilon_{ijk}, \mu_i + \beta_j + \gamma_{ij} + \epsilon_{ijk'}] \\ &= \text{cov}(\beta_j, \beta_j) + \text{cov}(\gamma_{ij}, \gamma_{ij}) \\ &= \text{V}(\beta_j) + \text{V}(\gamma_{ij}) = \sigma_\beta^2 + \sigma_\gamma^2.\end{aligned}$$

(c) For $i \neq i'$, for fixed $j$, and for $k \neq k'$, we have

$$\begin{aligned}\text{cov}(Y_{ijk}, Y_{i'jk'}) &= \text{cov}[\mu_i + \beta_j + \gamma_{ij} + \epsilon_{ijk}, \mu_{i'} + \beta_j + \gamma_{i'j} + \epsilon_{i'jk'}] \\ &= \text{cov}(\beta_j, \beta_j) = \text{V}(\beta_j) = \sigma_\beta^2.\end{aligned}$$

(d) Now,

$$\begin{aligned}\bar{Y}_{ij} &= n_{ij}^{-1} \sum_{k=1}^{n_{ij}} [\mu_i + \beta_j + \gamma_{ij} + \epsilon_{ijk}] \\ &= \mu_i + \beta_j + \gamma_{ij} + \bar{\epsilon}_{ij},\end{aligned}$$

where $\bar{\epsilon}_{ij} = n_{ij}^{-1} \sum_{k=1}^{n_{ij}} \epsilon_{ij}$.
So,

$$\text{E}(\bar{Y}_{ij}) = \mu_i \quad \text{and} \quad \text{V}(\bar{Y}_{ij}) = \sigma_\beta^2 + \sigma_\gamma^2 + \frac{\sigma_\epsilon^2}{n_{ij}}.$$

Also, for $i \neq i'$,

$$\begin{aligned}\text{cov}(\bar{Y}_{ij}, \bar{Y}_{i'j}) &= \text{cov}[\mu_i + \beta_j + \gamma_{ij} + \bar{\epsilon}_{ij}, \mu_{i'} + \beta_j + \gamma_{i'j} + \bar{\epsilon}_{i'j}] \\ &= \text{cov}(\beta_j, \beta_j) = \text{V}(\beta_j) = \sigma_\beta^2.\end{aligned}$$

(e) First,

$$\begin{aligned}L = \sum_{i=1}^{t} a_i \bar{Y}_i &= \sum_{i=1}^{t} a_i \left( c^{-1} \sum_{j=1}^{c} \bar{Y}_{ij} \right) = c^{-1} \sum_{i=1}^{t} a_i \sum_{j=1}^{c} \bar{Y}_{ij} \\ &= c^{-1} \sum_{i=1}^{t} a_i \sum_{j=1}^{c} [\mu_i + \beta_j + \gamma_{ij} + \bar{\epsilon}_{ij}] \\ &= \sum_{i=1}^{t} a_i \mu_i + c^{-1} \sum_{i=1}^{t} \sum_{j=1}^{c} a_i \gamma_{ij} + c^{-1} \sum_{i=1}^{t} \sum_{j=1}^{c} a_i \bar{\epsilon}_{ij},\end{aligned}$$

since $\left( \sum_{i=1}^{t} a_i \right) \left( \sum_{j=1}^{c} \beta_j \right) = 0$.

So, we clearly have

$$\mathrm{E}(L) = \sum_{i=1}^{t} a_i \mu_i.$$

And,

$$\mathrm{V}(L) = c^{-2}\sum_{i=1}^{t}\sum_{j=1}^{c} a_i^2\sigma_\gamma^2 + c^{-2}\sum_{i=1}^{t}\sum_{j=1}^{c} a_i^2\frac{\sigma_\epsilon^2}{n_{ij}}$$

$$= \frac{\sigma_\gamma^2}{c}\sum_{i=1}^{t} a_i^2 + \frac{\sigma_\epsilon^2}{c^2}\sum_{i=1}^{t} a_i^2 \sum_{j=1}^{c} n_{ij}^{-1}.$$

For the special case when $a_1 = +1, a_2 = -1, a_3 = a_4 = \cdots = a_t = 0$, we obtain

$$\mathrm{E}(L) = (\mu_1 - \mu_2),$$

which is the true difference in average effects for drug therapies 1 and 2; and,

$$\mathrm{V}(L) = \frac{2\sigma_\gamma^2}{c} + \frac{\sigma_\epsilon^2}{c^2}\left(\sum_{j=1}^{c} n_{1j}^{-1} + \sum_{j=1}^{c} n_{2j}^{-1}\right).$$

The random variable $L = \sum_{i=1}^{t} a_i \bar{Y}_i$ is called a *contrast* since $\sum_{i=1}^{t} a_i = 0$, and $L$ can be used to estimate unbiasedly important comparisons among the set $\{\mu_1, \mu_2, \ldots, \mu_t\}$ of $t$ drug therapy average effects. For example, if $a_1 = +1$, $a_2 = -\frac{1}{2}, a_3 = -\frac{1}{2}, a_4 = a_5 = \cdots = a_t = 0$, then $\mathrm{E}(L) = \mu_1 - \frac{1}{2}(\mu_2 + \mu_3)$, which is a comparison between the average effect of drug therapy 1 and the mean of the average effects of drug therapies 2 and 3.

**Solution 3.37***

(a) Now, $\mathrm{V}(S^2) = (n-1)^{-2}\mathrm{V}\left[\sum_{i=1}^{n}(X_i - \bar{X})^2\right]$. So,

$$\mathrm{V}\left[\sum_{i=1}^{n}(X_i - \bar{X})^2\right] = \mathrm{E}\left\{\left[\sum_{i=1}^{n}(X_i - \bar{X})^2\right]^2\right\} - \left\{\mathrm{E}\left[\sum_{i=1}^{n}(X_i - \bar{X})^2\right]\right\}^2$$

$$= \mathrm{E}\left\{\left[\sum_{i=1}^{n}(X_i - \mu)^2 - n(\bar{X} - \mu)^2\right]^2\right\} - (n-1)^2\sigma^4.$$

Now,

$$\mathrm{E}\left\{\left[\sum_{i=1}^{n}(X_i-\mu)^2-n(\bar{X}-\mu)^2\right]^2\right\}=\mathrm{E}\left\{\left[\sum_{i=1}^{n}(X_i-\mu)^2\right]^2\right\}+n^2\mathrm{E}\left[(\bar{X}-\mu)^4\right]$$
$$-2n\mathrm{E}\left[(\bar{X}-\mu)^2\sum_{i=1}^{n}(X_i-\mu)^2\right].$$

Now,

$$\mathrm{E}\left\{\left[\sum_{i=1}^{n}(X_i-\mu)^2\right]^2\right\}=\mathrm{E}\left[\sum_{i=1}^{n}(X_i-\mu)^4+2\sum_{i=1}^{n-1}\sum_{j=i+1}^{n}(X_i-\mu)^2(X_j-\mu)^2\right]$$
$$=n\mu_4+n(n-1)\sigma^4.$$

And,

$$\mathrm{E}\left[(\bar{X}-\mu)^4\right]=\mathrm{E}\left[\left(\frac{1}{n}\sum_{i=1}^{n}X_i-\mu\right)^4\right]=E\left\{\left[\frac{1}{n}\sum_{i=1}^{n}(X_i-\mu)\right]^4\right\}$$
$$=n^{-4}\mathrm{E}\left\{\left[\sum_{i=1}^{n}(X_i-\mu)\right]^4\right\}$$
$$=n^{-4}\mathrm{E}\left\{\sum\sum\cdots\sum\frac{4!}{(\prod_{i=1}^{n}\alpha_i!)}\prod_{i=1}^{n}(X_i-\mu)^{\alpha_i}\right\}$$
$$=n^{-4}\sum\sum\cdots\sum\frac{4!}{(\prod_{i=1}^{n}\alpha_i!)}\prod_{i=1}^{n}\mathrm{E}\left[(X_i-\mu)^{\alpha_i}\right],$$

where the notation $\sum\sum\cdots\sum$ denotes the summation over all nonnegative integer value choices for $\alpha_1, \alpha_2, \ldots, \alpha_n$ such that $\sum_{i=1}^{n}\alpha_i = 4$.

Noting that $\mathrm{E}\left[(X_i-\mu)^{\alpha_i}\right] = 0$ when $\alpha_i = 1$, we only have to consider two types of terms: i) $\alpha_i = 4$ for some $i$ and $\alpha_j = 0$ for all $j(\neq i)$; and, ii) $\alpha_i = 2$ and $\alpha_j = 2$ for $i \neq j$, and $\alpha_k = 0$ for all $k(\neq i \text{ or } j)$. There are $n$ of the former terms, each with expectation $\mu_4$, and there are $n(n-1)/2$ of the latter terms, each with expectation $6\sigma^4$. Thus,

$$\mathrm{E}\left[(\bar{X}-\mu)^4\right]=n^{-4}\left[n\mu_4+\frac{n(n-1)}{2}(6\sigma^4)\right]=n^{-3}\left[\mu_4+3(n-1)\sigma^4\right].$$

And,

$$\mathrm{E}\left[(\bar{X}-\mu)^2\sum_{i=1}^{n}(X_i-\mu)^2\right]=\mathrm{E}\left\{\left[\frac{1}{n}\sum_{i=1}^{n}(X_i-\mu)\right]^2\left[\sum_{i=1}^{n}(X_i-\mu)^2\right]\right\}$$

$$=\frac{1}{n^2}\mathrm{E}\left\{\left[\sum_{i=1}^{n}(X_i-\mu)^2\right]^2+2\left[\sum_{k=1}^{n}(X_k-\mu)^2\right]\sum_{i=1}^{n-1}\sum_{j=i+1}^{n}(X_i-\mu)(X_j-\mu)\right\}$$

$$=\frac{1}{n^2}\mathrm{E}\left\{\left[\sum_{i=1}^{n}(X_i-\mu)^2\right]^2\right\}=\frac{1}{n^2}\left[n\mu_4+n(n-1)\sigma^4\right]$$

$$=n^{-1}\left[\mu_4+(n-1)\sigma^4\right].$$

So, we have

$$\mathrm{V}\left[\sum_{i=1}^{n}(X_i-\bar{X})^2\right]=\left[n\mu_4+n(n-1)\sigma^4\right]+n^2\left\{n^{-3}\left[\mu_4+3(n-1)\sigma^4\right]\right\}$$

$$-2n\left\{n^{-1}\left[\mu_4+(n-1)\sigma^4\right]\right\}-(n-1)^2\sigma^4$$

$$=\frac{(n-1)^2}{n}\mu_4-\frac{(n-1)(n-3)}{n}\sigma^4.$$

Finally,

$$\mathrm{V}(S^2)=(n-1)^{-2}\mathrm{V}\left[\sum_{i=1}^{n}(X_i-\bar{X})^2\right]=\frac{1}{n}\left[\mu_4-\left(\frac{n-3}{n-1}\right)\sigma^4\right].$$

(b) For the POI($\lambda$) distribution, $\sigma^2=\lambda$ and $\mu_4=\lambda(1+3\lambda)$, giving

$$\mathrm{V}(S^2)=\frac{1}{n}\left\{[\lambda(1+3\lambda)]-\left(\frac{n-3}{n-1}\right)\lambda^2\right\}=\frac{\lambda}{n}\left[1+\left(\frac{2n}{n-1}\right)\lambda\right].$$

For the N($\mu,\sigma^2$) distribution, $\mu_4=3\sigma^4$, giving

$$\mathrm{V}(S^2)=\frac{1}{n}\left[3\sigma^4-\left(\frac{n-3}{n-1}\right)\sigma^4\right]=\frac{2\sigma^4}{(n-1)}.$$

**Solution 3.38***

(a) First,

$$\mathrm{cov}(\bar{X},S^2)=\mathrm{E}\{[\bar{X}-\mathrm{E}(\bar{X})][S^2-\mathrm{E}(S^2)]\}=\mathrm{E}(\bar{X}S^2)-\mathrm{E}(\bar{X})\mathrm{E}(S^2)$$

$$=\frac{1}{n(n-1)}\mathrm{E}\left[\left(\sum_{i=1}^{n}X_i\right)\left(\sum_{j=1}^{n}(X_j-\bar{X})^2\right)\right]-\mu\sigma^2.$$

Now,

$$\mathrm{E}\left[\left(\sum_{i=1}^{n} X_i\right)\left(\sum_{j=1}^{n}(X_j-\bar{X})^2\right)\right] = \mathrm{E}\left[\sum_{i=1}^{n} X_i(X_i-\bar{X})^2\right] + \mathrm{E}\left[\sum_{\text{all } i\neq j} X_i(X_j-\bar{X})^2\right].$$

And,

$$\begin{aligned}
X_i(X_i-\bar{X})^2 &= (X_i-\mu)[(X_i-\mu)-(\bar{X}-\mu)]^2 + \mu(X_i-\bar{X})^2 \\
&= (X_i-\mu)[(X_i-\mu)^2 - 2(\bar{X}-\mu)(X_i-\mu) \\
&\quad + (\bar{X}-\mu)^2] + \mu(X_i-\bar{X})^2 \\
&= (X_i-\mu)^3 - 2(\bar{X}-\mu)(X_i-\mu)^2 \\
&\quad + (\bar{X}-\mu)^2(X_i-\mu) + \mu(X_i-\bar{X})^2 \\
&= (X_i-\mu)^3 - \left[\frac{2}{n}\sum_{l=1}^{n}(X_l-\mu)\right](X_i-\mu)^2 \\
&\quad + \left[\frac{1}{n}\sum_{l=1}^{n}(X_l-\mu)\right]^2 (X_i-\mu) + \mu(X_i-\bar{X})^2 \\
&= (X_i-\mu)^3 - \frac{2}{n}(X_i-\mu)^3 - \frac{2}{n}(X_i-\mu)^2 \sum_{\text{all } l(\neq i)} (X_l-\mu) \\
&\quad + \frac{1}{n^2}\left[\sum_{l=1}^{n}(X_l-\mu)^2 + \sum_{\text{all } l\neq l'} (X_l-\mu)(X_{l'}-\mu)\right] \\
&\quad \times (X_i-\mu) + \mu(X_i-\bar{X})^2 \\
&= (X_i-\mu)^3 - \frac{2}{n}(X_i-\mu)^3 - \frac{2}{n}(X_i-\mu)^2 \sum_{\text{all } l(\neq i)} (X_l-\mu) \\
&\quad + \frac{1}{n^2}(X_i-\mu)^3 + \frac{1}{n^2}(X_i-\mu) \sum_{\text{all } l(\neq i)} (X_l-\mu)^2 \\
&\quad + \frac{1}{n^2}(X_i-\mu) \sum_{\text{all } l\neq l'} (X_l-\mu)(X_{l'}-\mu) + \mu(X_i-\bar{X})^2.
\end{aligned}$$

Finally,

$$\begin{aligned}
\mathrm{E}\left[\sum_{i=1}^{n} X_i(X_i-\bar{X})^2\right] &= n\mu_3 - 2\mu_3 - 0 + \frac{\mu_3}{n} + 0 + 0 + \mu(n-1)\sigma^2 \\
&= \left[\frac{(n-1)^2}{n}\right]\mu_3 + (n-1)\mu\sigma^2.
\end{aligned}$$

Also, for $i \neq j$,

$$
\begin{aligned}
X_i(X_j - \bar{X})^2 &= (X_i - \mu)[(X_j - \mu) - (\bar{X} - \mu)]^2 + \mu(X_j - \bar{X})^2 \\
&= (X_i - \mu)[(X_j - \mu)^2 - 2(\bar{X} - \mu)(X_j - \mu) \\
&\quad + (\bar{X} - \mu)^2] + \mu(X_j - \bar{X})^2 \\
&= (X_i - \mu)(X_j - \mu)^2 - 2(\bar{X} - \mu)(X_i - \mu)(X_j - \mu) \\
&\quad + (\bar{X} - \mu)^2(X_i - \mu) + \mu(X_j - \bar{X})^2 \\
&= (X_i - \mu)(X_j - \mu)^2 - \left[\frac{2}{n}\sum_{l=1}^{n}(X_l - \mu)\right] \\
&\quad \times (X_i - \mu)(X_j - \mu) + \left[\frac{1}{n}\sum_{l=1}^{n}(X_l - \mu)\right]^2 \\
&\quad \times (X_i - \mu) + \mu(X_j - \bar{X})^2 \\
&= (X_i - \mu)(X_j - \mu)^2 - \frac{2}{n}(X_i - \mu)(X_j - \mu) \\
&\quad \times \sum_{l=1}^{n}(X_l - \mu) + \frac{1}{n^2}\left[\sum_{l=1}^{n}(X_l - \mu)^2\right. \\
&\quad \left. + \sum_{\text{all } l \neq l'}(X_l - \mu)(X_{l'} - \mu)\right](X_i - \mu) + \mu(X_j - \bar{X})^2 \\
&= (X_i - \mu)(X_j - \mu)^2 - \frac{2}{n}(X_i - \mu)(X_j - \mu)\sum_{l=1}^{n}(X_l - \mu) \\
&\quad + \frac{1}{n^2}(X_i - \mu)^3 + \frac{1}{n^2}(X_i - \mu)\sum_{\text{all } l(\neq i)}(X_l - \mu)^2 \\
&\quad + (X_i - \mu)\sum_{\text{all } l \neq l'}(X_l - \mu)(X_{l'} - \mu) + \mu(X_j - \bar{X})^2.
\end{aligned}
$$

Hence, we have

$$
\begin{aligned}
E\left[\sum_{\text{all } i \neq j} X_i(X_j - \bar{X})^2\right] &= 0 - 0 + \frac{1}{n^2}[n(n-1)\mu_3] + 0 + 0 \\
&\quad + \mu(n-1)^2\sigma^2 \\
&= \left(\frac{n-1}{n}\right)\mu_3 + (n-1)^2\mu\sigma^2.
\end{aligned}
$$

Finally, we have

$$\begin{aligned}\text{cov}(\bar{X}, S^2) &= \frac{1}{n(n-1)}\left\{\left[\frac{(n-1)^2}{n}\right]\mu_3 + (n-1)\mu\sigma^2 \right.\\ &\quad \left. + \left(\frac{n-1}{n}\right)\mu_3 + (n-1)^2\mu\sigma^2\right\} - \mu\sigma^2.\\ &= \mu_3/n.\end{aligned}$$

(b) The joint distribution of $X_1$ and $X_2$ is equal to

$$\begin{aligned}p_{X_1,X_2}(x_1, x_2) &= p_{X_1}(x_1)p_{X_2}(x_2)\\ &= \left(\tfrac{1}{4}\right)^{|x_1|}\left(\tfrac{1}{2}\right)^{1-|x_1|}\left(\tfrac{1}{4}\right)^{|x_2|}\left(\tfrac{1}{2}\right)^{1-|x_2|}\\ &= \left(\tfrac{1}{2}\right)^{2+|x_1|+|x_2|}, \qquad x_1 = -1, 0, 1 \quad \text{and} \quad x_2 = -1, 0, 1.\end{aligned}$$

Hence, it follows directly that the following pairs of $(\bar{X}, S^2)$ values occur with the following probabilities: (−1,0) with probability 1/16, (−1/2,1/2) with probability 1/4, (0,0) with probability 1/4, (0,2) with probability 1/8, (1/2, 1/2) with probability 1/4, and (1,0) with probability 1/16.

Hence, it is easy to show by direct computation that $\text{cov}(\bar{X}, S^2) = 0$. However, since

$$\text{pr}(S^2 = 0|\bar{X} = 1) = 1 \neq \text{pr}(S^2 = 0) = \tfrac{3}{8},$$

it follows that $\bar{X}$ and $S^2$ are *dependent* random variables.

Clearly, $p_X(x)$ is a discrete distribution that is symmetric about $E(X) = 0$, so that $\mu_3 = 0$. Thus, it follows from part (a) that, as shown directly, $\text{cov}(\bar{X}, S^2) = 0$. More generally, the random variables $\bar{X}$ and $S^2$ are independent when selecting a random sample from a normally distributed parent population, but are generally dependent when selecting a random sample from a nonnormal parent population.

# 4

# Estimation Theory

## 4.1 Concepts and Notation

### 4.1.1 Point Estimation of Population Parameters

Let the random variables $X_1, X_2, \ldots, X_n$ constitute a sample of size $n$ from some population with properties depending on a row vector $\boldsymbol{\theta} = (\theta_1, \theta_2, \ldots, \theta_p)$ of $p$ unknown parameters, where the *parameter space* is the set $\Omega$ of all possible values of $\boldsymbol{\theta}$. In the most general situation, the $n$ random variables $X_1, X_2, \ldots, X_n$ are allowed to be mutually dependent and to have different distributions (e.g., different means and different variances).

A *point estimator* or a *statistic* is any scalar function $U(X_1, X_2, \ldots, X_n) \equiv U(\boldsymbol{X})$ of the random variables $X_1, X_2, \ldots, X_n$, but *not* of $\boldsymbol{\theta}$. A point estimator or statistic is itself a random variable since it is a function of the random vector $\boldsymbol{X} = (X_1, X_2, \ldots, X_n)$. In contrast, the corresponding *point estimate* or *observed statistic* $U(x_1, x_2, \ldots, x_n) \equiv U(\boldsymbol{x})$ is the realized (or observed) numerical value of the point estimator or statistic that is computed using the realized (or observed) numerical values $x_1, x_2, \ldots, x_n$ of $X_1, X_2, \ldots, X_n$ for the particular sample obtained.

Some popular methods for obtaining a row vector $\hat{\boldsymbol{\theta}} = (\hat{\theta}_1, \hat{\theta}_2, \ldots, \hat{\theta}_p)$ of point estimators of the elements of the row vector $\boldsymbol{\theta} = (\theta_1, \theta_2, \ldots, \theta_p)$, where $\hat{\theta}_j \equiv \hat{\theta}_j(\boldsymbol{X})$ for $j = 1, 2, \ldots, p$, are the following:

#### 4.1.1.1 *Method of Moments (MM)*

For $j = 1, 2, \ldots, p$, let

$$M_j = \frac{1}{n}\sum_{i=1}^{n} X_i^j \quad \text{and} \quad \text{E}(M_j) = \frac{1}{n}\sum_{i=1}^{n} \text{E}(X_i^j),$$

where $\text{E}(M_j)$, $j = 1, 2, \ldots, p$, is a function of the elements of $\boldsymbol{\theta}$.

Then, $\hat{\boldsymbol{\theta}}_{\text{mm}}$, the MM estimator of $\boldsymbol{\theta}$, is obtained as the solution of the $p$ equations

$$M_j = \text{E}(M_j), j = 1, 2, \ldots, p.$$

#### 4.1.1.2 Unweighted Least Squares (ULS)

Let $Q_u = \sum_{i=1}^{n}[X_i - E(X_i)]^2$. Then, $\hat{\boldsymbol{\theta}}_{uls}$, the ULS estimator of $\boldsymbol{\theta}$, is chosen to *minimize* $Q_u$ and is defined as the solution of the $p$ equations

$$\frac{\partial Q_u}{\partial \theta_j} = 0, \quad j = 1, 2, \ldots, p.$$

#### 4.1.1.3 Weighted Least Squares (WLS)

Let $Q_w = \sum_{i=1}^{n} w_i[X_i - E(X_i)]^2$, where $w_1, w_2, \ldots, w_n$ are weights. Then, $\hat{\boldsymbol{\theta}}_{wls}$, the WLS estimator of $\boldsymbol{\theta}$, is chosen to minimize $Q_w$ and is defined as the solution of the $p$ equations

$$\frac{\partial Q_w}{\partial \theta_j} = 0, \quad j = 1, 2, \ldots, p.$$

#### 4.1.1.4 Maximum Likelihood (ML)

Let $\mathcal{L}(\boldsymbol{x}; \boldsymbol{\theta})$ denote the *likelihood function*, which is often simply the joint distribution of the random variables $X_1, X_2, \ldots, X_n$. Then, $\hat{\boldsymbol{\theta}}_{ml}$, the ML estimator (MLE) of $\boldsymbol{\theta}$, is chosen to *maximize* $\mathcal{L}(\boldsymbol{x}; \boldsymbol{\theta})$ and is defined as the solution of the $p$ equations

$$\frac{\partial \ln \mathcal{L}(\boldsymbol{x}; \boldsymbol{\theta})}{\partial \theta_j} = 0, \quad j = 1, 2, \ldots, p.$$

If $\tau(\boldsymbol{\theta})$ is a scalar function of $\boldsymbol{\theta}$, then $\tau(\hat{\boldsymbol{\theta}}_{ml})$ is the MLE of $\tau(\boldsymbol{\theta})$; this is known as the *invariance property* of MLEs.

### 4.1.2 Data Reduction and Joint Sufficiency

The goal of any statistical analysis is to quantify the information contained in a sample of size $n$ by making valid and precise statistical inferences using the smallest possible number of point estimators or statistics. This data reduction goal leads to the concept of joint sufficiency.

#### 4.1.2.1 Joint Sufficiency

The statistics $U_1(\boldsymbol{X}), U_2(\boldsymbol{X}), \ldots, U_k(\boldsymbol{X}),\ k \geq p$, are jointly sufficient for the parameter vector $\boldsymbol{\theta}$ if and only if the conditional distribution of $\boldsymbol{X}$ given $U_1(\boldsymbol{X}) = U_1(\boldsymbol{x}), U_2(\boldsymbol{X}) = U_2(\boldsymbol{x}), \ldots, U_k(\boldsymbol{X}) = U_k(\boldsymbol{x})$ does not *in any way* depend on $\boldsymbol{\theta}$. More specifically, the phrase "in any way" means that the conditional distribution of $\boldsymbol{X}$, including the domain of $\boldsymbol{X}$, given the $k$ sufficient statistics is not a function of $\boldsymbol{\theta}$. In other words, the jointly sufficient statistics $U_1(\boldsymbol{X}), U_2(\boldsymbol{X}), \ldots, U_k(\boldsymbol{X})$ utilize all the information about $\boldsymbol{\theta}$ that is contained in the sample $\boldsymbol{X}$.

#### 4.1.2.2 Factorization Theorem

To demonstrate joint sufficiency, the *Factorization Theorem* (Halmos and Savage, 1949) is quite useful: Let $\boldsymbol{X}$ be a discrete or continuous random vector with distribution $\mathcal{L}(\boldsymbol{x};\boldsymbol{\theta})$. Then, $U_1(\boldsymbol{X}), U_2(\boldsymbol{X}), \ldots, U_k(\boldsymbol{X})$ are jointly sufficient for $\boldsymbol{\theta}$ if and only if there are nonnegative functions $g[U_1(\boldsymbol{x}), U_2(\boldsymbol{x}), \ldots, U_k(\boldsymbol{x});\boldsymbol{\theta}]$ and $h(\boldsymbol{x})$ such that

$$\mathcal{L}(\boldsymbol{x};\boldsymbol{\theta}) = g[U_1(\boldsymbol{x}), U_2(\boldsymbol{x}), \ldots, U_k(\boldsymbol{x});\boldsymbol{\theta}]h(\boldsymbol{x}),$$

where, given $U_1(\boldsymbol{X}) = U_1(\boldsymbol{x}), U_2(\boldsymbol{X}) = U_2(\boldsymbol{x}), \ldots, U_k(\boldsymbol{X}) = U_k(\boldsymbol{x})$, the function $h(\boldsymbol{x})$ in no way depends on $\boldsymbol{\theta}$. Also, any one-to-one function of a sufficient statistic is also a sufficient statistic.

As an important example, a *family* $\mathcal{F}_d = \{\mathrm{p}_X(x;\boldsymbol{\theta}), \boldsymbol{\theta} \in \Omega\}$ of discrete probability distributions is a member of the *exponential* family of distributions if $\mathrm{p}_X(x;\boldsymbol{\theta})$ can be written in the general form

$$\mathrm{p}_X(x;\boldsymbol{\theta}) = h(x)b(\boldsymbol{\theta})\mathrm{e}^{\sum_{j=1}^{k} w_j(\boldsymbol{\theta})v_j(x)},$$

where $h(x) \geq 0$ does not in any way depend on $\boldsymbol{\theta}$, $b(\boldsymbol{\theta}) \geq 0$ does not depend on $x$, $w_1(\boldsymbol{\theta}), w_2(\boldsymbol{\theta}), \ldots, w_k(\boldsymbol{\theta})$ are real-valued functions of $\boldsymbol{\theta}$ but not of $x$, and $v_1(x), v_2(x), \ldots, v_k(x)$ are real-valued functions of $x$ but not of $\boldsymbol{\theta}$. Then, if $X_1, X_2, \ldots, X_n$ constitute a random sample of size $n$ from $\mathrm{p}_X(x;\boldsymbol{\theta})$, so that $\mathrm{p}_{\boldsymbol{X}}(\boldsymbol{x};\boldsymbol{\theta}) = \prod_{i=1}^{n} \mathrm{p}_X(x_i;\boldsymbol{\theta})$, it follows that

$$\mathrm{p}_{\boldsymbol{X}}(\boldsymbol{x};\boldsymbol{\theta}) = \left\{[b(\boldsymbol{\theta})]^n \mathrm{e}^{\sum_{j=1}^{k} w_j(\boldsymbol{\theta})[\sum_{i=1}^{n} v_j(x_i)]}\right\}\left\{\prod_{i=1}^{n} h(x_i)\right\};$$

so, by the Factorization Theorem, the $p$ statistics $U_j(\boldsymbol{X}) = \sum_{i=1}^{n} v_j(X_i), j = 1, 2, \ldots, k$, are jointly sufficient for $\boldsymbol{\theta}$. The above results also hold when considering a family $\mathcal{F}_c = \{\mathrm{f}_X(x;\boldsymbol{\theta}), \boldsymbol{\theta} \in \Omega\}$ of continuous probability distributions. Many important families of distributions are members of the exponential family; these include the binomial, Poisson, and negative binomial families in the discrete case, and the normal, gamma, and beta families in the continuous case.

### 4.1.3 Methods for Evaluating the Properties of a Point Estimator

For now, consider the special case of one unknown parameter $\theta$.

#### 4.1.3.1 Mean-Squared Error (MSE)

The *mean-squared error* of $\hat{\theta}$ as an estimator of the parameter $\theta$ is defined as

$$\mathrm{MSE}(\hat{\theta}, \theta) = \mathrm{E}[(\hat{\theta} - \theta)^2] = \mathrm{V}(\hat{\theta}) + [\mathrm{E}(\hat{\theta}) - \theta]^2,$$

where $V(\hat{\theta})$ is the variance of $\hat{\theta}$ and $[E(\hat{\theta}) - \theta]^2$ is the squared-bias of $\hat{\theta}$ as an estimator of the parameter $\theta$. An estimator with small MSE has both a small variance and a small squared-bias.

Using MSE as the criterion for choosing among a class of possible estimators of $\theta$ is problematic because this class is too large. Hence, it is common practice to limit the class of possible estimators of $\theta$ to those estimators that are *unbiased* estimators of $\theta$. More formally, $\hat{\theta}$ is an *unbiased* estimator of the parameter $\theta$ if $E(\hat{\theta}) = \theta$ for all $\theta \in \Omega$. Then, if $\hat{\theta}$ is an unbiased estimator of $\theta$, we have $\text{MSE}(\hat{\theta}, \theta) = V(\hat{\theta})$, so that the criterion for choosing among competing unbiased estimators of $\theta$ is based solely on variance considerations.

#### 4.1.3.2 Cramér–Rao Lower Bound (CRLB)

Let $\mathcal{L}(x;\theta)$ denote the distribution of the random vector $\boldsymbol{X}$, and let $\hat{\theta}$ be *any* unbiased estimator of the parameter $\theta$. Then, under certain mathematical regularity conditions, it can be shown (Rao, 1945; Cramér, 1946) that

$$V(\hat{\theta}) \geq \frac{1}{E_x\left[(\partial \ln \mathcal{L}(x;\theta)/\partial\theta)^2\right]} = \frac{1}{-E_x\left[\partial^2 \ln \mathcal{L}(x;\theta)/\partial\theta^2\right]}.$$

In the important special case when $X_1, X_2, \ldots, X_n$ constitute a random sample of size $n$ from the discrete probability distribution $p_X(x;\theta)$, so that $\mathcal{L}(x;\theta) = \prod_{i=1}^{n} p_X(x_i;\theta)$, then we obtain

$$V(\hat{\theta}) \geq \frac{1}{nE_x\left\{\left(\partial \ln[p_X(x;\theta)]/\partial\theta\right)^2\right\}} = \frac{1}{-nE_x\left\{\partial^2 \ln[p_X(x;\theta)]/\partial\theta^2\right\}}.$$

A completely analogous result holds when $X_1, X_2, \ldots, X_n$ constitute a random sample of size $n$ from the density function $f_X(x;\theta)$. For further discussion, see Lehmann (1983).

#### 4.1.3.3 Efficiency

The *efficiency* of any unbiased estimator $\hat{\theta}$ of $\theta$ relative to the CRLB is defined as

$$\text{EFF}(\hat{\theta}, \theta) = \frac{\text{CRLB}}{V(\hat{\theta})}, \quad 0 \leq \text{EFF}(\hat{\theta}, \theta) \leq 1,$$

and the corresponding *asymptotic efficiency* is $\lim_{n\to\infty} \text{EFF}(\hat{\theta}, \theta)$.

There are situations when no unbiased estimator of $\theta$ achieves the CRLB. In such a situation, we can utilize the *Rao–Blackwell Theorem* (Rao, 1945; Blackwell, 1947) to aid in the search for that unbiased estimator with the smallest variance (i.e., the minimum variance unbiased estimator or MVUE).

First, we need to introduce the concept of a *complete* sufficient statistic:

#### 4.1.3.4 Completeness

The family $\mathcal{F}_u = \{\mathrm{p}_U(u;\theta), \theta \in \Omega\}$, or $\mathcal{F}_u = \{\mathrm{f}_U(u;\theta), \theta \in \Omega\}$, for the sufficient statistic $U$ is called *complete* (or, equivalently, $U$ is a complete sufficient statistic) if the condition $\mathrm{E}[g(U)] = 0$ for all $\theta \in \Omega$ implies that $\mathrm{pr}[g(U) = 0] = 1$ for all $\theta \in \Omega$.

As an important special case, for an exponential family with $U_j(\boldsymbol{X}) = \sum_{i=1}^{n} v_j(X_i)$ for $j = 1, 2, \ldots, k$, the vector of sufficient statistics

$$\boldsymbol{U}(\boldsymbol{X}) = [U_1(\boldsymbol{X}), U_2(\boldsymbol{X}), \ldots, U_k(\boldsymbol{X})]$$

is complete if $\{w_1(\boldsymbol{\theta}), w_2(\boldsymbol{\theta}), \ldots, w_k(\boldsymbol{\theta}) : \boldsymbol{\theta} \in \Omega\}$ contains an open set in $\Re^k$.

#### 4.1.3.5 Rao–Blackwell Theorem

Let $U^* \equiv U^*(\boldsymbol{X})$ be *any* unbiased point estimator of $\theta$, and let $U \equiv U(\boldsymbol{X})$ be a sufficient statistic for $\theta$. Then, $\hat{\theta} = \mathrm{E}(U^*|U = u)$ is an unbiased point estimator of $\theta$, and $\mathrm{V}(\hat{\theta}) \leq \mathrm{V}(U^*)$. If $U$ is a *complete* sufficient statistic for $\theta$, then $\hat{\theta}$ is the *unique* (with probability one) MVUE of $\theta$.

It is important to emphasize that the variance of the MVUE of $\theta$ may not achieve the CRLB.

### 4.1.4 Interval Estimation of Population Parameters

#### 4.1.4.1 Exact Confidence Intervals

An *exact* $100(1-\alpha)\%$ confidence interval (CI) for a parameter $\theta$ involves two random variables, $L$ (called the *lower limit*) and $U$ (called the *upper limit*), defined so that

$$\mathrm{pr}(L < \theta < U) = (1 - \alpha),$$

where typically $0 < \alpha \leq 0.10$.

The construction of exact CIs often involves the properties of statistics based on random samples from normal populations. Some illustrations are as follows.

#### 4.1.4.2 Exact CI for the Mean of a Normal Distribution

Let $X_1, X_2, \ldots, X_n$ constitute a random sample from a $\mathrm{N}(\mu, \sigma^2)$ parent population. The sample mean is $\bar{X} = n^{-1}\sum_{i=1}^{n} X_i$ and the sample variance is $S^2 = (n-1)^{-1}\sum_{i=1}^{n}(X_i - \bar{X})^2$.

Then,

$$\bar{X} \sim \mathrm{N}\left(\mu, \frac{\sigma^2}{n}\right),$$

$$\frac{(n-1)S^2}{\sigma^2} = \frac{\sum_{i=1}^{n}(X_i - \bar{X})^2}{\sigma^2} \sim \chi^2_{n-1},$$

and $\bar{X}$ and $S^2$ are independent random variables.

In general, if $Z \sim \mathrm{N}(0, 1), U \sim \chi^2_{\nu}$, and $Z$ and $U$ are independent random variables, then the random variable $T_{\nu} = Z/\sqrt{U/\nu} \sim t_{\nu}$; that is, $T_{\nu}$ has a $t$-distribution with $\nu$ degrees of freedom (df). Thus, the random variable

$$T_{n-1} = \frac{(\bar{X} - \mu)/(\sigma/\sqrt{n})}{\sqrt{[(n-1)S^2/\sigma^2]/(n-1)}} = \frac{\bar{X} - \mu}{S/\sqrt{n}} \sim t_{n-1}.$$

With $t_{n-1,1-\alpha/2}$ defined so that $\mathrm{pr}(T_{n-1} < t_{n-1,1-\alpha/2}) = 1 - \alpha/2$, we then have

$$\begin{aligned}
(1-\alpha) &= \mathrm{pr}(-t_{n-1,1-\alpha/2} < T_{n-1} < t_{n-1,1-\alpha/2}) \\
&= \mathrm{pr}\left[-t_{n-1,1-\alpha/2} < \frac{\bar{X} - \mu}{S/\sqrt{n}} < t_{n-1,1-\alpha/2}\right] \\
&= \mathrm{pr}\left[\bar{X} - t_{n-1,1-\alpha/2}\frac{S}{\sqrt{n}} < \mu < \bar{X} + t_{n-1,1-\alpha/2}\frac{S}{\sqrt{n}}\right].
\end{aligned}$$

Thus,

$$L = \bar{X} - t_{n-1,1-\alpha/2}\frac{S}{\sqrt{n}} \quad \text{and} \quad U = \bar{X} + t_{n-1,1-\alpha/2}\frac{S}{\sqrt{n}},$$

giving

$$\bar{X} \pm t_{n-1,1-\alpha/2}\frac{S}{\sqrt{n}}$$

as the *exact* $100(1-\alpha)\%$ CI for $\mu$ based on a random sample $X_1, X_2, \ldots, X_n$ of size $n$ from a $\mathrm{N}(\mu, \sigma^2)$ parent population.

#### 4.1.4.3 Exact CI for a Linear Combination of Means of Normal Distributions

More generally, for $i = 1, 2, \ldots, k$, let $X_{i1}, X_{i2}, \ldots, X_{in_i}$ constitute a random sample of size $n_i$ from a $\mathrm{N}(\mu_i, \sigma_i^2)$ parent population. Then,

i. For $i = 1, 2, \ldots, k, \bar{X}_i = n_i^{-1}\sum_{j=1}^{n_i} X_{ij} \sim \mathrm{N}\left(\mu_i, \frac{\sigma_i^2}{n_i}\right)$;

ii. For $i = 1, 2, \ldots, k$, $\frac{(n_i - 1)S_i^2}{\sigma_i^2} = \frac{\sum_{j=1}^{n_i}(X_{ij} - \bar{X}_i)^2}{\sigma_i^2} \sim \chi^2_{n_i - 1}$;

iii. The $2k$ random variables $\{\bar{X}_i, S_i^2\}_{i=1}^k$ are mutually independent.

Now, assuming $\sigma_i^2 = \sigma^2$ for all $i$ (i.e., assuming *variance homogeneity*), if $c_1, c_2, \ldots, c_k$ are known constants, then the random variable

$$\sum_{i=1}^{k} c_i \bar{X}_i \sim \mathrm{N}\left[\sum_{i=1}^{k} c_i \mu_i, \sigma^2 \left(\sum_{i=1}^{k} \frac{c_i^2}{n_i}\right)\right];$$

and, with $N = \sum_{i=1}^k n_i$, the random variable

$$\frac{\sum_{i=1}^{k}(n_i - 1)S_i^2}{\sigma^2} = \frac{\sum_{i=1}^{k}\sum_{j=1}^{n_i}(X_{ij} - \bar{X}_i)^2}{\sigma^2} \sim \chi^2_{N-k};$$

Thus, the random variable

$$T_{N-k} = \frac{\sum_{i=1}^{k} c_i \bar{X}_i - \sum_{i=1}^{k} c_i \mu_i}{S_p \sqrt{\sum_{i=1}^{k} \frac{c_i^2}{n_i}}} \sim t_{N-k},$$

where the pooled sample variance is $S_p^2 = \sum_{i=1}^k (n_i - 1)S_i^2/(N - k)$.

This gives

$$\sum_{i=1}^{k} c_i \bar{X}_i \pm t_{N-k, 1-\frac{\alpha}{2}} S_p \sqrt{\sum_{i=1}^{k} \frac{c_i^2}{n_i}}$$

as the *exact* $100(1-\alpha)\%$ CI for the parameter $\sum_{i=1}^k c_i \mu_i$.

In the special case when $k = 2, c_1 = +1$, and $c_2 = -1$, we obtain the well-known two-sample CI for $(\mu_1 - \mu_2)$, namely,

$$(\bar{X}_1 - \bar{X}_2) \pm t_{n_1+n_2-2, 1-\alpha/2} S_p \sqrt{\frac{1}{n_1} + \frac{1}{n_2}}.$$

#### 4.1.4.4 Exact CI for the Variance of a Normal Distribution

For $i = 1, 2, \ldots, k$, since $(n_i - 1)S_i^2/\sigma_i^2 \sim \chi^2_{n_i - 1}$, we have

$$(1 - \alpha) = \mathrm{pr}\left[\chi^2_{n_i - 1, \alpha/2} < \frac{(n_i - 1)S_i^2}{\sigma_i^2} < \chi^2_{n_i - 1, 1-\alpha/2}\right] = \mathrm{pr}(L < \sigma_i^2 < U),$$

where

$$L = \frac{(n_i - 1)S_i^2}{\chi^2_{n_i-1,1-\alpha/2}} \quad \text{and} \quad U = \frac{(n_i - 1)S_i^2}{\chi^2_{n_i-1,\alpha/2}},$$

and where $\chi^2_{n_i-1,\alpha/2}$ and $\chi^2_{n_i-1,1-\alpha/2}$ are, respectively, the $100\,(\alpha/2)$ and $100\,(1-\alpha/2)$ percentiles of the $\chi^2_{n_i-1}$ distribution.

#### 4.1.4.5 *Exact CI for the Ratio of Variances of Two Normal Distributions*

In general, if $U_1 \sim \chi^2_{\nu_1}, U_2 \sim \chi^2_{\nu_2}$, and $U_1$ and $U_2$ are independent random variables, then the random variable

$$F_{\nu_1,\nu_2} = \frac{U_1/\nu_1}{U_2/\nu_2} \sim f_{\nu_1,\nu_2};$$

that is, $F_{\nu_1,\nu_2}$ follows an $f$-distribution with $\nu_1$ numerator df and $\nu_2$ denominator df. As an example, when $k = 2$, the random variable

$$F_{n_1-1,n_2-1} = \frac{\left[[(n_1-1)S_1^2]/\sigma_1^2\right]/(n_1-1)}{\left[[(n_2-1)S_2^2]/\sigma_2^2\right]/(n_2-1)} = \left(\frac{S_1^2}{S_2^2}\right)\left(\frac{\sigma_2^2}{\sigma_1^2}\right) \sim f_{n_1-1,n_2-1}.$$

So, since $f_{n_1-1,n_2-1,\alpha/2} = f^{-1}_{n_2-1,n_1-1,1-\alpha/2}$, we have

$$\begin{aligned}(1-\alpha) &= \text{pr}\left[f^{-1}_{n_2-1,n_1-1,1-\alpha/2} < \left(\frac{S_1^2}{S_2^2}\right)\left(\frac{\sigma_2^2}{\sigma_1^2}\right) < f_{n_1-1,n_2-1,1-\alpha/2}\right] \\ &= \text{pr}\left[L < \left(\frac{\sigma_2^2}{\sigma_1^2}\right) < U\right],\end{aligned}$$

where

$$L = f^{-1}_{n_2-1,n_1-1,1-\alpha/2}\left(\frac{S_2^2}{S_1^2}\right) \quad \text{and} \quad U = f_{n_1-1,n_2-1,1-\alpha/2}\left(\frac{S_2^2}{S_1^2}\right),$$

and where $f_{n_1-1,n_2-1,1-\alpha/2}$ is the $100\,(1-\alpha/2)$ percentile of the $f$-distribution with $(n_1-1)$ numerator df and $(n_2-1)$ denominator df.

#### 4.1.4.6 *Large-Sample Approximate CIs*

By an *approximate* CI for a parameter $\theta$, we mean that the random variables $L$ and $U$ satisfy

$$\text{pr}(L < \theta < U) \approx (1-\alpha),$$

where typically $0 < \alpha \leq 0.10$.

The concepts of *convergence in distribution* (discussed in the front material for Chapter 3: Multivariate Distribution Theory) and *consistency*, coupled with the use of *Slutsky's Theorem* (see Serfling, 2002), are typically used for the development of ML-based approximate CIs.

#### 4.1.4.7 Consistency

A point estimator $\hat{\theta}$ is a *consistent* estimator of a parameter $\theta$ if, for every $\epsilon > 0$,

$$\lim_{n \to \infty} \mathrm{pr}(|\hat{\theta} - \theta| > \epsilon) = 0.$$

In this case, we say that $\hat{\theta}$ *converges in probability* to $\theta$, and we write $\hat{\theta} \xrightarrow{\mathrm{P}} \theta$. Two *sufficient* conditions so that $\hat{\theta} \xrightarrow{\mathrm{P}} \theta$ are

$$\lim_{n \to \infty} \mathrm{E}(\hat{\theta}) = \theta \text{ and } \lim_{n \to \infty} \mathrm{V}(\hat{\theta}) = 0.$$

#### 4.1.4.8 Slutsky's Theorem

If $V_n \xrightarrow{\mathrm{P}} c$, where $c$ is a constant, and if $W_n \xrightarrow{\mathrm{D}} W$, then

$$V_n W_n \xrightarrow{\mathrm{D}} cW \quad \text{and} \quad (V_n + W_n) \xrightarrow{\mathrm{D}} (c + W).$$

To develop ML-based large-sample approximate CIs, we make use of the following properties of the MLE $\hat{\boldsymbol{\theta}}_{\mathrm{ml}} \equiv \hat{\boldsymbol{\theta}}$ of $\boldsymbol{\theta}$, assuming $\mathcal{L}(\boldsymbol{x}; \boldsymbol{\theta})$ is the correct likelihood function and assuming that certain regularity conditions hold:

i. For $j = 1, 2, \ldots, p$, $\hat{\theta}_j$ is a consistent estimator of $\theta_j$. More generally, if the scalar function $\tau(\boldsymbol{\theta})$ is a continuous function of $\boldsymbol{\theta}$, then $\tau(\hat{\boldsymbol{\theta}})$ is a consistent estimator of $\tau(\boldsymbol{\theta})$.

ii.

$$\sqrt{n}(\hat{\boldsymbol{\theta}} - \boldsymbol{\theta}) \xrightarrow{\mathrm{D}} \mathrm{MVN}_p[\mathbf{0}, n\mathcal{I}^{-1}(\boldsymbol{\theta})],$$

where $\mathcal{I}(\boldsymbol{\theta})$ is the $(p \times p)$ *expected information matrix*, with $(j, j')$ element equal to

$$-\mathrm{E}_{\boldsymbol{x}} \left[ \frac{\partial^2 \ln \mathcal{L}(\boldsymbol{x}; \boldsymbol{\theta})}{\partial \theta_j \partial \theta_{j'}} \right],$$

and where $\mathcal{I}^{-1}(\boldsymbol{\theta})$ is the large-sample covariance matrix of $\hat{\boldsymbol{\theta}}$ based on expected information. In particular, the $(j, j')$ element of $\mathcal{I}^{-1}(\boldsymbol{\theta})$ is denoted $v_{jj'}(\boldsymbol{\theta}) = \mathrm{cov}(\hat{\theta}_j, \hat{\theta}_{j'})$, $j = 1, 2, \ldots, p$ and $j' = 1, 2, \ldots, p$.

#### 4.1.4.9 Construction of ML-Based CIs

As an illustration, properties (i) and (ii) will now be used to construct a large-sample ML-based approximate 100(1 − α)% CI for the parameter $\theta_j$.

First, with the $(j, j)$ diagonal element $v_{jj}(\boldsymbol{\theta})$ of $\mathcal{I}^{-1}(\boldsymbol{\theta})$ being the large-sample variance of $\hat{\theta}_j$ based on expected information, it follows that

$$\frac{\hat{\theta}_j - \theta_j}{\sqrt{v_{jj}(\boldsymbol{\theta})}} \xrightarrow{\text{D}} \text{N}(0,1) \quad \text{as } n \longrightarrow \infty.$$

Then, with $\mathcal{I}^{-1}(\hat{\boldsymbol{\theta}})$ denoting the *estimated* large-sample covariance matrix of $\hat{\boldsymbol{\theta}}$ based on expected information, and with the $(j, j)$ diagonal element $v_{jj}(\hat{\boldsymbol{\theta}})$ of $\mathcal{I}^{-1}(\hat{\boldsymbol{\theta}})$ being the estimated large-sample variance of $\hat{\theta}_j$ based on expected information, it follows by Sluksky's Theorem that

$$\frac{\hat{\theta}_j - \theta_j}{\sqrt{v_{jj}(\hat{\boldsymbol{\theta}})}} = \sqrt{\frac{v_{jj}(\boldsymbol{\theta})}{v_{jj}(\hat{\boldsymbol{\theta}})}} \left[ \frac{\hat{\theta}_j - \theta_j}{\sqrt{v_{jj}(\boldsymbol{\theta})}} \right] \xrightarrow{\text{D}} \text{N}(0,1) \quad \text{as } n \longrightarrow \infty$$

since $v_{jj}(\hat{\boldsymbol{\theta}})$ is a consistent estimator of $v_{jj}(\boldsymbol{\theta})$.

Thus, it follows from the above results that

$$\frac{\hat{\theta}_j - \theta_j}{\sqrt{v_{jj}(\hat{\boldsymbol{\theta}})}} \dot{\sim} \text{N}(0,1) \quad \text{for large } n.$$

Finally, with $Z_{1-\alpha/2}$ defined so that $\text{pr}(Z < Z_{1-\alpha/2}) = (1 - \alpha/2)$ when $Z \sim \text{N}(0,1)$, we have

$$\begin{aligned}
(1 - \alpha) &= \text{pr}(-Z_{1-\alpha/2} < Z < Z_{1-\alpha/2}) \\
&\approx \text{pr}\left[ -Z_{1-\alpha/2} < \frac{\hat{\theta}_j - \theta_j}{\sqrt{v_{jj}(\hat{\boldsymbol{\theta}})}} < Z_{1-\alpha/2} \right] \\
&= \text{pr}\left[ \hat{\theta}_j - Z_{1-\alpha/2}\sqrt{v_{jj}(\hat{\boldsymbol{\theta}})} < \theta_j < \hat{\theta}_j + Z_{1-\alpha/2}\sqrt{v_{jj}(\hat{\boldsymbol{\theta}})} \right].
\end{aligned}$$

Thus,

$$\hat{\theta}_j \pm Z_{1-\alpha/2}\sqrt{v_{jj}(\hat{\boldsymbol{\theta}})}$$

is the large-sample ML-based approximate 100(1 − α)% CI for the parameter $\theta_j$ based on expected information.

In practice, instead of the estimated expected information matrix, the estimated *observed* information matrix $\boldsymbol{I}(\boldsymbol{x};\hat{\boldsymbol{\theta}})$ is used, with its $(j,j')$ element equal to

$$-\left[\frac{\partial^2 \ln \mathcal{L}(\boldsymbol{x};\boldsymbol{\theta})}{\partial\theta_j \partial\theta_{j'}}\right]_{|\boldsymbol{\theta}=\hat{\boldsymbol{\theta}}}.$$

Then, with $\boldsymbol{I}^{-1}(\boldsymbol{x};\hat{\boldsymbol{\theta}})$ denoting the estimated large-sample covariance matrix of $\hat{\boldsymbol{\theta}}$ based on observed information, and with the $(j,j)$ diagonal element $v_{jj}(\boldsymbol{x};\hat{\boldsymbol{\theta}})$ of $\boldsymbol{I}^{-1}(\boldsymbol{x};\hat{\boldsymbol{\theta}})$ being the estimated large-sample variance of $\hat{\theta}_j$ based on observed information, it follows that

$$\hat{\theta}_j \pm Z_{1-\alpha/2}\sqrt{v_{jj}(\boldsymbol{x};\hat{\boldsymbol{\theta}})}$$

is the large-sample ML-based approximate $100(1-\alpha)\%$ CI for the parameter $\theta_j$ based on observed information.

#### 4.1.4.10 ML-Based CI for a Bernoulli Distribution Probability

As a simple one-parameter $(p = 1)$ example, let $X_1, X_2, \ldots, X_n$ constitute a random sample of size $n$ from the Bernoulli parent population

$$\mathrm{p}_X(x;\theta) = \theta^x(1-\theta)^{1-x}, \quad x = 0, 1 \quad \text{and} \quad 0 < \theta < 1,$$

and suppose that it is desired to develop a large-sample ML-based approximate $100(1-\alpha)\%$ CI for the parameter $\theta$. First, the appropriate likelihood function is

$$\mathcal{L}(\boldsymbol{x};\theta) = \prod_{i=1}^{n}\left[\theta^{x_i}(1-\theta)^{1-x_i}\right] = \theta^s(1-\theta)^{n-s},$$

where $s = \sum_{i=1}^{n} x_i$ is a sufficient statistic for $\theta$.

Now,

$$\ln \mathcal{L}(\boldsymbol{x};\theta) = s\ln\theta + (n-s)\ln(1-\theta),$$

so that the equation

$$\frac{\partial \ln \mathcal{L}(\boldsymbol{x};\theta)}{\partial\theta} = \frac{s}{\theta} - \frac{(n-s)}{(1-\theta)} = 0$$

gives $\hat{\theta} = \bar{X} = n^{-1}\sum_{i=1}^{n} X_i$ as the MLE of $\theta$.

And,

$$\frac{\partial^2 \ln \mathcal{L}(\boldsymbol{x};\theta)}{\partial\theta^2} = \frac{-s}{\theta^2} - \frac{(n-s)}{(1-\theta)^2},$$

so that

$$-\mathrm{E}\left[\frac{\partial^2 \ln \mathcal{L}(x;\theta)}{\partial\theta^2}\right] = \frac{n\theta}{\theta^2} + \frac{(n-n\theta)}{(1-\theta)^2} = \frac{n}{\theta(1-\theta)}.$$

Hence,

$$v_{11}(\hat{\theta}) = \left\{-\mathrm{E}\left[\frac{\partial^2 \ln \mathcal{L}(x;\theta)}{\partial\theta^2}\right]\right\}^{-1}_{|\theta=\hat{\theta}}$$

$$= v_{11}(x;\hat{\theta}) = \left\{-\frac{\partial^2 \ln \mathcal{L}(x;\theta)}{\partial\theta^2}\right\}^{-1}_{|\theta=\hat{\theta}}$$

$$= \frac{\bar{X}(1-\bar{X})}{n},$$

so that the large-sample ML-based approximate $100(1-\alpha)\%$ CI for $\theta$ is equal to

$$\bar{X} \pm Z_{1-\alpha/2}\sqrt{\frac{\bar{X}(1-\bar{X})}{n}}.$$

In this simple example, the same CI is obtained using either expected information or observed information. In more complicated situations, this will typically *not* happen.

#### 4.1.4.11 Delta Method

Let $Y = g(\mathbf{X})$, where $\mathbf{X} = (X_1, X_2, \ldots, X_k)$, $\boldsymbol{\mu} = (\mu_1, \mu_2, \ldots, \mu_k)$, $\mathrm{E}(X_i) = \mu_i$, $\mathrm{V}(X_i) = \sigma_i^2$, and $\mathrm{cov}(X_i, X_j) = \sigma_{ij}$ for $i \neq j, i = 1, 2, \ldots, k$ and $j = 1, 2, \ldots, k$. Then, a *first-order* (or *linear*) multivariate Taylor series approximation to $Y$ around $\boldsymbol{\mu}$ is

$$Y \approx g(\boldsymbol{\mu}) + \sum_{i=1}^{k} \frac{\partial g(\boldsymbol{\mu})}{\partial X_i}(X_i - \mu_i),$$

where

$$\frac{\partial g(\boldsymbol{\mu})}{\partial X_i} = \frac{\partial g(\mathbf{X})}{\partial X_i}_{|\mathbf{X}=\boldsymbol{\mu}}.$$

Thus, using the above linear approximation for $Y$, it follows that $\mathrm{E}(Y) \approx g(\boldsymbol{\mu})$ and that

$$\mathrm{V}(Y) \approx \sum_{i=1}^{k}\left[\frac{\partial g(\boldsymbol{\mu})}{\partial X_i}\right]^2 \sigma_i^2 + 2\sum_{i=1}^{k-1}\sum_{j=i+1}^{k}\left[\frac{\partial g(\boldsymbol{\mu})}{\partial X_i}\right]\left[\frac{\partial g(\boldsymbol{\mu})}{\partial X_j}\right]\sigma_{ij}.$$

The delta method for MLEs is as follows. For $q \leq p$, suppose that the $(1 \times q)$ row vector

$$\boldsymbol{\Phi}(\boldsymbol{\theta}) = [\tau_1(\boldsymbol{\theta}), \tau_2(\boldsymbol{\theta}), \ldots, \tau_q(\boldsymbol{\theta})]$$

involves $q$ scalar parametric functions of the parameter vector $\boldsymbol{\theta}$. Then,

$$\boldsymbol{\Phi}(\hat{\boldsymbol{\theta}}) = [\tau_1(\hat{\boldsymbol{\theta}}), \tau_2(\hat{\boldsymbol{\theta}}), \ldots, \tau_q(\hat{\boldsymbol{\theta}})]$$

is the MLE of $\boldsymbol{\Phi}(\boldsymbol{\theta})$.

Then, the $(q \times q)$ large-sample covariance matrix of $\boldsymbol{\Phi}(\hat{\boldsymbol{\theta}})$ based on expected information is

$$[\boldsymbol{\Delta}(\boldsymbol{\theta})]\mathcal{I}^{-1}(\boldsymbol{\theta})[\boldsymbol{\Delta}(\boldsymbol{\theta})]',$$

where the $(i, j)$ element of the $(q \times p)$ matrix $\boldsymbol{\Delta}(\boldsymbol{\theta})$ is equal to $\partial\tau_i(\boldsymbol{\theta})/\partial\theta_j, i = 1, 2, \ldots, q$ and $j = 1, 2, \ldots, p$.

Hence, the corresponding estimated large-sample covariance matrix of $\boldsymbol{\Phi}(\hat{\boldsymbol{\theta}})$ based on expected information is

$$[\boldsymbol{\Delta}(\hat{\boldsymbol{\theta}})]\mathcal{I}^{-1}(\hat{\boldsymbol{\theta}})[\boldsymbol{\Delta}(\hat{\boldsymbol{\theta}})]'.$$

Analogous expressions based on observed information are obtained by substituting $\boldsymbol{I}^{-1}(\boldsymbol{x}; \boldsymbol{\theta})$ for $\mathcal{I}^{-1}(\boldsymbol{\theta})$ and by substituting $\boldsymbol{I}^{-1}(\boldsymbol{x}; \hat{\boldsymbol{\theta}})$ for $\mathcal{I}^{-1}(\hat{\boldsymbol{\theta}})$ in the above two expressions.

The special case $q = p = 1$ gives

$$\mathrm{V}[\tau_1(\hat{\theta}_1)] \approx \left[\frac{\partial\tau_1(\theta_1)}{\partial\theta_1}\right]^2 \mathrm{V}(\hat{\theta}_1).$$

The corresponding large-sample ML-based approximate $100(1 - \alpha)\%$ CI for $\tau_1(\theta_1)$ based on expected information is equal to

$$\tau_1(\hat{\theta}_1) \pm Z_{1-\alpha/2}\sqrt{\left[\frac{\partial\tau_1(\theta_1)}{\partial\theta_1}\right]^2_{|\theta_1=\hat{\theta}_1} v_{11}(\hat{\theta}_1)}.$$

The corresponding CI based on observed information is obtained by substituting $v_{11}(\boldsymbol{x}; \hat{\theta}_1)$ for $v_{11}(\hat{\theta}_1)$ in the above expression.

#### *4.1.4.12 Delta Method CI for a Function of a Bernoulli Distribution Probability*

As a simple illustration, for the Bernoulli population example considered earlier, suppose that it is now desired to use the delta method to obtain a large-sample ML-based approximate $100(1 - \alpha)\%$ CI for the "odds"

$$\tau(\theta) = \frac{\theta}{(1 - \theta)} = \frac{\mathrm{pr}(X = 1)}{[1 - \mathrm{pr}(X = 1)]}.$$

So, by the invariance property, $\tau(\hat{\theta}) = \bar{X}/(1-\bar{X})$ is the MLE of $\tau(\theta)$ since $\hat{\theta} = \bar{X}$ is the MLE of $\theta$. And, via the delta method, the large-sample estimated variance of $\tau(\hat{\theta})$ is equal to

$$\hat{V}\left[\tau(\hat{\theta})\right] \approx \left[\frac{\partial\tau(\theta)}{\partial\theta}\right]^2_{|\theta=\hat{\theta}} \hat{V}(\hat{\theta})$$
$$= \left[\frac{1}{(1-\bar{X})^2}\right]^2 \left[\frac{\bar{X}(1-\bar{X})}{n}\right]$$
$$= \frac{\bar{X}}{n(1-\bar{X})^3}.$$

Finally, the large-sample ML-based approximate $100(1-\alpha)\%$ CI for $\tau(\theta) = \theta/(1-\theta)$ using the delta method is equal to

$$\frac{\bar{X}}{(1-\bar{X})} \pm Z_{1-\alpha/2}\sqrt{\frac{\bar{X}}{n(1-\bar{X})^3}}.$$

## EXERCISES

**Exercise 4.1.** Suppose that $Y_x \sim N(x\mu, x^3\sigma^2)$, $x = 1, 2, \ldots, n$. Further, assume that $\{Y_1, Y_2, \ldots, Y_n\}$ constitute a set of $n$ mutually independent random variables, and that $\sigma^2$ is a *known* positive constant. Consider the following three estimators of $\mu$:

1. $\hat{\mu}_1$, the method of moments estimator of $\mu$;
2. $\hat{\mu}_2$, the unweighted least squares estimator of $\mu$;
3. $\hat{\mu}_3$, the MLE of $\mu$.

(a) Derive expressions for $\hat{\mu}_1$, $\hat{\mu}_2$, and $\hat{\mu}_3$. (These expressions can involve summation signs.) Also, determine the exact distribution of each of these estimators of $\mu$.

(b) If $n = 5$, $\sigma^2 = 2$, and $y_x = (x+1)$ for $x = 1, 2, 3, 4$, and 5, construct what you believe to be the "best" exact 95% CI for $\mu$.

**Exercise 4.2.** An epidemiologist gathers data $(x_i, Y_i)$ on each of $n$ randomly chosen noncontiguous cities in the United States, where $x_i$ $(i = 1, 2, \ldots, n)$ is the known population size (in millions of people) in city $i$, and where $Y_i$ is the random variable denoting the number of people in city $i$ with liver cancer. It is reasonable to assume that $Y_i$ $(i = 1, 2, \ldots, n)$ has a Poisson distribution with mean $E(Y_i) = \theta x_i$, where $\theta$ $(>0)$ is an unknown parameter, and that $Y_1, Y_2, \ldots, Y_n$ constitute a set of mutually independent random variables.

(a) Find an explicit expression for the unweighted least-squares estimator $\hat{\theta}_{uls}$ of $\theta$. Also, find explicit expressions for $E(\hat{\theta}_{uls})$ and $V(\hat{\theta}_{uls})$.

(b) Find an explicit expression for the method of moments estimator $\hat{\theta}_{mm}$ of $\theta$. Also, find explicit expressions for E $(\hat{\theta}_{mm})$ and $V(\hat{\theta}_{mm})$.

(c) Find an explicit expression for the MLE $\hat{\theta}_{ml}$ of $\theta$. Also, find explicit expressions for $E(\hat{\theta}_{ml})$ and $V(\hat{\theta}_{ml})$.

(d) Find an explicit expression for the CRLB for the variance of any unbiased estimator of $\theta$. Which (if any) of the three estimators $\hat{\theta}_{uls}$, $\hat{\theta}_{mm}$, and $\hat{\theta}_{ml}$ achieve this lower bound?

**Exercise 4.3.** Suppose that $\hat{\theta}_1$ and $\hat{\theta}_2$ are two *unbiased* estimators of an unknown parameter $\theta$. Further, suppose that the variance of $\hat{\theta}_1$ is $\sigma_1^2$, that the variance of $\hat{\theta}_2$ is $\sigma_2^2$, and that $\text{corr}(\hat{\theta}_1, \hat{\theta}_2) = \rho, -1 < \rho < +1$. Define the parameter $\lambda = \sigma_1/\sigma_2$, and assume (without loss of generality) that $0 < \sigma_1 \leq \sigma_2 < +\infty$, so that $0 < \lambda \leq 1$. Consider the unbiased estimator of $\theta$ of the general form

$$\hat{\theta} = k\hat{\theta}_1 + (1 - k)\hat{\theta}_2,$$

where the quantity $k$ satisfies the inequality $-\infty < k < +\infty$.

(a) Develop an explicit expression (as a function of $\lambda$ and $\rho$) for that value of $k$ (say, $k^*$) that *minimizes* the variance of the unbiased estimator $\hat{\theta}$ of $\theta$. Discuss the special cases when $\rho > \lambda$ and when $\lambda = 1$.

(b) Let $\hat{\theta}^* = k^*\hat{\theta}_1 + (1 - k^*)\hat{\theta}_2$, where $k^*$ was determined in part (a). Develop a sufficient condition (as a function of $\lambda$ and $\rho$) for which

$$V(\hat{\theta}^*) < \sigma_1^2 = V(\hat{\theta}_1) \leq \sigma_2^2 = V(\hat{\theta}_2).$$

**Exercise 4.4.** Suppose that the random variable $X_i \sim N(\beta a_i, \sigma_i^2), i = 1, 2, \ldots, n$. Further, assume that $\{X_1, X_2, \ldots, X_n\}$ constitute a set of mutually independent random variables, that $\{a_1, a_2, \ldots, a_n\}$ constitute a set of known constants, and that $\{\sigma_1^2, \sigma_2^2, \ldots, \sigma_n^2\}$ constitute a set of known variances. A biostatistician suggests that the random variable

$$\hat{\beta} = \sum_{i=1}^{n} c_i X_i$$

would be an excellent estimator of the unknown parameter $\beta$ if the constants $c_1, c_2, \ldots, c_n$ are chosen so that the following two conditions simultaneously hold: (1) $E(\hat{\beta}) = \beta$; and, (2) $V(\hat{\beta})$ is a minimum.

Find explicit expressions for $c_1, c_2, \ldots, c_n$ (as functions of the $a_i$'s and $\sigma_i^2$'s) such that these two conditions simultaneously hold. Using these "optimal" choices of the $c_i$'s, what then is the exact distribution of this "optimal" estimator of $\beta$?

**Exercise 4.5.** For $i = 1, 2, \ldots, k$, let $Y_{i1}, Y_{12}, \ldots, Y_{in_i}$ constitute a random sample of size $n_i$ $(> 1)$ from a $N(\mu_i, \sigma^2)$ parent population. Further,

$$\bar{Y}_i = n_i^{-1} \sum_{j=1}^{n_i} Y_{ij} \quad \text{and} \quad S_i^2 = (n_i - 1)^{-1} \sum_{j=1}^{n_i} (Y_{ij} - \bar{Y}_i)^2$$

are, respectively, the sample mean and sample variance of the $n_i$ observations from this $N(\mu_i, \sigma^2)$ parent population. Further, let $N = \sum_{i=1}^{k} n_i$ denote the total number of observations.

(a) Consider estimating $\sigma^2$ with the estimator

$$\hat{\sigma}^2 = \sum_{i=1}^{k} w_i S_i^2,$$

where $w_i, w_2, \ldots, w_k$ are constants satisfying the constraint $\sum_{i=1}^{k} w_i = 1$. Prove rigorously that $E(\hat{\sigma}^2) = \sigma^2$, namely, that $\hat{\sigma}^2$ is an unbiased estimator of $\sigma^2$.

(b) Under the constraint $\sum_{i=1}^{k} w_i = 1$, find explicit expressions for $w_1, w_2, \ldots, w_k$ such that $V(\hat{\sigma}^2)$, the variance of $\hat{\sigma}^2$, is a *minimum*.

**Exercise 4.6.** Suppose that a professor in the Maternal and Child Health Department at the University of North Carolina at Chapel Hill administers a questionnaire (consisting of $k$ questions, each of which is to be answered "yes" or "no") to each of $n$ randomly selected mothers of infants less than 6 months of age in Chapel Hill. The purpose of this questionnaire is to assess the quality of maternal infant care in Chapel Hill, with "yes" answers indicating good care and "no" answers indicating bad care.

Suppose that this professor asks you, the consulting biostatistician on this research project, the following question: Is it possible for you to provide me with a "good" estimator of the probability that a randomly selected new mother in Chapel Hill will respond "yes" to all $k$ items on the questionnaire, reflecting "perfect care"?

As a start, assume that the number $X$ of "yes" answers to the questionnaire for a randomly chosen new mother in Chapel Hill follows a binomial distribution with sample size $k$ and probability parameter $\pi$, $0 < \pi < 1$. Then, the responses $X_1, X_2, \ldots, X_n$ of the $n$ randomly chosen mothers can be considered to be a random sample of size $n$ from this binomial distribution. Your task as the consulting biostatistician is to find the minimum variance unbiased estimator (MVUE) $\hat{\theta}$ of $\theta = \text{pr}(X = k) = \pi^k$. Once you have found an explicit expression for $\hat{\theta}$, demonstrate by direct calculation that $E(\hat{\theta}) = \theta$.

**Exercise 4.7.** Let $Y_1, Y_2, \ldots, Y_n$ constitute a random sample of size $n$ ($n \geq 2$) from a $N(0, \sigma^2)$ population.

(a) Develop an explicit expression for an unbiased estimator $\hat{\theta}$ of the unknown parameter $\theta = \sigma^r$ ($r$ a known positive integer) that is a function of a sufficient statistic for $\theta$.

(b) Derive an explicit expression for the CRLB for the variance of any unbiased estimator of the parameter $\theta = \sigma^r$. Find a particular value of $r$ for which the variance of $\hat{\theta}$ actually achieves the CRLB.

**Exercise 4.8.** In a certain laboratory experiment, the time $Y$ (in milliseconds) for a certain blood clotting agent to show an observable effect is assumed to have the negative exponential distribution

$$f_Y(y) = \alpha^{-1}e^{-y/\alpha}, \quad y > 0, \quad \alpha > 0.$$

Let $Y_1, Y_2, \ldots, Y_n$ constitute a random sample of size $n$ from $f_Y(y)$, and let $y_1, y_2, \ldots, y_n$ be the corresponding observed values (or realizations) of $Y_1, Y_2, \ldots, Y_n$. One can think of $y_1, y_2, \ldots, y_n$ as the set of observed times for the blood clotting agent to show an observable effect based on $n$ repetitions of the laboratory experiment.

It is of interest to make statistical inferences about the unknown parameter $\theta = \text{V}(Y) = \alpha^2$ using the available data $\boldsymbol{y} = \{y_1, y_2, \ldots, y_n\}$.

(a) Develop an explicit expression for the MLE $\hat{\theta}$ of $\theta$. If the observed value of $S = \sum_{i=1}^{n} Y_i$ is the value $s = 40$ when $n = 50$, compute an appropriate large-sample 95% CI for the parameter $\theta$.

(b) Develop an explicit expression for the MVUE $\hat{\theta}^*$ of $\theta$, and then develop an explicit expression for $\text{V}(\hat{\theta}^*)$, the variance of the MVUE of $\theta$.

(c) Does $\hat{\theta}^*$ achieve the CRLB for the variance of any unbiased estimator of $\theta$?

(d) For any finite value of $n$, develop explicit expressions for $\text{MSE}(\hat{\theta}, \theta)$ and $\text{MSE}(\hat{\theta}^*, \theta)$, the *mean squared errors* of $\hat{\theta}$ and $\hat{\theta}^*$ as estimators of the unknown parameter $\theta$. Using this MSE criterion, which estimator do you prefer for finite $n$, and which estimator do you prefer asymptotically (i.e., as $n \to +\infty$)?

**Exercise 4.9.** Suppose that a laboratory test is conducted on a blood sample from each of $n$ randomly chosen human subjects in a certain city in the United States. The purpose of the test is to detect the presence of a particular biomarker reflecting recent exposure to benzene, a known human carcinogen. Let $\pi, 0 < \pi < 1$, be the unknown probability that a randomly chosen subject in this city has been recently exposed to benzene. When a subject has been recently exposed to benzene, the biomarker will be *correctly* detected with known probability $\gamma, 0 < \gamma < 1$; when a subject has not been recently exposed to benzene, the biomarker will be *incorrectly* detected with known probability $\delta, 0 < \delta < \gamma < 1$. Let $X$ be the random variable denoting the number of the $n$ subjects who are classified as having been recently exposed to benzene (or, equivalently, who provide a blood sample in which the biomarker is detected).

(a) Find an *unbiased estimator* $\hat{\pi}$ of the parameter $\pi$ that is an explicit function of the random variable X, and also derive an explicit expression for $\text{V}(\hat{\pi})$, the variance of the estimator $\hat{\pi}$.

(b) If $n = 50, \alpha = 0.05, \beta = 0.90$, and if the observed value of $X$ is $x = 20$, compute an appropriate 95% large-sample CI for the unknown parameter $\pi$.

**Exercise 4.10.** A scientist at the National Institute of Environmental Health Sciences (NIEHS) is studying the teratogenic effects of a certain chemical by injecting a group of pregnant female rats with this chemical and then observing the number of abnormal (i.e., dead or malformed) fetuses in each litter.

Suppose that $\pi, 0 < \pi < 1$, is the probability that a fetus is abnormal. Further, for the $i$th of $n$ litters, each litter being of size *two*, let the random variable $X_{ij}$ take the value 1 if the $j$th fetus is abnormal and let $X_{ij}$ take the value 0 if the $j$th fetus is normal, $j = 1, 2$.

Since the two fetuses in each litter have experienced the same gestational conditions, the dichotomous random variables $X_{i1}$ and $X_{i2}$ are expected to be correlated. To allow for such a correlation, the following *correlated binomial* model is proposed: for $i = 1, 2, \ldots, n$,

$$\begin{aligned}
\text{pr}[(X_{i1} = 1) \cap (X_{i2} = 1)] &= \pi^2 + \theta, \\
\text{pr}[(X_{i1} = 1) \cap (X_{i2} = 0)] &= \text{pr}[(X_{i1} = 0) \cap (X_{i2} = 1)] \\
&= \pi(1 - \pi) - \theta,
\end{aligned}$$

and

$$\text{pr}[(X_{i1} = 0) \cap (X_{i2} = 0)] = (1 - \pi)^2 + \theta.$$

Here, $\text{cov}(X_{i1}, X_{i2}) = \theta, -\min[\pi^2, (1 - \pi)^2] \leq \theta \leq \pi(1 - \pi)$.

(a) Let the random variable $Y_{11}$ be the number of litters out of $n$ for which both fetuses are abnormal, and let the random variable $Y_{00}$ be the number of litters out of $n$ for which both fetuses are normal. Show that the MLEs $\hat{\pi}$ of $\pi$ and $\hat{\theta}$ of $\theta$ are, respectively,

$$\hat{\pi} = \frac{1}{2} + \frac{(Y_{11} - Y_{00})}{2n},$$

and

$$\hat{\theta} = \frac{Y_{11}}{n} - \hat{\pi}^2.$$

(b) Develop explicit expressions for $\text{E}(\hat{\pi})$ and $\text{V}(\hat{\pi})$.

(c) If $n = 30$, and if the observed values of $Y_{11}$ and $Y_{00}$ are $y_{11} = 3$ and $y_{00} = 15$, compute an appropriate large-sample 95% CI for $\pi$.

For a more general statistical treatment of such a correlated binomial model, see Kupper and Haseman (1978).

**Exercise 4.11.** A popular epidemiologic study design is the pair-matched case–control study design, where a case (i.e., a diseased person, denoted D) is "matched" (on covariates such as age, race, and sex) to a control (i.e., a nondiseased person, denoted $\bar{\text{D}}$). Each member of the pair is then interviewed as to the presence (E) or absence ($\bar{\text{E}}$) of a history of exposure to some potentially harmful substance (e.g., cigarette smoke, asbestos, benzene, etc.). The data from such a study involving $n$ case–control pairs can be presented in tabular form, as follows:

| | | $\bar{\text{D}}$ | | |
|---|---|---|---|---|
| | | E | $\bar{\text{E}}$ | |
| D | E | $Y_{11}$ | $Y_{10}$ | |
| | $\bar{\text{E}}$ | $Y_{01}$ | $Y_{00}$ | |
| | | | | $n$ |

Here, $Y_{11}$ is the number of pairs for which *both* the case *and* the control are exposed (i.e., both have a history of exposure), $Y_{10}$ is the number of pairs for which the case is exposed but the control is not, and so on. Clearly, $\sum_{i=0}^{1}\sum_{j=0}^{1} Y_{ij} = n$.

In what follows, assume that the $\{Y_{ij}\}$ have a multinomial distribution with sample size $n$ and associated cell probabilities $\{\pi_{ij}\}$, where

$$\sum_{i=0}^{1}\sum_{j=0}^{1} \pi_{ij} = 1.$$

For example, then, $\pi_{10}$ is the probability of obtaining a pair in which the case is exposed and its matched control is not. In such a study, the parameter measuring the association between exposure status and disease status is the odds ratio $\text{OR} = \pi_{10}/\pi_{01}$; the estimator of OR is $\widehat{\text{OR}} = Y_{10}/Y_{01}$.

(a) Under the assumed multinomial model for the $\{Y_{ij}\}$, use the delta method to develop an appropriate estimator $\hat{\text{V}}(\ln \widehat{\text{OR}})$ of $\text{V}(\ln \widehat{\text{OR}})$, the variance of the random variable $\ln \widehat{\text{OR}}$. What is the numerical value of your variance estimator when $n = 100$ and when the observed cell counts are $y_{11} = 15$, $y_{10} = 25$, $y_{01} = 15$, and $y_{00} = 45$?

(b) Assuming that

$$\frac{\ln \widehat{\text{OR}} - \ln \text{OR}}{\sqrt{\hat{\text{V}}(\ln \widehat{\text{OR}})}} \dot{\sim} \text{N}(0,1),$$

for large $n$, use the observed cell counts given in part (a) to construct an appropriate 95% CI for OR.

**Exercise 4.12.** Actinic keratoses are small skin lesions that serve as precursors for skin cancer. It has been theorized that adults who are residents of U.S. cities near the equator are more likely to develop actinic keratoses, and hence to be at greater risk for skin cancer, than are adults who are residents of U.S. cities distant from the equator. To test this theory, suppose that dermatology records for a random sample of $n_1$ adult residents of a particular U.S. city (say, City 1) near the equator are examined to determine the number of actinic keratoses that each of these $n_1$ adults has developed. In addition, dermatology records for a random sample of $n_0$ adult residents of a particular U.S. city (say, City 0) distant from the equator are examined to determine the number of actinic keratoses that each of these adults has developed.

As a statistical model for evaluating this theory, for adult resident $j$ $(j = 1, 2, \ldots, n_i)$ in City $i$ $(i = 0, 1)$, suppose that the random variable $Y_{ij} \sim \text{POI}(L_{ij}\lambda_i)$, where $L_{ij}$ is the length of time (in years) that adult $j$ has resided in City $i$ and where $\lambda_i$ is the rate of development of actinic keratoses per year (i.e., the expected number of actinic keratoses that develop per year) for an adult resident of City $i$. So, the pair $(L_{ij}, y_{ij})$ constitutes the observed information for adult resident $j$ in City $i$.

(a) Develop an appropriate ML-based large-sample $100(1-\alpha)\%$ CI for the log rate ratio $\ln \psi = \ln(\lambda_1/\lambda_0)$.

(b) If $n_1 = n_0 = 30, \sum_{j=1}^{30} y_{1j} = 40, \sum_{j=1}^{30} L_{1j} = 350, \sum_{j=1}^{30} y_{0j} = 35,$ and $\sum_{j=1}^{30} L_{0j} = 400$, compute a 95% CI for the rate ratio $\psi$. Comment on your findings.

**Exercise 4.13.** The time $T$ (in months) in remission for leukemia patients who have completed a certain type of chemotherapy treatment is assumed to have the negative exponential distribution

$$f_T(t;\theta) = \theta e^{-\theta t}, \quad t > 0, \ \theta > 0.$$

Suppose that monitoring a random sample of $n$ leukemia patients who have completed this chemotherapy treatment leads to the $n$ observed remission times $t_1, t_2, \ldots, t_n$. In formal statistical terms, $T_1, T_2, \ldots, T_n$ represent a random sample of size $n$ from $f_T(t;\theta)$, and $t_1, t_2, \ldots, t_n$ are the observed values (or realizations) of the $n$ random variables $T_1, T_2, \ldots, T_n$.

(a) Using the available data, derive an explicit expression for the large-sample variance (based on expected information) of the MLE $\hat{\theta}$ of $\theta$.

(b) A biostatistician responsible for analyzing this data set realizes that it is not possible to know with certainty the *exact* number of months that each patient is in remission after completing the chemotherapy treatment. So, this biostatistician suggests the following alternative procedure for estimating $\theta$: "After some specified time period (in months) of length $t^*$ (a known positive constant) after completion of the chemotherapy treatment, let $Y_i = 1$ if the $i$th patient is still in remission after $t^*$ months and let $Y_i = 0$ if not, where $\text{pr}(Y_i = 1) = \text{pr}(T_i > t^*), i = 1, 2, \ldots, n$. Then, use the $n$ mutually independent dichotomous random variables $Y_1, Y_2, \ldots, Y_n$ to find an alternative MLE $\hat{\theta}^*$ of the parameter $\theta$." Develop an explicit expression for $\hat{\theta}^*$.

(c) Use expected information to compare the large-sample variances of $\hat{\theta}$ and $\hat{\theta}^*$. Assuming $t^* \geq E(T)$, which of these two MLEs has the smaller variance, and why should this be the anticipated finding? Are there circumstances where the MLE with the larger variance might be preferred?

**Exercise 4.14.** For a typical woman in a certain high-risk population of women, suppose that the number $Y$ of lifetime events of domestic violence involving emergency room treatment is assumed to have the Poisson distribution

$$p_Y(y;\lambda) = \lambda^y e^{-\lambda}/y!, \quad y = 0, 1, \ldots, +\infty \quad \text{and} \quad \lambda > 0.$$

Let $Y_1, Y_2, \ldots, Y_n$ constitute a random sample of size $n$ (where $n$ is large) from this Poisson population (i.e., $n$ women from this high-risk population are randomly sampled and then each woman in the random sample is asked to recall the number of lifetime events of domestic violence involving emergency room treatment that she has experienced).

(a) Find an explicit expression for the CRLB for the variance of any unbiased estimator of parameter $\theta = \text{pr}(Y = 0)$. Does there exist an unbiased estimator of $\theta$ that achieves this CRLB for all finite values of $n$?

(b) Suppose that a certain domestic violence researcher believes that reported values of $Y$ greater than zero are not very accurate (although a reported value greater than zero almost surely indicates at least one domestic violence experience involving emergency room treatment), but that reported values of $Y$ equal to zero are accurate. Because of this possible data inaccuracy problem, this researcher wants to analyze the data by converting each $Y_i$ to a two-valued (or dichotomous) random variable $X_i$, where $X_i$ is defined as follows: if $Y_i \geq 1$, then $X_i = 1$; and, if $Y_i = 0$, then $X_i = 0$. Using the $n$ mutually independent dichotomous random variables $X_1, X_2, \ldots, X_n$, find an explicit expression for the MLE $\hat{\lambda}^*$ of $\lambda$ and then find an explicit expression for the large-sample variance of $\hat{\lambda}^*$.

(c) This domestic violence researcher is concerned that she may be doing something wrong by using the dichotomous variables $X_1, X_2, \ldots, X_n$ (instead of the original Poisson variables $Y_1, Y_2, \ldots, Y_n$) to estimate the unknown parameter $\lambda$. To address her concern, make a quantitative comparison between the properties of $\hat{\lambda}^*$ and $\hat{\lambda}$, where $\hat{\lambda}$ is the MLE of $\lambda$ obtained by using $Y_1, Y_2, \ldots, Y_n$. Also, comment on issues of validity (i.e., bias) and precision (i.e.,variability) as they relate to the choice between $\hat{\lambda}$ and $\hat{\lambda}^*$.

**Exercise 4.15.** For a certain African village, available data strongly suggest that the expected number of new cases of AIDS developing in any particular year is directly proportional to the expected number of new AIDS cases that developed during the immediately preceding year. An important statistical goal is to estimate the value of this unknown proportionality constant $\theta$ ($\theta > 1$), which is assumed not to vary from year to year, and then to find an appropriate 95% CI for $\theta$.

To accomplish this goal, the following statistical model is to be used: For $j = 0, 1, \ldots, n$ consecutive years of data, let $Y_j$ be the random variable denoting the number of new AIDS cases developing in year $j$. Further, suppose that the $(n+1)$ random variables $Y_0, Y_1, \ldots, Y_n$ are such that the conditional distribution of $Y_{j+1}$, given $Y_k = y_k$ for $k = 0, 1, \ldots, j$, depends only on $y_j$ and is *Poisson* with $E(Y_{j+1}|Y_j = y_j) = \theta y_j, j = 0, 1, \ldots, (n-1)$. Further, assume that the distribution of the random variable $Y_0$ is *Poisson* with $E(Y_0) = \theta$, where $\theta > 1$.

(a) Using all $(n+1)$ random variables $Y_0, Y_1, \ldots, Y_n$, develop an explicit expression for the MLE $\hat{\theta}$ of the unknown proportionality constant $\theta$.

(b) If $n = 25$ and $\hat{\theta} = 1.20$, compute an appropriate ML-based 95% CI for $\theta$.

**Exercise 4.16.** In a certain clinical trial, suppose that the outcome variable $X$ represents the 6-month change in cholesterol level (in milligrams per deciliter) for subjects in the treatment (T) group who will be given a certain cholesterol-lowering drug, and suppose that $Y$ represents this same outcome variable for subjects in the control (C) group who will be given a placebo. Further, suppose that it is reasonable to assume that $X \sim N(\mu_t, \sigma_t^2)$ and $Y \sim N(\mu_c, \sigma_c^2)$, and that $\sigma_t^2$ and $\sigma_c^2$ have *known* values such that $\sigma_t^2 \neq \sigma_c^2$.

Let $X_1, X_2, \ldots, X_{n_t}$ constitute a random sample of size $n_t$ from $N(\mu_t, \sigma_t^2)$; namely, these $n_t$ observations represent the set of outcomes to be measured on the $n_t$ subjects who have been randomly assigned to the T group. Similarly, let $Y_1, Y_2, \ldots, Y_{n_c}$ constitute a random sample of size $n_c$ from $N(\mu_c, \sigma_c^2)$; namely, these $n_c$ observations represent the set of outcomes to be measured on the $n_c$ subjects who have been randomly assigned to the C group.

Because of monetary and logistical constraints, suppose that a total of only $N$ subjects can participate in this clinical trial, so that $n_t$ and $n_c$ are constrained to satisfy the relationship $(n_t + n_c) = N$. Based on the stated assumptions (namely, random samples from two normal populations with known, but unequal, variances), determine the "optimal" partition of $N$ into values $n_t$ and $n_c$ that will produce the most "precise" *exact* 95% CI for $(\mu_t - \mu_c)$. When $N = 100, \sigma_t^2 = 4$, and $\sigma_c^2 = 9$, find the optimal choices for $n_t$ and $n_c$. Comment on your findings.

**Exercise 4.17.** Suppose that the random variable $Y = \ln(X)$, where $X$ is the ambient carbon monoxide (CO) concentration (in parts per million) in a certain highly populated U.S. city, is assumed to have a normal distribution with mean $E(Y) = \mu$ and variance $V(Y) = \sigma^2$. Let $Y_1, Y_2, \ldots, Y_n$ constitute a random sample from this $N(\mu, \sigma^2)$ population. Practically speaking, $Y_1, Y_2, \ldots, Y_n$ can be considered to be ln(CO concentration) readings taken on days $1, 2, \ldots, n$, where these $n$ days are spaced far enough apart so that $Y_1, Y_2, \ldots, Y_n$ can be assumed to be mutually independent random variables. It is of interest to be able to predict with some accuracy the value of the random variable $Y_{n+1}$, namely, the value of the random variable representing the ln(CO concentration) on day $(n+1)$, where day $(n+1)$ is far enough in time from day $n$ so that $Y_{n+1}$ can reasonably be assumed to be independent of the random variables $Y_1, Y_2, \ldots, Y_n$. Also, it can be further assumed, as well, that $Y_{n+1} \sim N(\mu, \sigma^2)$.

If $\bar{Y} = n^{-1} \sum_{i=1}^{n} Y_i$ and if $S^2 = (n-1)^{-1} \sum_{i=1}^{n} (Y_i - \bar{Y})^2$, determine explicit expressions for random variables $L$ and $U$ (involving $\bar{Y}$ and $S$) such that

$$\text{pr}[L < Y_{n+1} < U] = (1 - \alpha), 0 < \alpha \leq 0.10.$$

In other words, rigorously derive an exact $100(1-\alpha)\%$ *prediction interval* for the random variable $Y_{n+1}$. If $n = 5$, and if $Y_i = i, \; i = 1, 2, 3, 4, 5$, compute an exact 95% prediction interval for $Y_6$. As a hint, construct a statistic involving the random variable $(\bar{Y} - Y_{n+1})$ that has a $t$-distribution.

**Exercise 4.18.** Let $X_1, X_2, \ldots, X_n$ constitute a random sample of size $n$ from a $N(\mu, \sigma^2)$ population. Let

$$\bar{X} = n^{-1} \sum_{i=1}^{n} X_i \quad \text{and} \quad S^2 = (n-1)^{-1} \sum_{i=1}^{n} (X_i - \bar{X})^2.$$

Under the stated assumptions, the most appropriate $100(1-\alpha)\%$ CI for $\mu$ is

$$\bar{X} \pm t_{n-1,1-\alpha/2} S/\sqrt{n},$$

where $t_{n-1,1-\alpha/2}$ is the $100(1-\alpha/2)\%$ percentile point of Student's $t$-distribution with $(n-1)$ df. The width $W_n$ of this CI is

$$W_n = 2t_{n-1,1-\alpha/2} S/\sqrt{n}.$$

(a) Under the stated assumptions, derive an explicit expression for $E(W_n)$, the expected width of this CI. What is the exact numerical value of $E(W_n)$ if $n = 4$, $\alpha = 0.05$, and $\sigma^2 = 4$?

(b) Suppose that it is desired to find the smallest sample size $n^*$ such that

$$\text{pr}(W_{n^*} \leq \delta) = \text{pr}\{2t_{n^*-1,1-\alpha/2}S/\sqrt{n^*} \leq \delta\} \geq (1-\gamma),$$

where $\delta$ ($> 0$) and $\gamma$ ($0 < \gamma < 1$) are specified positive numbers.

Under the stated assumptions, prove rigorously that $n^*$ should be chosen to be the smallest positive integer satisfying the inequality

$$n^*(n^*-1) \geq \left(\frac{2\sigma}{\delta}\right)^2 \chi^2_{n^*-1,1-\gamma} f_{1,n^*-1,1-\alpha},$$

where $\chi^2_{n^*-1,1-\gamma}$ and $f_{1,n^*-1,1-\alpha}$ denote, respectively, $100(1-\gamma)$ and $100(1-\alpha)$ percentile points for a chi-square distribution with $(n^*-1)$ df and for an $f$-distribution with 1 numerator, and $(n^*-1)$ denominator, df.

**Exercise 4.19.** Suppose that an epidemiologist desires to make statistical inferences about the true mean diastolic blood pressure levels for adult residents in three rural North Carolina cities. As a starting model, suppose that she assumes that the true underlying distribution of diastolic blood pressure measurements for adults in each city is normal, and that these three normal distributions have a common variance (say, $\sigma^2$), but possibly different means (say, $\mu_1$, $\mu_2$, and $\mu_3$). This epidemiologist decides to obtain her blood pressure study data by randomly selecting $n_i$ adult residents from city $i$, $i = 1, 2, 3$, and then measuring their diastolic blood pressures.

Using more formal statistical notation, for $i = 1, 2, 3$, let $Y_{i1}, Y_{i2}, \ldots, Y_{in_i}$ constitute a random sample of size $n_i$ from a $N(\mu_i, \sigma^2)$ population. Define the random variables

$$\bar{Y}_i = n_i^{-1} \sum_{j=1}^{n_i} Y_{ij}, \quad i = 1, 2, 3,$$

and

$$S_i^2 = (n_i - 1)^{-1} \sum_{j=1}^{n_i} \left(Y_{ij} - \bar{Y}_i\right)^2, \quad i = 1, 2, 3.$$

(a) Consider the parameter

$$\theta = (2\mu_1 - 3\mu_2 + \mu_3).$$

Using *all* the available data (in particular, all three sample means and all three sample variances), construct a random variable that has a Student's $t$-distribution.

(b) If $n_1 = n_2 = n_3 = 4$, $\bar{y}_1 = 80$, $\bar{y}_2 = 75$, $\bar{y}_3 = 70$, $s_1^2 = 4$, $s_2^2 = 3$, and $s_3^2 = 5$, find an *exact* 95% CI for $\theta$ given the stated assumptions.

(c) Now, suppose that governmental reviewers of this study are skeptical about both the epidemiologist's assumptions of normality and homogeneous variance, claiming that her sample sizes were much too small to provide reliable information about the appropriateness of these assumptions or about the parameter θ. To address these criticisms, this epidemiologist goes back to these same three rural North Carolina cities and takes blood pressure measurements on *large* random samples of adult residents in each of the three cities; she obtains the following data:

$$n_1 = n_2 = n_3 = 50;\ \bar{y}_1 = 85, \bar{y}_2 = 82, \bar{y}_3 = 79;\ s_1^2 = 7, s_2^2 = 2, s_3^2 = 6.$$

Retaining the normality assumption for now, find an appropriate 95% CI for $\sigma_1^2/\sigma_2^2$, and then comment regarding the appropriateness of the homogeneous variance assumption.

(d) Using the data in part (c), compute an appropriate large-sample 95% CI for θ. Comment on the advantages of increasing the sizes of the random samples selected from each of the three populations.

**Exercise 4.20.** Let $X_1, X_2, \ldots, X_{n_1}$ constitute a random sample of size $n_1(>2)$ from a normal parent population with mean 0 and variance θ. Also, let $Y_1, Y_2, \ldots, Y_{n_2}$ constitute a random sample of size $n_2(>2)$ from a normal parent population with mean 0 and variance $\theta^{-1}$. The set of random variables $\{X_1, X_2, \ldots, X_{n_1}\}$ is independent of the set of random variables $\{Y_1, Y_2, \ldots, Y_{n_2}\}$, and $\theta(>0)$ is an unknown parameter.

(a) Derive an explicit expression for $\mathrm{E}(\sqrt{L})$ when $L = \sum_{i=1}^{n_1} X_i^2$.

(b) Using all $(n_1 + n_2)$ available observations, derive an explicit expression for an exact $100(1-\alpha)\%$ CI for the unknown parameter θ. If $n_1 = 8, n_2 = 5, \sum_{i=1}^{8} x_i^2 = 30$, and $\sum_{i=1}^{5} y_i^2 = 15$, compute a 95% confidence interval for θ.

**Exercise 4.21.** In certain types of studies called *crossover studies*, each of $n$ randomly chosen subjects is administered both a treatment T (e.g., a new drug pill) and a placebo P (e.g., a sugar pill). Typically, neither the subject nor the person administering the pills knows which pill is T and which pill is P (namely, the study is a so-called *double-blind* study). Also, the two possible pill administration orderings "first T, then P" and "first P, then T" are typically allocated randomly to subjects, and sufficient time is allowed between administrations to avoid so-called "carry-over" effects. One advantage of a crossover study is that a comparison between the effects of T and P can be made within (or specific to) each subject (since each subject supplies information on the effects of both T and P), thus eliminating subject-to-subject variability in each subject-specific comparison.

For the $i$th subject $(i = 1, 2, \ldots, n)$, suppose that $D_i = (Y_{\mathrm{T}i} - Y_{\mathrm{P}i})$ is the continuous random variable representing the difference between a continous response $(Y_{\mathrm{T}i})$ following T administration and a continuous response $(Y_{\mathrm{P}i})$ following P administration. So, $D_i$ is measuring the effect of T relative to P for subject $i$. Since $Y_{\mathrm{T}i}$ and $Y_{\mathrm{P}i}$ are responses for the same subject (namely, subject $i$), it is very sensible to expect that $Y_{\mathrm{T}i}$ and $Y_{\mathrm{P}i}$ will be correlated to some extent. To allow for this potential intra-subject response correlation, assume in what follows that $Y_{\mathrm{T}i}$ and $Y_{\mathrm{P}i}$ jointly follow a *bivariate normal* distribution with $\mathrm{E}(Y_{\mathrm{T}i}) = \mu_{\mathrm{T}}, \mathrm{V}(Y_{\mathrm{T}i}) = \sigma_{\mathrm{T}}^2, \mathrm{E}(Y_{\mathrm{P}i}) = \mu_{\mathrm{P}}, \mathrm{V}(Y_{\mathrm{P}i}) = \sigma_{\mathrm{P}}^2$, and with $\mathrm{corr}(Y_{\mathrm{T}i}, Y_{\mathrm{P}i}) = \rho, i = 1, 2, \ldots, n$. Further, assume that the $n$ differences $D_1, D_2, \ldots, D_n$ are mutually independent of one another.

(a) Assuming that $\sigma_T^2, \sigma_P^2$, and $\rho$ have *known* values, use the $n$ mutually independent random variables $D_1, D_2, \ldots, D_n$ to derive an exact $100(1-\alpha)\%$ CI for the unknown parameter $\theta = (\mu_T - \mu_P)$, the true difference between the expected responses for the T and P administrations. In particular, find explicit expressions for random variables $L$ and $U$ such that $\text{pr}(L < \theta < U) = (1-\alpha), 0 < \alpha \leq 0.10$. If there are available data for which $n = 10, \bar{y}_T = 15.0, \bar{y}_P = 14.0, \sigma_T^2 = 2.0, \sigma_P^2 = 3.0$, $\rho = 0.30$, and $\alpha = 0.05$, use this numerical information to compute exact numerical values for $L$ and $U$. Interpret these numerical results with regard to whether or not the available data provide statistical evidence that $\mu_T$ and $\mu_P$ have different values.

(b) Now, assume that treatment effectiveness is equivalent to the inequality $\theta > 0$ (or, equivalently, $\mu_T > \mu_P$). If $\alpha = 0.05, \sigma_T^2 = 2.0, \sigma_P^2 = 3.0, \rho = 0.30$, and $\theta = 1.0$ (so that T is truly effective compared to P), what is the *minimum number* $n^*$ of subjects that should be enrolled in this crossover study so that the random variable $L$ determined in part (a) exceeds the value zero with probability at least equal to 0.95? The motivation for finding $n^*$ is that, if the treatment is truly effective, it is highly desirable for the lower limit $L$ of the CI for $\theta$ to have a high probability of exceeding zero in value, thus providing statistical evidence in favor of a real treatment effect relative to the placebo effect.

**Exercise 4.22.** For $i = 1, 2, \ldots, n$, let the random variables $X_i$ and $Y_i$ denote, respectively, the diastolic blood pressure (DBP) and systolic blood pressure (SBP) for the $i$th of $n$ (>1) randomly chosen hypertensive adult males. Assume that the pairs $(X_i, Y_i)$, $i = 1, 2, \ldots, n$, constitute a random sample of size $n$ from a bivariate normal population, where $E(X_i) = \mu_x$, $E(Y_i) = \mu_y$, $V(X_i) = V(Y_i) = \sigma^2$, and $\text{corr}(X_i, Y_i) = \rho$. The goal is to develop an exact 95% CI for the correlation coefficient $\rho$. To accomplish this goal, consider the following random variables. Let $U_i = (X_i + Y_i)$ and $V_i = (X_i - Y_i)$, $i = 1, 2, \ldots, n$. Further, let $n\bar{U} = \sum_{i=1}^n U_i$, $n\bar{V} = \sum_{i=1}^n V_i$, $(n-1)S_u^2 = \sum_{i=1}^n (U_i - \bar{U})^2$, and $(n-1)S_v^2 = \sum_{i=1}^n (V_i - \bar{V})^2$.

(a) Derive explicit expressions for the means and variances of the random variables $U_i$ and $V_i$, $i = 1, 2, \ldots, n$.

(b) Prove rigorously that $\text{cov}(U_i, V_i) = 0, i = 1, 2, \ldots, n$, so that, in this situation, it will follow that $U_i$ and $V_i$ are independent random variables, $i = 1, 2, \ldots, n$.

(c) Use rigorous arguments to prove that the random variable

$$W = \frac{(1-\rho)S_u^2}{(1+\rho)S_v^2}$$

has an $f$-distribution.

(d) If $n = 10$, and if the realized values of $S_u^2$ and $S_v^2$ are 1.0 and 2.0, respectively, use these data, along with careful arguments, to compute an exact 95% CI for $\rho$.

**Exercise 4.23.** An economist postulates that the distribution of income (in thousands of dollars) in a certain large U.S. city can be modeled by the Pareto density function

$$f_Y(y; \gamma, \theta) = \theta\gamma^\theta y^{-(\theta+1)}, \quad 0 < \gamma < y < \infty \quad \text{and} \quad 2 < \theta < \infty,$$

where $\gamma$ and $\theta$ are unknown parameters. Let $Y_1, Y_2, \ldots, Y_n$ constitute a random sample of size $n$ from $f_Y(y; \gamma, \theta)$.

(a) If $n = 50, \bar{y} = n^{-1}\sum_{i=1}^{n} y_i = 30$, and $s^2 = (n-1)^{-1}\sum_{i=1}^{n}(y_i - \bar{y})^2 = 10$, find exact numerical values for the method of moments estimators $\hat{\gamma}_{mm}$ and $\hat{\theta}_{mm}$, respectively, of $\gamma$ and $\theta$.

(b) A consulting biostatistician suggests that the smallest order statistic $Y_{(1)} = \min\{Y_1, Y_2, \ldots, Y_n\}$ is also a possible estimator for $\gamma$. Is $Y_{(1)}$ a consistent estimator of $\gamma$?

(c) Now, assume that $\theta = 3$, so that the only unknown parameter is $\gamma$. It is desired to use the random variable $Y_{(1)}$ to compute an exact upper one-sided CI for $\gamma$. In particular, derive an explicit expression for a random variable $U = cY_{(1)}, 0 < c < 1$, such that $\text{pr}(\gamma < U) = (1 - \alpha), 0 < \alpha \leq 0.10$. If $n = 5$, $\alpha = 0.10$, and the observed value of $Y_{(1)}$ is $y_{(1)} = 20$, use this information to compute an upper one-sided 90% CI for the unknown parameter $\gamma$.

**Exercise 4.24.** Let $X_1, X_2, \ldots, X_n$ constitute a random sample of size $n$ from the parent population $f_X(x), -\infty < x < +\infty$. Further, let $X_{(1)}, X_{(2)}, \ldots, X_{(n)}$ be the set of corresponding order statistics, where $-\infty < X_{(1)} < X_{(2)} < \cdots < X_{(n-1)} < X_{(n)} < +\infty$.

(a) Let $U_r$ be the random variable defined as

$$U_r = \text{pr}\left[X \leq X_{(r)}\right] = \int_{-\infty}^{X_{(r)}} f_X(x)\,dx = F_X(X_{(r)}), \quad r = 1, 2, \ldots, n,$$

so that $U_r$ is the amount of area under $f_X(x)$ to the left of $X_{(r)}$. Develop an explicit expression for $E(U_r)$.

(b) For $0 < p < 1$, define the $p$th *quantile* of $f_X(x)$ to be $\theta_p = F_X^{-1}(p)$; in particular, $\theta_p$ is that value of $x$ such that an amount $p$ of area under $f_X(x)$ is to the left of $x$. Describe how the result in part (a) can be used to develop a reasonable estimator of $\theta_p$.

**Exercise 4.25.** Suppose that $X_1, X_2, \ldots, X_n$ constitute a random sample of size $n$ from the density function $f_X(x; \theta)$, where $-\infty < x < \infty$. It is desired to construct an appropriate CI for the *median* $\xi$ of $f_X(x; \theta)$, where $\xi$ is defined as the population parameter satisfying the relationship $\int_{-\infty}^{\xi} f_X(x; \theta)\,dx = \frac{1}{2}$. As one possible CI for $\xi$, consider using $X_{(1)} = \min\{X_1, X_2, \ldots, X_n\}$ for the lower limit and $X_{(n)} = \max\{X_1, X_2, \ldots, X_n\}$ for the upper limit.

(a) If $f_X(x; \theta) = \theta x^{\theta - 1}, 0 < x < 1$ and $\theta > 0$, derive an explicit expression for the expected value of the width $W$ of the proposed CI $[X_{(1)}, X_{(n)}]$.

(b) Now, suppose that the structure of $f_X(x; \theta)$ is completely unknown. Again consider the proposed CI $[X_{(1)}, X_{(n)}]$. Derive an explicit expression for $\text{pr}[X_{(1)} < \xi < X_{(n)}]$, and then comment on this result with regard to the utility of this particular CI for $\xi$.

**Exercise 4.26.** Let $X_1, X_2, \ldots, X_n$ constitute a random sample from the uniform density $f_X(x) = 1, 0 < x < 1$. Then, $G = \left(\prod_{i=1}^{n} X_i\right)^{1/n}$ is the *geometric mean*. Develop explicit expressions for $E(G)$ and $V(G)$, and then use these results to determine to what quantity $G$ converges in probability.

**Exercise 4.27.** Let $Y$ be a continuous random variable with density $f_Y(y) = e^{-y}, y > 0$. Consider the sequence of random variables $X_n = e^n I(Y > n)$, $n = 1, 2, \ldots,$ where the indicator function $I(Y > n)$ takes the value 1 if $Y > n$ and takes the value 0 otherwise. Working *directly* with the definition of "convergence in probability," prove that $X_n$ converges in probability to the value 0.

**Exercise 4.28.** Suppose that a continuous response $Y$ is to be measured on each of $n$ subjects during a two-group clinical trial comparing a new drug therapy to a standard drug therapy. Without loss of generality, suppose that the first $n_1$ subjects $(i = 1, 2, \ldots, n_1)$ constitute the treatment group (i.e., the group of subjects receiving the new drug therapy) and the remaining $n_0 = (n - n_1)$ subjects $(i = n_1 + 1, n_1 + 2, \ldots, n)$ constitute the comparison group (i.e., the group of subjects receiving the standard therapy).

Further, suppose that the following *multiple linear regression* model defines the *true* underlying relationship between the continuous response and relevant covariates:

$$E(Y_i|T_i, A_i) = \alpha + \beta T_i + \gamma A_i, \quad i = 1, 2, \ldots, n,$$

where $T_i$ equals 1 if the $i$th subject is a member of the treatment group and equals 0 if the $i$th subject is a member of the comparison group, where $A_i$ is the age of the $i$th subject, and where $\alpha \neq 0, \beta \neq 0$, and $\gamma \neq 0$. Note that the key parameter of interest is $\beta$, which measures the effect of the new drug therapy relative to the standard drug therapy, adjusting for the possible confounding effect of the differing ages of study subjects.

Consider the unfortunate situation where the researchers running the clinical trial lose that subset $\{A_i\}_{i=1}^{n}$ of the complete data set $\{Y_i, T_i, A_i\}_{i=1}^{n}$ which gives the age of each subject. Suppose that these researchers then decide to fit the alternative *incorrect* straight-line model $E(Y_i|T_i) = \alpha^* + \beta^* T_i, i = 1, 2, \ldots, n,$ to the available data by the method of unweighted least squares, thus obtaining

$$\hat{\beta}^* = \frac{\sum_{i=1}^{n}(T_i - \bar{T})(Y_i - \bar{Y})}{\sum_{i=1}^{n}(T_i - \bar{T})^2}$$

as their suggested estimator of $\beta$, where $\bar{T} = n^{-1}\sum_{i=1}^{n} T_i$ and $\bar{Y} = n^{-1}\sum_{i=1}^{n} Y_i$.

Rigorously derive an explicit expression for $E(\hat{\beta}^*|\{T_i\}, \{A_i\})$, and then provide a sufficient condition involving $A_1, A_2, \ldots, A_n$ such that this conditional expected value is equal to $\beta$. Although these researchers do not know the ages of the $n$ subjects in the clinical trial, suppose that they did decide to assign these subjects randomly to the treatment and comparison groups. Discuss how such a randomization procedure could possibly affect the degree of bias in $\hat{\beta}^*$ as an estimator of $\beta$.

For further details about multiple linear regression, see Kleinbaum et al. (2008) and Kutner et al. (2004).

**Exercise 4.29.** For $i = 1, 2, \ldots, n$, suppose that the dichotomous random variable $Y_i$ takes the value 1 if the $i$th subject in a certain clinical trial experiences a particular outcome of interest and takes the value 0 if not, and assume that $Y_1, Y_2, \ldots, Y_n$ constitute a set of $n$ mutually independent random variables. Further, given $p$ covariate values $x_{i0}(\equiv 1), x_{i1}, \ldots, x_{ip}$ associated with the $i$th subject, make the assumption that $\pi_i = \text{pr}(Y_i = 1|x_{ij}, j = 0, 1, \ldots, p)$ has the *logistic* model form, namely,

$$\pi_i = \text{pr}(Y_i = 1|x_{ij}, j = 0, 1, \ldots, p) = \frac{e^{\sum_{j=0}^{p} \beta_j x_{ij}}}{1 + e^{\sum_{j=0}^{p} \beta_j x_{ij}}}.$$

(a) If $\mathcal{L}(y; \boldsymbol{\beta})$ denotes the appropriate likelihood function for the random vector $\boldsymbol{Y}' = (Y_1, Y_2, \ldots, Y_n)$, where $\boldsymbol{y}' = (y_1, y_2, \ldots, y_n)$ and $\boldsymbol{\beta}' = (\beta_0, \beta_1, \ldots, \beta_p)$, show that the $(p+1)$ equations that need to be simultaneously solved to obtain the vector $\hat{\boldsymbol{\beta}} = (\hat{\beta}_0, \hat{\beta}_1, \ldots, \hat{\beta}_p)'$ of MLEs of $\boldsymbol{\beta}$ can be compactly in matrix notation as

$$\mathbf{X}'\left[\boldsymbol{y} - \text{E}(\boldsymbol{Y})\right] = \mathbf{0},$$

where $\text{E}(\boldsymbol{Y}) = [\text{E}(Y_1), \text{E}(Y_2), \ldots, \text{E}(Y_n)]'$, where $\mathbf{0}$ is a $[(p+1) \times 1]$ column vector of zeros, and where $\boldsymbol{X}$ is an appropriately specified $[n \times (p+1)]$ matrix.

(b) For the likelihood function $\mathcal{L}(y; \boldsymbol{\beta})$, show that both the observed and expected information matrices are identical and that each can be written as the same function of $\boldsymbol{X}$ and $\boldsymbol{V}$, where $\boldsymbol{V}$ is the covariance matrix for the random vector $\boldsymbol{Y}$. Then, use this result to describe how to obtain an estimate of the covariance matrix of $\hat{\boldsymbol{\beta}}$.

**Exercise 4.30***. Research investigators from the Division of Marine Fisheries in a certain U.S. state are interested in evaluating possible causes of ulcerative lesions in fish inhabiting a large coastal estuary. The investigators hypothesize that fish born in nesting sites rich in Pfiesteria, a toxic alga, are more susceptible to such lesions than are fish born in nesting sites without Pfiesteria. The goal of the research is to estimate the mean number of lesions for fish born in each of these two types of nesting sites, as well as the proportion $\pi$ of coastal estuary fish actually born in Pfiesteria-rich sites. The only available data to estimate these three parameters consist of lesion counts on $n$ randomly chosen young adult fish residing in this estuary, each of which is known to have been born in one of these two types of nesting sites. Unfortunately, for each of these $n$ fish, the type of nesting site (Pfiesteria-rich or non-Pfiesteria) in which that fish was born is not known.

To analyze these data in order to estimate the three parameters of interest, a biostatistician consulting with these investigators proposes the following statistical model. For fish born in Pfiesteria-rich sites, the number $Y$ of lesions is assumed to follow a Poisson distribution with mean $\mu_1$; for fish born in sites without Pfiesteria, $Y$ is assumed to follow a Poisson distribution with mean $\mu_2$. The statistical goal is to estimate the unknown parameters $\pi, \mu_1$, and $\mu_2$ using the data set $\boldsymbol{y} = (y_1, y_2, \ldots, y_n)'$, where the type of birth site for the $i$th young adult fish with observed lesion count $y_i$ is unknown, $i = 1, 2, \ldots, n$.

This consulting biostatistician recommends the following method for obtaining the MLEs of the model parameters:

(1) Introduce an *unobserved* (or "latent") indicator variable $Z_i$ that takes the value 1 (with probability $\pi$) if the $i$th of the $n$ fish was born in a Pfiesteria-rich site, and takes the value 0 otherwise;

(2) Define $\mathcal{L}_c(y, z; \pi, \mu_1, \mu_2)$ to be the joint (or "complete-data") likelihood for $y = (y_1, y_2, \ldots, y_n)'$ and $z = (z_1, z_2, \ldots, z_n)'$; and

(3) Use this complete-data likelihood, along with the *expectation-maximization* (EM) algorithm described below (Dempster et al., 1977), to derive the desired MLEs.

Starting from well-chosen initial values (specified at iteration $t = 0$), the EM algorithm computes the MLEs by iterating between two steps: the "E-step," which evaluates the conditional expectation of the complete-data log-likelihood with respect to the unobservable vector $\mathbf{Z} = (Z_1, Z_2, \ldots, Z_n)'$, given the observed data $y$ and the current parameter estimates; and, the "M-step," in which this conditional expectation is maximized with respect to the model parameters. Under certain regularity conditions, the EM algorithm will converge to (at least) a local maximum of the observed-data likelihood $\mathcal{L}(y; \pi, \mu_1, \mu_2)$, which, if the vector $\mathbf{Z}$ was known, could be used directly to estimate the three unknown parameters of interest.

Develop explicit expressions (as functions of $y$, $\pi$, $\mu_1$, and $\mu_2$) for the quantities obtained for the E-step and for the M-step at iteration $t$ $(t \geq 1)$. In particular, for the E-step at iteration $t$, derive an explicit expression for

$$\begin{aligned} \mathcal{Q}^{(t)}(y; \pi, \mu_1, \mu_2) &\equiv \mathcal{Q}^{(t)} \\ &= \mathrm{E}_Z\left\{\ln[\mathcal{L}_c(y, z; \pi, \mu_1, \mu_2)] \,\middle|\, y, \hat{\pi}^{(t-1)}, \hat{\mu}_1^{(t-1)}, \hat{\mu}_2^{(t-1)}\right\}. \end{aligned}$$

Then, for the M-step, use $\mathcal{Q}^{(t)}$ to find the MLEs of $\pi$, $\mu_1$, and $\mu_2$ at iteration $t$.

**Exercise 4.31*.** The number $X$ of colds per year for a resident in Alaska is assumed to have the discrete distribution $\mathrm{p}_X(x; \theta) = \theta^{-1}$, $x = 1, 2, \ldots, \theta$, where the parameter $\theta$ is an unknown positive integer. It is desired to find a reasonable candidate for the minimum variance unbiased estimator (MVUE) of $\theta$ using the information contained in a random sample $X_1, X_2, \ldots, X_n$ from $\mathrm{p}_X(x; \theta)$.

(a) Prove that $U = \max\{X_1, X_2, \ldots, X_n\}$ is a sufficient statistic for the parameter $\theta$. Also, show that $U^* = (2X_1 - 1)$ is an unbiased estimator of the parameter $\theta$.

(b) Given that $U$ is a complete sufficient statistic for $\theta$, use the Rao–Blackwell Theorem to derive an explicit expression for the MVUE $\hat{\theta}$ of $\theta$, where $\hat{\theta} = \mathrm{E}(U^* | U = u)$. Then, show directly that $\mathrm{E}(\hat{\theta}) = \theta$. Do you notice any undesirable properties of the estimator $\hat{\theta}$?

**Exercise 4.32*.** For children with autism, it is postulated that the time $X$ (in minutes) for such children to complete a certain manual dexterity test follows the distribution

$$\mathrm{f}_X(x) = 1, 0 < \theta < x < (\theta + 1) < +\infty.$$

Let $X_1, X_2, \ldots, X_n$ constitute a random sample of size $n(>1)$ from $\mathrm{f}_X(x; \theta)$. Let $X_{(1)} = \min\{X_1, X_2, \ldots, X_n\}$ and let $X_{(n)} = \max\{X_1, X_2, \ldots, X_n\}$. Then, consider the following two estimators of the unknown parameter $\theta$:

$$\hat{\theta}_1 = \frac{1}{2}\left[X_{(1)} + X_{(n)} - 1\right]$$

and

$$\hat{\theta}_2 = \frac{1}{(n-1)}\left[nX_{(1)} - X_{(n)}\right].$$

(a) Show that $\hat{\theta}_1$ and $\hat{\theta}_2$ are both unbiased estimators of the parameter $\theta$ and find explicit expressions for $V(\hat{\theta}_1)$ and $V(\hat{\theta}_2)$.

(b) More generally, consider the linear function $W = (c_0 + c_1U_1 + c_2U_2)$, where $V(U_1) = V(U_2) = \sigma^2$, where $\text{cov}(U_1, U_2) = \sigma_{12}$, and where $c_0, c_1$, and $c_2$ are constants with $(c_1 + c_2) = 1$. Determine values for $c_1$ and $c_2$ that minimize $V(W)$, and explain how this general result relates to a comparison of the variance expressions obtained in part (a).

(c) Show that $X_{(1)}$ and $X_{(n)}$ constitute a set of jointly sufficient statistics for $\theta$. Do $X_{(1)}$ and $X_{(n)}$ constitute a set of *complete* sufficient statistics for $\theta$?

**Exercise 4.33*.** Reliable estimation of the numbers of subjects in the United States living with different types of medical conditions is important to both public health and health policy professionals. In the United States, disease-specific registries have been established for a variety of medical conditions including birth defects, tuberculosis, HIV, and cancer. Such registries are very often only partially complete, meaning that the number of registry records for a particular medical condition generally provides an under-estimate of the actual number of subjects with that particular medical condition.

When two registries exist for the same medical condition, statistical models can be used to estimate the degree of under-ascertainment for each registry and to produce an improved estimate of the actual number of subjects having the medical condition of interest. The simplest statistical model for this purpose is based on the assumption that membership status for one registry is statistically independent of membership status for the other registry.

Let the parameter $N$ denote the true unknown number of subjects who have a certain medical condition of interest. Define the random variables

$$\begin{aligned}
X_{yy} &= \text{number of subjects listed in both Registry 1 and Registry 2,}\\
X_{yn} &= \text{number of subjects listed in Registry 1 but not in Registry 2,}\\
X_{ny} &= \text{number of subjects listed in Registry 2 but not in Registry 1,}\\
X_{nn} &= \text{number of subjects listed in neither of the two registries,}
\end{aligned}$$

and the corresponding probabilities

$$\begin{aligned}
\pi_{yy} &= \text{pr(a subject is listed in both Registry 1 and Registry 2),}\\
\pi_{yn} &= \text{pr(a subject is listed in Registry 1 only),}\\
\pi_{ny} &= \text{pr(a subject is listed in Registry 2 only),}\\
\pi_{nn} &= \text{pr(a patient is listed in neither Registry).}
\end{aligned}$$

It is reasonable to assume that the data arise from a multinomial distribution of the form

$$p_{X_{yy},X_{yn},X_{ny},X_{nn}}(x_{yy}, x_{yn}, x_{ny}, x_{nn}) = \frac{N!}{x_{yy}!x_{yn}!x_{ny}!x_{nn}!}\pi_{yy}^{x_{yy}}\pi_{yn}^{x_{yn}}\pi_{ny}^{x_{ny}}\pi_{nn}^{x_{nn}},$$

where $0 \leq x_{yy} \leq N, 0 \leq x_{yn} \leq N, 0 \leq x_{ny} \leq N, 0 \leq x_{nn} \leq N$, and $(x_{yy} + x_{yn} + x_{ny} + x_{nn}) = N$.

It is important to note that the random variable $X_{nn}$ is *not* observable.

(a) Let $\pi_1 = (\pi_{yy} + \pi_{yn})$ denote the marginal probability that a patient is listed in Registry 1, and let $\pi_2 = (\pi_{yy} + \pi_{ny})$ denote the marginal probability that a patient is listed in Registry 2. Under the assumption of statistical independence [i.e., $\pi_{yy} = \pi_1\pi_2, \pi_{yn} = \pi_1(1-\pi_2)$, etc.], develop an estimator $\hat{N}$ of $N$ by equating observed cell counts to their expected values under the assumed model. What is the numerical value of $\hat{N}$ when $x_{yy} = 12,000$, $x_{yn} = 6,000$, and $x_{ny} = 8,000$?

(b) For $j = 1, 2$, let $E_j$ denote the event that a subject with the medical condition is listed in Registry $j$, and let $\bar{E}_j$ denote the event that this subject is not listed in Registry $j$. In part (a), it was assumed that the events $E_1$ and $E_2$ are independent. As an alternative to this independence assumption, assume that membership in one of the two registries increases or decreases the odds of membership in the other registry by a factor of $k$; in other words,

$$\frac{\text{odds}(E_1 \mid E_2)}{\text{odds}(E_1 \mid \bar{E}_2)} = \frac{\text{odds}(E_2 \mid E_1)}{\text{odds}(E_2 \mid \bar{E}_1)} = k, \quad 0 < k < +\infty,$$

where, for two events A and B, odds(A|B) = pr(A|B)/[1 − pr(A|B)]. Note that $k > 1$ implies a positive association between the events $E_1$ and $E_2$, that $k < 1$ implies a negative association between the events $E_1$ and $E_2$, and that $k = 1$ implies no association (i.e., independence) between the events $E_1$ and $E_2$.

Although $k$ is not known in practice, it is of interest to determine whether estimates of $N$ would meaningfully change when plugging in various plausible values for $k$. Toward this end, develop an explicit expression for the method-of-moments estimator $\tilde{N}(k)$ of $N$ that would be obtained under the assumption that $k$ is a known constant. Using the data from part (a), calculate numerical values of $\tilde{N}(1/2)$, $\tilde{N}(2)$, and $\tilde{N}(4)$. Comment on your findings. In particular, is the estimate of $N$ sensitive to different assumptions about the direction and magnitude of the association between membership status for the two registries (i.e., to the value of $k$)?

**Exercise 4.34***. University researchers are conducting a study involving $n$ infants to assess whether infants placed in day care facilities are more likely to be overweight than are infants receiving care at home. Infants are defined as "overweight" if they fall within the 85th or higher percentile on the official Centers for Disease Control and Prevention (CDC) age-adjusted and sex-adjusted body mass index (BMI) growth chart.

Let $Y_i = 1$ if the $i$th infant ($i = 1, 2, \ldots, n$) is overweight, and let $Y_i = 0$ otherwise. It is assumed that $Y_i$ has the *Bernoulli* distribution

$$p_{Y_i}(y_i; \pi_i) = \pi_i^{y_i}(1-\pi_i)^{1-y_i}, \quad y_i = 0, 1 \quad \text{and} \quad 0 < \pi_i < 1.$$

Also, $Y_1, Y_2, \ldots, Y_n$ are assumed to be mutually independent random variables.

To make statistical inferences about the association between type of care and the probability of being overweight, the researchers propose the following *logistic regression model:*

$$\pi_i \equiv \pi(x_i) = \text{pr}(Y_i = 1|x_i) = \frac{e^{\alpha+\beta x_i}}{(1+e^{\alpha+\beta x_i})}, \text{ or equivalently,}$$

$$\text{logit}[\pi(x_i)] = \ln\left[\frac{\pi(x_i)}{1-\pi(x_i)}\right] = \alpha + \beta x_i, \quad i = 1, \ldots, n,$$

where $x_i = 1$ if the $i$th infant is in day care and $x_i = 0$ if the $i$th infant is at home, and where $\alpha$ and $\beta$ are unknown parameters to be estimated. Here, the parameter $\alpha$ represents the "log odds" of being overweight for infants in home care ($x_i = 0$), and the parameter $\beta$ represents the *difference* in log odds (or "log odds ratio") of being overweight for infants placed in day care ($x_i = 1$) compared to infants receiving care at home ($x_i = 0$).

Suppose that $n$ pairs $(y_1, x_1 = 0), (y_2, x_2 = 0), \ldots, (y_{n_0}, x_{n_0} = 0)$, $(y_{n_0+1}, x_{n_0+1} = 1), (y_{n_0+2}, x_{n_0+2} = 1) \ldots, (y_n, x_n = 1)$ of observed data are collected during the study, where the first $n_0$ data pairs are associated with the infants receiving home care, where the last $n_1$ data pairs are associated with the infants placed in day care, and where $(n_0 + n_1) = n$.

(a) Show that the MLEs of $\alpha$ and $\beta$ are

$$\hat{\alpha} = \ln\left(\frac{p_0}{1-p_0}\right) \quad \text{and} \quad \hat{\beta} = \ln\left[\frac{p_1/(1-p_1)}{p_0/(1-p_0)}\right],$$

where $p_0 = n_0^{-1}\sum_{i=1}^{n_0} y_i$ is the sample proportion of overweight infants receiving home care and $p_1 = n_1^{-1}\sum_{i=n_0+1}^{n} y_i$ is the sample proportion of overweight infants in day care.

(b) Develop an explicit expression for the large-sample variance–covariance matrix of $\hat{\alpha}$ and $\hat{\beta}$ based on both expected and observed information.

(c) Suppose that there are 100 infants receiving home care, 18 of whom are overweight, and that there are 100 infants in day care, 26 of whom are overweight. Use these data to compute large-sample 95% CIs for $\alpha$ and $\beta$. Based on these CI results, do the data supply statistical evidence that infants placed in day care facilities are more likely to be overweight than are infants receiving care at home?

For further details about logistic regression, see Breslow and Day (1980), Hosmer and Lemeshow (2000), Kleinbaum and Klein (2002), and Kleinbaum et al. (1982).

**Exercise 4.35*.** In April of 1986, a reactor exploded at the Chernobyl Nuclear Power Plant in Chernobyl, Russia. There were roughly 14,000 permanent residents of Chernobyl who were exposed to varying levels of radioactive iodine, as well as to other radioactive substances. It took about 3 days before these permanent residents, and other persons living in nearby areas, could be evacuated. As a result, many children and adults have since developed various forms of cancer.

In particular, many young children developed thyroid cancer. As a model for the development of thyroid cancer in such children, the following statistical model is proposed. Let $T$ be a continuous random variable representing the time (in years) from childhood radioactive iodine exposure caused by the Chernobyl explosion to the diagnosis of thyroid cancer, and let the continuous random variable $X$ be the level (in Joules per kilogram) of radioactive iodine exposure. Then, it is assumed that the conditional distribution of $T$ given $X = x$ is

$$f_T(t|X = x) = \theta x e^{-\theta x t}, \quad t > 0, \; x > 0, \; \theta > 0.$$

Further, assume that the distribution of $X$ is GAMMA$(\alpha = 1, \beta)$, so that

$$f_X(x) = \frac{x^{\beta-1}e^{-x}}{\Gamma(\beta)}, \quad x > 0, \; \beta > 1.$$

Suppose that an epidemiologist locates $n$ children with thyroid cancer who were residents of Cherynobyl at the time of the explosion. For each of these children, this epidemiologist determines the time in years (i.e., the so-called latency period) from exposure to the diagnosis of thyroid cancer. In particular, let $t_1, t_2, \ldots, t_n$ denote these observed latency periods. Since it is impossible to determine the true individual level of radioactive iodine exposure for each of these $n$ children, the only data available to this epidemiologist are the $n$ observed latency periods.

Based on the use of the observed latency periods for a random sample of $n = 300$ children, if the MLEs of $\theta$ and $\beta$ are $\hat{\theta} = 0.32$ and $\hat{\beta} = 1.50$, compute an appropriate large-sample 95% CI for $\gamma = \mathrm{E}(T)$, the true average latency period for children who developed thyroid cancer as a result of the Chernobyl nuclear reactor explosion.

**Exercise 4.36*.** Using $n$ mutually independent data pairs of the general form $(x, Y)$, it is desired to use the method of unweighted least squares to fit the model

$$Y = \beta_0 + \beta_1 x + \beta_2 x^2 + \epsilon,$$

where $\mathrm{E}(\epsilon) = 0$ and $\mathrm{V}(\epsilon) = \sigma^2$. Suppose that the three possible values of the predictor $x$ are $-1, 0,$ and $+1$. What proportion of the $n$ data points should be assigned to each of the three values of $x$ so as to minimize $\mathrm{V}(\hat{\beta}_2)$, the variance of the unweighted least-squares estimator of $\beta_2$?

To proceed, assume that $n = n\pi_1 + n\pi_2 + n\pi_3$, where $\pi_1 (0 < \pi_1 < 1)$ is the proportion of the $n$ observations to be assigned to the $x$-value of $-1$, where $\pi_2 (0 < \pi_2 < 1)$ is the proportion of the $n$ observations to be assigned to the $x$-value of 0, and where $\pi_3 (0 < \pi_3 < 1)$ is the proportion of the $n$ observations to be assigned to the $x$-value of $+1$. Further, assume that $n$ can be chosen so that $n_1 = n\pi_1, n_2 = n\pi_2,$ and $n_3 = n\pi_3$ are positive integers.

(a) With

$$\hat{\boldsymbol{\beta}} = (\hat{\beta}_0, \hat{\beta}_1, \hat{\beta}_2)' = (\mathbf{X}'\mathbf{X})^{-1}\mathbf{X}'\mathbf{Y},$$

show that $\mathbf{X}'\mathbf{X}$ can be written in the form

$$\mathbf{X}'\mathbf{X} = n \begin{bmatrix} 1 & b & a \\ b & a & b \\ a & b & a \end{bmatrix},$$

where $a = (\pi_1 + \pi_3)$ and $b = (\pi_3 - \pi_1)$.

(b) Show that

$$V(\hat{\beta}_2) = \left\{ \frac{[(\pi_1 + \pi_3) - (\pi_3 - \pi_1)^2]}{4n\pi_1\pi_2\pi_3} \right\} \sigma^2.$$

(c) Use the result from part (b) to find the values of $\pi_1, \pi_2$ and $\pi_3$ that minimize $V(\hat{\beta}_2)$ subject to the constraint $(\pi_1 + \pi_2 + \pi_3) = 1$.

**Exercise 4.37*.** Given appropriate data, one possible (but *not* necessarily optimal) algorithm for deciding whether or not there is statistical evidence that $p(\geq 2)$ population means are not all equal to the same value is the following: compute a $100(1-\alpha)\%$ CI for each population mean and decide that there is no statistical evidence that these population means are not all equal to the same value if these $p$ CIs have at least one value in common (i.e., if there is at least one value that is simultaneously contained in all $p$ CIs); otherwise, decide that there is statistical evidence that these $p$ population means are not all equal to the same value. To evaluate some statistical properties of this proposed algorithm, consider the following scenario.

For $i = 1, 2, \ldots, p$, let $X_{i1}, X_{i2}, \ldots, X_{in}$ constitute a random sample of size $n$ from a $N(\mu_i, \sigma^2)$ population. Given the stated assumptions, the appropriate exact $100(1-\alpha)\%$ CI for $\mu_i$, using only the data $\{X_{i1}, X_{i2}, \ldots, X_{in}\}$ from the $i$th population, involves the $t$-distribution with $(n-1)$ df and takes the form

$$\bar{X}_i \pm k\frac{S_i}{\sqrt{n}}, \quad \text{where } k = t_{(n-1),1-\alpha/2},$$

where

$$\bar{X}_i = n^{-1}\sum_{j=1}^{n} X_{ij} \quad \text{and} \quad S_i^2 = (n-1)^{-1}\sum_{j=1}^{n}(X_{ij} - \bar{X}_i)^2,$$

and where, for $0 < \alpha < 0.50$,

$$\text{pr}\left(T_\nu > t_{\nu,1-\alpha/2}\right) = \frac{\alpha}{2}$$

when the random variable $T_\nu$ has a $t$-distribution with $\nu$ df.

For notational convenience, let $I_i$ denote the set of values included in the $i$th computed CI; and, for $i \neq i'$, let the event $E_{ii'} = I_i \cap I_{i'} = \emptyset$, the empty (or null) set; in other words, $E_{ii'}$ is the event that the CIs $\bar{X}_i \pm kS_i/\sqrt{n}$ and $\bar{X}_{i'} \pm kS_{i'}/\sqrt{n}$ have no values in common (i.e., do not overlap).

(a) Show that

$$\pi_{ii'} = \text{pr}(E_{ii'}) = \text{pr}\left[|\bar{X}_i - \bar{X}_{i'}| > k\left(\frac{S_i}{\sqrt{n}} + \frac{S_{i'}}{\sqrt{n}}\right)\right].$$

(b) Under the condition (say, $C_p$) that all $p$ population means are actually equal to the same value (i.e., $\mu_1 = \mu_2 = \cdots = \mu_p = \mu$, say), use the result from part (a) to show that, for $i \neq i'$,

$$\pi_{ii'}^* = \text{pr}(E_{ii'}|C_p) \leq \text{pr}\left[|T_{2(n-1)}| > k|C_p\right] \leq \alpha.$$

(c) When $p = 3$ and under the condition $C_3$ that $\mu_1 = \mu_2 = \mu_3 = \mu$, say, find a crude upper bound for the probability that there are no values common to all three CIs. Comment on this finding and, in general, on the utility of this algorithm.

**Exercise 4.38*.** A highway safety researcher theorizes that the number $Y$ of automobile accidents per year occurring on interstate highways in the United States is linearly related to a certain measure $x$ of traffic density. To evaluate his theory, this highway safety researcher gathers appropriate data from $n$ independently chosen locations across the United States. More specifically, for the $i$th of $n$ independently chosen locations $(i = 1, 2, \ldots, n)$, the data point $(x_i, y_i)$ is recorded, where $x_i$ (the measure of traffic density at location $i$) is assumed to be a known positive constant and where $y_i$ is the observed value (or "realization") of the random variable $Y_i$. Here, the random variable $Y_i$ is assumed to have a Poisson distribution with $E(Y_i) = E(Y_i|x_i) = \theta_0 + \theta_1 x_i$. You can assume that $E(Y_i) > 0$ for all $i$ and that the set $\{Y_1, Y_2, \ldots, Y_n\}$ constitutes a set of $n$ mutually independent Poisson random variables. The goal is to use the available $n$ pairs of data points $(x_i, y_i), i = 1, 2, \ldots, n$, to make statistical inferences about the unknown parameters $\theta_0$ and $\theta_1$.

(a) Derive explicit expressions for the unweighted least-squares (ULS) estimators $\tilde{\theta}_0$ and $\tilde{\theta}_1$, respectively, of $\theta_0$ and $\theta_1$. Also, derive expressions for the expected values and variances of these two ULS estimators.

(b) For a set of $n = 100$ data pairs $(x_i, y_i), i = 1, 2, \ldots, 100$, suppose that each of 25 data pairs has an $x$ value equal to 1.0, that each of 25 data pairs has an $x$ value equal to 2.0, that each of 25 data pairs has an $x$ value equal to 3.0, and that each of 25 data pairs has an $x$ value of 4.0. If the MLEs of $\theta_0$ and $\theta_1$ are, respectively, $\hat{\theta}_0 = 2.00$ and $\hat{\theta}_1 = 4.00$, compute an appropriate large-sample 95% CI for the parameter

$$\psi = E(Y|x = 2.5) = \theta_0 + (2.5)\theta_1.$$

**Exercise 4.39*.** Let $X_1, X_2, \ldots, X_n$ constitute a random sample of size $n(> 1)$ from an $N(\mu, \sigma^2)$ population, and let $Y_1, Y_2, \ldots, Y_n$ constitute a random sample of size $n(> 1)$ from a completely different $N(\mu, \sigma^2)$ population. Hence, the set $\{X_1, X_2, \ldots, X_n; Y_1, Y_2, \ldots, Y_n\}$ is made up of a total of $2n$ mutually independent random variables, with each random variable in the set having a $N(\mu, \sigma^2)$ distribution. Consider the following random variables:

$$\bar{X} = n^{-1} \sum_{i=1}^{n} X_i, \quad S_x^2 = (n-1)^{-1} \sum_{i=1}^{n} (X_i - \bar{X})^2,$$

$$\bar{Y} = n^{-1} \sum_{i=1}^{n} Y_i, \quad S_y^2 = (n-1)^{-1} \sum_{i=1}^{n} (Y_i - \bar{Y})^2.$$

(a) For any particular value of $i(1 \leq i \leq n)$, determine the exact distribution of the random variable $D_i = (X_i - \bar{X})$.

(b) For particular values of $i(1 \leq i \leq n)$ and $j(1 \leq j \leq n)$, where $i \neq j$, derive an explicit expression for $\text{corr}(D_i, D_j)$, the correlation between the random variables $D_i$ and

$D_j$. Also, find the limiting value of $\text{corr}(D_i, D_j)$ as $n \to \infty$, and then provide an argument as to why this result makes sense.

(c) Prove rigorously that the density function of the random variable $R = S_x^2/S_y^2$ is

$$f_R(r) = [\Gamma(n-1)]\{\Gamma[(n-1)/2]\}^{-2} r^{[(n-3)/2]}(1+r)^{-(n-1)},$$
$$0 < r < \infty.$$

Also, find an explicit expression for $E(R)$.

(d) Now, consider the following two estimators of the unknown parameter $\mu$:

(i) $\hat{\mu}_1 = (\bar{X} + \bar{Y})/2$;

(ii) $\hat{\mu}_2 = [(\bar{X})(S_y^2) + (\bar{Y})(S_x^2)]/(S_x^2 + S_y^2)$.

Prove rigorously that both $\hat{\mu}_1$ and $\hat{\mu}_2$ are unbiased estimators of $\mu$.

(e) Derive explicit expressions for $V(\hat{\mu}_1)$ and $V(\hat{\mu}_2)$. Which estimator, $\hat{\mu}_1$ or $\hat{\mu}_2$, do you prefer and why?

**Exercise 4.40*.** A certain company manufactures stitches for coronary bypass graft surgeries. The distribution of the length $Y$ (in feet) of defect-free stitches manufactured by this company is assumed to have the uniform density

$$f_Y(y;\theta) = \theta^{-1}, \quad 0 < y < \theta, \; \theta > 0.$$

Clearly, the larger is $\theta$, the better is the quality of the manufacturing process. Suppose that $Y_1, Y_2, \ldots, Y_n$ constitute a random sample of size $n (n > 2)$ from $f_Y(y;\theta)$. A statistician proposes three estimators of the parameter $\mu = E(Y) = \theta/2$, the true average length of defect-free stitches manufactured by this company. These three estimators are as follows:

(1) $\hat{\mu}_1 = k_1\bar{Y} = k_1 n^{-1}\sum_{i=1}^{n} Y_i$, where $k_1$ is to be chosen so that $E(\hat{\mu}_1) = \mu$;

(2) $\hat{\mu}_2 = k_2 Y_{(n)}$, where $Y_{(n)}$ is the largest order statistic based on this random sample and where $k_2$ is to be chosen so that $E(\hat{\mu}_2) = \mu$;

(3) $\hat{\mu}_3 = k_3[Y_{(1)} + Y_{(n)}]/2$, the so-called "midpoint" of the data, where $Y_{(1)}$ is the smallest order statistic based on this random sample and where $k_3$ is to be chosen so that $E(\hat{\mu}_3) = \mu$.

(a) Find the value of $k_1$, and then find $V(\hat{\mu}_1)$.

(b) Find the value of $k_2$, and then find $V(\hat{\mu}_2)$.

(c) Find the value of $k_3$, and then find $V(\hat{\mu}_3)$.

(d) Compare the variances of $\hat{\mu}_1, \hat{\mu}_2$, and $\hat{\mu}_3$ for both finite $n$ and as $n \to \infty$. Which estimator do you prefer and why?

**Exercise 4.41*.** For adult males with incurable malignant melanoma who have lived at least 25 consecutive years in Arizona, an epidemiologist theorizes that the true mean time (in years) to death differs between those adult males with a family history of skin cancer and those adult males without a family history of skin cancer.

To test this theory, this epidemiologist selects a random sample of $n$ adult males in Arizona with incurable malignant melanoma, each of whom has lived in Arizona for at least 25 consecutive years. Then, this epidemiologist and a collaborating biostatistician agree to consider the following statistical model for two random variables $X$ and $Y$. Here, $X$ is a *dichotomous* random variable taking the value 1 for an adult male in the random sample *without* a family history of skin cancer and taking the value 0 for an adult male in the random sample *with* a family history of skin cancer; and, $Y$ is a *continuous* random variable representing the time (in months) to death for an adult male in the random sample with incurable malignant melanoma. More specifically, assume that the *marginal* distribution of $X$ is Bernoulli (or point-binomial), namely,

$$\mathrm{p}_X(x;\theta) = \theta^x(1-\theta)^{(1-x)}, \quad x = 0,1;\ 0 < \theta < 1.$$

Moreover, assume that the *conditional* distribution of $Y$, given $X = x$, is negative exponential with conditional mean $\mathrm{E}(Y|X = x) = \mu(x) = \mathrm{e}^{\alpha+\beta x}$, namely,

$$\mathrm{f}_Y(y|X = x;\alpha,\beta) = [\mu(x)]^{-1}\mathrm{e}^{-y/\mu(x)}, \quad 0 < y < +\infty.$$

This two-variable model involves three unknown parameters, namely, $\theta(0 < \theta < 1), \alpha(-\infty < \alpha < +\infty)$, and $\beta(-\infty < \beta < +\infty)$.

Let $(X_1, Y_1), (X_2, Y_2), \ldots, (X_n, Y_n)$ constitute a random sample of size $n$ from the joint distribution $\mathrm{f}_{X,Y}(x, y;\theta,\alpha,\beta)$ of the random variables $X$ and $Y$, where this joint distribution is given by the product

$$\mathrm{f}_{X,Y}(x, y;\theta,\alpha,\beta) = \mathrm{p}_X(x;\theta)\mathrm{f}_Y(y|X = x;\alpha,\beta).$$

Now, suppose that the available data contain $n_1$ adult males without a family history of skin cancer and $n_0 = (n - n_1)$ adult males with a family history of skin cancer. Further, assume (without loss of generality) that the $n$ observed data pairs (i.e., the $n$ realizations) are arranged (for notational simplicity) so that the first $n_1$ pairs $(1, y_1), (1, y_2), \ldots, (1, y_{n_1})$ are the observed data for the $n_1$ adult males without a family history of skin cancer, and the remaining $n_0$ data pairs $(0, y_{n_1+1}), (0, y_{n_1+2}), \ldots, (0, y_n)$ are the observed data for the $n_0$ adult males with a family history of skin cancer.

(a) Develop explicit expressions for the MLEs $\hat{\theta}, \hat{\alpha}$, and $\hat{\beta}$ of $\theta, \alpha$, and $\beta$, respectively. In particular, show that these three ML estimators can be written as explicit functions of one or more of the sample means $\bar{x} = n_1/n, \bar{y}_1 = n_1^{-1}\sum_{i=1}^{n_1} y_i$, and $\bar{y}_0 = n_0^{-1}\sum_{i=(n_1+1)}^{n} y_i$.

(b) Using *expected* information, develop an explicit expression for the $(3\times3)$ large-sample covariance matrix $\mathcal{I}^{-1}$ for the three ML estimators $\hat{\theta}, \hat{\alpha}$, and $\hat{\beta}$.

(c) If $n = 50, \hat{\theta} = 0.60, \hat{\alpha} = 0.50$, and $\hat{\beta} = 0.40$, use appropriate CI calculations to determine whether this numerical information supplies statistical evidence that the true mean time to death differs between adult males with a family history of skin cancer and adult males without a family history of skin cancer, all of whom developed incurable malignant melanoma and lived at least 25 consecutive years in Arizona.

**Exercise 4.42*.** For a certain laboratory experiment, the concentration $Y_x$ (in milligrams per cubic centimeter) of a certain pollutant produced via a chemical reaction taking place at temperature $x$ (conveniently scaled so that $-1 \leq x \leq +1$) has a *normal* distribution with mean $E(Y_x) = \theta_x = (\beta_0 + \beta_1 x + \beta_2 x^2)$ and variance $V(Y_x) = \sigma^2$. Also, the temperature $x$ is nonstochastic (i.e., is not a random variable) and is known without error.

Suppose that an environmental scientist runs this experiment $N$ times, with each run involving a different temperature setting. Further, suppose that these $N$ runs produce the $N$ pairs of data $(x_1, Y_{x_1}), (x_2, Y_{x_2}), \ldots, (x_N, Y_{x_N})$. Assume that the random variables $Y_{x_1}, Y_{x_2}, \ldots, Y_{x_N}$ constitute a set of mutually independent random variables, and that $x_1, x_2, \ldots, x_N$ constitute a set of known constants.

Further, let $\mu_k = N^{-1} \sum_{i=1}^{N} x_i^k, k = 1, 2, 3$; and, assume that the environmental scientist chooses the $N$ temperature values $x_1, x_2, \ldots, x_N$ so that $\mu_1 = \mu_3 = 0$.

Suppose that this environmental scientist decides to estimate the parameter $\theta_x$ using the *straight-line estimator* $\hat{\theta}_x = (B_0 + B_1 x)$, where $B_0 = N^{-1} \sum_{i=1}^{N} Y_{x_i}$ and where $B_1 = \sum_{i=1}^{N} x_i Y_{x_i} / \sum_{i=1}^{N} x_i^2$. Note that $\hat{\theta}_x$ is a straight-line function of $x$, but that the true model relating the expected value of $Y_x$ to $x$ actually involves a squared term in $x$. Hence, the wrong model is being fit to the available data.

(a) Develop explicit expressions for $E(\hat{\theta}_x)$ and $V(\hat{\theta}_x)$. What is the exact distribution of the estimator $\hat{\theta}_x$?

(b) Consider the expression

$$Q = \int_{-1}^{1} [E(\hat{\theta}_x) - \theta_x]^2 \, dx.$$

Since $[E(\hat{\theta}_x) - \theta_x]^2$ is the *squared bias* when $\hat{\theta}_x$ is used to estimate $\theta_x$ at temperature setting $x$, $Q$ is called the *integrated squared bias*. The quantity $Q$ can be interpreted as being the cumulative bias over all values of $x$ such that $-1 \leq x \leq +1$. It is desirable to choose the temperature settings $x_1, x_2, \ldots, x_N$ to make $Q$ as small as possible. More specifically, find the numerical value of $\mu_2$ that minimizes $Q$. Then, given this result, if $N = 4$, find a set of values for $x_1, x_2, x_3$, and $x_4$ such that $Q$ is minimized and that $\mu_1 = \mu_3 = 0$.

**Exercise 4.43*.** For the $i$th of $n$ $(i = 1, 2, \ldots, n)$ U.S. military families, suppose that there are $y_{i1}$ events of child abuse during a period of $L_{i1}$ months when the soldier-father is not at home (i.e., is deployed to a foreign country), and suppose that there are $y_{i0}$ events of child abuse during a period of $L_{i0}$ months when the soldier-father is at home (i.e., is not deployed). To assess whether the rate of child abuse when the soldier-father is deployed is different from the rate of child abuse when the soldier-father is not deployed, the following statistical model is proposed.

Let $\alpha_i \sim N(0, \sigma_\alpha^2)$. For $j = 0, 1$, and given $\alpha_i$ fixed, suppose that the random variable $Y_{ij}$, with realization (or observed value) $y_{ij}$, is assumed to have a Poisson distribution with conditional mean $E(Y_{ij}|\alpha_i) = L_{ij}\lambda_{ij}$, where

$$\ln(\lambda_{ij}) = \alpha_i + \beta D_{ij} + \sum_{l=1}^{p} \gamma_l C_{il};$$

here, $\lambda_{i0}$ and $\lambda_{i1}$ denote the respective nondeployment and deployment rates of child abuse per month for the $i$th family, $D_{ij}$ takes the value 1 if $j = 1$ (i.e., if the soldier-father is deployed) and $D_{ij}$ takes the value 0 if $j = 0$ (i.e., if the soldier-father is not deployed), and $C_{i1}, C_{i2}, \ldots, C_{ip}$ are the values of $p$ covariates $C_1, C_2, \ldots, C_p$ specific to the $i$th family. Further, conditional on $\alpha_i$ being fixed, $Y_{i0}$ and $Y_{i1}$ are assumed to be independent random variables, $i = 1, 2, \ldots, n$.

(a) Find an explicit expression for $\text{cov}(Y_{i0}, Y_{i1})$, and comment on the rationale for including the random effect $\alpha_i$ in the proposed statistical model.

(b) Let the random variable $Y_i = (Y_{i0} + Y_{i1})$ be the total number of child abuse events for the $i$th family. Show that the conditional distribution $\text{p}_{Y_{i1}}(y_{i1}|Y_i = y_i, \alpha_i)$ of $Y_{i1}$, given $Y_i = y_i$ and $\alpha_i$ fixed, is $\text{BIN}(y_i, \pi_i)$, where

$$\pi_i = \frac{L_{i1}\theta}{L_{i0} + L_{i1}\theta};$$

here,

$$\theta = \frac{\lambda_{i1}}{\lambda_{i0}} = e^{\beta}$$

is the rate ratio comparing the deployment and non-deployment rates of child abuse per month for the $i$th family. Note that the rate ratio parameter $\theta$ does not vary with $i$ (i.e., does not vary across families), even though the individual rate parameters are allowed to vary with $i$.

(c) Under the reasonable assumption that families behave independently of one another, use the conditional likelihood function

$$\mathcal{L} = \prod_{i=1}^{n} \text{p}_{Y_{i1}}(y_{i1}|Y_i = y_i, \alpha_i)$$

to show that the conditional MLE $\hat{\theta}$ of $\theta$ satisfies the equation

$$\hat{\theta} \sum_{i=1}^{n} \left( \frac{y_i L_{i1}}{L_{i0} + L_{i1}\hat{\theta}} \right) = \sum_{i=1}^{n} y_{i1}.$$

(d) Using *expected* information, develop a general expression for a large-sample 95% CI for the rate ratio parameter $\theta$.

**Exercise 4.44*.** Let $Y$ be a continuous response variable and let $X$ be a continuous predictor variable. Also, assume that

$$\text{E}(Y|X = x) = (\beta_0 + \beta_1 x) \quad \text{and} \quad X \sim \text{N}(\mu_x, \sigma_x^2).$$

Further, suppose that the predictor variable $X$ is very expensive to measure, but that a *surrogate* variable $X^*$ is available and can be measured fairly inexpensively. Further, for $i = 1, 2, \ldots, n$, assume that $X_i^*$ and $X_i$ are related by the *measurement error model* $X_i^* = (X_i + U_i)$, where $U_i \sim \text{N}(0, \sigma_u^2)$ and where $X_i$ and $U_i$ are independent random variables. Suppose that it is decided to use $X^*$ instead of $X$ as the predictor variable when estimating the slope parameter $\beta_1$ by fitting a straight-line regression model via unweighted least squares. In particular, suppose that the $n$ mutually independent pairs $(X_i^*, Y_i) = (X_i + U_i, Y_i), i = 1, 2, \ldots, n$, are used to construct an estimator $\hat{\beta}_1^*$ of $\beta_1$ of the form

$$\hat{\beta}_1^* = \frac{\sum_{i=1}^{n}(X_i^* - \bar{X}^*)Y_i}{\sum_{i=1}^{n}(X_i^* - \bar{X}^*)^2},$$

where $\bar{X}^* = n^{-1}\sum_{i=1}^{n} X_i^*$.

Using conditional expectation theory, derive an explicit expression for $E(\hat{\beta}_1^*)$, and then comment on how $E(\hat{\beta}_1^*)$ varies as a function of the ratio $\lambda = \sigma_u^2/\sigma_x^2, 0 < \lambda < \infty$. In your derivation, use the fact that $X_i$ and $X_i^*$ have a bivariate normal distribution and employ the assumption that $E(Y_i|X_i = x_i, X_i^* = x_i^*) = E(Y_i|X_i = x_i), i = 1, 2, \ldots, n$. This assumption is known as the *nondifferential error assumption* and states that $X_i^*$ contributes no further information regarding $Y_i$ if $X_i$ is available.

For an excellent book on measurement error and its effects on the validity of statistical analyses, see Fuller (2006).

**Exercise 4.45*.** Let the random variable $Y$ take the value 1 if a person develops a certain rare disease, and let $Y$ take the value 0 if not. Consider the following exponential regression model, namely,

$$\text{pr}(Y = 1|X, C) = e^{(\beta_0 + \beta_1 X + \gamma' C)},$$

where $X$ is a continuous exposure variable, $C' = (C_1, C_2, \ldots, C_p)$ is a row vector of $p$ covariates, and $\gamma' = (\gamma_1, \gamma_2, \ldots, \gamma_p)$ is a row vector of $p$ regression coefficients. Here, $\beta_1(>0)$ is the key parameter of interest; in particular, $\beta_1$ measures the effect of the exposure $X$ on the probability (or risk) of developing the rare disease after adjusting for the effects of the covariates $C_1, C_2, \ldots, C_p$. Since the disease in question is rare, it is reasonable to assume that

$$\text{pr}(Y = 1|X, C) = e^{(\beta_0 + \beta_1 X + \gamma' C)} < 1.$$

Now, suppose that the exposure variable $X$ is very expensive to measure, but that a *surrogate* variable $X^*$ for $X$ is available and can be measured fairly inexpensively. Further, assume that $X$ and $X^*$ are related via the *Berkson measurement error model* (Berkson, 1950)

$$X = \alpha_0 + \alpha_1 X^* + \delta' C + U,$$

where $\alpha_1 > 0$, where $U \sim N(0, \sigma_u^2)$, where $\delta' = (\delta_1, \delta_2, \ldots, \delta_p)$ is a row vector of $p$ regression coefficients, and where the random variables $U$ and $X^*$ are independent given $C$.

(a) Show that $\text{corr}(X, X^*|C) < 1$, in which case $X^*$ is said to be an imperfect surrogate for $X$ (since it is not perfectly correlated with $X$).

(b) Determine the structure of $f_X(x|X^*, C)$, the conditional density function of $X$ given $X^*$ and $C$.

(c) Now, suppose that an epidemiologist decides to use $X^*$ instead of $X$ for the exposure variable in the exponential regression model given above. To assess the implications of this decision, show that $\text{pr}(Y = 1|X^*, C)$ has the structure

$$\text{pr}(Y = 1|X^*, C) = e^{(\theta_0 + \theta_1 X^* + \boldsymbol{\xi}' C)},$$

where $\theta_0, \theta_1$, and $\boldsymbol{\xi}'$ are specific parametric functions of one or more of the quantities $\beta_0, \beta_1, \alpha_0, \alpha_1, \sigma_u^2, \boldsymbol{\gamma}'$, and $\boldsymbol{\delta}'$. In your derivation, assume that

$$\text{pr}(Y = 1|X, X^*, C) = \text{pr}(Y = 1|X, C);$$

this is known as the *nondifferential error assumption* and states that $X^*$ contributes no further information regarding $Y$ if $X$ is available. In particular, show that $\theta_1 \neq \beta_1$, and then comment on the implication of this result with regard to the estimation of $\beta_1$ using $X^*$ instead of $X$ in the stated exponential regression model.

For an application of this methodology, see Horick et al. (2006).

**Exercise 4.46*.** Suppose that $Y$ is a dichotomous outcome variable taking the values 0 and 1, and that $X$ is a dichotomous predictor variable also taking the values 0 and 1. Further, for $x = 0, 1$, let

$$\mu_x = \text{pr}(Y = 1|X = x) \quad \text{and} \quad \text{let } \delta = \text{pr}(X = 1).$$

(a) Suppose that $X$ is unobservable, and that a *surrogate* dichotomous variable $X^*$ is used in place of $X$. Further, assume that $X$ and $X^*$ are related via the *misclassification probabilities*

$$\pi_{xx^*} = \text{pr}(X^* = x^*|X = x), \quad x = 0, 1 \quad \text{and} \quad x^* = 0, 1.$$

Find an explicit expression for $\text{corr}(X, X^*)$. For what values of $\pi_{00}, \pi_{10}, \pi_{01}$, and $\pi_{11}$ will $\text{corr}(X, X^*) = 1$, in which case $X^*$ is said to be a *perfect* surrogate for $X$? Comment on your findings.

(b) Now, consider the *risk difference* parameter $\theta = (\mu_1 - \mu_0)$. With $\mu^*_{x^*} = \text{pr}(Y = 1|X^* = x^*)$, prove that $|\theta^*| \leq |\theta|$, where

$$\theta^* = (\mu^*_1 - \mu^*_0).$$

Then, comment on this finding with regard to the *misclassification bias* resulting from the use of $X^*$ instead of $X$ for estimating $\theta$. In your proof, assume that

$$\text{pr}[Y = 1|(X = x) \cap (X^* = x^*)] = \text{pr}(Y = 1|X = x);$$

this *nondifferential error assumption* states that $X^*$ contributes no further information regarding $Y$ if $X$ is available.

## SOLUTIONS

### Solution 4.1

(a) Method of Moments:

$$\bar{Y} = \frac{1}{n}\sum_{x=1}^{n} Y_x \text{ is equated to } \mathrm{E}(\bar{Y}) = \frac{1}{n}\sum_{x=1}^{n}\mathrm{E}(Y_x) = \frac{\mu}{n}\sum_{x=1}^{n} x,$$

so that

$$\hat{\mu}_1 = \frac{n\bar{Y}}{\sum_{x=1}^{n} x} = \frac{n\bar{Y}}{\left[\frac{n(n+1)}{2}\right]} = \left(\frac{2}{n+1}\right)\bar{Y}.$$

Unweighted Least Squares:

$$Q = \sum_{x=1}^{n}[Y_x - \mathrm{E}(Y_x)]^2 = \sum_{x=1}^{n}(Y_x - x\mu)^2.$$

So,

$$\frac{\partial Q}{\partial \mu} = 2\sum_{x=1}^{n}(Y_x - x\mu)(-x) = 0 \Longrightarrow \mu\sum_{x=1}^{n} x^2 = \sum_{x=1}^{n} xY_x,$$

so that

$$\hat{\mu}_2 = \frac{\sum_{x=1}^{n} xY_x}{\sum_{x=1}^{n} x^2} = \frac{\sum_{x=1}^{n} xY_x}{\left[\frac{n(n+1)(2n+1)}{6}\right]} = \frac{6\sum_{x=1}^{n} xY_x}{n(n+1)(2n+2)}.$$

Maximum Likelihood:

$$\mathcal{L} = \prod_{x=1}^{n}\left\{\frac{1}{\sqrt{2\pi}(r^3\sigma^2)^{1/2}}\exp\left[\frac{-(y_x - x\mu)^2}{2x^3\sigma^2}\right]\right\}$$

$$= (2\pi)^{-n/2}\sigma^{-n}\left(\prod_{x=1}^{n} x^{-3/2}\right)\exp\left[-\frac{1}{2\sigma^2}\sum_{x=1}^{n} x^{-3}(y_x - x\mu)^2\right].$$

So,

$$\ln\mathcal{L} = -\frac{n}{2}\ln(2\pi) - n\ln(\sigma) - \frac{3}{2}\sum_{x=1}^{n}\ln x - \frac{1}{2\sigma^2}\sum_{x=1}^{n} x^{-3}(y_x - x\mu)^2.$$

Thus,

$$\frac{\partial \ln\mathcal{L}}{\partial \mu} = \frac{-1}{\sigma^2}\sum_{x=1}^{n} x^{-3}(y_x - x\mu)(-x) = 0$$

gives

$$\mu \sum_{x=1}^{n} x^{-1} = \sum_{x=1}^{n} x^{-2} y_x,$$

so that

$$\hat{\mu}_3 = \frac{\sum_{x=1}^{n} x^{-2} Y_x}{\sum_{x=1}^{n} x^{-1}}.$$

Now, since $\hat{\mu}_1$, $\hat{\mu}_2$, and $\hat{\mu}_3$ are each a linear combination of mutually independent normal random variables, all three of these estimators are normally distributed. Now,

$$\begin{aligned}
\mathrm{E}(\hat{\mu}_1) &= \left(\frac{2}{n+1}\right) \mathrm{E}(\bar{Y}) = \left(\frac{2}{n+1}\right) \frac{1}{n} \sum_{x=1}^{n} \mathrm{E}(Y_x) \\
&= \left(\frac{2}{n+1}\right) \frac{\mu}{n} \sum_{x=1}^{n} x = \mu, \\
\mathrm{V}(\hat{\mu}_1) &= \left(\frac{2}{n+1}\right)^2 \frac{1}{n^2} \sum_{x=1}^{n} x^3 \sigma^2 = \frac{4\sigma^2 \sum_{x=1}^{n} x^3}{n^2(n+1)^2} = \sigma^2. \\
\mathrm{E}(\hat{\mu}_2) &= \frac{\sum_{x=1}^{n} x(x\mu)}{\sum_{x=1}^{n} x^2} = \mu, \\
\mathrm{V}(\hat{\mu}_2) &= \frac{\sum_{x=1}^{n} x^2(x^3\sigma^2)}{\left(\sum_{x=1}^{n} x^2\right)^2} = \frac{36\sigma^2}{n^2(n+1)^2(2n+1)^2} \sum_{x=1}^{n} x^5. \\
\mathrm{E}(\hat{\mu}_3) &= \frac{\sum_{x=1}^{n} x^{-2}(x\mu)}{\sum_{x=1}^{n} x^{-1}} = \mu, \\
\mathrm{V}(\hat{\mu}_3) &= \frac{\sum_{x=1}^{n} x^{-4}(x^3\sigma^2)}{\left(\sum_{x=1}^{n} x^{-1}\right)^2} = \frac{\sigma^2}{\left(\sum_{x=1}^{n} x^{-1}\right)}.
\end{aligned}$$

(b) Clearly, since all these estimators are unbiased estimators of $\mu$, we want to use the estimator with the smallest variance. We could analytically compare $\mathrm{V}(\hat{\mu}_1)$, $\mathrm{V}(\hat{\mu}_2)$, and $\mathrm{V}(\hat{\mu}_3)$, but there is a more direct way. Since

$$\frac{\partial \ln \mathcal{L}}{\partial \mu} = \frac{1}{\sigma^2} \sum_{x=1}^{n} x^{-2} y_x - \frac{\mu}{\sigma^2} \sum_{x=1}^{n} x^{-1},$$

and since

$$\mathrm{E}\left(\frac{\partial^2 \ln \mathcal{L}}{\partial \mu \, \partial \sigma^2}\right) = 0,$$

so that the expected information matrix is a diagonal matrix, the Cramér–Rao lower bound for the variance of any unbiased estimator of $\mu$ using $\{Y_1, Y_2, \ldots, Y_n\}$ is

$$\frac{1}{-\mathrm{E}\left(\frac{\partial^2 \ln \mathcal{L}}{\partial \mu^2}\right)} = \frac{\sigma^2}{\sum_{x=1}^{n} x^{-1}},$$

which is achieved by $\hat{\mu}_3$ for any finite $n$. So, the "best" exact 95% CI for $\mu$ should be based on $\hat{\mu}_3$, the minimum variance bound unbiased estimator (MVBUE) of $\mu$. Since

$$\frac{\hat{\mu}_3 - \mu}{\sqrt{\mathrm{V}(\hat{\mu}_3)}} \sim \mathrm{N}(0,1),$$

the "best" exact 95% CI for $\mu$ is $\hat{\mu}_3 \pm 1.96\sqrt{\mathrm{V}(\hat{\mu}_3)}$. For the given data,

$$\hat{\mu}_3 = \frac{\sum_{x=1}^{5} x^{-2}(x+1)}{\sum_{x=1}^{5} x^{-1}} = \frac{\left(\frac{2}{1} + \frac{3}{4} + \frac{4}{9} + \frac{5}{16} + \frac{6}{25}\right)}{\left(1 + \frac{1}{2} + \frac{1}{3} + \frac{1}{4} + \frac{1}{5}\right)} = 1.641,$$

and

$$\mathrm{V}(\hat{\mu}_3) = \frac{(2)}{(2.283)} = 0.876,$$

so that the computed exact 95% CI for $\mu$ is

$$1.641 \pm 1.96\sqrt{0.876} = 1.641 \pm 1.835 = (-0.194, 3.476).$$

**Solution 4.2**

(a) The unweighted least-squares estimator $\hat{\theta}_{\text{uls}}$ is the value of $\theta$ that minimizes

$$Q = \sum_{i=1}^{n} (Y_i - \theta x_i)^2.$$

Solving

$$\frac{\partial Q}{\partial \theta} = -2\sum_{i=1}^{n} x_i(Y_i - \theta x_i) = 0$$

yields

$$\hat{\theta}_{\text{uls}} = \sum_{i=1}^{n} x_i Y_i \Big/ \sum_{i=1}^{n} x_i^2.$$

Since

$$\frac{\partial^2 Q}{\partial \theta^2} = 2\sum_{i=1}^{n} x_i^2 > 0,$$

$\hat{\theta}_{\text{uls}}$ minimizes $Q$. Also,

$$\text{E}(\hat{\theta}_{\text{uls}}) = \frac{\sum_{i=1}^{n} x_i \text{E}(Y_i)}{\sum_{i=1}^{n} x_i^2} = \frac{\sum_{i=1}^{n} x_i(\theta x_i)}{\sum_{i=1}^{n} x_i^2} = \theta,$$

and

$$\text{V}(\hat{\theta}_{\text{uls}}) = \frac{\sum_{i=1}^{n} x_i^2 \text{V}(Y_i)}{(\sum_{i=1}^{n} x_i^2)^2} = \frac{\sum_{i=1}^{n} x_i^2(\theta x_i)}{(\sum_{i=1}^{n} x_i^2)^2}$$

$$= \theta \sum_{i=1}^{n} x_i^3 \Bigg/ \left(\sum_{i=1}^{n} x_i^2\right)^2.$$

(b) The method of moments estimator is obtained by solving for $\theta$ using the equation

$$\bar{Y} = \text{E}(\bar{Y}),$$

where

$$\bar{Y} = n^{-1} \sum_{i=1}^{n} Y_i$$

and

$$\text{E}(\bar{Y}) = n^{-1} \sum_{i=1}^{n} \text{E}(Y_i) = n^{-1} \sum_{i=1}^{n} (\theta x_i)$$

$$= \theta n^{-1} \sum_{i=1}^{n} x_i = \theta \bar{x}.$$

Hence, the equation

$$\bar{Y} = \text{E}(\bar{Y}) = \theta \bar{x}$$

gives

$$\hat{\theta}_{\text{mm}} = \bar{Y}/\bar{x}.$$

Obviously, $\text{E}(\hat{\theta}_{\text{mm}}) = \theta$, and

$$\text{V}(\hat{\theta}_{\text{mm}}) = \frac{\text{V}(\bar{Y})}{(\bar{x})^2} = \frac{\text{V}\left[n^{-1} \sum_{i=1}^{n} Y_i\right]}{(\bar{x})^2}$$

$$= \frac{n^{-2} \sum_{i=1}^{n} \text{V}(Y_i)}{(\bar{x})^2} = \frac{n^{-2} \sum_{i=1}^{n} (\theta x_i)}{(\bar{x})^2}$$

$$= \frac{n^{-1} \theta \bar{x}}{(\bar{x})^2} = \frac{\theta}{n \bar{x}} = \frac{\theta}{\sum_{i=1}^{n} x_i}.$$

(c) Now, with $y = (y_1, y_2, \ldots, y_n)$, we have

$$\mathcal{L}(y;\theta) = \prod_{i=1}^{n}\left\{\frac{(\theta x_i)^{y_i} e^{-\theta x_i}}{y_i!}\right\}$$

$$= \frac{\theta^{\sum_{i=1}^{n} y_i}\left(\prod_{i=1}^{n} x_i^{y_i}\right) e^{-\theta \sum_{i=1}^{n} x_i}}{\prod_{i=1}^{n} y_i!},$$

so that

$$\ln \mathcal{L}(y;\theta) = \left(\sum_{i=1}^{n} y_i\right) \ln \theta + \sum_{i=1}^{n} y_i \ln x_i - \theta \sum_{i=1}^{n} x_i - \sum_{i=1}^{n} \ln y_i!.$$

So,

$$\frac{\partial \ln \mathcal{L}(y;\theta)}{\partial \theta} = \frac{\left(\sum_{i=1}^{n} y_i\right)}{\theta} - \sum_{i=1}^{n} x_i = 0$$

gives

$$\hat{\theta}_{\text{ml}} = \sum_{i=1}^{n} Y_i \Big/ \sum_{i=1}^{n} x_i = \frac{\bar{Y}}{\bar{x}} \ (= \hat{\theta}_{\text{mm}}).$$

So,

$$\mathrm{E}(\hat{\theta}_{\text{ml}}) = \mathrm{E}(\hat{\theta}_{\text{mm}}) = \theta \quad \text{and} \quad \mathrm{V}(\hat{\theta}_{\text{ml}}) = \mathrm{V}(\hat{\theta}_{\text{mm}}) = \frac{\theta}{n\bar{x}}.$$

Note that one can use exponential family theory to show that $\hat{\theta}_{\text{ml}} (= \hat{\theta}_{\text{mm}})$ is the MVBUE of $\theta$. In particular,

$$\mathcal{L}(y;\theta) = \exp\left\{\hat{\theta}_{\text{ml}}\left(\sum_{i=1}^{n} x_i\right)(\ln \theta) - \theta \sum_{i=1}^{n} x_i + \ln\left[\frac{\prod_{i=1}^{n} x_i^{y_i}}{\prod_{i=1}^{n} y_i!}\right]\right\}.$$

(d) From part (c),

$$\frac{\partial^2 \ln \mathcal{L}(y;\theta)}{\partial \theta^2} = \frac{-\sum_{i=1}^{n} y_i}{\theta^2},$$

so that

$$-\mathrm{E}_y\left[\frac{\partial^2 \ln \mathcal{L}(y;\theta)}{\partial \theta^2}\right] = \frac{\sum_{i=1}^{n} \mathrm{E}(Y_i)}{\theta^2} = \frac{\theta \sum_{i=1}^{n} x_i}{\theta^2} = \frac{n\bar{x}}{\theta}.$$

Hence,

$$\text{CRLB} = \frac{1}{(n\bar{x}/\theta)} = \frac{\theta}{n\bar{x}},$$

which is achieved by the estimators $\hat{\theta}_{ml}$ and $\hat{\theta}_{mm}$ (which are identical), but which is not achieved by $\hat{\theta}_{uls}$.

**Solution 4.3**

(a) Since $V(\hat{\theta}) = k^2\sigma_1^2 + (1-k)^2\sigma_2^2 + 2k(1-k)\rho\sigma_1\sigma_2$, it follows that

$$\frac{\partial V(\hat{\theta})}{\partial k} = 2k\sigma_1^2 - 2(1-k)\sigma_2^2 + 2(1-2k)\rho\sigma_1\sigma_2 = 0.$$

Solving the above equation gives

$$k^* = \frac{\sigma_2^2 - \rho\sigma_1\sigma_2}{\sigma_1^2 + \sigma_2^2 - 2\rho\sigma_1\sigma_2} = \frac{\dfrac{\sigma_2}{\sigma_1} - \rho}{\dfrac{\sigma_1}{\sigma_2} + \dfrac{\sigma_2}{\sigma_1} - 2\rho} = \frac{(1-\rho\lambda)}{(1+\lambda^2-2\rho\lambda)}, \quad k^* > 0,$$

which minimizes $V(\hat{\theta})$.

Interestingly, if $\rho > \lambda$, then $k^* > 1$, so that the unbiased estimator $\hat{\theta}_2$ gets *negative* weight. And, when $\lambda = 1$, so that $\sigma_1 = \sigma_2$, $k^* = \frac{1}{2}$, regardless of the value of $\rho$.

(b) In general, since $\sigma_2 = \sigma_1/\lambda$,

$$\begin{aligned} V(\hat{\theta}) &= k^2\sigma_1^2 + (1-k)^2\frac{\sigma_1^2}{\lambda^2} + 2k(1-k)\rho\frac{\sigma_1^2}{\lambda} \\ &= \sigma_1^2\left[k^2 + (1-k)^2\frac{1}{\lambda^2} + 2k(1-k)\frac{\rho}{\lambda}\right]. \end{aligned}$$

So, after substituting $k^*$ for $k$ in the above expression and doing some algebraic simplification, we obtain

$$V(\hat{\theta}^*) = \sigma_1^2\left[1 - \frac{(\lambda-\rho)^2}{(1-2\rho\lambda+\lambda^2)}\right].$$

Thus, if $\lambda \neq \rho$, $V(\hat{\theta}^*) < \sigma_1^2$.
For further discussion, see Samuel-Cahn (1994).

**Solution 4.4.** Since

$$E(\hat{\beta}) = \sum_{i=1}^{n} c_i E(X_i) = \sum_{i=1}^{n} c_i(\beta a_i) = \beta\sum_{i=1}^{n} c_i a_i,$$

we require that $\sum_{i=1}^{n} c_i a_i = 1$. Now,

$$V(\hat{\beta}) = \sum_{i=1}^{n} c_i^2 V(X_i) = \sum_{i=1}^{n} c_i^2\sigma_i^2.$$

So, we need to minimize $\sum_{i=1}^{n} c_i^2\sigma_i^2$ subject to the constraint $\sum_{i=1}^{n} c_i a_i = 1$. Although the Lagrange Multiplier Method could be used, we will do this minimization directly. So,

$$\begin{aligned} V(\hat{\beta}) &= \sum_{i=1}^{n} c_i^2\sigma_i^2 \\ &= \sum_{i=1}^{n-1} c_i^2\sigma_i^2 + (c_n a_n)^2\left(\frac{\sigma_n^2}{a_n^2}\right) \\ &= \sum_{i=1}^{n-1} c_i^2\sigma_i^2 + \left(1 - \sum_{i=1}^{n-1} c_i a_i\right)^2\left(\frac{\sigma_n^2}{a_n^2}\right). \end{aligned}$$

So, for $i = 1, 2, \ldots, (n-1)$,

$$\begin{aligned} 0 = \frac{dV(\hat{\beta})}{dc_i} &= 2c_i\sigma_i^2 + 2\left(1 - \sum_{i=1}^{n-1} c_i a_i\right)(-a_i)\left(\frac{\sigma_n^2}{a_n^2}\right) \\ &= 2c_i\sigma_i^2 + 2(c_n a_n)(-a_i)\left(\frac{\sigma_n^2}{a_n^2}\right) \\ \Rightarrow 0 &= c_i\sigma_i^2 - \frac{a_i c_n \sigma_n^2}{a_n} \\ \Rightarrow 0 &= a_i c_i - \frac{a_i^2 c_n}{a_n}\left(\frac{\sigma_n^2}{\sigma_i^2}\right) \\ \Rightarrow 0 &= \sum_{i=1}^{n} a_i c_i - \sum_{i=1}^{n} \frac{a_i^2 c_n}{a_n}\left(\frac{\sigma_n^2}{\sigma_i^2}\right) \\ \Rightarrow 0 &= 1 - \frac{c_n\sigma_n^2}{a_n}\sum_{i=1}^{n}\frac{a_i^2}{\sigma_i^2}, \end{aligned}$$

so that

$$c_n = \frac{(a_n/\sigma_n^2)}{\sum_{i=1}^{n}(a_i^2/\sigma_i^2)}.$$

Substituting this result into the above equations yields

$$c_i = \frac{(a_i/\sigma_i^2)}{\sum_{i=1}^{n}(a_i/\sigma_i^2)}, \quad i = 1, 2, \ldots, n.$$

For this choice of the $c_i$'s,

$$\hat{\beta} = \sum_{i=1}^{n}\left[\frac{(a_i/\sigma_i^2)}{\sum_{i=1}^{n}(a_i/\sigma_i^2)}\right]X_i.$$

Since $\hat{\beta}$ is a linear combination of mutually independent normal variates,

$$\hat{\beta} \sim \mathrm{N}[\mathrm{E}(\hat{\beta}), \mathrm{V}(\hat{\beta})],$$

with

$$\mathrm{E}(\hat{\beta}) = \beta$$

and with

$$\begin{aligned} \mathrm{V}(\hat{\beta}) &= \sum_{i=1}^{n} \left[ \frac{(a_i/\sigma_i^2)}{\sum_{i=1}^{n}(a_i/\sigma_i^2)} \right]^2 \sigma_i^2 \\ &= \frac{\sum_{i=1}^{n}(a_i^2/\sigma_i^2)}{[\sum_{i=1}^{n}(a_i^2/\sigma_i^2)]^2} \\ &= \left[ \sum_{i=1}^{n}(a_i^2/\sigma_i^2) \right]^{-1}. \end{aligned}$$

**Solution 4.5**

(a) Since

$$\frac{(n_i - 1)S_i^2}{\sigma^2} \sim \chi^2_{n_i - 1} = \mathrm{GAMMA}\left[\alpha = 2, \beta_i = \frac{(n_i - 1)}{2}\right],$$

we have

$$\mathrm{E}\left[\frac{(n_i - 1)S_i^2}{\sigma^2}\right] = 2 \cdot \frac{(n_i - 1)}{2},$$

so that

$$\frac{(n_i - 1)}{\sigma^2}\mathrm{E}(S_i^2) = (n_i - 1),$$

and hence $\mathrm{E}(S_i^2) = \sigma^2$, $i = 1, 2, \ldots, k$.

Thus,

$$\begin{aligned} \mathrm{E}(\hat{\sigma}^2) &= \mathrm{E}\left[\sum_{i=1}^{k} w_i S_i^2\right] = \sum_{i=1}^{k} w_i \mathrm{E}(S_i^2) \\ &= \sigma^2 \left(\sum_{i=1}^{k} w_i\right) = \sigma^2. \end{aligned}$$

(b) Now, since $S_1^2, S_2^2, \ldots, S_k^2$ constitute a set of $k$ mutually independent random variables, we have

$$\mathrm{V}(\hat{\sigma}^2) = \sum_{i=1}^{k} w_i^2 \mathrm{V}(S_i^2).$$

And, since

$$\mathrm{V}\left[\frac{(n_i-1)S_i^2}{\sigma^2}\right] = \frac{(n_i-1)^2}{\sigma^4}\mathrm{V}(S_i^2)$$

$$= (2)^2\frac{(n_i-1)}{2} = 2(n_i-1),$$

it follows that $\mathrm{V}(S_i^2) = 2\sigma^4/(n_i-1)$, so that

$$\mathrm{V}(\hat{\sigma}^2) = \sum_{i=1}^{k} w_i^2 \frac{2\sigma^4}{(n_i-1)}.$$

So,

$$\mathrm{V}(\hat{\sigma}^2) \propto \sum_{i=1}^{k-1} w_i^2(n_i-1)^{-1} + \left(1-\sum_{i=1}^{k-1} w_i\right)^2 (n_k-1)^{-1}.$$

Thus,

$$\frac{\partial \mathrm{V}(\hat{\sigma}^2)}{\partial w_i} = \frac{2w_i}{(n_i-1)} - \frac{2(1-\sum_{i=1}^{k-1} w_i)}{(n_k-1)} = 0, \quad i = 1, 2, \ldots, (k-1),$$

so that

$$\frac{w_i}{(n_i-1)} - \frac{w_k}{(n_k-1)} = 0, \quad i = 1, 2, \ldots, k.$$

Hence,

$$(n_k-1)\sum_{i=1}^{k} w_i = w_k \sum_{i=1}^{k}(n_i-1),$$

or

$$w_k = \frac{(n_k-1)}{(N-k)}.$$

And, since $w_i = [(n_i-1)/(n_k-1)]w_k$, we have, in general,

$$w_i = \frac{(n_i-1)}{(N-k)} = \frac{(n_i-1)}{\sum_{i=1}^{k}(n_i-1)}, \quad i = 1, 2, \ldots, k.$$

Using these optimal choices for the weights $w_1, w_2, \ldots, w_k$, the estimator $\hat{\sigma}^2$ takes the specific form

$$\hat{\sigma}^2 = \sum_{i=1}^{k}\left[\frac{(n_i-1)}{\sum_{i=1}^{k}(n_i-1)}S_i^2\right] = \frac{\sum_{i=1}^{k}\sum_{j=1}^{n_i}(Y_{ij}-\bar{Y}_i)^2}{(N-k)},$$

which is recognizable as a pooled variance estimator often encountered when using analysis of variance (ANOVA) methods.

**Solution 4.6.** The joint distribution of $X_1, X_2, \ldots, X_n$ is

$$\begin{aligned} p_{X_1,X_2,\ldots,X_n}(x_1, x_2, \ldots, x_n; \pi) &= \prod_{i=1}^{n} \left\{ C_{x_i}^{k} \pi^{x_i}(1-\pi)^{k-x_i} \right\} \\ &= \left( \prod_{i=1}^{n} C_{x_i}^{k} \right) \pi^{\sum_{i=1}^{n} x_i}(1-\pi)^{nk-\sum_{i=1}^{n} x_i}. \end{aligned}$$

Substituting $\theta = \pi^k$ and $u = \sum_{i=1}^{n} x_i$ in the above expression, we have

$$p_{X_1,X_2,\ldots,X_n}(x_1, x_2, \ldots, x_n; \theta) = \left[ (\theta^{1/k})^u (1-\theta^{1/k})^{nk-u} \right] \left( \prod_{i=1}^{n} C_{x_i}^{k} \right),$$

which has the form $g(u;\theta) \cdot h(x_1, x_2, \ldots, x_n)$, where $h(x_1, x_2, \ldots, x_n)$ does not (in any way) depend on $\theta$. Hence, by the Factorization Theorem, $U = \sum_{i=1}^{n} X_i$ is sufficient for $\theta$. Note that $U \sim \text{BIN}(nk, \pi)$. To show that this binomial distribution represents a complete family of distributions, let $g(U)$ denote a generic function of $U$, and note that

$$\begin{aligned} \text{E}[g(U)] &= \sum_{u=0}^{nk} g(u) C_u^{nk} \pi^u (1-\pi)^{nk-u} \\ &= (1-\pi)^{nk} \sum_{u=0}^{nk} \left[ g(u) C_u^{nk} \right] \left( \frac{\pi}{1-\pi} \right)^u. \end{aligned}$$

Using this result and appealing to the theory of polynomials, we find that the condition

$$\text{E}[g(U)] = 0 \ \forall \pi, \quad 0 < \pi < 1,$$

implies that $g(u) = 0,\ u = 0, 1, \ldots, nk$. Hence, $U$ is a complete sufficient statistic for $\theta$.

$$\text{Let } U^* = \begin{cases} 1 & \text{if } X_1 = k, \\ 0 & \text{otherwise.} \end{cases}$$

Then, $\text{E}(U^*) = \pi^k$. Thus, by the Rao–Blackwell Theorem,

$$\hat{\theta} = \text{E}(U^* | U = u) = \text{pr}(U^* = 1 | U = u) = \text{pr}(X_1 = k | U = u)$$

is the MVUE of $\theta$. Clearly, since $U = \sum_{i=1}^{n} X_i, \hat{\theta} = 0$ for $u = 0, 1, \ldots, (k-1)$. So, for $u = k, (k+1), \ldots, nk$,

$$\begin{aligned} \hat{\theta} &= \text{E}\left( U^* | U = u \right) \\ &= \text{pr}(U^* = 1 | U = u) \end{aligned}$$

$$
\begin{aligned}
&= \text{pr}(X_1 = k|U = u) \\
&= \frac{\text{pr}[(X_1 = k) \cap (U = u)]}{\text{pr}(U = u)} \\
&= \frac{\text{pr}(X_1 = k) \times \text{pr}(\sum_{i=2}^{n} X_i = u - k)}{\text{pr}(\sum_{i=1}^{n} X_i = u)} \\
&= \frac{(\pi^k) \times C_{u-k}^{k(n-1)} \pi^{u-k}(1-\pi)^{k(n-1)-(u-k)}}{C_u^{nk}\pi^u(1-\pi)^{nk-u}} \\
&= \frac{C_{u-k}^{k(n-1)}}{C_u^{nk}},
\end{aligned}
$$

where the next-to-last line follows because $\sum_{i=2}^{n} X_i \sim \text{BIN}[k(n-1), \pi]$ and $U \sim \text{BIN}(nk, \pi)$.

$$
\text{So, } \hat{\theta} = \begin{cases} 0, & u = 0, 1, 2, \ldots, (k-1), \\ \dfrac{C_{u-k}^{k(n-1)}}{C_u^{nk}}, & u = k, (k+1), \ldots, nk, \end{cases}
$$

where $u = \sum_{i=1}^{n} x_i$.

To demonstrate that $\text{E}(\hat{\theta}) = \theta$, note that

$$
\begin{aligned}
\text{E}(\hat{\theta}) &= \sum_{u=k}^{nk} \frac{C_{u-k}^{k(n-1)}}{C_u^{nk}} C_u^{nk}\pi^u(1-\pi)^{nk-u} \\
&= \sum_{u=k}^{nk} C_{u-k}^{k(n-1)}\pi^u(1-\pi)^{nk-u} \\
&= \sum_{z=0}^{k(n-1)} C_z^{k(n-1)}\pi^{z+k}(1-\pi)^{nk-(z+k)} \\
&= \pi^k \sum_{z=0}^{k(n-1)} C_z^{k(n-1)}\pi^{z}(1-\pi)^{k(n-1)-z} \\
&= \pi^k[\pi + (1-\pi)]^{k(n-1)} = \pi^k.
\end{aligned}
$$

**Solution 4.7**

(a) Now,

$$
\mathcal{L}(y; \sigma^r) = \prod_{i=1}^{n} \left\{ \frac{1}{\sqrt{2\pi}\sigma} e^{-y_i^2/2\sigma^2} \right\} = (2\pi)^{-n/2}\theta^{-n/r} \exp\left\{ -\frac{\sum_{i=1}^{n} y_i^2}{2\theta^{2/r}} \right\},
$$

so that $U = \sum_{i=1}^{n} Y_i^2$ is a sufficient statistic for $\theta = \sigma^r$. Also,

$$\frac{U}{\sigma^2} = \sum_{i=1}^{n}\left(\frac{Y_i}{\sigma}\right)^2 = \sum_{i=1}^{n} Z_i^2 \sim \chi_n^2 = \text{GAMMA}(\alpha = 2, \beta = \frac{n}{2}),$$

since $Z_i \sim \text{N}(0,1)$ and the $\{Z_i\}_{i=1}^n$ are mutually independent. So, $\text{E}\left(U/\sigma^2\right) = n$, so that $\text{E}(U) = n\sigma^2$. So, we might consider some function of $U^{r/2}$. Thus,

$$\begin{aligned}\text{E}\left[\left(\frac{U}{\sigma^2}\right)^{r/2}\right] = \frac{\text{E}\left(U^{r/2}\right)}{\sigma^r} &= \frac{\Gamma(n/2 + r/2)}{\Gamma(n/2)}2^{r/2} \\ &= \frac{\Gamma[(n+r)/2]}{\Gamma(\frac{n}{2})}2^{r/2}.\end{aligned}$$

So,

$$\hat{\theta} = 2^{-r/2}\frac{\Gamma(n/2)}{\Gamma[(n+r)/2]}U^{r/2}$$

is a function of a sufficient statistic (namely, $U$) that is an unbiased estimator of $\theta$. As a special case, when $r = 2$,

$$\hat{\theta} = 2^{-1}\frac{\Gamma(n/2)}{\Gamma(n/2+1)}U = 2^{-1}\left(\frac{2}{n}\right)U = \frac{U}{n},$$

as expected.

(b) Since

$$\mathcal{L}(y;\theta) \equiv \mathcal{L} = (2\pi)^{-n/2}\theta^{-n/r}\exp\left\{-\frac{\sum_{i=1}^{n} y_i^2}{2\theta^{2/r}}\right\},$$

we have

$$\ln\mathcal{L} = -\frac{n}{2}\ln(2\pi) - \frac{n}{r}\ln\theta - \frac{u}{2\theta^{2/r}}, \quad \text{where } u = \sum_{i=1}^{n} y_i^2.$$

So,

$$\frac{\partial \ln\mathcal{L}}{\partial\theta} = -\frac{n}{r\theta} - \left(\frac{-2}{r}\right)\frac{\theta^{-2/r-1}u}{2} = \frac{-n}{r\theta} + \frac{\theta^{-2/r-1}u}{r},$$

and

$$\frac{\partial^2 \ln\mathcal{L}}{\partial\theta^2} = \frac{n}{r\theta^2} + \left(\frac{-2}{r} - 1\right)\frac{\theta^{-2/r-2}u}{r} = \frac{n}{r\theta^2} - \frac{(2+r)\theta^{-2/r-2}u}{r^2}.$$

Thus,

$$-\mathrm{E}\left(\frac{\partial^2 \ln \mathcal{L}}{\partial\theta^2}\right) = \frac{-n}{r\theta^2} + \frac{(2+r)\theta^{-2/r-2}(n\sigma^2)}{r^2}$$
$$= \frac{-n}{r\theta^2} + \frac{(2+r)\theta^{-2/r-2}n\theta^{2/r}}{r^2}$$
$$= \frac{-n}{r\theta^2} + \frac{n(2+r)}{r^2\theta^2} = \frac{2n}{r^2\theta^2}.$$

So, the CRLB is

$$\text{CRLB} = \frac{r^2\theta^2}{2n} = \frac{r^2\sigma^{2r}}{2n}.$$

When $r = 2$, we obtain

$$\text{CRLB} = \frac{4\sigma^4}{2n} = \frac{2\sigma^4}{n},$$

which is achieved by $\hat{\theta}$ since

$$\mathrm{V}\left(\hat{\theta}\right) = \mathrm{V}\left(\frac{U}{n}\right) = \mathrm{V}\left(\frac{\sigma^2}{n}\cdot\frac{U}{\sigma^2}\right) = \frac{\sigma^4}{n^2}\mathrm{V}(\chi_n^2) = \frac{\sigma^4}{n^2}(2n) = \frac{2\sigma^4}{n}.$$

**Solution 4.8**

(a) First,

$$\mathcal{L}(y;\theta) = \prod_{i=1}^{n}[\theta^{-1/2}\mathrm{e}^{-y_i/\sqrt{\theta}}] = \theta^{-n/2}\mathrm{e}^{-s/\sqrt{\theta}},$$

where $s = \sum_{i=1}^{n} y_i$. Solving for $\theta$ in the equation

$$\frac{\partial \ln \mathcal{L}(y;\theta)}{\partial\theta} = \frac{-n}{2\theta} + \frac{s}{2\theta^{3/2}} = 0,$$

yields the MLE $\hat{\theta} = \bar{Y}^2$. Now,

$$\frac{\partial^2 \ln \mathcal{L}(y;\theta)}{\partial\theta^2} = \frac{n}{2\theta^2} - \frac{3s}{4\theta^{5/2}},$$

so that

$$-\mathrm{E}_y\left[\frac{\partial^2 \ln \mathcal{L}(y;\theta)}{\partial\theta^2}\right] = \frac{-n}{2\theta^2} + \frac{3(n\theta^{1/2})}{4\theta^{5/2}} = \frac{-n}{2\theta^2} + \frac{3n}{4\theta^2} = \frac{n}{4\theta^2}.$$

So, an appropriate large-sample 95% CI for $\theta$ is

$$\bar{Y}^2 \pm 1.96\sqrt{\frac{4\hat{\theta}^2}{n}},$$

or, equivalently,

$$\bar{Y}^2 \pm \frac{3.92\bar{Y}^2}{\sqrt{n}}.$$

When $s = 40$ and $n = 50$, this CI is

$$\left(\frac{40}{50}\right)^2 \pm \frac{(3.92)\left(\frac{40}{50}\right)^2}{\sqrt{50}}$$

or (0.285, 0.995).

(b) Clearly, $S = \sum_{i=1}^{n} Y_i$ is a sufficient statistic for $\theta = \alpha^2$. And, $\text{E}(S) = n\sqrt{\theta}$ and $\text{V}(S) = n\theta$. Since $\text{E}(S^2) = n\theta + n^2\theta = n(n+1)\theta$, it follows that

$$\frac{S^2}{n(n+1)} = \hat{\theta}^*$$

is the MVUE of $\theta$ (because $S$ is a complete sufficient statistic for $\theta$ from exponential family theory). Now,

$$\text{V}(\hat{\theta}^*) = \frac{\text{V}(S^2)}{n^2(n+1)^2} = \frac{\text{E}(S^4) - [\text{E}(S^2)]^2}{n^2(n+1)^2}.$$

Since $S \sim \text{GAMMA}(\alpha = \sqrt{\theta}, \beta = n)$, it follows that

$$\text{E}(S^r) = \frac{\Gamma(n+r)}{\Gamma(n)}\alpha^r = \frac{\Gamma(n+r)}{\Gamma(n)}\theta^{r/2}, \quad r \geq 0.$$

So,

$$\text{E}(S^4) = \frac{\Gamma(n+4)}{\Gamma(n)}\theta^2 = n(n+1)(n+2)(n+3)\theta^2.$$

So,

$$\begin{aligned}
\text{V}(\hat{\theta}^*) &= \frac{n(n+1)(n+2)(n+3)\theta^2 - n^2(n+1)^2\theta^2}{n^2(n+1)^2} \\
&= \frac{\theta^2}{n(n+1)}[n^2 + 5n + 6 - n^2 - n] \\
&= \frac{\theta^2}{n(n+1)}(4n+6) \\
&= \frac{2(2n+3)}{n(n+1)}\theta^2.
\end{aligned}$$

(c) From part (a),

$$-\mathrm{E}_y\left[\frac{\partial^2 \ln \mathcal{L}(y;\theta)}{\partial\theta^2}\right] = \frac{n}{4\theta^2},$$

so that the CRLB is

$$\text{CRLB} = \frac{4\theta^2}{n}.$$

However,

$$\mathrm{V}(\hat{\theta}^*) = \frac{2(2n+3)\theta^2}{n(n+1)} = \left[\frac{2n+3}{2n+2}\right]\left[\frac{4\theta^2}{n}\right] > \frac{4\theta^2}{n},$$

so $\hat{\theta}^*$ does *not* achieve the CRLB.

(d) In general,

$$\begin{aligned}\text{MSE}(\hat{\theta}^*, \theta) &= \mathrm{V}(\hat{\theta}^*) + [\mathrm{E}(\hat{\theta}^*) - \theta]^2 \\ &= \mathrm{V}(\hat{\theta}^*) + 0 \\ &= \frac{2(2n+3)}{n(n+1)}\theta^2.\end{aligned}$$

Since $\hat{\theta} = \bar{Y}^2$,

$$\mathrm{E}\left(\bar{Y}^2\right) = \mathrm{V}\left(\bar{Y}\right) + \left[\mathrm{E}\left(\bar{Y}\right)\right]^2 = \frac{\theta}{n} + \theta.$$

And,

$$\begin{aligned}\mathrm{V}\left(\bar{Y}^2\right) &= \frac{\mathrm{V}(S^2)}{n^4} \\ &= \frac{n(n+1)(n+2)(n+3)\theta^2 - n^2(n+1)^2\theta^2}{n^4} \\ &= \frac{(n+1)\theta^2}{n^3}[n^2 + 5n + 6 - n^2 - n] \\ &= \frac{2(n+1)(2n+3)\theta^2}{n^3}.\end{aligned}$$

So,

$$\begin{aligned}\text{MSE}(\hat{\theta}, \theta) &= \frac{2(n+1)(2n+3)\theta^2}{n^3} + \frac{\theta^2}{n^2} \\ &= \frac{\theta^2}{n^3}[2(n+1)(2n+3) + n] \\ &= \frac{(4n^2 + 11n + 6)}{n^3}\theta^2.\end{aligned}$$

Since

$$\hat{\theta}^* = \frac{S^2}{n(n+1)} = \left(\frac{n}{n+1}\right)\hat{\theta},$$

we have

$$\text{MSE}(\hat{\theta}^*, \theta) = \text{V}(\hat{\theta}^*) = \left(\frac{n}{n+1}\right)^2 \text{V}(\hat{\theta}) < \text{V}(\hat{\theta}) < \text{MSE}(\hat{\theta}, \theta),$$

for all finite $n$, so that $\hat{\theta}^*$ is preferable to $\hat{\theta}$ for finite $n$. However,

$$\lim_{n\to\infty}\left[\frac{\text{MSE}(\hat{\theta}^*, \theta)}{\text{MSE}(\hat{\theta}, \theta)}\right] = 1,$$

so that there is no difference asymptotically.

**Solution 4.9**

(a) Let C be the event that a subject is *classified* as having been recently exposed to benzene, and let E be the event that a subject has *truly* been recently exposed to benzene. Then, $\text{pr(C)} = \text{pr(C|E)pr(E)} + \text{pr(C|}\bar{\text{E}}\text{)pr(}\bar{\text{E}}\text{)}$, so that $\text{pr(C)} = \gamma\pi + \delta(1 - \pi)$. Since $X$ has a binomial distribution with mean $\text{E}(X) = n[\text{pr(C)}]$, equating $X$ to $\text{E}(X)$ via the method of moments gives

$$\hat{\pi} = \frac{\frac{X}{n} - \delta}{\gamma - \delta},$$

as the unbiased estimator of $\pi$.

Since $\text{V}(X) = n[\text{pr(C)}][1 - \text{pr(C)}]$, the variance of the estimator $\hat{\pi}$ is

$$\begin{aligned} V(\hat{\pi}) &= \frac{\text{V}(X/n)}{(\gamma - \delta)^2} \\ &= \frac{[\gamma\pi + \delta(1-\pi)][1 - \gamma\pi - \delta(1-\pi)]}{n(\gamma - \delta)^2}. \end{aligned}$$

(b) Since $n$ is large, the standardized random variable $(\hat{\pi} - \pi)/\sqrt{\hat{\text{V}}(\hat{\pi})} \mathrel{\dot\sim} \text{N}(0, 1)$ by Slutsky's Theorem, where

$$\hat{\text{V}}(\hat{\pi}) = \frac{[\gamma\hat{\pi} + \delta(1-\hat{\pi})][1 - \gamma\hat{\pi} - \delta(1-\hat{\pi})]}{n(\gamma - \delta)^2}.$$

Thus, an appropriate large-sample 95% CI for $\pi$ is

$$\hat{\pi} \pm 1.96\sqrt{\hat{\text{V}}(\hat{\pi})}.$$

When $n = 50, \delta = 0.05, \gamma = 0.90$, and $x = 20$, the computed 95% interval for $\pi$ is $0.412 \pm 1.96(0.0815) = (0.252, 0.572)$.

**Solution 4.10**

(a) Since

$$(Y_{11}, Y_{00}, n - Y_{11} - Y_{00}) \\ \sim \text{MULT}\left\{n; (\pi^2 + \theta), [(1-\pi)^2 + \theta], 2[\pi(1-\pi) - \theta]\right\},$$

it follows directly that

$$(\hat{\pi}^2 + \hat{\theta}) = \frac{Y_{11}}{n} \quad \text{and} \quad [(1-\hat{\pi})^2 + \hat{\theta}] = \frac{Y_{00}}{n}.$$

Solving these two equations simultaneously gives the desired expressions for $\hat{\pi}$ and $\hat{\theta}$.

(b) Appealing to properties of the multinomial distribution, we have

$$\begin{aligned} \mathrm{E}(\hat{\pi}) &= \frac{1}{2} + \frac{[\mathrm{E}(Y_{11}) - \mathrm{E}(Y_{00})]}{2n} \\ &= \frac{1}{2} + \frac{n(\pi^2 + \theta) - n[(1-\pi)^2 + \theta]}{2n} \\ &= \frac{1}{2} + \frac{1}{2}(\pi^2 - 1 + 2\pi - \pi^2) = \pi, \end{aligned}$$

so that $\hat{\pi}$ is an unbiased estimator of the parameter $\pi$.
And, with $\beta_{11} = (\pi^2 + \theta)$ and $\beta_{00} = [(1-\pi)^2 + \theta]$, it follows that

$$\begin{aligned} \mathrm{V}(\hat{\pi}) &= (4n^2)^{-1}[\mathrm{V}(Y_{11}) + \mathrm{V}(Y_{00}) - 2\mathrm{cov}(Y_{11}, Y_{00})] \\ &= (4n^2)^{-1}[n\beta_{11}(1-\beta_{11}) + n\beta_{00}(1-\beta_{00}) - 2n\beta_{11}\beta_{00}] \\ &= (4n)^{-1}[\beta_{11}(1-\beta_{11}) + \beta_{00}(1-\beta_{00}) - 2\beta_{11}\beta_{00}]. \end{aligned}$$

(c) Since $\hat{\beta}_{11} = Y_{11}/n$ and $\hat{\beta}_{00} = Y_{00}/n$, it follows that the estimator $\hat{\mathrm{V}}(\hat{\pi})$ of $\mathrm{V}(\hat{\pi})$ is equal to

$$\hat{\mathrm{V}}(\hat{\pi}) = (4n)^{-1}\left[\frac{Y_{11}}{n}\left(1 - \frac{Y_{11}}{n}\right) + \frac{Y_{00}}{n}\left(1 - \frac{Y_{00}}{n}\right) - 2\left(\frac{Y_{11}}{n}\right)\left(\frac{Y_{00}}{n}\right)\right].$$

When $n = 30, y_{11} = 3$, and $y_{00} = 15$, then the estimated value $\hat{\pi}$ of $\pi$ is equal to

$$\hat{\pi} = \frac{1}{2} + \frac{(3-15)}{30} = 0.50 - 0.40 = 0.10.$$

And, the estimated variance of $\hat{\pi}$ is equal to

$$\begin{aligned} \hat{\mathrm{V}}(\hat{\pi}) &= [4(30)]^{-1}\left[\left(\frac{3}{30}\right)\left(\frac{27}{30}\right) + \left(\frac{15}{30}\right)\left(\frac{15}{30}\right) - 2\left(\frac{3}{30}\right)\left(\frac{15}{30}\right)\right] \\ &= 0.0020. \end{aligned}$$

Thus, the computed 95% CI for $\pi$ is equal to

$$\hat{\pi} \pm 1.96\sqrt{\hat{V}(\hat{\pi})} = 0.10 \pm 1.96\sqrt{0.0020} = 0.10 \pm 0.0877,$$

or (0.0123, 0.1877).

**Solution 4.11**

(a) With $Y = g(\boldsymbol{X})$, where $\boldsymbol{X}' = (X_1, X_2, \ldots, X_k)$ and $\boldsymbol{\mu}' = (\mu_1, \mu_2, \ldots, \mu_k)$, and with $\mathrm{E}(X_i) = \mu_i, \mathrm{V}(X_i) = \sigma_i^2$, and $\mathrm{cov}(X_i, X_j) = \sigma_{ij}$ for all $i \neq j, i = 1, 2, \ldots, k$ and $j = 1, 2, \ldots, k$, then the delta method gives $\mathrm{E}(Y) \approx g(\boldsymbol{\mu})$ and

$$\mathrm{V}(Y) \approx \sum_{i=1}^{k} \left[\frac{\partial g(\boldsymbol{\mu})}{\partial X_i}\right]^2 \sigma_i^2 + 2\sum_{i=1}^{k-1} \sum_{j=i+1}^{k} \left[\frac{\partial g(\boldsymbol{\mu})}{\partial X_i}\right]\left[\frac{\partial g(\boldsymbol{\mu})}{\partial X_j}\right] \sigma_{ij},$$

where

$$\frac{\partial g(\boldsymbol{\mu})}{\partial X_i} = \frac{\partial g(\boldsymbol{X})}{\partial X_i}\bigg|_{\boldsymbol{X}=\boldsymbol{\mu}}.$$

In our particular situation, $k = 2$; and, with $X_1 \equiv Y_{10}$ and $X_2 \equiv Y_{01}$, then $Y = g(X_1, X_2) = \ln(X_1/X_2) = \ln\widehat{\mathrm{OR}} = \ln X_1 - \ln X_2$. So,

$$\frac{\partial g(X_1, X_2)}{\partial X_1} = \frac{1}{X_1} \quad \text{and} \quad \frac{\partial g(X_1, X_2)}{\partial X_2} = -\frac{1}{X_2}.$$

Now, $\mathrm{E}(X_1) = n\pi_{10}, \mathrm{E}(X_2) = n\pi_{01}, \mathrm{V}(X_1) = n\pi_{10}(1 - \pi_{10})$, and $\mathrm{V}(X_2) = n\pi_{01}(1 - \pi_{01})$. Also, $\mathrm{cov}(X_1, X_2) = -n\pi_{10}\pi_{01}$. Finally,

$$\begin{aligned}
\mathrm{V}(\ln\widehat{\mathrm{OR}}) &\approx \left(\frac{1}{n\pi_{10}}\right)^2 n\pi_{10}(1 - \pi_{10}) + \left(\frac{-1}{n\pi_{01}}\right)^2 n\pi_{01}(1 - \pi_{01}) \\
&\quad + 2\left(\frac{1}{n\pi_{10}}\right)\left(\frac{-1}{n\pi_{01}}\right)(-n\pi_{10}\pi_{01}) \\
&= \frac{(1 - \pi_{10})}{n\pi_{10}} + \frac{(1 - \pi_{01})}{n\pi_{01}} + \frac{2}{n} \\
&= \frac{1}{n\pi_{10}} - \frac{1}{n} + \frac{1}{n\pi_{01}} - \frac{1}{n} + \frac{2}{n} \\
&= \frac{1}{n\pi_{10}} + \frac{1}{n\pi_{01}}.
\end{aligned}$$

Since $\mathrm{E}(Y_{10}) = n\pi_{10}$ and $\mathrm{E}(Y_{01}) = n\pi_{01}$, we have

$$\hat{\mathrm{V}}(\ln\widehat{\mathrm{OR}}) \mathrel{\dot{\approx}} \frac{1}{Y_{10}} + \frac{1}{Y_{01}}.$$

For the given set of data, the estimate of the variance of $\ln\widehat{\text{OR}}$ is

$$\hat{\text{V}}(\ln\widehat{\text{OR}}) \approx \frac{1}{25} + \frac{1}{15} = 0.107.$$

(b) Assume $\text{Z} \sim \text{N}(0, 1)$. Then,

$$\begin{aligned}
0.95 &= \text{pr}\{-1.96 < \text{Z} < +1.96\} \\
&\approx \text{pr}\left\{-1.96 < \frac{\ln\widehat{\text{OR}} - \ln\text{OR}}{\sqrt{\hat{\text{V}}(\ln\widehat{\text{OR}})}} < 1.96\right\} \\
&= \text{pr}\left\{\ln\widehat{\text{OR}} - 1.96\sqrt{\hat{\text{V}}(\ln\widehat{\text{OR}})} < \ln\text{OR}\right. \\
&\qquad \left. < \ln\widehat{\text{OR}} + 1.96\sqrt{\hat{\text{V}}(\ln\widehat{\text{OR}})}\right\} \\
&= \text{pr}\left\{(\widehat{\text{OR}})\text{e}^{-1.96\sqrt{\hat{\text{V}}(\ln\widehat{\text{OR}})}} < \text{OR} < (\widehat{\text{OR}})\text{e}^{+1.96\sqrt{\hat{\text{V}}(\ln\widehat{\text{OR}})}}.\right\}
\end{aligned}$$

So, the 95% CI for OR is

$$\left[(\widehat{\text{OR}})\text{e}^{-1.96\sqrt{\hat{\text{V}}(\ln\widehat{\text{OR}})}}, (\widehat{\text{OR}})\text{e}^{+1.96\sqrt{\hat{\text{V}}(\ln\widehat{\text{OR}})}}\right].$$

For the data in part (a), we obtain

$$\left[\left(\frac{25}{15}\right)\text{e}^{-1.96\sqrt{0.107}}, \left(\frac{25}{15}\right)\text{e}^{+1.96\sqrt{0.107}}\right] = (0.878, 3.164).$$

**Solution 4.12**

(a) The appropriate likelihood function $\mathcal{L}$ is

$$\mathcal{L} = \prod_{i=0}^{1}\prod_{j=1}^{n_i}\left[\frac{\left(L_{ij}\lambda_i\right)^{y_{ij}}\text{e}^{-L_{ij}\lambda_i}}{y_{ij}!}\right],$$

so that $\ln\mathcal{L}$ can be written as

$$\ln\mathcal{L} = \sum_{i=0}^{1}\left[\sum_{j=1}^{n_i} y_{ij}\ln L_{ij} + \left(\sum_{j=1}^{n_i} y_{ij}\right)\ln\lambda_i - \lambda_i\sum_{j=1}^{n_i} L_{ij} - \sum_{j=1}^{n_i} y_{ij}!\right].$$

So, for $i = 0, 1$,

$$\frac{\partial\ln\mathcal{L}}{\partial\lambda_i} = \frac{\sum_{j=1}^{n_1} y_{ij}}{\lambda_i} - \sum_{j=1}^{n_i} L_{ij} = 0$$

gives

$$\hat{\lambda}_i = \frac{\sum_{j=1}^{n_i} Y_{ij}}{\sum_{j=1}^{n_i} L_{ij}}$$

as the MLE of $\lambda_i$.

Also, for $i = 0, 1$,

$$\frac{\partial^2 \ln \mathcal{L}}{\partial \lambda_i^2} = \frac{-\sum_{j=1}^{n_1} y_{ij}}{\lambda_i^2},$$

so that, with $\text{E}(Y_{ij}) = L_{ij}\lambda_i$ and $\partial^2 \ln \mathcal{L}/\partial\lambda_1\partial\lambda_2 = 0$, we have

$$\text{V}(\hat{\lambda}_i) = \left\{-\text{E}\left(\frac{\partial^2 \ln \mathcal{L}}{\partial \lambda_i^2}\right)\right\}^{-1} = \frac{\lambda_i}{\sum_{j=1}^{n_i} L_{ij}}.$$

Now, by the invariance principle, the MLE $\ln \hat{\psi}$ of $\ln \psi$ is

$$\ln \hat{\psi} = \ln \hat{\lambda}_1 - \ln \hat{\lambda}_0.$$

And, using the delta method, we have

$$\begin{aligned}\text{V}(\ln \hat{\psi}) &= \text{V}(\ln \hat{\lambda}_1) + \text{V}(\ln \hat{\lambda}_0) \\ &\approx \left(\frac{1}{\lambda_1}\right)^2 \text{V}(\hat{\lambda}_1) + \left(\frac{1}{\lambda_0}\right)^2 \text{V}(\hat{\lambda}_0) \\ &= \frac{1}{\lambda_1 \sum_{j=1}^{n_1} L_{1j}} + \frac{1}{\lambda_0 \sum_{j=1}^{n_0} L_{0j}}.\end{aligned}$$

Hence, from ML theory, the random variable

$$\frac{\ln \hat{\psi} - \ln \psi}{\sqrt{\hat{\text{V}}(\ln \hat{\psi})}} = \frac{\ln \hat{\psi} - \ln \psi}{\left(\frac{1}{\sum_{j=1}^{n_1} y_{1j}} + \frac{1}{\sum_{j=1}^{n_0} y_{0j}}\right)^{1/2}} \stackrel{.}{\sim} \text{N}(0, 1) \text{ for large samples,}$$

so that a ML-based large-sample $100(1-\alpha)\%$ CI for $\ln \psi$ is

$$\begin{aligned}\ln \hat{\psi} \pm Z_{1-\alpha/2}\sqrt{\hat{\text{V}}(\ln \hat{\psi})} &= (\ln \hat{\lambda}_1 - \ln \hat{\lambda}_0) \\ &\pm Z_{1-\alpha/2}\left(\frac{1}{\sum_{j=1}^{n_1} y_{1j}} + \frac{1}{\sum_{j=1}^{n_0} y_{0j}}\right)^{1/2},\end{aligned}$$

where $\text{pr}(Z > Z_{1-\alpha/2}) = \alpha/2$ when $Z \sim \text{N}(0, 1)$.

(b) Based on the CI for $\ln\psi$ developed in part (a), an appropriate ML-based large-sample $100(1-\alpha)\%$ CI for the rate ratio $\psi$ is

$$(\hat{\psi})\exp\left[\pm Z_{1-\alpha/2}\left(\frac{1}{\sum_{j=1}^{n_1} y_{1j}}+\frac{1}{\sum_{j=1}^{n_0} y_{0j}}\right)^{1/2}\right].$$

For the given data, the computed 95% CI for $\psi$ is

$$\left(\frac{40/350}{35/400}\right)\exp\left[\pm 1.96\left(\frac{1}{40}+\frac{1}{35}\right)^{1/2}\right]=(1.306)e^{\pm 0.454},$$

or (0.829, 2.056).

Since the number 1 is contained in this 95% CI, these data provide no evidence in favor of the proposed theory. Of course, there could be several reasons why there were no significant findings. In particular, important individual-specific risk factors for skin cancer and related skin conditions were not considered, some of these important risk factors being skin color (i.e., having fair skin), having a family history of skin cancer, having had a previous skin cancer, being older, being male, and so on.

**Solution 4.13**

(a) The likelihood function $\mathcal{L}(t_1,t_2,\ldots,t_n)\equiv\mathcal{L}$ is

$$\mathcal{L}=\prod_{i=1}^{n} f_T(t_i;\theta)=\prod_{i=1}^{n}\left[\theta e^{-\theta t_i}\right]=\theta^n e^{-\theta\sum_{i=1}^{n} t_i}.$$

So,

$$\ln\mathcal{L}=n\ln\theta-\theta\sum_{i=1}^{n} t_i,$$

$$\frac{\partial\mathcal{L}}{\partial\theta}=\frac{n}{\theta}-\sum_{i=1}^{n} t_i,$$

and

$$\frac{\partial^2\mathcal{L}}{\partial\theta^2}=\frac{-n}{\theta^2}.$$

Thus, the large-sample variance of $\hat{\theta}$ is

$$V(\hat{\theta})=\left[-E\left(\frac{\partial^2\ln\mathcal{L}}{\partial\theta^2}\right)\right]^{-1}=\frac{\theta^2}{n}.$$

(b) Now,

$$\text{pr}(T_i > t^*) = \int_{t^*}^{\infty} \theta \mathrm{e}^{-\theta t_i} \, \mathrm{d}t_i = \left[ -\mathrm{e}^{-\theta t_i} \right]_{t^*}^{\infty} = \mathrm{e}^{-\theta t^*}.$$

So, the likelihood function $\mathcal{L}^*(y_1, y_2, \ldots, y_n) \equiv \mathcal{L}^*$ is

$$\begin{aligned}
\mathcal{L}^* &= \prod_{i=1}^{n} \left\{ \left( \mathrm{e}^{-\theta t^*} \right)^{y_i} \left( 1 - \mathrm{e}^{-\theta t^*} \right)^{1-y_i} \right\} \\
&= \mathrm{e}^{-\theta t^* \sum_{i=1}^{n} y_i} (1 - \mathrm{e}^{-\theta t^*})^{n - \sum_{i=1}^{n} y_i}.
\end{aligned}$$

So,

$$\ln \mathcal{L}^* = -\theta t^* n\bar{y} + n(1 - \bar{y}) \ln(1 - \mathrm{e}^{-\theta t^*}),$$

and

$$\frac{\partial \ln \mathcal{L}^*}{\partial \theta} = -t^* n\bar{y} + n(1 - \bar{y}) \frac{t^* \mathrm{e}^{-\theta t^*}}{(1 - \mathrm{e}^{-\theta t^*})}.$$

So,

$$\begin{aligned}
\frac{\partial \ln \mathcal{L}^*}{\partial \theta} = 0 &\Rightarrow n(1 - \bar{y}) t^* \mathrm{e}^{-\theta t^*} = n t^* \bar{y} (1 - \mathrm{e}^{-\theta t^*}) \\
&\Rightarrow (1 - \bar{y}) \mathrm{e}^{-\theta t^*} = \bar{y} (1 - \mathrm{e}^{-\theta t^*}) \\
&\Rightarrow \mathrm{e}^{-\theta t^*} = \bar{y} \\
&\Rightarrow \hat{\theta}^* = \frac{-\ln \bar{y}}{t^*} = \frac{1}{t^*} \ln \left( \frac{1}{\bar{y}} \right).
\end{aligned}$$

(c) Now,

$$\begin{aligned}
\frac{\partial^2 \ln \mathcal{L}^*}{\partial \theta^2} &= n t^* (1 - \bar{y}) \left[ \frac{-t^* \mathrm{e}^{-\theta t^*} (1 - \mathrm{e}^{-\theta t^*}) - \mathrm{e}^{-\theta t^*} (t^* \mathrm{e}^{-\theta t^*})}{(1 - \mathrm{e}^{-\theta t^*})^2} \right] \\
&= \frac{-n t^* (1 - \bar{y})}{(1 - \mathrm{e}^{-\theta t^*})^2} (t^* \mathrm{e}^{-\theta t^*}),
\end{aligned}$$

so that

$$\begin{aligned}
-\mathrm{E} \left( \frac{\partial^2 \ln \mathcal{L}^*}{\partial \theta^2} \right) &= \frac{n (t^*)^2 \mathrm{e}^{-\theta t^*} \mathrm{E}(1 - \bar{Y})}{(1 - \mathrm{e}^{-\theta t^*})^2} \\
&= \frac{n (t^*)^2 \mathrm{e}^{-\theta t^*}}{(1 - \mathrm{e}^{-\theta t^*})^2} (1 - \mathrm{e}^{-\theta t^*}) \\
&= \frac{n (t^*)^2}{(\mathrm{e}^{\theta t^*} - 1)}.
\end{aligned}$$

So, the large-sample variance of $\hat{\theta}^*$ is $(e^{\theta t^*} - 1)/n(t^*)^2$.

Hence, with $t^* \geq E(T) = \theta^{-1}$, we have

$$\frac{V(\hat{\theta})}{V(\hat{\theta}^*)} = \frac{\theta^2/n}{(e^{\theta t^*} - 1)/n(t^*)^2} = \frac{\theta^2(t^*)^2}{(e^{\theta t^*} - 1)} < 1,$$

so that $\hat{\theta}$ is preferred based solely on large-sample variance considerations. This finding reflects the fact that we have lost information by categorizing $\{T_1, T_2, \ldots, T_n\}$ into dichotomous data $\{Y_1, Y_2, \ldots, Y_n\}$. However, if the remission times are measured with error, then $\hat{\theta}^*$ would be preferred to $\hat{\theta}$ on validity grounds; in other words, if the remission times are measured with error, then $\hat{\theta}$ would be an *asymptotically biased* estimator of the unknown parameter $\theta$.

**Solution 4.14**

(a) The parameter of interest is

$$\theta = \text{pr}(Y = 0) = e^{-\lambda},$$

so that

$$\lambda = -\ln\theta.$$

Now, with $y = (y_1, y_2, \ldots, y_n)$,

$$\mathcal{L}(y; \theta) = \prod_{i=1}^{n} \left\{ \frac{(-\ln\theta)^{y_i}\theta}{y_i!} \right\} = \frac{\theta^n(-\ln\theta)^s}{\prod_{i=1}^{n} y_i!},$$

where $s = \sum_{i=1}^{n} y_i$. So,

$$\ln\mathcal{L}(y; \theta) = n\ln\theta + s\ln(-\ln\theta) - \sum_{i=1}^{n} \ln(y_i!);$$

$$\frac{\partial \ln\mathcal{L}(y; \theta)}{\partial\theta} = \frac{n}{\theta} + \frac{s}{\theta\ln\theta};$$

$$\frac{\partial^2 \ln\mathcal{L}(y; \theta)}{\partial\theta^2} = \frac{-n}{\theta^2} - \frac{s(\ln\theta + 1)}{(\theta\ln\theta)^2}.$$

So, since $S \sim \text{POI}(n\lambda)$,

$$-E\left[\frac{\partial^2 \ln\mathcal{L}(y; \theta)}{\partial\theta^2}\right] = \frac{n}{\theta^2} + \frac{(-n\ln\theta)(\ln\theta + 1)}{(\theta\ln\theta)^2}$$

$$= \frac{n}{\theta^2} - \frac{n}{\theta^2} - \frac{n}{\theta^2\ln\theta}$$

$$= \frac{-n}{\theta^2\ln\theta}.$$

So, the CRLB is

$$\text{CRLB} = \frac{\theta^2(-\ln\theta)}{n} = \frac{e^{-2\lambda}\lambda}{n} = \frac{\lambda}{ne^{2\lambda}}.$$

Consider the estimator

$$\hat{\theta} = \left(\frac{n-1}{n}\right)^S.$$

Since $\hat{\theta}$ is an unbiased estimator of $\theta$ and is a function of a complete sufficient statistic for $\theta$, it is the MVUE of $\theta$. Since

$$V(\hat{\theta}) = \frac{(e^{\lambda/n}-1)}{e^{2\lambda}} > \frac{\lambda}{ne^{2\lambda}},$$

there is no unbiased estimator that attains the CRLB for all finite values of $n$.

(b) Note that

$$\text{pr}(X_i = 0) = \text{pr}(Y_i = 0) = e^{-\lambda},$$

and that

$$\text{pr}(X_i = 1) = \text{pr}(Y_i \geq 1) = 1 - \text{pr}(Y_i = 0) = 1 - e^{-\lambda}.$$

So, with $\boldsymbol{x} = (x_1, x_2, \ldots, x_n)$,

$$p_{X_i}(x_i; \lambda) = (1 - e^{-\lambda})^{x_i}(e^{-\lambda})^{1-x_i}, x_i = 0, 1.$$

Thus,

$$\begin{aligned}\mathcal{L}(\boldsymbol{x}; \lambda) &= \prod_{i=1}^{n}\left\{(1-e^{-\lambda})^{x_i}e^{-\lambda(1-x_i)}\right\} \\ &= (1-e^{-\lambda})^{n\bar{x}}e^{-n\lambda(1-\bar{x})},\end{aligned}$$

where

$$\bar{x} = n^{-1}\sum_{i=1}^{n} x_i.$$

So,

$$\ln \mathcal{L}(\boldsymbol{x}; \lambda) = n\bar{x}\,\ln(1-e^{-\lambda}) - n\lambda(1-\bar{x}).$$

The equation

$$\begin{aligned}\frac{\partial \ln \mathcal{L}(\boldsymbol{x}; \lambda)}{\partial\lambda} &= \frac{n\bar{x}e^{-\lambda}}{(1-e^{-\lambda})} - n(1-\bar{x}) = 0 \\ &\Rightarrow n\bar{x}e^{-\lambda} - n(1-\bar{x})(1-e^{-\lambda}) = 0 \\ &\Rightarrow \bar{x}e^{-\lambda} - 1 + e^{-\lambda} + \bar{x} - \bar{x}e^{-\lambda} = 0\end{aligned}$$

$$\Rightarrow e^{-\lambda} = (1 - \bar{x}) \Rightarrow -\lambda = \ln(1 - \bar{x})$$
$$\Rightarrow \hat{\lambda}^* = -\ln(1 - \bar{x}).$$

This result also follows because $\bar{X}$ is the MLE of $\text{pr}(X_i = 1) = (1 - e^{-\lambda})$. And,

$$\frac{\partial^2 \ln \mathcal{L}(\boldsymbol{x}; \lambda)}{\partial \lambda^2} = n\bar{x}\left[\frac{-e^{-\lambda}(1 - e^{-\lambda}) - e^{-\lambda}e^{-\lambda}}{(1 - e^{-\lambda})^2}\right]$$
$$= \frac{-n\bar{x}e^{-\lambda}}{(1 - e^{-\lambda})^2}.$$

So,

$$-\text{E}\left[\frac{\partial^2 \ln \mathcal{L}(\boldsymbol{x}; \lambda)}{\partial \lambda^2}\right] = \frac{ne^{-\lambda}\text{E}(\bar{X})}{(1 - e^{-\lambda})^2}$$
$$= \frac{ne^{-\lambda}(1 - e^{-\lambda})}{(1 - e^{-\lambda})^2}$$
$$= \frac{ne^{-\lambda}}{(1 - e^{-\lambda})}.$$

Thus, for large $n$,

$$\text{V}(\hat{\lambda}^*) = \frac{(1 - e^{-\lambda})}{ne^{-\lambda}} = \frac{(e^{\lambda} - 1)}{n}.$$

(c) There are two scenarios to consider:

*Scenario* 1: Assume that $Y_1, Y_2, \ldots, Y_n$ are accurate. Then, $\hat{\lambda} = \bar{Y}$ is the MLE (and MVBUE) of $\lambda$, with $\text{E}(\hat{\lambda}) = \lambda$ and $\text{V}(\hat{\lambda}) = \lambda/n$. Since, for large $n$, $\hat{\lambda}^*$ is essentially unbiased, a comparison of variances is appropriate. Now,

$$\text{EFF}(\hat{\lambda}^*, \hat{\lambda}) = \frac{\lambda/n}{(e^{\lambda} - 1)/n} = \frac{\lambda}{(e^{\lambda} - 1)} = \frac{\lambda}{\lambda + \sum_{j=2}^{\infty} \frac{\lambda^j}{j!}} < 1,$$

so that $\hat{\lambda}^*$ always has a larger variance than $\hat{\lambda}$ (which is an expected result since we are losing information by categorizing $Y_i$ into the dichotomous variable $X_i$). In fact,

$$\lim_{\lambda \to \infty} \text{EFF}(\hat{\lambda}^*, \hat{\lambda}) = 0,$$

so the loss in efficiency gets worse as $\lambda$ gets larger (and this loss of information is not affected by increasing $n$).

*Scenario* 2: Assume that $Y_1, Y_2, \ldots, Y_n$ are inaccurate. In this case, using $\hat{\lambda} = \bar{Y}$ to estimate $\lambda$ could lead to a severe bias problem. Assuming that $X_1, X_2, \ldots, X_n$ are accurate, then $\hat{\lambda}^*$ is essentially unbiased for large $n$ and so would be the preferred estimator. Since validity takes preference over precision, $\hat{\lambda}^*$ would be preferred to $\hat{\lambda}$ when $\{Y_1, Y_2, \ldots, Y_n\}$ are inaccurate but $\{X_1, X_2, \ldots, X_n\}$ are correct.

**Solution 4.15**

(a) For the assumed statistical model, and with $\boldsymbol{y} = (y_0, y_1, \ldots, y_n)$, the corresponding likelihood function $\mathcal{L}(\boldsymbol{y}; \theta) \equiv \mathcal{L}$ is

$$\begin{aligned}\mathcal{L} &= \mathrm{p}_{Y_0}(y_0;\theta)\prod_{j=0}^{n-1}\mathrm{p}_{Y_{j+1}}(y_{j+1}|Y_k = y_k,\ k = 0, 1, \ldots, j; \theta)\\ &= \mathrm{p}_{Y_0}(y_0;\theta)\prod_{j=0}^{n-1}\mathrm{p}_{Y_{j+1}}(y_{j+1}|Y_j = y_j;\theta)\\ &= \left(\frac{\theta^{y_0}\mathrm{e}^{-\theta}}{y_0!}\right)\prod_{j=0}^{n-1}\frac{(\theta y_j)^{y_{j+1}}\mathrm{e}^{\theta y_j}}{y_{j+1}!}.\end{aligned}$$

Thus,

$$\ln(\mathcal{L}) \sim \left(\sum_{j=0}^{n} y_j\right)\ln(\theta) - \theta\left(1 + \sum_{j=0}^{n-1} y_j\right),$$

so that the equation

$$\frac{\partial \ln(\mathcal{L})}{\partial\theta} = \theta^{-1}\sum_{j=0}^{n} y_j - \left(1 + \sum_{j=0}^{n-1} y_j\right) = 0$$

gives

$$\hat{\theta} = \frac{\sum_{j=0}^{n} Y_j}{1 + \sum_{j=0}^{n-1} Y_j}$$

as the MLE of $\theta$.

(b) Now,

$$\frac{\partial^2 \ln(\mathcal{L})}{\partial\theta^2} = \frac{-\sum_{j=0}^{n} y_j}{\theta^2}, \text{ so that } -\mathrm{E}\left(\frac{\partial^2 \ln(\mathcal{L})}{\partial\theta^2}\right) = \frac{\sum_{j=0}^{n}\mathrm{E}(Y_j)}{\theta^2}.$$

And, $\mathrm{E}(Y_0) = \theta, \mathrm{E}(Y_1) = \mathrm{E}_{y_0}[\mathrm{E}(Y_1|Y_0 = y_0)] = E_{y_0}(\theta y_0) = \theta^2, \mathrm{E}(Y_2) = \mathrm{E}_{y_1}[\mathrm{E}(Y_2|Y_1 = y_1)] = \mathrm{E}_{y_1}(\theta y_1) = \theta^3$ and so on, so that, in general, $\mathrm{E}(Y_j) = \theta^{(j+1)}, j = 0, 1, \ldots, n$. Finally,

$$-\mathrm{E}\left(\frac{\partial^2 \ln(\mathcal{L})}{\partial\theta^2}\right) = \frac{\sum_{j=0}^{n}\theta^{(j+1)}}{\theta^2} = \theta^{-1}\sum_{j=0}^{n}\theta^j = \theta^{-1}\left(\frac{1-\theta^{(n+1)}}{1-\theta}\right).$$

So, for large $n$, $V(\hat{\theta}) \doteq [\theta(1-\theta)]/[1-\theta^{(n+1)}]$, and a ML-based 95% CI for $\theta$ is

$$\hat{\theta} \pm 1.96\sqrt{\hat{V}(\hat{\theta})} = \hat{\theta} \pm 1.96\sqrt{\frac{\hat{\theta}(1-\hat{\theta})}{1-\hat{\theta}^{(n+1)}}}.$$

When $n = 25$ and $\hat{\theta} = 1.20$, the computed 95% CI for $\theta$ is (1.11, 1.29).

**Solution 4.16.** Given the stated assumptions, the appropriate CI for $(\mu_t - \mu_c)$ using $(\bar{X} - \bar{Y})$ is:

$$(\bar{X} - \bar{Y}) \pm 1.96\sqrt{\frac{\sigma_t^2}{n_t} + \frac{\sigma_c^2}{n_c}},$$

where $\bar{X} = n_t^{-1}\sum_{i=1}^{n_t} X_i$ and $\bar{Y} = n_c^{-1}\sum_{i=1}^{n_c} Y_i$.

The *optimal* choices for $n_t$ and $n_c$, subject to the constraint $(n_t + n_c) = N$, would minimize the width of the above CI.

So, we want to minimize the function $\left(\sigma_t^2/n_t + \sigma_t^2/n_c\right)$ subject to the constraint $(n_t + n_c) = N$, or, equivalently, we want to minimize the function

$$Q = \frac{\sigma_t^2}{n_t} + \frac{\sigma_c^2}{(N - n_t)},$$

with respect to $n_t$.

So,

$$\frac{dQ}{dn_t} = \frac{-\sigma_t^2}{n_t^2} + \frac{\sigma_c^2}{(N - n_t)^2} = 0$$

$$\Rightarrow (\sigma_t^2 - \sigma_c^2)n_t^2 - 2N\sigma_t^2 n_t + N^2\sigma_t^2 = 0.$$

So, via the quadratic formula, the two roots of the above quadratic equation are

$$\frac{2N\sigma_t^2 \pm \sqrt{4N^2\sigma_t^4 - 4(\sigma_t^2 - \sigma_c^2)N^2\sigma_t^2}}{2(\sigma_t^2 - \sigma_c^2)} = N\left[\frac{\sigma_t(\sigma_t \pm \sigma_c)}{(\sigma_t + \sigma_c)(\sigma_t - \sigma_c)}\right].$$

If the positive sign is used, the possible answer is $N\sigma_t/(\sigma_t - \sigma_c)$, which cannot be correct. If the negative sign is used, the answer is

$$n_t = N\left(\frac{\sigma_t}{\sigma_t + \sigma_c}\right),$$

so that

$$n_c = N\left(\frac{\sigma_c}{\sigma_t + \sigma_c}\right).$$

This choice for $n_t$ minimizes $Q$ since $\left.\frac{dQ^2}{dn_t^2}\right|_{n_t = \frac{N\sigma_t}{(\sigma_t + \sigma_c)}} > 0.$

When $N = 100$, $\sigma_t^2 = 4$, and $\sigma_c^2 = 9$, then $n_t = 40$ and $n_c = 60$. Note that these answers make sense, since more data are required from the more variable population.

**Solution 4.17.** Consider the random variable $(\bar{Y} - Y_{n+1})$, which is a linear combination of independent $\mathrm{N}(\mu, \sigma^2)$ variates. Since

$$\mathrm{E}(\bar{Y} - Y_{n+1}) = \mathrm{E}(\bar{Y}) - \mathrm{E}(Y_{n+1}) = \mu - \mu = 0,$$

and since

$$\mathrm{V}(\bar{Y} - Y_{n+1}) = \mathrm{V}(\bar{Y}) + \mathrm{V}(Y_{n+1}) = \frac{\sigma^2}{n} + \sigma^2 = \left(\frac{n+1}{n}\right)\sigma^2,$$

it follows that

$$(\bar{Y} - Y_{n+1}) \sim \mathrm{N}\left[0, \left(\frac{n+1}{n}\right)\sigma^2\right].$$

Hence,

$$\frac{(\bar{Y} - Y_{n+1})}{\sqrt{\left(\frac{n+1}{n}\right)\sigma^2}} \sim \mathrm{N}(0,1).$$

Also, we know that

$$\frac{(n-1)S^2}{\sigma^2} \sim \chi^2_{n-1}.$$

So,

$$\frac{\left[\dfrac{(\bar{Y} - Y_{n+1})}{\sqrt{\left(\frac{n+1}{n}\right)\sigma^2}}\right]}{\sqrt{\dfrac{(n-1)S^2}{\sigma^2}\Big/(n-1)}} = \frac{(\bar{Y} - Y_{n+1})}{S\sqrt{\dfrac{(n+1)}{n}}} \sim t_{n-1},$$

since $(\bar{Y} - Y_{n+1})$ and $S^2$ are independent random variables. So,

$$(1-\alpha) = \mathrm{pr}\left\{-t_{n-1,1-\alpha/2} < \frac{(\bar{Y} - Y_{n+1})}{S\sqrt{\dfrac{(n+1)}{n}}} < t_{n-1,1-\alpha/2}\right\}$$

$$= \mathrm{pr}\left\{\bar{Y} - t_{n-1,1-\alpha/2}S\sqrt{\frac{(n+1)}{n}} < Y_{n+1} < \bar{Y} + t_{n-1,1-\alpha/2}S\sqrt{\frac{(n+1)}{n}}\right\}.$$

Thus,

$$L = \bar{Y} - t_{n-1,1-\alpha/2}S\sqrt{\frac{(n+1)}{n}}$$

and

$$U = \bar{Y} + t_{n-1,1-\alpha/2} S \sqrt{\frac{(n+1)}{n}}.$$

For the given data, the realized values of $\bar{Y}$ and $S^2$ are $\bar{y} = 3$ and $s^2 = 2.50$, so that the computed 95% prediction interval for the random variable $Y_6$ is

$$\begin{aligned}\bar{y} \pm t_{n-1,1-\alpha/2} s \sqrt{\frac{(n+1)}{n}} &= 3 \pm t_{0.975,4}\sqrt{2.50}\sqrt{\frac{6}{5}} \\ &= 3 \pm 2.776\sqrt{3} \\ &= (-1.8082, 7.8082).\end{aligned}$$

**Solution 4.18**

(a) We know that

$$U = \frac{(n-1)S^2}{\sigma^2} \sim \chi^2_{n-1} = \text{GAMMA}\left(\alpha = 2, \beta = \frac{n-1}{2}\right).$$

If $Y \sim \text{GAMMA}(\alpha, \beta)$, then

$$\text{E}(Y^r) = \int_0^\infty y^r \frac{y^{\beta-1} e^{-y/\alpha}}{\Gamma(\beta)\alpha^\beta} dy = \frac{\Gamma(\beta + r)}{\Gamma(\beta)} \alpha^r, \quad (\beta + r) > 0.$$

So,

$$\text{E}(U^r) = \frac{\Gamma\left[(n-1)/2 + r\right]}{\Gamma\left[(n-1)/2\right]} 2^r.$$

So,

$$\begin{aligned}\text{E}\left[U^{1/2}\right] &= \text{E}\left[\sqrt{\frac{(n-1)S^2}{\sigma^2}}\right] = \frac{\sqrt{n-1}}{\sigma}\text{E}(S) \\ &= \frac{\Gamma[(n-1)/2 + 1/2]}{\Gamma[(n-1)/2]} 2^{1/2} = \frac{\Gamma(n/2)}{\Gamma[(n-1)/2]}\sqrt{2} \\ \Rightarrow \quad \text{E}(S) &= \frac{\Gamma(n/2)}{\Gamma[(n-1)/2]}\sqrt{\frac{2}{(n-1)}}\sigma \\ \Rightarrow \quad \text{E}(W) &= 2t_{n-1,1-\alpha/2}\frac{\text{E}(S)}{\sqrt{n}} = 2^{3/2} t_{n-1,1-\alpha/2} \frac{\Gamma(n/2)}{\Gamma[(n-1)/2]} \frac{\sigma}{\sqrt{n(n-1)}}.\end{aligned}$$

If $\alpha = 0.05$, $n = 4$, and $\sigma^2 = 4$, then

$$\begin{aligned}\text{E}(W) &= 2^{3/2} t_{3,.975} \frac{\Gamma(2)}{\Gamma(3/2)} \frac{2}{\sqrt{4(4-1)}} \\ &= 2^{3/2}(3.182) \frac{1}{(\sqrt{\pi}/2)} \frac{1}{\sqrt{3}} = 5.8633.\end{aligned}$$

(b)

$$(1-\gamma) \le \text{pr}\left\{2t_{n^*-1,1-\frac{\alpha}{2}}\frac{S}{\sqrt{n^*}} \le \delta\right\} = \text{pr}\left\{4f_{1,n^*-1,1-\alpha}\frac{S^2}{n^*} \le \delta^2\right\}$$

$$= \text{pr}\left\{S^2 \le \frac{n^*\delta^2}{4f_{1,n^*-1,1-\alpha}}\right\}$$

$$= \text{pr}\left\{\frac{(n^*-1)S^2}{\sigma^2} \le \frac{n^*(n^*-1)\delta^2}{4\sigma^2 f_{1,n^*-1,1-\alpha}}\right\}$$

$$= \text{pr}\left\{\chi^2_{n^*-1} \le \frac{n^*(n^*-1)\delta^2}{4\sigma^2 f_{1,n^*-1,1-\alpha}}\right\}.$$

So, we require

$$\frac{n^*(n^*-1)\delta^2}{4\sigma^2 f_{1,n^*-1,1-\alpha}} \ge \chi^2_{n^*-1,1-\gamma},$$

or

$$n^*(n^*-1) \ge \left(\frac{2\sigma}{\delta}\right)^2 \chi^2_{n^*-1,1-\gamma} f_{1,n^*-1,1-\alpha}.$$

For further details, see Kupper and Hafner (1989).

**Solution 4.19**

(a) Let $\hat{\theta} = 2\bar{Y}_1 - 3\bar{Y}_2 + \bar{Y}_3$; so, $E(\hat{\theta}) = \theta$, and

$$V(\hat{\theta}) = 4\left(\frac{\sigma^2}{n_1}\right) + 9\left(\frac{\sigma^2}{n_2}\right) + \left(\frac{\sigma^2}{n_3}\right) = \sigma^2\left(\frac{4}{n_1} + \frac{9}{n_2} + \frac{1}{n_3}\right).$$

So,

$$Z = \frac{\hat{\theta} - E(\hat{\theta})}{\sqrt{V(\hat{\theta})}} = \frac{(2\bar{Y}_1 - 3\bar{Y}_2 + \bar{Y}_3) - (2\mu_1 - 3\mu_2 + \mu_3)}{\sigma\sqrt{4/n_1 + 9/n_2 + 1/n_3}} \sim N(0,1).$$

Now, $(n_i - 1)S_i^2/\sigma^2 \sim \chi^2_{n_i-1}$, $i = 1, 2, 3$, and the $S_i^2$'s are mutually independent random variables. Thus, by the additivity property of mutually independent gamma random variables,

$$U = \frac{(n_1-1)S_1^2 + (n_2-1)S_2^2 + (n_3-1)S_3^2}{\sigma^2} \sim \chi^2_{(n_1+n_2+n_3-3)}$$

and

$$E(U) = E\left[\frac{\sum_{i=1}^3 (n_i-1)S_i^2}{(n_1+n_2+n_3-3)}\right] = \sigma^2,$$

where $\sum_{i=1}^{3}(n_i - 1)S_i^2/(n_1 + n_2 + n_3 - 3)$ is called a "pooled estimator" of $\sigma^2$.

So, noting that the numerators and denominators in each of the following expressions are independent, we have

$$T_{(n_1+n_2+n_3-3)} = \frac{Z}{\sqrt{U/(n_1+n_2+n_3-3)}}$$

$$= \frac{(\hat{\theta} - \mathrm{E}(\hat{\theta}))/\sqrt{\mathrm{V}(\hat{\theta})}}{\sqrt{\frac{\sum_{i=1}^{3}(n_i-1)S_i^2}{\sigma^2} \Big/ (n_1+n_2+n_3-3)}}$$

$$= \frac{(2\bar{Y}_1 - 3\bar{Y}_2 + \bar{Y}_3) - \theta}{\sqrt{\frac{\sum_{i=1}^{3}(n_i-1)S_i^2}{(n_1+n_2+n_3-3)}}\sqrt{\frac{4}{n_1} + \frac{9}{n_2} + \frac{1}{n_3}}} \sim t_{\left(\sum_{i=1}^{3} n_i - 3\right)}.$$

(b) Let $\hat{\theta} = 2\bar{Y}_1 - 3\bar{Y}_2 + \bar{Y}_3$ and $S_p^2 = \sum_{i=1}^{3}(n_i - 1)\,S_i^2/(n_1 + n_2 + n_3 - 3)$. From part (a),

$$\frac{\hat{\theta} - \theta}{S_p\sqrt{4/n_1 + 9/n_2 + 1/n_3}} \sim t_{(n_1+n_2+n_3-3)}.$$

So,

$$(1-\alpha) = \mathrm{pr}\left\{-t_{\left(\sum_{i=1}^{3} n_i-3,1-\alpha/2\right)} < \frac{\hat{\theta} - \theta}{S_p\sqrt{4/n_1 + 9/n_2 + 1/n_3}} < t_{\left(\sum_{i=1}^{3} n_i-3,1-\alpha/2\right)}\right\},$$

and hence an *exact* $100(1-\alpha)\%$ CI for $\theta$ is

$$\hat{\theta} \pm t_{\left(\sum_{i=1}^{3} n_i-3,1-\alpha/2\right)} S_p\sqrt{\frac{4}{n_1} + \frac{9}{n_2} + \frac{1}{n_3}}.$$

For these data, we have:

$$[2(80) - 3(75) + 70] \pm t_{9,0.975}\sqrt{\frac{3(4+3+5)}{9}}\sqrt{\frac{4}{4} + \frac{9}{4} + \frac{1}{4}} = 5 \pm 8.46,$$

or $(-3.46, 13.46)$.

(c) An *exact* $100(1-\alpha)\%$ CI for $\sigma_1^2/\sigma_2^2$ is

$$(1-\alpha) = \mathrm{pr}\left\{\frac{S_1^2/S_2^2}{f_{n_1-1,n_2-1,1-\alpha/2}} < \frac{\sigma_1^2}{\sigma_2^2} < \frac{S_1^2/S_2^2}{1/f_{n_2-1,n_1-1,1-\alpha/2}}\right\}.$$

Now, $f_{49,49,0.975} = 1.76$. So,

$$\text{lower limit} = \frac{7/2}{1.76} = 1.99,$$

and

$$\text{upper limit} = \left(\frac{7}{2}\right)(1.76) = 6.16;$$

thus, our 95% CI for $\sigma_1^2/\sigma_2^2$ is (1.99, 6.16). Note that the value 1 is *not* included in this interval, suggesting variance heterogeneity.

(d) Consider the statistic

$$\frac{\hat{\theta} - \theta}{\sqrt{4S_1^2/n_1 + 9S_2^2/n_2 + S_3^2/n_3}} = \left[\frac{4\sigma_1^2/n_1 + 9\sigma_2^2/n_2 + \sigma_3^2/n_3}{4S_1^2/n_1 + 9S_2^2/n_2 + S_3^2/n_3}\right]^{1/2} \times \left[\frac{\hat{\theta} - \theta}{\sqrt{4\sigma_1^2/n_1 + 9\sigma_2^2/n_2 + \sigma_3^2/n_3}}\right].$$

The expression in the first set of brackets converges to 1, since $S_i^2$ is consistent for $\sigma_i^2$, $i = 1, 2, 3$, while the expression in the second set of brackets converges in distribution to N(0,1) by the Central Limit Theorem. So, by Slutsky's Theorem,

$$\frac{\hat{\theta} - \theta}{\sqrt{4S_1^2/n_1 + 9S_2^2/n_2 + S_3^2/n_3}} \mathrel{\dot\sim} \mathrm{N}(0,1) \quad \text{for large } n_1,\ n_2,\ n_3.$$

Thus, an approximate large-sample 95% CI for $\theta$ is

$$\hat{\theta} \pm 1.96\sqrt{\frac{4S_1^2}{n_1} + \frac{9S_2^2}{n_2} + \frac{S_3^2}{n_3}}.$$

For the data in part (c), $\hat{\theta} = 2(85) - 3(82) + 79 = 3$. So, our large-sample CI is

$$3 \pm 1.96\sqrt{\frac{4(7)}{50} + \frac{9(2)}{50} + \frac{6}{50}} = 3 \pm 2.00 \text{ or } (1.00, 5.00).$$

The advantage of selecting large random samples from each of the three populations is that the assumptions of exactly normally distributed populations and homogenous variance across populations can both be relaxed.

**Solution 4.20**

(a) Since $X_i/\sqrt{\theta} \sim \mathrm{N}(0,1)$, then $L/\theta \sim \chi^2_{n_1}$, or equivalently GAMMA $(2, n_1/2)$, since $X_1, X_2, \ldots, X_{n_1}$ constitute a set of mutually independent random variables. If

$U \sim \text{GAMMA}(\alpha, \beta)$, then $\text{E}(U^r) = [\Gamma(\beta + r)/\Gamma(\beta)]\alpha^r$, $(\beta + r) > 0$. Thus, for $r = \frac{1}{2}$, we have

$$\text{E}\left(\sqrt{\frac{L}{\theta}}\right) = \theta^{-1/2}\text{E}(\sqrt{L}) = \frac{\Gamma(n_1/2 + 1/2)}{\Gamma(n_1/2)} 2^{1/2},$$

so that

$$\text{E}(\sqrt{L}) = \frac{\Gamma[(n_1 + 1)/2]}{\Gamma(n_1/2)} \sqrt{2\theta}.$$

(b) The random variable

$$F_{n_1,n_2} = \frac{\sum_{i=1}^{n_1} \left(X_i/\sqrt{\theta}\right)^2 / n_1}{\sum_{i=1}^{n_2} \left(\sqrt{\theta} Y_i\right)^2 / n_2} = \theta^{-2}\left(\frac{n_2}{n_1}\right)\left(\frac{\sum_{i=1}^{n_1} X_i^2}{\sum_{i=1}^{n_2} Y_i^2}\right) \sim f_{n_1,n_2}.$$

So,

$$\begin{aligned}(1 - \alpha) &= \text{pr}\left(f_{n_1,n_2,\alpha/2} < F_{n_1,n_2} < f_{n_1,n_2,1-\alpha/2}\right) \\ &= \text{pr}\left(f^{-1}_{n_1,n_2,1-\alpha/2} < F^{-1}_{n_1,n_2} < f_{n_2,n_1,1-\alpha/2}\right) = \text{pr}(L < \theta < U),\end{aligned}$$

where

$$L = \left(\frac{n_2}{n_1}\right)^{1/2} \left(\frac{\sum_{i=1}^{n_1} X_i^2}{\sum_{i=1}^{n_2} Y_i^2}\right)^{1/2} f^{-1/2}_{n_1,n_2,1-\alpha/2}$$

and

$$U = \left(\frac{n_2}{n_1}\right)^{1/2} \left(\frac{\sum_{i=1}^{n_1} X_i^2}{\sum_{i=1}^{n_2} Y_i^2}\right)^{1/2} f^{1/2}_{n_2,n_1,1-\alpha/2}.$$

For the available data, since $f_{8,5,0.975} = 6.76$ and $f_{5,8,0.975} = 4.82$, the computed exact 95% CI for $\theta$ is (0.430,2.455).

**Solution 4.21**

(a) The best point estimator of $\theta$ is $\bar{D} = n^{-1}\sum_{i=1}^{n} D_i$. Since $\text{E}(D_i) = \text{E}(Y_{\text{T}i} - Y_{\text{P}i}) = (\mu_\text{T} - \mu_\text{P}) = \theta$, and $\text{V}(D_i) = \text{V}(Y_{\text{T}i}) + \text{V}(Y_{\text{P}i}) - 2\rho\sqrt{\text{V}(Y_{\text{T}i})\text{V}(Y_{\text{P}i})} = (\sigma_\text{T}^2 + \sigma_\text{P}^2 - 2\rho\sigma_\text{T}\sigma_\text{P})$, it follows that $\text{E}(\bar{D}) = \theta$ and $\text{V}(\bar{D}) = (\sigma_\text{T}^2 + \sigma_\text{P}^2 - 2\rho\sigma_\text{T}\sigma_\text{P})/n$. Since $[\bar{D} - \text{E}(\bar{D})]/\sqrt{\text{V}(\bar{D})} \sim \text{N}(0, 1)$, it follows that

$$\text{pr}\left[-z_{1-\alpha/2} < \frac{\bar{D} - \text{E}(\bar{D})}{\sqrt{\text{V}(\bar{D})}} < z_{1-\alpha/2}\right] = (1 - \alpha) = \text{pr}(L < \theta < U)$$

where

$$L = \bar{D} - z_{1-\alpha/2}\sqrt{\text{V}(\bar{D})} \quad \text{and} \quad U = \bar{D} + z_{1-\alpha/2}\sqrt{\text{V}(\bar{D})}.$$

Given the available data, the realized value of $L$ is 0.02, and the realized value of $U$ is 1.98, so that the computed 95% CI for $\theta$ is (0.02, 0.98). This computed 95% CI does not include the value zero, indicating that there is statistical evidence that $\theta \neq 0$ (or, equivalently, that $\mu_T \neq \mu_P$).

(b) Now,

$$\begin{aligned}
\text{pr}(L > 0|\theta = 1.0) &= \text{pr}\left[\bar{D} - z_{0.975}\sqrt{\text{V}(\bar{D})} > 0|\theta = 1.0\right] \\
&= \text{pr}\left[\frac{\bar{D} - 1.0}{\sqrt{\text{V}(\bar{D})}} > \frac{1.96\sqrt{\text{V}(\bar{D})} - 1.0}{\sqrt{\text{V}(\bar{D})}}\right] \\
&= \text{pr}\left[Z > 1.96 - \frac{1.0}{\sqrt{\text{V}(\bar{D})}}\right],
\end{aligned}$$

where $Z \sim \text{N}(0, 1)$.
So, to achieve $\text{pr}(L > 0|\theta = 1.0) \geq 0.95$, we require $1.96 - 1.0/\sqrt{\text{V}(\bar{D})} \leq -1.645$, or, equivalently,

$$\frac{1.0}{\sqrt{\text{V}(\bar{D})}} = \frac{1.0\sqrt{n}}{\sqrt{\sigma_T^2 + \sigma_P^2 - 2\rho\sigma_T\sigma_P}} \geq (1.96 + 1.645),$$

which gives $n^* = 46$.

## Solution 4.22

(a)

$$\begin{aligned}
\text{E}(U_i) &= \text{E}(X_i + Y_i) = \text{E}(X_i) + \text{E}(Y_i) = (\mu_x + \mu_y), \\
\text{E}(V_i) &= \text{E}(X_i - Y_i) = \text{E}(X_i) - \text{E}(Y_i) = (\mu_x - \mu_y), \\
\text{V}(U_i) &= \text{V}(X_i + Y_i) = \text{V}(X_i) + \text{V}(Y_i) + 2\text{cov}(X_i, Y_i) \\
&= \sigma^2 + \sigma^2 + 2\rho\sigma^2 = 2\sigma^2(1 + \rho), \text{ and} \\
\text{V}(V_i) &= \text{V}(X_i - Y_i) = \text{V}(X_i) + \text{V}(Y_i) - 2\text{cov}(X_i, Y_i) \\
&= \sigma^2 + \sigma^2 - 2\rho\sigma^2 = 2\sigma^2(1 - \rho).
\end{aligned}$$

(b)

$$\begin{aligned}
\text{cov}(U_i, V_i) &= \text{E}(U_iV_i) - \text{E}(U_i)\text{E}(V_i) \\
&= \text{E}[(X_i + Y_i)(X_i - Y_i)] - (\mu_x + \mu_y)(\mu_x - \mu_y) \\
&= \text{E}(X_i^2 - Y_i^2) - (\mu_x^2 - \mu_y^2) \\
&= [\text{E}(X_i^2) - \mu_x^2] - [\text{E}(Y_i^2) - \mu_y^2] = \sigma^2 - \sigma^2 = 0.
\end{aligned}$$

(c) Given the bivariate normal assumption, it follows that $U_1, U_2, \ldots, U_n$ are i.i.d. $\mathrm{N}[(\mu_x + \mu_y), 2\sigma^2(1+\rho)]$ random variables; and, $V_1, V_2, \ldots, V_n$ are i.i.d. $\mathrm{N}[(\mu_x - \mu_y), 2\sigma^2(1-\rho)]$ random variables. Hence,

$$\frac{(n-1)S_u^2}{2\sigma^2(1+\rho)} \sim \chi^2_{(n-1)}, \quad \frac{(n-1)S_v^2}{2\sigma^2(1-\rho)} \sim \chi^2_{(n-1)},$$

and $S_u^2$ and $S_v^2$ are independent random variables because of the result in part (b). So,

$$\frac{\left[\frac{(n-1)S_u^2}{2\sigma^2(1+\rho)}\right] \Big/ (n-1)}{\left[\frac{(n-1)S_v^2}{2\sigma^2(1-\rho)}\right] \Big/ (n-1)} = \frac{(1-\rho)S_u^2}{(1+\rho)S_v^2} \sim f_{(n-1),(n-1)}.$$

(d) If $f_{n-1,n-1,1-\frac{\alpha}{2}}$ is defined such that

$$\mathrm{pr}\left[F_{n-1,n-1} > f_{n-1,n-1,1-\alpha/2}\right] = \frac{\alpha}{2},$$

then

$$\begin{aligned}(1-\alpha) &= \mathrm{pr}\left[f_{n-1,n-1,\alpha/2} < F_{n-1,n-1} < f_{n-1,n-1,1-\alpha/2}\right] \\ &= \mathrm{pr}\left[\frac{1}{f_{n-1,n-1,1-\alpha/2}} < \frac{(1-\rho)S_u^2}{(1+\rho)S_v^2} < f_{n-1,n-1,1-\alpha/2}\right] \\ &= \mathrm{pr}\left[\left(\frac{S_v^2}{S_u^2}\right)\frac{1}{f_{n-1,n-1,1-\alpha/2}} < \frac{2}{(1+\rho)} - 1 \right. \\ &\qquad \left. < \left(\frac{S_v^2}{S_u^2}\right) f_{n-1,n-1,1-\alpha/2}\right] \\ &= \mathrm{pr}\left[\left(\frac{2}{W} - 1\right) < \rho < \left(\frac{2}{V} - 1\right)\right],\end{aligned}$$

where

$$V = \left[1 + \left(\frac{S_v^2}{S_u^2}\right)\frac{1}{f_{n-1,n-1,1-\alpha/2}}\right]$$

and

$$W = \left[1 + \left(\frac{S_v^2}{S_u^2}\right) f_{n-1,n-1,1-\alpha/2}\right].$$

In our situation, $n = 10$, $s_u^2 = 1$, $s_v^2 = 2$, $\alpha = 0.05$, and $f_{9,9,0.975} = 4.03$. So,

$$v = 1 + \frac{2}{1}(4.03)^{-1} = 1.4963,$$
$$w = 1 + \frac{2}{1}(4.03) = 9.06,$$
$$\left(\frac{2}{w} - 1\right) = \frac{2}{9.06} - 1 = -0.7792,$$
$$\left(\frac{2}{v} - 1\right) = \frac{2}{1.4963} - 1 = 0.3366.$$

Hence, the computed exact 95% CI for $\rho$ is

$$(-0.7792, 0.3366).$$

**Solution 4.23**

(a) First, note that

$$\mu_r' = \mathrm{E}(Y^r) = \int_\gamma^\infty y^r \theta\gamma^\theta y^{-(\theta+1)}\, dy = \theta\gamma^\theta \left[\frac{y^{r-\theta}}{(r-\theta)}\right]_\gamma^\infty = \frac{\theta\gamma^r}{(\theta - r)}, \quad \theta > r.$$

The method of moments estimators are found by solving for $\gamma$ and $\theta$ using the following two equations:

$$\hat{\mu}_1' = \bar{y} = \frac{\theta\gamma}{(\theta - 1)}$$

and

$$\hat{\mu}_2' = \frac{1}{n}\sum_{i=1}^n y_i^2 = \mathrm{E}(Y^2) = \frac{\theta\gamma^2}{(\theta - 2)}.$$

The above equations imply that

$$\frac{\hat{\mu}_2'}{\bar{y}^2} = \frac{\theta\gamma^2/(\theta - 2)}{\theta^2\gamma^2/(\theta - 1)^2} = \frac{(\theta - 1)^2}{\theta(\theta - 2)}.$$

Hence,

$$\frac{(\theta - 1)^2}{\theta(\theta - 2)} - 1 = \frac{1}{\theta(\theta - 2)} = \frac{\hat{\mu}_2'}{\bar{y}^2} - 1 = \frac{(\hat{\mu}_2' - \bar{y}^2)}{\bar{y}^2}$$
$$= \frac{\frac{1}{n}\sum_{i=1}^n (y_i - \bar{y})^2}{\bar{y}^2} = \left(\frac{n-1}{n}\right)\frac{s^2}{\bar{y}^2}.$$

So,

$$\theta(\theta-2)=\left(\frac{n}{n-1}\right)\frac{\bar{y}^2}{s^2}=\left(\frac{50}{49}\right)\left(\frac{900}{10}\right)=91.8367.$$

The roots of the quadratic equation $\theta^2-2\theta-91.8367=0$ are

$$\frac{2\pm\sqrt{(-2)^2+4(91.8367)}}{2},$$

or $-8.6352$ and $10.6352$. Since $\theta>2$, we take the positive root and use $\hat{\theta}_{\mathrm{mm}}=10.6352$. Finally,

$$\hat{\gamma}_{\mathrm{mm}}=\frac{(\hat{\theta}_{\mathrm{mm}}-1)}{\hat{\theta}_{\mathrm{mm}}}\bar{y}=\left(\frac{9.6352}{10.6352}\right)(30)=27.1793.$$

So, $\hat{\gamma}_{\mathrm{mm}}=27.1793$.

(b) Now,

$$\begin{aligned}F(y;\gamma,\theta)&=\int_{\gamma}^{y}\theta\gamma^{\theta}t^{-(\theta+1)}\,dt=\gamma^{\theta}\left[-t^{-\theta}\right]_{\gamma}^{y}\\&=\gamma^{\theta}\left[\gamma^{-\theta}-y^{-\theta}\right]=1-\left(\frac{\gamma}{y}\right)^{\theta},\quad 0<\gamma<y<\infty.\end{aligned}$$

So,

$$\begin{aligned}f_{Y_{(1)}}(y_{(1)};\gamma,\theta)&=n\left[1-F_Y(y_{(1)};\gamma,\theta)\right]^{n-1}f_Y(y_{(1)};\gamma,\theta)\\&=n\left[\left(\frac{\gamma}{y_{(1)}}\right)^{\theta}\right]^{n-1}\theta\gamma^{\theta}y_{(1)}{}^{-(\theta+1)}\\&=n\theta\gamma^{n\theta}y_{(1)}{}^{-(n\theta+1)},\quad 0<\gamma<y_{(1)}<\infty.\end{aligned}$$

Using this density, we have

$$E\left[Y_{(1)}{}^{r}\right]=\int_{\gamma}^{\infty}y_{(1)}{}^{r}n\theta\gamma^{n\theta}y_{(1)}{}^{-(n\theta+1)}\,dy_{(1)}=\frac{n\theta\gamma^{r}}{(n\theta-r)},\quad n\theta>r.$$

So,

$$E\left[Y_{(1)}\right]=\frac{n\theta\gamma}{(n\theta-1)},$$

and

$$\lim_{n\to\infty}E\left[Y_{(1)}\right]=\lim_{n\to\infty}\frac{\theta\gamma}{(\theta-\frac{1}{n})}=\frac{\theta\gamma}{\theta}=\gamma,$$

so that $Y_{(1)}$ is an asymptotically unbiased estimator of $\gamma$. Also,

$$\begin{aligned} V\left[Y_{(1)}\right] &= \frac{n\theta\gamma^2}{(n\theta-2)} - \left[\frac{n\theta\gamma}{(n\theta-1)}\right]^2 \\ &= n\theta\gamma^2\left[\frac{1}{(n\theta-2)} - \frac{n\theta}{(n\theta-1)^2}\right] \\ &= \frac{n\theta\gamma^2}{(n\theta-1)^2(n\theta-2)}. \end{aligned}$$

Since $\lim_{n\to\infty} V\left[Y_{(1)}\right] = 0$, and since $Y_{(1)}$ is asymptotically unbiased, it follows that $Y_{(1)}$ is a consistent estimator of $\gamma$.

(c) For $0 < c < 1$, we wish to find $c$ such that $\text{pr}\left[\gamma < cY_{(1)}\right] = (1-\alpha)$. Now,

$$\begin{aligned} \text{pr}\left[\gamma < cY_{(1)}\right] &= \text{pr}\left[\frac{\gamma}{c} < Y_{(1)}\right] = \int_{\gamma/c}^{\infty} n\theta\gamma^{n\theta} y_{(1)}{}^{-(n\theta+1)}\,dy_{(1)} \\ &= \gamma^{n\theta}\left[-y_{(1)}{}^{-n\theta}\right]_{\gamma/c}^{\infty} = \gamma^{n\theta}\left(\frac{\gamma}{c}\right)^{-n\theta} = c^{n\theta} = (1-\alpha). \end{aligned}$$

So, $c = (1-\alpha)^{1/(n\theta)}$. Thus, since $\theta = 3$, we have $U = cY_{(1)} = (1-\alpha)^{1/3n}Y_{(1)}$. When $n = 5, \alpha = 0.10$, and $y_{(1)} = 20$, the computed value of $U$ is $u = (1-0.10)^{1/15}(20) = 19.860$. So, the upper 90% CI for $\gamma$ is $(0, 19.860)$.

**Solution 4.24**

(a) From order statistics theory, we know that, for $r = 1, 2, \ldots, n$,

$$f_{X_{(r)}}(x_{(r)}) = nC_{r-1}^{n-1}\left[F_X(x_{(r)})\right]^{r-1}\left[1 - F_X(x_{(r)})\right]^{n-r} f_X(x_{(r)}), \quad -\infty < x_{(r)} < +\infty.$$

Hence, letting $u = F_X(x_{(r)})$, so that $du = f_X(x_{(r)})\,dx_{(r)}$, and then appealing to properties of the beta distribution, we have

$$\begin{aligned} E(U_r) &= \int_{-\infty}^{\infty}\left[F_X(X_{(r)})\right] nC_{r-1}^{n-1}\left[F_X(x_{(r)})\right]^{r-1}\left[1 - F_X(x_{(r)})\right]^{n-r} f_X(x_{(r)})\,dx_{(r)} \\ &= \int_0^1 \frac{n!}{(r-1)!(n-r)!}u^r(1-u)^{n-r}\,du \\ &= \int_0^1 \frac{\Gamma(n+1)}{\Gamma(r)\Gamma(n-r+1)}u^r(1-u)^{n-r}\,du \\ &= \left[\frac{\Gamma(n+1)}{\Gamma(r)\Gamma(n-r+1)}\right]\left[\frac{\Gamma(r+1)\Gamma(n-r+1)}{\Gamma(n+2)}\right] \\ &= \frac{r}{(n+1)}. \end{aligned}$$

(b) For any particular value of $p$, we can pick a pair of values for $r$ and $n$ such that

$$\mathrm{E}(U_r) = \mathrm{E}\left[\mathrm{F_X}(\mathrm{X}_{(r)})\right] = \frac{r}{(n+1)} \approx p.$$

For these particular choices for $r$ and $n$, the amount of area under $\mathrm{f_X}(x)$ to the left of $X_{(r)}$ is, on average (i.e., on expectation), equal to $p$.

Thus, for these values of $r$ and $n$, it is reasonable to use $\mathrm{X}_{(r)}$ as the estimator of the $p$th quantile $\theta_p$; in particular, $\mathrm{X}_{(r)}$ is called the $p$th *sample* quantile.

**Solution 4.25**

(a) $\mathrm{E}(W) = \mathrm{E}[\mathrm{X}_{(n)} - \mathrm{X}_{(1)}] = \mathrm{E}[\mathrm{X}_{(n)}] - \mathrm{E}[\mathrm{X}_{(1)}]$. Since $\mathrm{f_X}(x;\theta) = \theta x^{\theta-1}$, $\mathrm{F_X}(x;\theta) = x^{\theta}$, $0 < x < 1$, $\theta > 0$.

So,

$$\mathrm{f}_{\mathrm{X}_{(1)}}(x_{(1)};\theta) = n\left(1 - x_{(1)}^{\theta}\right)^{n-1} \theta x_{(1)}^{\theta-1}, \quad 0 < x_{(1)} < 1$$

and

$$\mathrm{f}_{\mathrm{X}_{(n)}}(x_{(n)};\theta) = n\left[x_{(n)}^{\theta}\right]^{n-1} \theta x_{(n)}^{\theta-1}, \quad 0 < x_{(n)} < 1.$$

So,

$$\mathrm{E}[\mathrm{X}_{(n)}] = \int_0^1 x_{(n)} n x_{(n)}^{\theta(n-1)} \theta x_{(n)}^{\theta-1}\, \mathrm{d}x_{(n)} = \left(\frac{n\theta}{n\theta+1}\right).$$

And, with $u = x_{(1)}^{\theta}$ and $\mathrm{d}u = \theta x_{(1)}^{\theta-1}\,\mathrm{d}x_{(1)}$, we have

$$\begin{aligned}
\mathrm{E}(X_{(1)}) &= \int_0^1 x_{(1)} n\left(1 - x_{(1)}^{\theta}\right)^{n-1} \theta x_{(1)}^{\theta-1}\, \mathrm{d}x_{(1)} \\
&= n\int_0^1 u^{1/\theta}(1-u)^{n-1}\, du \\
&= n\int_0^1 u^{\left(\frac{1}{\theta}+1\right)-1}(1-u)^{n-1}\, du \\
&= \frac{n\Gamma(1/\theta+1)\Gamma(n)}{\Gamma(1/\theta+1+n)} = \frac{\Gamma(1/\theta+1)\Gamma(n+1)}{\Gamma(1/\theta+1+n)}.
\end{aligned}$$

So,

$$\mathrm{E}(W) = \left(\frac{n\theta}{n\theta+1}\right) - \frac{\Gamma(1/\theta+1)\Gamma(n+1)}{\Gamma(1/\theta+1+n)}.$$

(b) Let A be the event "$\mathrm{X}_{(1)} < \xi$," and let B be the event "$\mathrm{X}_{(n)} > \xi$." Then,

$$\begin{aligned}
\mathrm{pr}\left[\mathrm{X}_{(1)} < \xi < \mathrm{X}_{(n)}\right] &= \mathrm{pr}\left\{[\mathrm{X}_{(1)} < \xi] \cap [\mathrm{X}_{(n)} > \xi]\right\} \\
&= \mathrm{pr}(\mathrm{A} \cap \mathrm{B}) = 1 - \mathrm{pr}(\overline{\mathrm{A} \cap \mathrm{B}}) = 1 - \mathrm{pr}(\bar{\mathrm{A}} \cup \bar{\mathrm{B}}) \\
&= 1 - [\mathrm{pr}(\bar{\mathrm{A}}) + \mathrm{pr}(\bar{\mathrm{B}}) - \mathrm{pr}(\bar{\mathrm{A}} \cap \bar{\mathrm{B}})].
\end{aligned}$$

Now,

$$\mathrm{pr}(\bar{\mathrm{A}}) = \mathrm{pr}[X_{(1)} > \xi] = \mathrm{pr}[\cap_{i=1}^{n}(X_i > \xi)] = \prod_{i=1}^{n} \mathrm{pr}(X_i > \xi) = \left(\frac{1}{2}\right)^n;$$

similarly,

$$\mathrm{pr}(\bar{\mathrm{B}}) = \mathrm{pr}[X_{(n)} < \xi] = \mathrm{pr}[\cap_{i=1}^{n}(X_i < \xi)] = \prod_{i=1}^{n} \mathrm{pr}(X_i < \xi) = \left(\frac{1}{2}\right)^n,$$

and

$$\mathrm{pr}(\bar{\mathrm{A}} \cap \bar{\mathrm{B}}) = \mathrm{pr}\left[(X_{(1)} > \xi) \cap (X_{(n)} < \xi)\right] = 0.$$

So,

$$\mathrm{pr}\left[X_{(1)} < \xi < X_{(n)}\right] = 1 - 2\left(\frac{1}{2}\right)^n = 1 - \frac{1}{2^{n-1}}.$$

So, the confidence coefficient for the interval $[X_{(1)}, X_{(n)}]$ varies with $n$, which is a highly undesirable property.

**Solution 4.26.** First, if $X$ has a uniform distribution on the interval $(0, 1)$, then, for $r \geq 0$, we have

$$\mathrm{E}(X^r) = \int_0^1 x^r(1)\,\mathrm{d}x = (1+r)^{-1}.$$

So,

$$\mathrm{E}(G) = \mathrm{E}\left[\left(\prod_{i=1}^{n} X_i\right)^{1/n}\right] = \prod_{i=1}^{n} \mathrm{E}\left(X_i^{1/n}\right)$$
$$= \prod_{i=1}^{n}\left(1 + \frac{1}{n}\right)^{-1} = \left[\left(1 + \frac{1}{n}\right)^n\right]^{-1},$$

so that $\lim_{n\to\infty} \mathrm{E}(G) = \mathrm{e}^{-1}$.

And, similarly,

$$\mathrm{E}(G^2) = \mathrm{E}\left[\left(\prod_{i=1}^{n} X_i\right)^{2/n}\right] = \prod_{i=1}^{n} \mathrm{E}\left(X_i^{2/n}\right)$$
$$= \prod_{i=1}^{n}\left(1 + \frac{2}{n}\right)^{-1} = \left[\left(1 + \frac{2}{n}\right)^n\right]^{-1},$$

so that $\lim_{n\to\infty} \mathrm{E}(G^2) = \mathrm{e}^{-2}$.

Thus,

$$\lim_{n\to\infty} \mathrm{V}(G) = \lim_{n\to\infty} \left\{\mathrm{E}(G^2) - [\mathrm{E}(G)]^2\right\} = \mathrm{e}^{-2} - (\mathrm{e}^{-1})^2 = 0.$$

Hence, since $\lim_{n\to\infty} \mathrm{E}(G) = \mathrm{e}^{-1}$ and $\lim_{n\to\infty} \mathrm{V}(G) = 0$, it follows that the random variable $G$ converges in probability to (i.e., is a consistent estimator of) the quantity $\mathrm{e}^{-1} = 0.368$.

**Solution 4.27.** We wish to prove that $\lim_{n\to\infty} \mathrm{pr}\{|X_n - 0| > \epsilon\} = 0 \,\forall \epsilon > 0$. Since $\mathrm{pr}(Y > n) = \int_n^\infty \mathrm{e}^{-y}\, dy = \mathrm{e}^{-n}$, we have

$$X_n = \mathrm{e}^n \mathrm{I}(Y > n) = \begin{cases} \mathrm{e}^n & \text{with probability } \mathrm{e}^{-n}, \\ 0 & \text{with probability } (1 - \mathrm{e}^{-n}). \end{cases}$$

Thus, for any $\epsilon > 0$,

$$\mathrm{p}\{|X_n| > \epsilon\} = \mathrm{pr}\{X_n > \epsilon\} = \begin{cases} 0 & \text{if } \mathrm{e}^n \le \epsilon, \\ \mathrm{e}^{-n} & \text{if } \mathrm{e}^n > \epsilon. \end{cases}$$

So,

$$\lim_{n\to\infty} \mathrm{pr}\{|X_n| > \epsilon\} \le \lim_{n\to\infty} \mathrm{e}^{-n} = 0, \quad \forall \epsilon > 0.$$

Note that $\mathrm{E}(X_n) = 1$ and $\mathrm{V}(X_n) = (\mathrm{e}^n - 1)$, so that $\lim_{n\to\infty} \mathrm{V}(X_n) = +\infty$; hence, a direct proof of convergence in probability is required.

**Solution 4.28.** Now,

$$\hat{\beta}^* = \frac{\sum_{i=1}^n (T_i - \bar{T})Y_i}{\sum_{i=1}^n (T_i - \bar{T})^2},$$

where $\bar{T} = n^{-1} \sum_{i=1}^n T_i = n_1/n$. Also, define $\bar{A}_1 = n_1^{-1} \sum_{i=1}^{n_1} A_i$, and $\bar{A}_0 = n_0^{-1} \sum_{i=(n_1+1)}^n A_i$.

Now, since $\mathrm{E}(Y_i | T_i, A_i) = \alpha + \beta T_i + \gamma A_i$, it follows that

$$\mathrm{E}(\hat{\beta}^* | \{T_i\}, \{A_i\}) = \frac{\sum_{i=1}^n (T_i - \bar{T})(\alpha + \beta T_i + \gamma A_i)}{\sum_{i=1}^n (T_i - \bar{T})^2}$$

$$= \beta + \frac{\gamma \sum_{i=1}^n (T_i - \frac{n_1}{n})A_i}{\sum_{i=1}^n (T_i - \frac{n_1}{n})^2}.$$

Now,

$$\sum_{i=1}^n (T_i - \frac{n_1}{n})A_i = (1 - \frac{n_1}{n})n_1\bar{A}_1 + (0 - \frac{n_1}{n})n_0\bar{A}_0 = \frac{n_0 n_1}{n}(\bar{A}_1 - \bar{A}_0).$$

And,

$$\sum_{i=1}^n (T_i - \frac{n_1}{n})^2 = n_1(1 - \frac{n_1}{n})^2 + n_0(-\frac{n_1}{n})^2 = \frac{n_0 n_1}{n}.$$

So,

$$E(\hat{\beta}^* | \{T_i\}, \{A_i\}) = \beta + \gamma(\bar{A}_1 - \bar{A}_0).$$

Thus, since $\gamma \neq 0$, a sufficient condition for $E(\hat{\beta}^* | \{T_i\}, \{A_i\}) = \beta$ is $\bar{A}_1 = \bar{A}_0$ (i.e., the average age of the $n_1$ subjects in the treatment group is equal to the average age of the $n_0$ subjects in the comparison group).

If the $n$ subjects are randomly selected from a large population of subjects and then randomization is employed in assigning these $n$ subjects to the treatment and comparison groups, then it follows that $E(\bar{A}_0) = E(\bar{A}_1)$, so that

$$E(\hat{\beta}^* | \{T_i\}) = E_{\{A_i\}}[E(\hat{\beta}^* | \{T_i\}, \{A_i\}] = \beta + \gamma E(\bar{A}_1 - \bar{A}_0) = \beta.$$

So, on expectation, randomization is sufficient to insure that $\hat{\beta}^*$ is an unbiased estimator of $\beta$.

**Solution 4.29**

(a) For $i = 1, 2, \ldots, n$, note that $E(Y_i) = \pi_i$, $V(Y_i) = \pi_i(1 - \pi_i)$, and

$$\mathcal{L}(\boldsymbol{y}; \boldsymbol{\beta}) = \prod_{i=1}^{n} \pi_i^{y_i}(1 - \pi_i)^{1-y_i},$$

so that

$$\ln \mathcal{L}(\boldsymbol{y}; \boldsymbol{\beta}) = \sum_{i=1}^{n} \left[ y_i \ln \pi_i + (1 - y_i) \ln(1 - \pi_i) \right].$$

So, for $j = 0, 1, \ldots, p$, we have

$$\frac{\partial \ln \mathcal{L}(\boldsymbol{y}; \boldsymbol{\beta})}{\partial \beta_j} = \sum_{i=1}^{n} \left[ \left( \frac{y_i}{\pi_i} \right) \frac{\partial \pi_i}{\partial \beta_j} - \frac{(1 - y_i)}{(1 - \pi_i)} \frac{\partial \pi_i}{\partial \beta_j} \right]$$
$$= \sum_{i=1}^{n} \frac{\partial \pi_i}{\partial \beta_j} \left[ \frac{(y_i - \pi_i)}{\pi_i(1 - \pi_i)} \right].$$

And,

$$\frac{\partial \pi_i}{\partial \beta_j} = \frac{x_{ij} e^{\sum_{j=0}^{p} \beta_j x_{ij}} \left( 1 + e^{\sum_{j=0}^{p} \beta_j x_{ij}} \right) - x_{ij} e^{2 \sum_{j=0}^{p} \beta_j x_{ij}}}{\left( 1 + e^{\sum_{j=0}^{p} \beta_j x_{ij}} \right)^2}$$
$$= \frac{x_{ij} e^{\sum_{j=0}^{p} \beta_j x_{ij}}}{\left( 1 + e^{\sum_{j=0}^{p} \beta_j x_{ij}} \right)^2} = x_{ij} \pi_i (1 - \pi_i).$$

Thus,

$$\frac{\partial \ln \mathcal{L}(y;\beta)}{\partial \beta_j} = \sum_{i=1}^{n} x_{ij}\pi_i(1-\pi_i)\left[\frac{(y_i-\pi_i)}{\pi_i(1-\pi_i)}\right]$$

$$= \sum_{i=1}^{n} x_{ij}\left[y_i - \mathrm{E}(Y_i)\right] = x_j'\left[y - \mathrm{E}(Y)\right],$$

where $x_j' = (x_{1j}, x_{2j}, \ldots, x_{nj})$.

Finally, with the $[n\mathrm{x}(p+1)]$ matrix $X$ defined so that its $i$th row is $x_i' = (1, x_{i1}, x_{i2}, \ldots, x_{ip}), i = 1, 2, \ldots, n$, and with the $[(p+1)\times 1]$ column vector $[\partial \ln \mathcal{L}(y;\beta)]/\partial\beta$ defined as

$$\frac{\partial \ln \mathcal{L}(y;\beta)}{\partial \beta} = \left[\frac{\partial \ln \mathcal{L}(y;\beta)}{\partial \beta_0}, \frac{\partial \ln \mathcal{L}(y;\beta)}{\partial \beta_1}, \ldots, \frac{\partial \ln \mathcal{L}(y;\beta)}{\partial \beta_p}\right]',$$

we have

$$\frac{\partial \ln \mathcal{L}(y;\beta)}{\partial \beta} = X'\left[y - \mathrm{E}(Y)\right],$$

which gives the desired result.

(b) Since $\partial \ln \mathcal{L}(y;\beta)/\partial\beta_j = \sum_{i=1}^{n} x_{ij}(y_i - \pi_i)$, it follows that

$$-\frac{\partial^2 \ln \mathcal{L}(y;\beta)}{\partial\beta_j\partial\beta_{j'}} = \sum_{i=1}^{n} x_{ij}\frac{\partial \pi_i}{\partial \beta_{j'}}$$

$$= \sum_{i=1}^{n} x_{ij}x_{ij'}\pi_i(1-\pi_i)$$

$$= \sum_{i=1}^{n} x_{ij}x_{ij'}\mathrm{V}(Y_i),$$

which does *not* functionally depend on $Y$.

Since $Y$ has a diagonal covariance matrix of the simple form

$$V = diag\left[\mathrm{V}(Y_1), \mathrm{V}(Y_2), \ldots, \mathrm{V}(Y_n)\right]$$
$$= diag\left[\pi_1(1-\pi_1), \pi_2(1-\pi_2), \ldots, \pi_n(1-\pi_n)\right],$$

it follows directly that the observed information matrix $I(y;\beta)$ equals the expected information matrix $\mathcal{I}(\beta)$, which can be written in matrix notation as

$$\mathcal{I}(\beta) = X'VX.$$

Finally, the estimated covariance matrix $\hat{V}(\hat{\beta})$ of $\hat{\beta}$ is equal to

$$\hat{V}(\hat{\beta}) = \mathcal{I}^{-1}(\hat{\beta}) = \left(X'\hat{V}X\right)^{-1},$$

where

$$\hat{V} = diag\left[\hat{\pi}_1(1-\hat{\pi}_1), \hat{\pi}_2(1-\hat{\pi}_2), \ldots, \hat{\pi}_n(1-\hat{\pi}_n)\right],$$

and where

$$\hat{\pi}_i = \frac{e^{\sum_{j=0}^{p}\hat{\beta}_j x_{ij}}}{1+e^{\sum_{j=0}^{p}\hat{\beta}_j x_{ij}}}, \quad i = 1, 2, \ldots, n.$$

**Solution 4.30*.** At iteration $t$, the E-step requires that we evaluate

$$\mathcal{Q}^{(t)}(y;\pi,\mu_1,\mu_2) \equiv \mathcal{Q}^{(t)} = \mathrm{E}_{\mathbf{Z}}\left\{\ln[\mathcal{L}_c(y,z;\pi,\mu_1,\mu_2)] \,\middle|\, y, \hat{\pi}^{(t-1)}, \hat{\mu}_1^{(t-1)}, \hat{\mu}_2^{(t-1)}\right\},$$

where the complete-data likelihood is given by

$$\mathcal{L}_c(y,z;\pi,\mu_1,\mu_2) = \prod_{i=1}^{n}[\pi \mathrm{p}_{Y_i}(y_i;\mu_1)]^{z_i}[(1-\pi)\mathrm{p}_{Y_i}(y_i;\mu_2)]^{(1-z_i)}.$$

So,

$$\begin{aligned}\mathcal{Q}^{(t)}(y;\pi,\mu_1,\mu_2) &= \sum_{i=1}^{n}\mathrm{E}_{Z_i}\left\{z_i \ln[\hat{\pi}^{(t-1)}\mathrm{p}_{Y_i}(y_i;\hat{\mu}_1^{(t-1)})]\right.\\ &\quad \left. + (1-z_i)\ln[(1-\hat{\pi}^{(t-1)})\mathrm{p}_{Y_i}(y_i;\hat{\mu}_2^{(t-1)})] \,\middle|\, y_i\right\}\\ &= \sum_{i=1}^{n}\left[(C_{1i}-C_{2i})\mathrm{E}(Z_i|y_i) + C_{2i}\right],\end{aligned}$$

where $C_{1i} = \ln[\hat{\pi}^{(t-1)}\mathrm{p}_{Y_i}(y_i;\hat{\mu}_1^{(t-1)})]$ and $C_{2i} = \ln[(1-\hat{\pi}^{(t-1)})\mathrm{p}_{Y_i}(y_i;\hat{\mu}_2^{(t-1)})]$ are constants with respect to $Z_i$.

Now,

$$\begin{aligned}\mathrm{E}(Z_i|y_i) &= \mathrm{pr}(Z_i = 1|y_i)\\ &= \frac{\left[\mathrm{p}_{Y_i}(y_i;\hat{\mu}_1^{(t-1)})\right]\mathrm{pr}(Z_i=1)}{\left[\mathrm{p}_{Y_i}(y_i;\hat{\mu}_1^{(t-1)})\right]\mathrm{pr}(Z_i=1) + \left[\mathrm{p}_{Y_i}(y_i;\hat{\mu}_2^{(t-1)})\right]\mathrm{pr}(Z_i=0)}\\ &= \frac{\hat{\pi}^{(t-1)}\left[\mathrm{p}_{Y_i}(y_i;\hat{\mu}_1^{(t-1)})\right]}{\hat{\pi}^{(t-1)}\left[\mathrm{p}_{Y_i}(y_i;\hat{\mu}_1^{(t-1)})\right] + (1-\hat{\pi}^{(t-1)})\left[\mathrm{p}_{Y_i}(y_i;\hat{\mu}_2^{(t-1)})\right]} = \hat{Z}_i^{(t)}, \quad \text{say.}\end{aligned}$$

Note that $\hat{Z}_i^{(t)}$ is the $t$th iteration estimate of the probability that the $i$th fish was born in a Pfiesteria-rich site. Also, when $t = 1$, $\hat{\pi}^{(0)}$, $\hat{\mu}_1^{(0)}$, and $\hat{\mu}_2^{(0)}$ are the well-chosen initial values that must be specified to start the EM algorithm iteration process.

Thus,

$$\hat{\mathcal{Q}}^{(t)}(y;\pi,\mu_1,\mu_2) \equiv \hat{\mathcal{Q}}^{(t)} = \sum_{i=1}^{n}\left[(C_{1i}-C_{2i})\hat{Z}_i^{(t)} + C_{2i}\right]$$

$$= \sum_{i=1}^{n}\left\{\hat{Z}_i^{(t)}\ln[\hat{\pi}^{(t-1)}\mathrm{p}_{Y_i}(y_i;\hat{\mu}_1^{(t-1)})] + (1-\hat{Z}_i^{(t)})\right.$$

$$\left.\times\ln[(1-\hat{\pi}^{(t-1)})\mathrm{p}_{Y_i}(y_i;\hat{\mu}_2^{(t-1)})]\right\}.$$

For the M-step, maximizing $\hat{\mathcal{Q}}^{(t)}$ with respect to $\pi$ yields

$$\frac{\partial\hat{\mathcal{Q}}^{(t)}}{\partial\pi} = \partial\left(\sum_{i=1}^{n}\{\hat{Z}_i^{(t)}[\ln\pi + \ln\mathrm{p}_{Y_i}(y_i;\mu_1)] + (1-\hat{Z}_i^{(t)})[\ln(1-\pi)\right.$$

$$\left.+\ln\mathrm{p}_{Y_i}(y_i;\mu_2)]\}\right)\Big/\partial\pi$$

$$\Rightarrow \frac{\sum_{i=1}^{n}\hat{Z}_i^{(t)}}{\pi} - \frac{\left[n-\sum_{i=1}^{n}\hat{Z}_i^{(t)}\right]}{1-\pi} = 0$$

$$\Rightarrow \hat{\pi}^{(t)} = \frac{\sum_{i=1}^{n}\hat{Z}_i^{(t)}}{n}.$$

Thus, $\hat{\pi}^{(t)}$ is the sample average estimated probability that a randomly selected fish was born in a Pfiesteria-rich site.

And,

$$\frac{\partial\hat{\mathcal{Q}}^{(t)}}{\partial\mu_1} = \partial\left(\sum_{i=1}^{n}\left\{\hat{Z}_i^{(t)}\left[\ln\pi + \ln\mathrm{p}_{Y_i}(y_i;\mu_1)\right] + (1-\hat{Z}_i^{(t)})\right.\right.$$

$$\left.\left.\times\left[\ln(1-\pi) + \ln\mathrm{p}_{Y_i}(y_i;\mu_2)\right]\right\}\right)\Big/\partial\mu_1$$

$$= \frac{\partial\left\{\sum_{i=1}^{n}\hat{Z}_i^{(t)}\left[y_i\ln\mu_1 - \mu_1 - \ln y_i!\right]\right\}}{\partial\mu_1}$$

$$= \frac{\sum_{i=1}^{n}\hat{Z}_i^{(t)}y_i}{\mu_1} - \sum_{i=1}^{n}\hat{Z}_i^{(t)} = 0$$

$$\Rightarrow \hat{\mu}_1^{(t)} = \frac{\sum_{i=1}^{n}\hat{Z}_i^{(t)}y_i}{\sum_{i=1}^{n}\hat{Z}_i^{(t)}}.$$

Note that $\hat{\mu}_1^{(t)}$ is a weighted estimate of the average number of ulcerative lesions for fish born in Pfiesteria-rich sites.

Similarly, it can be shown that $\hat{\mu}_2^{(t)} = \sum_{i=1}^{n}(1-\hat{Z}_i^{(t)})y_i / \sum_{i=1}^{n}(1-\hat{Z}_i^{(t)})$ is a weighted estimate of the average number of ulcerative lesions for fish born in Pfiesteria-free sites.

**Solution 4.31***

(a) Let $\mathrm{I_E}(x)$ be the indicator function for the set E, so that $\mathrm{I_E}(x)$ equals 1 if $x \in \mathrm{E}$ and $\mathrm{I_E}(x)$ equals 0 otherwise. Then, letting A= $\{1,2,\ldots,\theta\}$ and letting B=$\{1,2,\ldots,\infty\}$, we have

$$\begin{aligned}
\mathrm{p}_{X_1,X_2,\ldots,X_n}(x_1,x_2,\ldots,x_n;\theta) &= \prod_{i=1}^{n}\left\{\theta^{-1}\mathrm{I_A}(x_i)\right\} \\
&= [(\theta^{-n})\mathrm{I_A}(x_{(n)})]\cdot\left[\prod_{i=1}^{n}\mathrm{I_B}(x_i)\right] \\
&= \mathrm{g}(u;\theta)\cdot \mathrm{h}(x_1,x_2,\ldots,x_n),
\end{aligned}$$

where $u = x_{(n)} = \max\{x_1,x_2,\ldots,x_n\}$. And, given $X_{(n)} = x_{(n)}, \mathrm{h}(x_1,x_2,\ldots,x_n)$ does not in any way depend on $\theta$, so that $X_{(n)}$ is a sufficient statistic for $\theta$.

Also,

$$\begin{aligned}
\mathrm{E}(U^*) &= \mathrm{E}[(2X_1-1)] = 2\mathrm{E}(X_1)-1 \\
&= 2\sum_{x_1=1}^{\theta} x_1\theta^{-1} - 1 = \frac{2}{\theta}\sum_{x_1=1}^{\theta} x_1 - 1 \\
&= \frac{2}{\theta}\left[\frac{\theta(\theta+1)}{2}\right] - 1 \\
&= \theta.
\end{aligned}$$

(b) For notational ease, let $X_{(n)} = U$. Now, $\hat{\theta} = \mathrm{E}(U^*|U=u) = \mathrm{E}(2X_1-1|U=u) = 2\mathrm{E}(X_1|U=u)-1$, so we need to evaluate $\mathrm{E}(X_1|U=u)$. To do so, we need to first find

$$\mathrm{p}_{X_1}(x_1|U=u) = \frac{\mathrm{p}_{X_1,U}(x_1,u)}{\mathrm{p}_U(u)} = \frac{\mathrm{pr}[(X_1=x_1)\cap(U=u)]}{\mathrm{p}_U(u)}.$$

Now,

$$\begin{aligned}
\mathrm{pr}(U=u) &= \mathrm{pr}(U\le u) - \mathrm{pr}(U\le u-1) \\
&= \mathrm{pr}\left[\bigcap_{i=1}^{n}(X_i\le u)\right] - \mathrm{pr}\left[\bigcap_{i=1}^{n}(X_i\le u-1)\right] \\
&= \prod_{i=1}^{n}\mathrm{pr}(X_i\le u) - \prod_{i=1}^{n}\mathrm{pr}(X_i\le u-1) \\
&= \left(\frac{u}{\theta}\right)^n - \left(\frac{u-1}{\theta}\right)^n, \quad u=1,2,\ldots,\theta.
\end{aligned}$$

And,

$$\mathrm{pr}[(X_1 = x_1) \cap (U = u)] = \begin{cases} 0, & x_1 > u, \\ \frac{1}{\theta}\left(\frac{u}{\theta}\right)^{n-1}, & x_1 = u, \\ \frac{1}{\theta}\left[\left(\frac{u}{\theta}\right)^{n-1} - \left(\frac{u-1}{\theta}\right)^{n-1}\right], & x_1 < u. \end{cases}$$

In the above expression, note that the equality "$x_1 = u$" implies consideration of the event $\{(X_1 = u) \cap [\bigcap_{i=2}^{n}(X_i \leq u)]\}$, and that the inequality "$x_1 < u$" implies consideration of the event

$$\{(X_1 = x_1) \cap [\max(X_2, X_2, \ldots, X_n) = u]\}.$$

So,

$$\mathrm{p}_{X_1}(x_1|U = u) = \begin{cases} 0, & x_1 > u, \\ \frac{u^{n-1}}{u^n - (u-1)^n}, & x_1 = u, \\ \frac{u^{n-1} - (u-1)^{n-1}}{u^n - (u-1)^n}, & x_1 = 1, 2, \ldots u - 1, \end{cases}$$

which cannot (and does not) depend in any way on $\theta$ by the sufficiency principle.
So,

$$\begin{aligned} \mathrm{E}(X_1|U = u) &= \sum_{x_1=1}^{u-1} x_1 \left[\frac{u^{n-1} - (u-1)^{n-1}}{u^n - (u-1)^n}\right] + u\left[\frac{u^{n-1}}{u^n - (u-1)^n}\right] \\ &= \left[\frac{(u-1)u}{2}\right]\left[\frac{u^{n-1} - (u-1)^{n-1}}{u^n - (u-1)^n}\right] + \left[\frac{u^n}{u^n - (u-1)^n}\right] \\ &= \frac{u^{n+1} - u(u-1)^n + u^n}{2\left[u^n - (u-1)^n\right]}. \end{aligned}$$

Thus,

$$\begin{aligned} 2\mathrm{E}(X_1|U = u) - 1 &= \frac{u^{n+1} - u(u-1)^n + u^n}{u^n - (u-1)^n} - 1 \\ &= \frac{u^{n+1} - (u-1)^{n+1}}{u^n - (u-1)^n}, \end{aligned}$$

so that the MVUE of $\theta$ is

$$\hat{\theta} = \frac{U^{n+1} - (U-1)^{n+1}}{U^n - (U-1)^n}.$$

As a simple check, when $n = 1$, so that $U = X_1$, we obtain

$$\hat{\theta} = \frac{U^2 - (U-1)^2}{U - (U-1)} = 2U - 1 = 2X_1 - 1, \quad \text{as desired}.$$

Note that

$$\begin{aligned} \mathrm{E}(\hat{\theta}) &= \sum_{u=1}^{\theta} \hat{\theta}\,\mathrm{pr}(U = u) = \sum_{u=1}^{\theta} \left[\frac{u^{n+1} - (u-1)^{n+1}}{u^n - (u-1)^n}\right]\left[\frac{u^n - (u-1)^n}{\theta^n}\right] \\ &= \frac{1}{\theta^n} \sum_{u=1}^{\theta} \left[u^{n+1} - (u-1)^{n+1}\right] \\ &= \frac{1}{\theta^n}\{(1-0) + (2^{n+1} - 1) + (3^{n+1} - 2^{n+1}) + \cdots + [\theta^{n+1} - (\theta-1)^{n+1}]\} \\ &= \frac{\theta^{n+1}}{\theta^n} = \theta. \end{aligned}$$

As a numerical example, if $n = 5$ and $u = 2$, then $\hat{\theta} = (2^6 - 1^6)/(2^5 - 1^5) = 2.0323$. So, one disadvantage of $\hat{\theta}$ is that it does not necessarily take positive integer values, even though the parameter $\theta$ is a positive integer.

**Solution 4.32***

(a) First, note that

$$\mathrm{F}_X(x) = \int_{\theta}^{x} (1)dx = (x - \theta), 0 < \theta < x < (\theta + 1) < +\infty.$$

Hence, from order statistics theory, it follows directly that

$$\begin{aligned} \mathrm{f}_{X_{(1)}}(x_{(1)}) &= n\left[1 - \mathrm{F}_X(x_{(1)})\right]^{n-1} \mathrm{f}_X(x_{(1)}) \\ &= n\left[1 - (x_{(1)} - \theta)\right]^{n-1}, \quad 0 < \theta < x_{(1)} < (\theta + 1) < +\infty. \end{aligned}$$

Then, with $u = (1 + \theta) - x_{(1)}$, so that $\mathrm{d}u = -\mathrm{d}x_{(1)}$, we have, for $r$ a non-negative integer,

$$\begin{aligned} \mathrm{E}\left[X_{(1)}^r\right] &= \int_{\theta}^{\theta+1} x_{(1)}^r n[(1+\theta) - x_{(1)}]^{n-1}\,\mathrm{d}x_{(1)} \\ &= \int_0^1 [(1+\theta) - u]^r n u^{n-1} du \\ &= n \int_0^1 \left[\sum_{j=0}^{r} \mathrm{C}_j^r (1+\theta)^j (-u)^{r-j}\right] u^{n-1}\,\mathrm{d}u \end{aligned}$$

$$= n\sum_{j=0}^{r} C_j^r (1+\theta)^j (-1)^{r-j} \int_0^1 u^{n+r-j-1}\, du$$

$$= n\sum_{j=0}^{r} C_j^r (1+\theta)^j (-1)^{r-j} (n+r-j)^{-1}.$$

When $r = 1$, we obtain $E(X_{(1)}) = \theta + 1/(n+1)$. Also, for $r = 2$, we obtain

$$E(X_{(1)}^2) = n\left[\frac{1}{(n+2)} + 2(1+\theta)(-1)\left(\frac{1}{n+1}\right) + (1+\theta)^2\left(\frac{1}{n}\right)\right],$$

so that

$$V(X_{(1)}) = E(X_{(1)}^2) - [E(X_{(1)})]^2 = \frac{n}{(n+1)^2(n+2)}.$$

By symmetry,

$$E(X_{(n)}) = (\theta+1) - \frac{1}{(n+1)} \quad \text{and} \quad V(X_{(n)}) = \frac{n}{(n+1)^2(n+2)}.$$

Or, more directly, one can use $f_{X_{(n)}}(x_{(n)}) = n\left[x_{(n)} - \theta\right]^{n-1}$, $0 < \theta < x_{(n)} < (\theta+1) < +\infty$, to show that

$$E(X_{(n)}{}^r) = n\sum_{j=0}^{r} C_j^r \theta^{r-j}(n+j)^{-1}.$$

Thus,

$$E(\hat{\theta}_1) = \frac{1}{2}\left[E(X_{(1)}) + E(X_{(n)}) - 1\right]$$

$$= \frac{1}{2}\left\{\left[\theta + \frac{1}{(n+1)}\right] + \left[(\theta+1) - \frac{1}{(n+1)}\right] - 1\right\} = \theta.$$

And,

$$E(\hat{\theta}_2) = \frac{1}{(n-1)}\left[nE(X_{(1)}) - E(X_{(n)})\right]$$

$$= \frac{1}{(n-1)}\left\{n\left[\theta + \frac{1}{(n+1)}\right] - \left[(\theta+1) - \frac{1}{(n+1)}\right]\right\} = \theta.$$

To find the variances of the estimators $\hat{\theta}_1$ and $\hat{\theta}_2$, we need to find $\text{cov}[X_{(1)}, X_{(n)}]$. If we let $Y_i = (X_i - \theta), i = 1, 2, \ldots, n$, then $f_{Y_i}(y_i) = 1, 0 < y_i < 1$. Also, $Y_{(1)} = \min\{Y_1, Y_2, \ldots, Y_n\} = (X_{(1)} - \theta)$ and $Y_{(n)} = \max\{Y_1, Y_2, \ldots, Y_n\} = (X_{(n)} - \theta)$, so that $\text{cov}[X_{(1)}, X_{(n)}] = \text{cov}[Y_{(1)}, Y_{(n)}]$.

Now, since $f_{Y_{(1)},Y_{(n)}}(y_{(1)},y_{(n)}) = n(n-1)(y_{(n)} - y_{(1)})^{n-2}, 0 < y_{(1)} < y_{(n)} < 1$, we have

$$E(Y_{(1)}Y_{(n)}) = \int_0^1 \int_{y_{(1)}}^1 [y_{(1)}y_{(n)}]n(n-1)(y_{(n)} - y_{(1)})^{n-2}\, dy_{(n)}\, dy_{(1)}.$$

So, using the relationship $w = (y_{(n)} - y_{(1)})$, so that $dw = dy_{(n)}$, and appealing to properties of the beta distribution, we obtain

$$\begin{aligned}
E(Y_{(1)}Y_{(n)}) &= n(n-1)\int_0^1 \int_0^{1-y_{(1)}} y_{(1)}(w + y_{(1)})w^{n-2}\, dw\, dy_{(1)} \\
&= n(n-1)\int_0^1 \int_0^{1-y_{(1)}} \left[y_{(1)}w^{n-1} + y_{(1)}^2 w^{n-2}\right] dw\, dy_{(1)} \\
&= n(n-1)\int_0^1 \left[\frac{y_{(1)}(1-y_{(1)})^n}{n} + \frac{y_{(1)}^2(1-y_{(1)})^{n-1}}{(n-1)}\right] dy_{(1)} \\
&= n(n-1)\left[\frac{\Gamma(2)\Gamma(n+1)/\Gamma(n+3)}{n} + \frac{\Gamma(3)\Gamma(n)/\Gamma(n+3)}{(n-1)}\right] \\
&= n(n-1)\left[\frac{1}{n(n+1)(n+2)} + \frac{2}{(n-1)n(n+1)(n+2)}\right] \\
&= \frac{1}{(n+2)}.
\end{aligned}$$

Finally,

$$\begin{aligned}
\text{cov}(X_{(1)}, X_{(n)}) &= \text{cov}(Y_{(1)}, Y_{(n)}) = E(Y_{(1)}Y_{(n)}) - E(Y_{(1)})E(Y_{(n)}) \\
&= \frac{1}{(n+2)} - \left(\frac{1}{n+1}\right)\left(\frac{n}{n+1}\right) = \frac{1}{(n+1)^2(n+2)}.
\end{aligned}$$

So,

$$\begin{aligned}
V(\hat{\theta}_1) &= \left(\frac{1}{2}\right)^2 \left[V(X_{(1)}) + V(X_{(n)}) + 2\text{cov}(X_{(1)}, X_{(n)})\right] \\
&= \frac{1}{4}\left[\frac{n}{(n+1)^2(n+2)} + \frac{n}{(n+1)^2(n+2)} + 2\left(\frac{1}{(n+1)^2(n+2)}\right)\right] \\
&= \frac{1}{2(n+1)(n+2)}.
\end{aligned}$$

And,

$$\begin{aligned}
V(\hat{\theta}_2) &= \frac{1}{(n-1)^2}\left[n^2 V(X_{(1)}) + V(X_{(n)}) - 2n\text{cov}(X_{(1)}, X_{(n)})\right] \\
&= \frac{1}{(n-1)^2}\left[\frac{n^3}{(n+1)^2(n+2)} + \frac{n}{(n+1)^2(n+2)} - \frac{2n}{(n+1)^2(n+2)}\right] \\
&= \frac{n}{(n-1)(n+1)(n+2)} = \frac{n}{(n^2-1)(n+2)}.
\end{aligned}$$

Thus, $V(\hat{\theta}_1) < V(\hat{\theta}_2)$, $n > 1$.

(b) Now,

$$V(W) = c_1^2\sigma^2 + c_2^2\sigma^2 + 2c_1c_2\sigma_{12} = [c_1^2 + (1-c_1)^2]\sigma^2 + 2c_1(1-c_1)\sigma_{12}.$$

So,

$$\frac{\partial V(W)}{\partial c_1} = [2c_1 - 2(1-c_1)] + 2(1-2c_1)\sigma_{12} = 0$$

gives

$$2(2c_1 - 1)(\sigma^2 - \sigma_{12}) = 0.$$

Thus, if $c_1 = c_2 = \frac{1}{2}$, then $V(W)$ is minimized as long as $\sigma^2 > \sigma_{12}$. Note that these conditions are met by the estimator $\hat{\theta}_1$, but not by the estimator $\hat{\theta}_2$. Also, another drawback associated with the estimator $\hat{\theta}_2$ is that it can take a negative value, even though $\theta > 0$.

(c) Let $I_A(x)$ be the indicator function for the set A, so that $I_A(x)$ equals 1 if $x \in A$ and $I_A(x)$ equals 0 otherwise. Then, with A equal to the open interval $(\theta, \theta+1)$, the joint distribution of $X_1, X_2, \ldots, X_n$ can be written in the form

$$(1)^n \prod_{i=1}^{n} I_{(\theta,\theta+1)}(x_i) = (1)^n \left\{I_{(\theta,\theta+1)}[x_{(1)}]\right\}\left\{I_{(\theta,\theta+1)}[x_{(n)}]\right\},$$

since $0 < \theta < x_{(1)} \leq x_i \leq x_{(n)} < (\theta+1) < +\infty$, $i = 1, 2, \ldots, n$. Hence, by the Factorization Theorem, $X_{(1)}$ and $X_{(n)}$ are jointly sufficient for $\theta$.
However, $X_{(1)}$ and $X_{(n)}$ do *not* constitute a set of *complete* sufficient statistics for $\theta$ since $E\left[g(X_{(1)}, X_{(n)})\right] = 0$ for all $\theta, 0 < \theta < +\infty$, where

$$g(X_{(1)}, X_{(n)}) = X_{(1)} - X_{(n)} + \left(\frac{n-1}{n+1}\right).$$

This finding raises the interesting question about whether or not the estimator $\hat{\theta}_1$ could be the MVUE of $\theta$, even though it is not a function of complete sufficient statistics. For more general discussion about this issue, see Bondesson (1983).

**Solution 4.33***

(a) Under the independence assumption,

$$E\begin{pmatrix} X_{yy} \\ X_{yn} \\ X_{ny} \end{pmatrix} = \begin{pmatrix} N\pi_1\pi_2 \\ N\pi_1(1-\pi_2) \\ N(1-\pi_1)\pi_2 \end{pmatrix}.$$

Hence, the method-of-moments estimator of $N$ is obtained by solving for $N$ using the three equations

$$X_{yy} = N\pi_1\pi_2, \tag{4.1}$$

$$X_{yn} = N\pi_1(1-\pi_2), \tag{4.2}$$

$$X_{ny} = (1-\pi_1)\pi_2. \tag{4.3}$$

The operations [(Equation 4.1) + (Equation 4.2)] and [(Equation 4.1) + (Equation 4.3)] give

$$(X_{yy} + X_{yn}) = N\pi_1, \tag{4.4}$$

$$(X_{yy} + X_{ny}) = N\pi_2. \tag{4.5}$$

Finally, the operation [(Equation 4.4) × (Equation 4.5)] / (Equation 4.1) produces

$$\hat{N} = \frac{(X_{yy} + X_{yn})(X_{yy} + X_{ny})}{X_{yy}}.$$

Note that $\hat{N}$ does not necessarily take integer values. Also, $E[\hat{N}]$ and $V[\hat{N}]$ are undefined since $X_{yy} = 0$ occurs with non-zero probability. When $x_{yy} = 12{,}000$, $x_{yn} = 6000$, and $x_{ny} = 8000$, we have

$$\hat{N} = \frac{[(12{,}000) + (6000)][(12{,}000) + (8000)]}{(12{,}000)} = 30{,}000.$$

(b) Since

$$\text{odds}(E_1|E_2) = \frac{\text{pr}(E_1|E_2)}{1 - \text{pr}(E_1|E_2)} = \frac{\pi_{yy}/(\pi_{yy} + \pi_{ny})}{\pi_{ny}/(\pi_{yy} + \pi_{ny})} = \frac{\pi_{yy}}{\pi_{ny}}$$

and

$$\text{odds}(E_1|\bar{E}_2) = \frac{\text{pr}(E_1|\bar{E}_2)}{1 - \text{pr}(E_1|\bar{E}_2)} = \frac{\pi_{yn}/(\pi_{yn} + \pi_{nn})}{\pi_{nn}/(\pi_{yn} + \pi_{nn})} = \frac{\pi_{yn}}{\pi_{nn}},$$

the assumption that

$$\frac{\text{odds}(E_1 \mid E_2)}{\text{odds}(E_1 \mid \bar{E}_2)} = k$$

implies that

$$\frac{\pi_{yy}/\pi_{ny}}{\pi_{yn}/\pi_{nn}} = k$$

or, equivalently,

$$\pi_{yy}(1 - \pi_{yy} - \pi_{yn} - \pi_{ny}) = k\pi_{yn}\pi_{ny}.$$

So, a method-of-moments estimator of $N$ is obtained by simultaneously solving the four equations

$$X_{yy} = N\pi_{yy}, \tag{4.6}$$

$$X_{yn} = N\pi_{yn}, \tag{4.7}$$

$$X_{ny} = N\pi_{ny}, \tag{4.8}$$

$$\pi_{yy}(1 - \pi_{yy} - \pi_{yn} - \pi_{ny}) = k\pi_{yn}\pi_{ny}. \tag{4.9}$$

Equations 4.6, 4.7, and 4.8 imply that $\hat{\pi}_{yy} = X_{yy}/N$, $\hat{\pi}_{yn} = X_{yn}/N$, and $\hat{\pi}_{ny} = X_{ny}/N$. Substituting these expressions into Equation 4.9 yields

$$\left(\frac{X_{yy}}{N}\right)\left[\frac{(N - X_{yy} - X_{yn} - X_{ny})}{N}\right] = k\left(\frac{X_{yn}}{N}\right)\left(\frac{X_{ny}}{N}\right),$$

giving

$$\tilde{N}(k) = \frac{X_{yy}^2 + X_{yy}X_{yn} + X_{yy}X_{ny} + kX_{yn}X_{ny}}{X_{yy}}.$$

When $x_{yy} = 12{,}000$, $x_{yn} = 6000$, and $x_{ny} = 8000$, we have

$$\begin{aligned}
\tilde{N}(1/2) &= \frac{(12{,}000)^2 + (12{,}000)(6000) + (12{,}000)(8000) + (1/2)(6000)(8000)}{12{,}000} \\
&= 28{,}000, \\
\tilde{N}(2) &= \frac{(12{,}000)^2 + (12{,}000)(6000) + (12{,}000)(8000) + (2)(6000)(8000)}{12{,}000} \\
&= 34{,}000, \\
\tilde{N}(4) &= \frac{(12{,}000)^2 + (12{,}000)(6000) + (12{,}000)(8000) + (4)(6000)(8000)}{12{,}000} \\
&= 42{,}000.
\end{aligned}$$

Apparently, the estimator $\hat{N}$, which assumes independence, has a tendency to over-estimate $N$ when $k < 1$ and to under-estimate $N$ when $k > 1$.

**Solution 4.34***

(a) Let $\pi(x_i) \equiv \pi_i = e^{\alpha+\beta x_i}/(1 + e^{\alpha+\beta x_i})$. The likelihood function $\mathcal{L}$ is equal to

$$\mathcal{L} = \prod_{i=1}^{n} \pi_i^{y_i}(1 - \pi_i)^{1-y_i},$$

so that

$$\ln \mathcal{L} = \sum_{i=1}^{n} y_i \ln(\pi_i) + (1 - y_i)\ln(1 - \pi_i).$$

By the chain rule,

$$\frac{\partial \ln \mathcal{L}}{\partial \alpha} = \sum_{i=1}^{n} \frac{\partial \ln \mathcal{L}}{\partial \pi_i} \cdot \frac{\partial \pi_i}{\partial \alpha}$$

and

$$\frac{\partial \ln \mathcal{L}}{\partial \beta} = \sum_{i=1}^{n} \frac{\partial \ln \mathcal{L}}{\partial \pi_i} \cdot \frac{\partial \pi_i}{\partial \beta}.$$

Now,

$$\frac{\partial \pi_i}{\partial \alpha} = \frac{\partial \left[ \dfrac{e^{\alpha+\beta x_i}}{(1+e^{\alpha+\beta x_i})} \right]}{\partial \alpha} = \frac{e^{\alpha+\beta x_i}}{(1+e^{\alpha+\beta x_i})^2} = \pi_i(1-\pi_i)$$

and

$$\frac{\partial \pi_i}{\partial \beta} = \frac{\partial \left[ \dfrac{e^{\alpha+\beta x_i}}{(1+e^{\alpha+\beta x_i})} \right]}{\partial \beta} = \frac{x_i e^{\alpha+\beta x_i}}{(1+e^{\alpha+\beta x_i})^2} = x_i \pi_i(1-\pi_i).$$

Thus,

$$\frac{\partial \ln \mathcal{L}}{\partial \alpha} = \sum_{i=1}^{n} \frac{\partial \ln \mathcal{L}}{\partial \pi_i} \cdot \frac{\partial \pi_i}{\partial \alpha} = \sum_{i=1}^{n} \left[ \frac{y_i}{\pi_i} - \frac{1-y_i}{1-\pi_i} \right] \pi_i(1-\pi_i)$$

$$= \sum_{i=1}^{n} (y_i - \pi_i) = 0;$$

$$\Rightarrow \sum_{i=1}^{n} y_i = \sum_{i=1}^{n} \pi_i = \sum_{i=1}^{n} \frac{e^{\alpha+\beta x_i}}{(1+e^{\alpha+\beta x_i})}$$

$$= \sum_{i=1}^{n_0} \frac{e^{\alpha}}{(1+e^{\alpha})} + \sum_{i=n_0+1}^{n} \frac{e^{\alpha+\beta}}{(1+e^{\alpha+\beta})}$$

$$\Rightarrow \sum_{i=1}^{n} y_i = n_0 \frac{e^{\alpha}}{(1+e^{\alpha})} + n_1 \frac{e^{\alpha+\beta}}{(1+e^{\alpha+\beta})}.$$

Similarly,

$$\frac{\partial \ln \mathcal{L}}{\partial \beta} = \sum_{i=1}^{n} \frac{\partial \ln \mathcal{L}}{\partial \pi_i} \cdot \frac{\partial \pi_i}{\partial \beta} = \sum_{i=1}^{n} \left[ \frac{y_i}{\pi_i} - \frac{1-y_i}{1-\pi_i} \right] x_i \pi_i(1-\pi_i)$$

$$= \sum_{i=1}^{n} x_i (y_i - \pi_i) = \sum_{i=n_0+1}^{n} (y_i - \pi_i) = 0;$$

$$\Rightarrow \sum_{i=n_0+1}^{n} y_i = \sum_{i=n_0+1}^{n} \pi_i = \sum_{i=n_0+1}^{n} \frac{e^{\alpha+\beta}}{(1+e^{\alpha+\beta})} = n_1 \frac{e^{\alpha+\beta}}{(1+e^{\alpha+\beta})}.$$

Via subtraction, we obtain

$$\sum_{i=1}^{n} y_i - \sum_{i=n_0+1}^{n} y_i = n_0 \frac{e^{\alpha}}{(1+e^{\alpha})}$$

$$\Rightarrow \sum_{i=1}^{n_0} y_i = n_0 \frac{e^{\alpha}}{(1+e^{\alpha})}$$

$$\Rightarrow \hat{\alpha} = \ln\left(\frac{p_0}{1-p_0}\right),$$

where $p_0 = n_0^{-1} \sum_{i=1}^{n_0} y_i$ is the sample proportion of overweight infants receiving home care.

Then, it follows directly that

$$\sum_{i=n_0+1}^{n} y_i = n_1 \frac{e^{\hat{\alpha}+\hat{\beta}}}{1+e^{\hat{\alpha}+\hat{\beta}}}$$

$$\Rightarrow \hat{\beta} = \ln\left(\frac{p_1}{1-p_1}\right) - \hat{\alpha}$$

$$= \ln\left[\frac{p_1}{1-p_1}\right] - \ln\left[\frac{p_0}{1-p_0}\right]$$

$$= \ln\left[\frac{p_1/(1-p_1)}{p_0/(1-p_0)}\right],$$

where $p_1 = n_1^{-1} \sum_{i=n_0+1}^{n} y_i$ is the sample proportion of overweight infants in day care.

Thus, $\hat{\alpha}$ is the sample log odds of being overweight for infants receiving home care, while $\hat{\beta}$ is the sample log odds ratio comparing the estimated odds of being overweight for infants in day care to the estimated odds of being overweight for infants receiving home care. The estimators $\hat{\alpha}$ and $\hat{\beta}$ make intuitive sense, since they are the sample counterparts of the population parameters $\alpha$ and $\beta$.

(b)

$$-\frac{\partial^2 \ln \mathcal{L}}{\partial \alpha^2} = \frac{-\partial \sum_{i=1}^{n}(y_i - \pi_i)}{\partial \pi_i} \cdot \frac{\partial \pi_i}{\partial \alpha}$$

$$= \sum_{i=1}^{n} \pi_i(1-\pi_i) = \sum_{i=1}^{n} \frac{e^{\alpha+\beta x_i}}{(1+e^{\alpha+\beta x_i})^2}$$

$$= \sum_{i=1}^{n_0} \frac{e^{\alpha}}{(1+e^{\alpha})^2} + \sum_{i=n_0+1}^{n} \frac{e^{\alpha+\beta}}{(1+e^{\alpha+\beta})^2}$$

$$= n_0\pi_0(1-\pi_0) + n_1\pi_1(1-\pi_1),$$

where $\pi_0 = e^{\alpha}/(1+e^{\alpha})$ and $\pi_1 = e^{\alpha+\beta}/(1+e^{\alpha+\beta})$.

Also,

$$
\begin{aligned}
-\frac{\partial^2 \ln \mathcal{L}}{\partial \beta^2} &= \frac{-\partial \sum_{i=1}^{n} x_i(y_i - \pi_i)}{\partial \pi_i} \cdot \frac{\partial \pi_i}{\partial \beta} \\
&= \sum_{i=1}^{n} x_i^2 \pi_i(1-\pi_i) \\
&= \sum_{i=n_0+1}^{n} \frac{e^{\alpha+\beta}}{(1+e^{\alpha+\beta})^2} = n_1\pi_1(1-\pi_1).
\end{aligned}
$$

Finally,

$$
\begin{aligned}
-\frac{\partial^2 \ln \mathcal{L}}{\partial \alpha \partial \beta} &= \frac{-\partial \sum_{i=1}^{n} (y_i - \pi_i)}{\partial \pi_i} \cdot \frac{\partial \pi_i}{\partial \beta} \\
&= \sum_{i=1}^{n} x_i \pi_i(1-\pi_i) \\
&= n_1\pi_1(1-\pi_1) = -\frac{\partial^2 \ln \mathcal{L}}{\partial \beta \partial \alpha}.
\end{aligned}
$$

So, with $\boldsymbol{y}' = (y_1, y_2, \ldots, y_n)$, it follows that

$$
\begin{aligned}
\boldsymbol{I}(\boldsymbol{y};\alpha,\beta) &= \mathcal{I}(\alpha,\beta) \\
&= \begin{bmatrix} n_0\pi_0(1-\pi_0) + n_1\pi_1(1-\pi_1) & n_1\pi_1(1-\pi_1) \\ n_1\pi_1(1-\pi_1) & n_1\pi_1(1-\pi_1) \end{bmatrix}.
\end{aligned}
$$

Hence, using either observed or expected information, the large-sample variance–covariance matrix for $\hat{\alpha}$ and $\hat{\beta}$ is equal to

$$
\mathcal{I}^{-1}(\alpha,\beta) = \begin{Bmatrix} \left[n_0 \frac{e^{\alpha}}{(1+e^{\alpha})^2}\right]^{-1} & -\left[n_0 \frac{e^{\alpha}}{(1+e^{\alpha})^2}\right]^{-1} \\ -\left[n_0 \frac{e^{\alpha}}{(1+e^{\alpha})^2}\right]^{-1} & \left[n_0 \frac{e^{\alpha}}{(1+e^{\alpha})^2}\right]^{-1} + \left[n_1 \frac{e^{\alpha+\beta}}{(1+e^{\alpha+\beta})^2}\right]^{-1} \end{Bmatrix}.
$$

(c) The estimated large-sample 95% CI for $\alpha$ is

$$
\hat{\alpha} \pm 1.96 \sqrt{\left[n_0 \frac{e^{\hat{\alpha}}}{(1+e^{\hat{\alpha}})^2}\right]^{-1}},
$$

and the estimated large-sample 95% CI for $\beta$ is

$$
\hat{\beta} \pm 1.96 \sqrt{\left[n_0 \frac{e^{\hat{\alpha}}}{(1+e^{\hat{\alpha}})^2}\right]^{-1} + \left[n_1 \frac{e^{\hat{\alpha}+\hat{\beta}}}{(1+e^{\hat{\alpha}+\hat{\beta}})^2}\right]^{-1}}.
$$

For the given data, $\hat{\alpha} = -1.52$ and $\hat{\beta} = 0.47$, so that the corresponding large-sample 95% CIs for $\alpha$ and $\beta$ are $(-2.03, -1.01)$ and $(-0.21, 1.15)$, respectively. Since the CI for $\beta$ includes the value 0, there is no statistical evidence using these data that infants placed in day care are more likely to be overweight than are infants receiving home care.

**Solution 4.35*.** First, the parameter of interest is

$$\gamma = \mathrm{E}(T) = \mathrm{E}_x[\mathrm{E}(T|X = x)] = \mathrm{E}_x[(\theta x)^{-1}] = \frac{1}{\theta}\mathrm{E}(X^{-1}) = \frac{1}{\theta(\beta - 1)}, \quad \beta > 1.$$

Since information on the random variable $X$ is unavailable, the marginal distribution $f_T(t)$ of the random variable $T$ must be used to make ML-based inferences about the parameters $\theta$, $\beta$, and $\gamma$. In particular, the observed latency periods $t_1, t_2, \ldots, t_n$ can then be considered to be the realizations of a random sample $T_1, T_2, \ldots, T_n$ of size $n$ from

$$\begin{aligned}
f_T(t) &= \int_0^\infty f_{T,X}(t, x)\,dx = \int_0^\infty f_T(t|X = x)f_X(x)\,dx \\
&= \int_0^\infty \theta x e^{-\theta x t} \cdot \frac{x^{\beta-1}e^{-x}}{\Gamma(\beta)}\,dx \\
&= \frac{\theta}{\Gamma(\beta)} \int_0^\infty x^{\beta} e^{-(1+\theta t)x}\,dx \\
&= \frac{\theta}{\Gamma(\beta)}\Gamma(\beta + 1)(1 + \theta t)^{-(\beta+1)} \\
&= \theta\beta(1 + \theta t)^{-(\beta+1)}, \quad t > 0,\ \theta > 0,\ \beta > 1.
\end{aligned}$$

To produce a large-sample CI for the unknown parameter $\gamma = [\theta(\beta - 1)]^{-1}$, we will employ the delta method. First, the appropriate likelihood function $\mathcal{L}$ is

$$\mathcal{L} = \prod_{i=1}^{n}\left[\theta\beta(1 + \theta t_i)^{-(\beta+1)}\right] = \theta^n\beta^n \prod_{i=1}^{n}(1 + \theta t_i)^{-(\beta+1)},$$

so that

$$\ln \mathcal{L} = n \ln \theta + n \ln \beta - (\beta + 1)\sum_{i=1}^{n} \ln(1 + \theta t_i).$$

So, we have

$$\frac{\partial \ln L}{\partial \theta} = \frac{n}{\theta} - (\beta + 1)\sum_{i=1}^{n} \frac{t_i}{(1 + \theta t_i)},$$

and

$$\frac{\partial \ln L}{\partial \beta} = \frac{n}{\beta} - \sum_{i=1}^{n} \ln(1 + \theta t_i),$$

so that

$$\frac{\partial^2 \ln \mathcal{L}}{\partial \theta^2} = \frac{-n}{\theta^2} + (\beta+1)\sum_{i=1}^{n} \frac{t_i^2}{(1+\theta t_i)^2},$$

$$\frac{\partial^2 \ln \mathcal{L}}{\partial \beta^2} = \frac{-n}{\beta^2},$$

and

$$\frac{\partial^2 \ln \mathcal{L}}{\partial \theta \partial \beta} = \frac{\partial^2 \ln \mathcal{L}}{\partial \beta \partial \theta} = -\sum_{i=1}^{n} \frac{t_i}{(1+\theta t_i)}.$$

Now, using integration-by-parts with $u = t, du = dt, dv = \theta(1+\theta t)^{-(\beta+2)}\,dt$, and $v = -(1+\theta t)^{-(\beta+1)}/(\beta+1)$, we have

$$\mathrm{E}\left[\frac{T}{(1+\theta T)}\right] = \int_0^{\infty} \frac{t}{(1+\theta t)} \theta\beta(1+\theta t)^{-(\beta+1)} dt = \frac{1}{\theta(\beta+1)}.$$

And, using integration-by-parts with $u = t^2, du = 2t\,dt, dv = \theta(1+\theta t)^{-(\beta+3)}\,dt$, and $v = -(1+\theta t)^{-(\beta+2)}/(\beta+2)$, we have

$$\mathrm{E}\left[\frac{T^2}{(1+\theta T)^2}\right] = \int_0^{\infty} \frac{t^2}{(1+\theta t)^2} \theta\beta(1+\theta t)^{-(\beta+1)}\,dt = \frac{2}{\theta^2(\beta+1)(\beta+2)}.$$

Thus, it follows directly that the expected information matrix $\mathcal{I}$ is equal to

$$\mathcal{I} = \begin{bmatrix} -\mathrm{E}\left(\frac{\partial^2 \ln \mathcal{L}}{\partial \theta^2}\right) & -\mathrm{E}\left(\frac{\partial^2 \ln \mathcal{L}}{\partial \theta \partial \beta}\right) \\ -\mathrm{E}\left(\frac{\partial^2 \ln \mathcal{L}}{\partial \theta \partial \beta}\right) & -\mathrm{E}\left(\frac{\partial^2 \ln \mathcal{L}}{\partial \beta^2}\right) \end{bmatrix}$$

$$= \begin{bmatrix} \frac{\beta n}{(\beta+2)\theta^2} & \frac{n}{\theta(\beta+1)} \\ \frac{n}{\theta(\beta+1)} & \frac{n}{\beta^2} \end{bmatrix},$$

and hence that

$$\mathcal{I}^{-1} = \begin{bmatrix} \frac{\theta^2(\beta+1)^2(\beta+2)}{\beta n} & \frac{-\theta\beta(\beta+1)(\beta+2)}{n} \\ \frac{-\theta\beta(\beta+1)(\beta+2)}{n} & \frac{\beta^2(\beta+1)^2}{n} \end{bmatrix}.$$

Now, with

$$\boldsymbol{\delta}' = \left[\frac{\partial \gamma}{\partial \theta}, \frac{\partial \gamma}{\partial \beta}\right]$$

$$= \left[\frac{-1}{\theta^2(\beta-1)}, \frac{-1}{\theta(\beta-1)^2}\right],$$

use of the delta method gives, for large $n$,

$$\begin{aligned}
V(\hat{\gamma}) &\approx \boldsymbol{\delta}'\mathcal{I}^{-1}\boldsymbol{\delta} \\
&= \frac{(\beta+1)}{n\theta^2(\beta-1)^2}\left[\frac{(\beta+1)(\beta+2)}{\beta} - \frac{2\beta(\beta+2)}{(\beta-1)} + \frac{\beta^2(\beta+1)}{(\beta-1)^2}\right].
\end{aligned}$$

Then, with $n = 300, \hat{\theta} = 0.32$, and $\hat{\beta} = 1.50$, we obtain $\hat{\gamma} = [0.32(1.50-1)]^{-1} = 6.25$ and $\hat{V}(\hat{\gamma}) = 2.387$, so that the 95% large-sample CI for $\gamma$ is $6.25 \pm 1.96\sqrt{2.387} = 6.25 \pm 3.03$, or (3.22,9.28).

**Solution 4.36***

(a) For $i = 1, 2, 3$, let

$$\mathbf{1}_{n_i} = (1, 1, \ldots, 1)'$$

denote the $(n_i \times 1)$ column vector of ones, and let

$$\mathbf{0}_{n_2} = (0, 0, \ldots, 0)'$$

denote the $(n_2 \times 1)$ column vector of zeros.

Then, the $(n \times 3)$ design matrix $\mathbf{X}$ can be written as

$$\mathbf{X} = \begin{bmatrix} \mathbf{1}_{n_1} & -\mathbf{1}_{n_1} & \mathbf{1}_{n_1} \\ \mathbf{1}_{n_2} & \mathbf{0}_{n_2} & \mathbf{0}_{n_2} \\ \mathbf{1}_{n_3} & \mathbf{1}_{n_3} & \mathbf{1}_{n_3} \end{bmatrix},$$

so that

$$\begin{aligned}
\mathbf{X}'\mathbf{X} &= \begin{bmatrix} n & (n_3 - n_1) & (n_1 + n_3) \\ (n_3 - n_1) & (n_1 + n_3) & (n_3 - n_1) \\ (n_1 + n_3) & (n_3 - n_1) & (n_1 + n_3) \end{bmatrix} \\
&= n \begin{bmatrix} 1 & b & a \\ b & a & b \\ a & b & a \end{bmatrix},
\end{aligned}$$

where $a = (\pi_1 + \pi_3)$ and $b = (\pi_3 - \pi_1)$.

(b) From standard unweighted least-squares theory, we know that $V(\hat{\beta}_2) = c_{22}\sigma^2$, where $c_{22}$ is the last diagonal entry in the matrix $(\mathbf{X}'\mathbf{X})^{-1} = ((c_{ll'})), l = 0, 1, 2$ and $l' = 0, 1, 2$. We can use the theory of cofactors to find an explicit expression for $c_{22}$. In particular, the cofactor needed for determining $c_{22}$ is equal to

$$(-1)^{(2+2)}n^2 \begin{vmatrix} 1 & b \\ b & a \end{vmatrix} = n^2(a - b^2),$$

so that $c_{22} = n^2(a - b^2)/|\mathbf{X}'\mathbf{X}|$. And,

$$\begin{aligned}
|\mathbf{X}'\mathbf{X}| &= n^3(a^2 + ab^2 + ab^2 - a^3 - b^2 - ab^2) = n^3(1-a)(a^2 - b^2) \\
&= n^3[1 - (\pi_1 + \pi_3)][(\pi_1 + \pi_3)^2 - (\pi_3 - \pi_1)^2] = 4n^3\pi_1\pi_2\pi_3.
\end{aligned}$$

Finally,

$$V(\hat{\beta}_2) = c_{22}\sigma^2 = \frac{n^2(a-b^2)\sigma^2}{|\mathbf{X}'\mathbf{X}|}$$
$$= \left\{\frac{[(\pi_1+\pi_3)-(\pi_3-\pi_1)^2]}{4n\pi_1\pi_2\pi_3}\right\}\sigma^2.$$

(c) We wish to find values for $\pi_1, \pi_2$, and $\pi_3$ that minimize $V(\hat{\beta}_2)$ subject to the constraint $(\pi_1 + \pi_2 + \pi_3) = 1$. Instead of considering $V(\hat{\beta}_2)$, we can equivalently consider the quantity

$$Q = \frac{[(\pi_1+\pi_3)-(\pi_3-\pi_1)^2]}{\pi_1\pi_2\pi_3},$$

which can be rewritten as

$$Q = \frac{(\pi_1+\pi_3)-[(\pi_1+\pi_3)^2-4\pi_1\pi_3]}{\pi_1\pi_2\pi_3}$$
$$= \frac{(\pi_1+\pi_3)[1-(\pi_1+\pi_3)]}{\pi_1\pi_2\pi_3} + \frac{4\pi_1\pi_3}{\pi_1\pi_2\pi_3}$$
$$= \frac{(\pi_1+\pi_3)}{\pi_1\pi_3} + \frac{4}{(1-\pi_1-\pi_3)},$$

since $\pi_2 = (1 - \pi_1 - \pi_3)$.

Now,

$$\frac{\partial Q}{\partial \pi_1} = 0 \quad \text{gives } (1-\pi_1-\pi_3)^2 = 4\pi_1^2$$

and

$$\frac{\partial Q}{\partial \pi_3} = 0 \quad \text{gives } (1-\pi_1-\pi_3)^2 = 4\pi_3^2.$$

Since $\pi_1, \pi_2$, and $\pi_3$ are positive, these two equations imply that $\pi_1 = \pi_3$. Then, from the equation $(1 - \pi_1 - \pi_3)^2 = 4\pi_1^2$, we obtain $(1 - 2\pi_1)^2 = 4\pi_1^2$, or $\pi_1 = \frac{1}{4}$. Thus, the values for $\pi_1, \pi_2, \text{and} \pi_3$ that minimize $V(\hat{\beta}_2)$ subject to the constraint $\sum_{i=1}^{3} \pi_i = 1$ are

$$\pi_1 = \frac{1}{4}, \quad \pi_2 = \frac{1}{2} \quad \text{and} \quad \pi_3 = \frac{1}{4}.$$

Note that the Lagrange Multiplier method can also be used to obtain this answer.

**Solution 4.37***

(a) For the CIs $\bar{X}_i \pm kS_i/\sqrt{n}$ and $\bar{X}_{i'} \pm kS_{i'}/\sqrt{n}$ *not* to have at least one value in common (i.e., *not* to overlap), it is required that either

$$\left(\bar{X}_i + k\frac{S_i}{\sqrt{n}}\right) < \left(\bar{X}_{i'} - k\frac{S_{i'}}{\sqrt{n}}\right), \quad \text{which implies } (\bar{X}_i - \bar{X}_{i'}) < -k\left(\frac{S_i}{\sqrt{n}} + \frac{S_{i'}}{\sqrt{n}}\right)$$

or

$$\left(\bar{X}_i - k\frac{S_i}{\sqrt{n}}\right) > \left(\bar{X}_{i'} + k\frac{S_{i'}}{\sqrt{n}}\right), \quad \text{which implies } (\bar{X}_i - \bar{X}_{i'}) > k\left(\frac{S_i}{\sqrt{n}} + \frac{S_{i'}}{\sqrt{n}}\right).$$

Thus, these two inequalities together can be written succinctly as the event $E_{ii'} = \{|\bar{X}_i - \bar{X}_{i'}| > k(S_i/\sqrt{n} + S_{i'}/\sqrt{n})\}$, which gives the desired result.

(b) First, note that

$$\left(\frac{S_i}{\sqrt{n}} + \frac{S_{i'}}{\sqrt{n}}\right)^2 = \left(\frac{S_i}{\sqrt{n}}\right)^2 + \left(\frac{S_{i'}}{\sqrt{n}}\right)^2 + 2\left(\frac{S_i}{\sqrt{n}}\right)\left(\frac{S_{i'}}{\sqrt{n}}\right) \geq \left(\frac{S_i}{\sqrt{n}}\right)^2 + \left(\frac{S_{i'}}{\sqrt{n}}\right)^2,$$

so that $(S_i/\sqrt{n} + S_{i'}/\sqrt{n}) \geq \sqrt{(S_i/\sqrt{n})^2 + (S_{i'}/\sqrt{n})^2}$. So, using the result from part (a), we have

$$\begin{aligned}
\pi^*_{ii'} &= \text{pr}\left[|\bar{X}_i - \bar{X}_{i'}| > k\left(\frac{S_i}{\sqrt{n}} + \frac{S_{i'}}{\sqrt{n}}\right)\Big|C_p\right] \\
&\leq \text{pr}\left[|\bar{X}_i - \bar{X}_{i'}| > k\sqrt{\left(\frac{S_i}{\sqrt{n}}\right)^2 + \left(\frac{S_{i'}}{\sqrt{n}}\right)^2}\Big|C_p\right] \\
&= \text{pr}\left[\frac{|\bar{X}_i - \bar{X}_{i'}|}{\sqrt{\left(\frac{S_i}{\sqrt{n}}\right)^2 + \left(\frac{S_{i'}}{\sqrt{n}}\right)^2}} > k\Big|C_p\right].
\end{aligned}$$

Now,

$$Z = \frac{(\bar{X}_i - \bar{X}_{i'})}{\sqrt{2\sigma^2/n}} \sim N(0, 1),$$

$$U = \frac{(n-1)(S_i^2 + S_{i'}^2)}{\sigma^2} \sim \chi^2_{2(n-1)},$$

and $Z$ and $U$ are independent random variables.

So, since

$$\frac{Z}{\sqrt{U/2(n-1)}} = \frac{(\bar{X}_i - \bar{X}_{i'})}{\sqrt{\left(\frac{S_i}{\sqrt{n}}\right)^2 + \left(\frac{S_{i'}}{\sqrt{n}}\right)^2}} \sim t_{2(n-1)},$$

and since

$$\text{pr}[T_{2(n-1)} > k] \leq \frac{\alpha}{2},$$

it follows that

$$\pi^*_{ii'} \leq \text{pr}\left[|T_{2(n-1)}| > k|C_p\right] \leq \alpha.$$

(c) For $p = 3$ and given condition $C_3$, let $\theta_3$ be the conditional probability that there are no values common to all three CIs, or equivalently, that at least two of the three CIs have no values in common. Hence,

$$\begin{aligned}\theta_3 &= \text{pr}(E_{12} \cup E_{13} \cup E_{23} | C_3) \\ &\leq \text{pr}(E_{12}|C_3) + \text{pr}(E_{13}|C_3) + \text{pr}(E_{23}|C_3) = 3\pi^*_{ii'}.\end{aligned}$$

Finally, from part (b), since $\pi^*_{ii'} \leq \alpha$, we obtain $\theta_3 \leq 3\alpha$.

Note that $\theta_3$ is the probability of incorrectly deciding statistically that the three population means are not equal to the same value when, in fact, they are equal to the same value; that is, $\theta_3$ is analogous to a Type I error rate when testing the null hypothesis $H_0$ that all three population means are equal to the same value versus the alternative hypothesis $H_1$ that they are all not equal to the same value. Since $3\alpha > \alpha$, this CI-based algorithm can lead to an inflated Type I error rate. For example, when $\alpha = 0.05$, then this Type I error rate could theoretically be as high as 0.15. Given the stated assumptions, one-way analysis of variance would be an appropriate method for testing $H_0$ versus $H_1$.

**Solution 4.38***

(a) We wish to choose $\tilde{\theta}_0$ and $\tilde{\theta}_1$ to minimize

$$Q = \sum_{i=1}^{n} [Y_i - (\theta_0 + \theta_1 x_i)]^2.$$

Now, the equation

$$\frac{\partial Q}{\partial \theta_0} = -2\sum_{i=1}^{n} [Y_i - (\theta_0 + \theta_1 x_i)] = 0$$

implies that

$$\tilde{\theta}_0 = \bar{Y} - \tilde{\theta}_1 \bar{x},$$

where

$$\bar{Y} = \frac{1}{n}\sum_{i=1}^{n} Y_i \quad \text{and} \quad \bar{x} = \frac{1}{n}\sum_{i=1}^{n} x_i.$$

And,

$$\frac{\partial Q}{\partial \theta_1} = -2\sum_{i=1}^{n} x_i [Y_i - (\theta_0 + \theta_1 x_i)] = 0$$

implies that

$$\sum_{i=1}^{n} x_i Y_i = \tilde{\theta}_0 \sum_{i=1}^{n} x_i + \tilde{\theta}_1 \sum_{i=1}^{n} x_i^2 = (\bar{Y} - \tilde{\theta}_1 \bar{x}) \sum_{i=1}^{n} x_i + \tilde{\theta}_1 \sum_{i=1}^{n} x_i^2.$$

Hence,

$$\widetilde{\theta}_1 = \frac{\sum_{i=1}^n x_i Y_i - \bar{Y}\sum_{i=1}^n x_i}{\sum_{i=1}^n x_i^2 - \bar{x}\sum_{i=1}^n x_i}$$
$$= \frac{\sum_{i=1}^n (x_i - \bar{x})(Y_i - \bar{Y})}{\sum_{i=1}^n (x_i - \bar{x})^2}$$
$$= \frac{\sum_{i=1}^n (x_i - \bar{x})Y_i}{\sum_{i=1}^n (x_i - \bar{x})^2}.$$

Now,

$$\mathrm{E}(\widetilde{\theta}_1) = \frac{\sum_{i=1}^n (x_i - \bar{x})(\theta_0 + \theta_1 x_i)}{\sum_{i=1}^n (x_i - \bar{x})^2}$$
$$= \theta_0 \frac{\sum_{i=1}^n (x_i - \bar{x})}{\sum_{i=1}^n (x_i - \bar{x})^2} + \theta_1 \frac{\sum_{i=1}^n (x_i - \bar{x})x_i}{\sum_{i=1}^n (x_i - \bar{x})^2}$$
$$= \theta_1.$$

And,

$$\mathrm{V}(\widetilde{\theta}_1) = \frac{\sum_{i=1}^n (x_i - \bar{x})^2(\theta_0 + \theta_1 x_i)}{\left[\sum_{i=1}^n (x_i - \bar{x})^2\right]^2}$$
$$= \frac{\theta_0}{\sum_{i=1}^n (x_i - \bar{x})^2} + \frac{\theta_1 \sum_{i=1}^n x_i (x_i - \bar{x})^2}{\left[\sum_{i=1}^n (x_i - \bar{x})^2\right]^2}.$$

Also,

$$\mathrm{E}(\widetilde{\theta}_0) = \mathrm{E}(\bar{Y}) - \bar{x}\mathrm{E}(\widetilde{\theta}_1)$$
$$= \frac{1}{n}\sum_{i=1}^n (\theta_0 + \theta_1 x_i) - \bar{x}\theta_1$$
$$= \theta_0 + \theta_1 \bar{x} - \bar{x}\theta_1 = \theta_0.$$

And, since

$$\widetilde{\theta}_0 = \frac{1}{n}\sum_{i=1}^n Y_i - \frac{\bar{x}\sum_{i=1}^n (x_i - \bar{x})Y_i}{\sum_{i=1}^n (x_i - \bar{x})^2} = \sum_{i=1}^n c_i Y_i,$$

where

$$c_i = \left[\frac{1}{n} - \frac{\bar{x}(x_i - \bar{x})}{\sum_{i=1}^n (x_i - \bar{x})^2}\right],$$

and where the $\{Y_i\}$ are mutually independent,

$$V(\widetilde{\theta}_0) = \sum_{i=1}^{n} c_i^2 V(Y_i) = \sum_{i=1}^{n} c_i^2(\theta_0 + \theta_1 x_i)$$
$$= \theta_0 \sum_{i=1}^{n} c_i^2 + \theta_1 \sum_{i=1}^{n} c_i^2 x_i,$$

where $c_i$ is defined as above.

(b) The likelihood function $\mathcal{L}$ has the structure

$$\mathcal{L} = \prod_{i=1}^{n} \left\{ (\theta_0 + \theta_1 x_i)^{y_i} e^{-(\theta_0+\theta_1 x_i)} / y_i! \right\}$$
$$= \left\{ \prod_{i=1}^{n} (\theta_0 + \theta_1 x_i)^{y_i} \right\} \left\{ e^{-\sum_{i=1}^{n}(\theta_0+\theta_1 x_i)} \right\} \left\{ \prod_{i=1}^{n} (y_i!)^{-1} \right\};$$

$$ln\mathcal{L} = \sum_{i=1}^{n} y_i \ln(\theta_0 + \theta_1 x_i) - \sum_{i=1}^{n} (\theta_0 + \theta_1 x_i) + \sum_{i=1}^{n} \ln\left[(y_i!)^{-1}\right];$$

$$\frac{\partial \ln \mathcal{L}}{\partial \theta_0} = \sum_{i=1}^{n} \frac{y_i}{(\theta_0 + \theta_1 x_i)} - n;$$

$$\frac{\partial^2 \ln \mathcal{L}}{\partial \theta_0^2} = -\sum_{i=1}^{n} \frac{y_i}{(\theta_0 + \theta_1 x_i)^2};$$

$$-E\left[\frac{\partial^2 \ln \mathcal{L}}{\partial \theta_0^2}\right] = \sum_{i=1}^{n} \frac{(\theta_0 + \theta_1 x_i)}{(\theta_0 + \theta_1 x_i)^2} = \sum_{i=1}^{n} (\theta_0 + \theta_1 x_i)^{-1} = A;$$

$$\frac{\partial^2 \ln \mathcal{L}}{\partial \theta_0 \partial \theta_1} = -\sum_{i=1}^{n} \frac{x_i y_i}{(\theta_0 + \theta_1 x_i)^2};$$

$$-E\left[\frac{\partial^2 \ln \mathcal{L}}{\partial \theta_0 \partial \theta_1}\right] = \sum_{i=1}^{n} x_i (\theta_0 + \theta_1 x_i)^{-1} = B;$$

$$\frac{\partial \ln \mathcal{L}}{\partial \theta_1} = \sum_{i=1}^{n} \frac{x_i y_i}{(\theta_0 + \theta_1 x_i)} - \sum_{i=1}^{n} x_i;$$

$$\frac{\partial^2 \ln \mathcal{L}}{\partial \theta_1^2} = -\sum_{i=1}^{n} \frac{x_i^2 y_i}{(\theta_0 + \theta_1 x_i)^2};$$

$$-E\left[\frac{\partial^2 \ln \mathcal{L}}{\partial \theta_1^2}\right] = \sum_{i=1}^{n} x_i^2 (\theta_0 + \theta_1 x_i)^{-1} = C.$$

So, the expected information matrix is

$$\mathcal{I}(\theta_0, \theta_1) = \begin{bmatrix} A & B \\ B & C \end{bmatrix}.$$

For the available data, we compute $\mathcal{I}(\hat{\theta}_0, \hat{\theta}_1)$ using $\hat{A}$, $\hat{B}$, and $\hat{C}$ as follows:

$$\hat{A} = 25\left\{[2+4(1)]^{-1} + [2+4(2)]^{-1} + [2+4(3)]^{-1} + [2+4(4)]^{-1}\right\} = 9.8425,$$

$$\hat{B} = 25\left[\frac{1}{6} + \frac{2}{10} + \frac{3}{14} + \frac{4}{18}\right] = 20.0800,$$

$$\hat{C} = 25\left[\frac{1}{6} + \frac{4}{10} + \frac{9}{14} + \frac{16}{18}\right] = 52.4625.$$

So,

$$\mathcal{I}(\hat{\theta}_0, \hat{\theta}_1) = \begin{bmatrix} 9.8425 & 20.0800 \\ 20.0800 & 52.4625 \end{bmatrix} = \begin{bmatrix} \hat{A} & \hat{B} \\ \hat{B} & \hat{C} \end{bmatrix},$$

and hence

$$\begin{aligned}
\mathcal{I}^{-1}(\hat{\theta}_0, \hat{\theta}_1) &= (\hat{A}\hat{C} - \hat{B}^2)^{-1}\begin{bmatrix} \hat{C} & -\hat{B} \\ -\hat{B} & \hat{A} \end{bmatrix} \\
&= [(9.8425)(52.4625) - (20.0800)^2]^{-1}\begin{bmatrix} 52.4625 & -20.0800 \\ -20.0800 & 9.8425 \end{bmatrix} \\
&= \begin{bmatrix} 0.4636 & -0.1775 \\ -0.1775 & 0.0870 \end{bmatrix}.
\end{aligned}$$

Now,

$$\hat{\psi} = \hat{\theta}_0 + (2.5)\hat{\theta}_1,$$

with

$$\hat{\mathrm{V}}(\hat{\psi}) = \hat{\mathrm{V}}(\hat{\theta}_0) + (2.5)^2\hat{\mathrm{V}}(\hat{\theta}_1) + 2(1)(2.5)\widehat{\mathrm{cov}}(\hat{\theta}_0, \hat{\theta}_1).$$

Since

$$\frac{\hat{\psi} - \psi}{\sqrt{\hat{\mathrm{V}}(\hat{\psi})}} \mathrel{\dot\sim} \mathrm{N}(0, 1)$$

for large $n$, our 95% CI for $\psi$ is

$$\begin{aligned}
\hat{\psi} \pm 1.96\sqrt{\hat{\mathrm{V}}(\hat{\psi})} &= [2 + (2.5)(4)] \pm 1.96\,[0.4636 + 6.250(0.0870) + 5(-0.1775)]^{1/2} \\
&= 12 \pm 1.96\sqrt{0.1199} \\
&= 12 \pm 0.6787 \\
&= (11.3213, 12.6787).
\end{aligned}$$

**Solution 4.39***

(a)

$$D_i = (X_i - \bar{X}) = X_i - \frac{1}{n}\sum_{i=1}^{n} X_i = \left(1 - \frac{1}{n}\right) X_i - \frac{1}{n}\sum_{j \neq i} X_j.$$

Since $X_i \sim \mathrm{N}(\mu, \sigma^2) \forall i$ and the $\{X_i\}$ are mutually independent, then $D_i$ is itself normal since $D_i$ is a linear combination of mutually independent normal variates. Also,

$$\mathrm{E}(D_i) = \mathrm{E}(X_i - \bar{X}) = \mathrm{E}(X_i) - \mathrm{E}(\bar{X}) = \mu - \mu = 0,$$

and

$$\begin{aligned}
\mathrm{V}(D_i) &= \left(1 - \frac{1}{n}\right)^2 \sigma^2 + \frac{(n-1)}{n^2}\sigma^2 \\
&= \left[\frac{(n-1)^2}{n^2} + \frac{(n-1)}{n^2}\right]\sigma^2 = \frac{(n-1)}{n^2}[(n-1)+1]\sigma^2 \\
&= \left(\frac{n-1}{n}\right)\sigma^2.
\end{aligned}$$

So,

$$D_i \sim \mathrm{N}\left[0, \left(\frac{n-1}{n}\right)\sigma^2\right].$$

(b) Now,

$$\begin{aligned}
\mathrm{cov}(D_i, D_j) &= \mathrm{E}(D_i D_j) \\
&= \mathrm{E}[(X_i - \bar{X})(X_j - \bar{X})] \\
&= \mathrm{E}(X_i X_j) - \mathrm{E}(X_i \bar{X}) - \mathrm{E}(X_j \bar{X}) + \mathrm{E}(\bar{X}^2) \\
&= \mu^2 - \mathrm{E}(X_i \bar{X}) - \mathrm{E}(X_j \bar{X}) + \left(\frac{\sigma^2}{n} + \mu^2\right).
\end{aligned}$$

Now,

$$\begin{aligned}
\mathrm{E}(X_i \bar{X}) &= \mathrm{E}\left[X_i \left(\frac{1}{n}\sum_{i=1}^{n} X_i\right)\right] \\
&= \frac{1}{n}\mathrm{E}[X_1 X_i + X_2 X_i + \cdots + X_{i-1}X_i + X_i^2 + X_{i+1}X_i + \ldots + X_n X_i] \\
&= \frac{1}{n}[(n-1)\mu^2 + (\mu^2 + \sigma^2)] \\
&= \frac{1}{n}(n\mu^2 + \sigma^2) \\
&= \mu^2 + \frac{\sigma^2}{n}.
\end{aligned}$$

An identical argument shows that $E(X_j\bar{X}) = \mu^2 + \sigma^2/n$. So

$$\text{cov}(D_i, D_j) = \mu^2 - \left(\mu^2 + \frac{\sigma^2}{n}\right) - \left(\mu^2 + \frac{\sigma^2}{n}\right) + \left(\frac{\sigma^2}{n} + \mu^2\right) = \frac{-\sigma^2}{n}.$$

Finally,

$$\begin{aligned} \text{corr}(D_i, D_j) &= \frac{\text{cov}(D_i, D_j)}{\sqrt{V(D_i) \cdot V(D_j)}} \\ &= \frac{-\sigma^2/n}{\sqrt{\left(\frac{n-1}{n}\right)\sigma^2 \cdot \left(\frac{n-1}{n}\right)\sigma^2}} \\ &= \frac{-1}{(n-1)}. \end{aligned}$$

Clearly,

$$\lim_{n\to\infty} [\text{corr}(D_i, D_j)] = 0.$$

Since $V(\bar{X}) \to 0$ as $n \to +\infty$, $\text{corr}(D_i, D_j) \to \text{corr}(X_i, X_j) = 0$ as $n \to +\infty$.

(c)

$$R = \frac{S_x^2}{S_y^2} = \frac{(n-1)S_x^2/\sigma^2}{(n-1)S_y^2/\sigma^2} = \frac{U_x}{U_y},$$

where $U_x \sim \chi^2_{n-1}$, $U_y \sim \chi^2_{n-1}$, and $U_x$ and $U_y$ are independent.
So,

$$f_{U_x,U_y}(u_x, u_y) = \frac{u_x^{[(n-1)/2]-1} e^{-u_x/2}}{\Gamma[(n-1)/2] \cdot 2^{[(n-1)/2]}} \cdot \frac{u_y^{[(n-1)/2]-1} e^{-u_y/2}}{\Gamma[(n-1)/2] \cdot 2^{[(n-1)/2]}},$$

$$u_x > 0, \;\; u_y > 0.$$

Let $R = U_x/U_y$ and $S = U_y$, so that $U_x = RS$ and $U_y = S$; and, $R > 0, S > 0$. Also,

$$J = \begin{vmatrix} \frac{\partial U_x}{\partial R} & \frac{\partial U_x}{\partial S} \\ \frac{\partial U_y}{\partial R} & \frac{\partial U_y}{\partial S} \end{vmatrix} = \begin{vmatrix} S & R \\ 0 & 1 \end{vmatrix} = S.$$

So,

$$\begin{aligned} f_{R,S}(r, s) &= f_{U_x,U_y}(rs, s) \times |J| \\ &= \frac{(rs)^{[(n-3)/2]} e^{-rs/2}}{\Gamma[(n-1)/2] \cdot 2^{[(n-1)/2]}} \cdot \frac{s^{[(n-3)/2]} e^{-s/2}}{\Gamma[(n-1)/2] \cdot 2^{[(n-1)/2]}} \cdot s \\ &= \frac{r^{[(n-3)/2]} s^{(n-2)} e^{-(1+r)s/2}}{[\Gamma[(n-1)/2]]^2 \cdot 2^{(n-1)}}, \quad r > 0, \;\; s > 0. \end{aligned}$$

So,

$$\begin{aligned}
f_R(r) &= \frac{r^{[(n-3)/2]}}{[\Gamma[(n-1)/2]]^2 \cdot 2^{(n-1)}} \int_0^{\infty} s^{(n-1)-1} e^{-s/[\frac{2}{(1+r)}]}\,ds \\
&= \frac{r^{[(n-3)/2]}}{[\Gamma[(n-1)/2]]^2 \cdot 2^{(n-1)}} \cdot \Gamma(n-1)\left[\frac{2}{(1+r)}\right]^{(n-1)} \\
&= [\Gamma(n-1)]\left[\Gamma\left(\frac{n-1}{2}\right)\right]^{-2} r^{[(n-3)/2]}(1+r)^{-(n-1)}, \quad r > 0.
\end{aligned}$$

(d) Clearly,

$$\mathrm{E}(\hat{\mu}_1) = \frac{1}{2}[\mathrm{E}(\bar{X}) + \mathrm{E}(\bar{Y})] = \frac{(\mu + \mu)}{2} = \mu.$$

And,

$$\mathrm{E}(\hat{\mu}_2) = \mathrm{E}\left[\frac{\bar{X}S_y^2 + \bar{Y}S_x^2}{S_x^2 + S_y^2}\right] = \mathrm{E}\left[\frac{\bar{X} + R\bar{Y}}{(1+R)}\right],$$

where

$$R = S_x^2/S_y^2 \sim \mathrm{F}_{(n-1),(n-1)}.$$

Now, since $\bar{X}, \bar{Y}, S_x^2$, and $S_y^2$ are mutually independent random variables,

$$\begin{aligned}
\mathrm{E}(\hat{\mu}_2) = \mathrm{E}_r\{\mathrm{E}(\hat{\mu}_2|R=r)\} &= \mathrm{E}_r\left\{\frac{\mathrm{E}(\bar{X}|R=r) + r\mathrm{E}(\bar{Y}|R=r)}{(1+r)}\right\} \\
&= \mathrm{E}_r\left\{\frac{\mathrm{E}(\bar{X}) + r\mathrm{E}(\bar{Y})}{(1+r)}\right\} \\
&= \mathrm{E}_r\left\{\frac{\mu + r\mu}{(1+r)}\right\} \\
&= \mathrm{E}_r\left\{\frac{\mu(1+r)}{(1+r)}\right\} \\
&= \mathrm{E}_r(\mu) = \mu.
\end{aligned}$$

(e)

$$\begin{aligned}
V(\hat{\mu}_1) &= V\left[\frac{1}{2}(\bar{X}+\bar{Y})\right] = \frac{1}{4}[V(\bar{X}) + V(\bar{Y})] \\
&= \frac{1}{4}\left[\frac{\sigma^2}{n} + \frac{\sigma^2}{n}\right] = \frac{\sigma^2}{2n}. \\
V(\hat{\mu}_2) &= V_r\{E(\hat{\mu}_2|R=r)\} + E_r\{V(\hat{\mu}_2)|R=r\} \\
&= V_r(\mu) + E_r\left\{V\left[\frac{\bar{X}+R\bar{Y}}{(1+R)}\middle| R=r\right]\right\} \\
&= 0 + E_r\left\{\frac{V(\bar{X}|R=r) + r^2 V(\bar{Y}|R=r)}{(1+r)^2}\right\} \\
&= E_r\left\{\frac{V(\bar{X}) + r^2 V(\bar{Y})}{(1+r)^2}\right\} \\
&= E_r\left\{\frac{(\sigma^2/n) + r^2(\sigma^2/n)}{(1+r)^2}\right\} \\
&= \frac{\sigma^2}{n} E_r\left[\frac{(1+r^2)}{(1+r)^2}\right].
\end{aligned}$$

So, to find $V(\hat{\mu}_2)$, we need to find $E\left[(1+R^2)/(1+R)^2\right]$, where $f_R(r)$ is given in part (c). So,

$$E\left[\frac{(1+R^2)}{(1+R)^2}\right] = E\left[1 - 2\frac{R}{(1+R)^2}\right] = 1 - 2E\left[\frac{R}{(1+R)^2}\right].$$

Now, with $u = r/(1+r)$,

$$\begin{aligned}
E\left[\frac{R}{(1+R)^2}\right] &= \int_0^{\infty} \frac{r}{(1+r)^2} \frac{\Gamma(n-1)}{[\Gamma[(n-1)/2]]^2} r^{[(n-3)/2]}(1+r)^{-(n-1)}\, dr \\
&= \frac{\Gamma(n-1)}{[\Gamma[(n-1)/2]]^2} \int_0^{\infty} \frac{r^{[(n-1)/2]}(1+r)^{-(n-1)}}{(1+r)^2}\, dr \\
&= \frac{\Gamma(n-1)}{[\Gamma(n-1/2)]^2} \int_0^1 u^{[(n+1)/2]-1}(1-u)^{[(n+1)/2]-1}\, du \\
&= \frac{\Gamma(n-1)}{[\Gamma[(n-1)/2]]^2} \cdot \frac{[\Gamma[(n+1)/2]]^2}{\Gamma(n+1)} = \frac{[(n-1)/2]^2}{n(n-1)} \\
&= \frac{(n-1)}{4n}.
\end{aligned}$$

Finally,

$$\begin{aligned}
V(\hat{\mu}_2) &= \frac{\sigma^2}{n}\left\{1 - 2E\left[\frac{R}{(1+R)^2}\right]\right\} \\
&= \frac{\sigma^2}{n}\left\{1 - 2\left[\frac{(n-1)}{4n}\right]\right\} \\
&= \frac{\sigma^2}{n}\left[1 - \frac{(n-1)}{2n}\right] \\
&= \frac{\sigma^2}{n}\left(\frac{2n-n+1}{2n}\right) \\
&= \left(\frac{n+1}{2n^2}\right)\sigma^2.
\end{aligned}$$

Since $V(\hat{\mu}_1) = (1/2n)\,\sigma^2 < V(\hat{\mu}_2) = \left[(n+1)/2n^2\right]\sigma^2$ when $n > 1$, we prefer $\hat{\mu}_1$ (a result which actually follows from the theory of sufficiency).

**Solution 4.40***

(a) Now,

$$E(\hat{\mu}_1) = E(k_1\bar{Y}) = k_1\mu,$$

so that $k_1 = 1$. Then,

$$V(\hat{\mu}_1) = V(\bar{Y}) = \frac{(\theta^2/12)}{n} = \frac{\theta^2}{12n}.$$

(b) Since

$$f_{Y_{(n)}}(y_{(n)};\theta) = n\left[\frac{y_{(n)}}{\theta}\right]^{n-1}\theta^{-1} = n\theta^{-n}y_{(n)}^{n-1}, \quad 0 < y_{(n)} < \theta,$$

we have, for $r \geq 0$,

$$E[Y_{(n)}^r] = n\theta^{-n}\int_0^\theta y_{(n)}^{(n+r)-1}\,dy_{(n)} = \left(\frac{n}{n+r}\right)\theta^r.$$

So,

$$E[Y_{(n)}] = \left(\frac{n}{n+1}\right)\theta.$$

Thus,

$$\begin{aligned}
E(\hat{\mu}_2) &= E[k_2Y_{(n)}] \\
&= k_2\left(\frac{n}{n+1}\right)\theta \\
&= 2k_2\left(\frac{n}{n+1}\right)\mu,
\end{aligned}$$

so that $k_2 = (n+1)/2n$. Since

$$\begin{aligned} V[Y_{(n)}] &= E[Y_{(n)}^2] - \{E[Y_{(n)}]\}^2 \\ &= \left(\frac{n}{n+2}\right)\theta^2 - \left(\frac{n}{n+1}\right)^2\theta^2 \\ &= \frac{n\theta^2}{(n+1)^2(n+2)}, \end{aligned}$$

it follows that

$$\begin{aligned} V(\hat{\mu}_2) &= V(k_2 Y_{(n)}) \\ &= k_2^2 V(Y_{(n)}) \\ &= \frac{(n+1)^2}{4n^2} \cdot \frac{n\theta^2}{(n+1)^2(n+2)} \\ &= \frac{\theta^2}{4n(n+2)}. \end{aligned}$$

(c) Since

$$f_{Y_{(1)},Y_{(n)}}(y_{(1)}, y_{(n)}; \theta) = n(n-1)\theta^{-n}(y_{(n)} - y_{(1)})^{n-2}, \quad 0 < y_{(1)} < y_{(n)} < \theta,$$

we have, for $r \geq 0$ and $s \geq 0$,

$$\begin{aligned} E\left[Y_{(1)}^r Y_{(n)}^s\right] &= n(n-1)\theta^{-n} \int_0^\theta \int_0^{y_{(n)}} y_{(1)}^r y_{(n)}^s (y_{(n)} - y_{(1)})^{n-2}\, dy_{(1)}\, dy_{(n)} \\ &= n(n-1)\theta^{-n} \int_0^\theta y_{(n)}^s \left[\int_0^{y_{(n)}} y_{(1)}^r (y_{(n)} - y_{(1)})^{n-2}\, dy_{(1)}\right] dy_{(n)} \\ &= n(n-1)\theta^{-n} \int_0^\theta y_{(n)}^s \left[\int_0^1 (y_{(n)}u)^r (y_{(n)} - y_{(n)}u)^{n-2} y_{(n)}\, du\right] dy_{(n)} \\ &= n(n-1)\theta^{-n} \int_0^\theta y_{(n)}^{(n+r+s)-1} \left[\int_0^1 u^{(r+1)-1}(1-u)^{(n-1)-1}\, du\right] dy_{(n)} \\ &= \frac{n(n-1)}{(n+r+s)} \cdot \frac{\Gamma(r+1)\Gamma(n-1)}{\Gamma(n+r)} \theta^{(r+s)}. \end{aligned}$$

Since

$$E[Y_{(1)}] = \frac{n(n-1)}{(n+1+0)} \cdot \frac{\Gamma(1+1)\Gamma(n-1)}{\Gamma(n+1)} \theta^{(1+0)} = \frac{\theta}{(n+1)},$$

it follows that

$$\begin{aligned} E(\hat{\mu}_3) &= \frac{k_3}{2}[E(Y_{(1)}) + E(Y_{(n)})] \\ &= \frac{k_3}{2}\left[\frac{\theta}{(n+1)} + \frac{n\theta}{(n+1)}\right] \\ &= k_3\left(\frac{\theta}{2}\right) = k_3\mu, \end{aligned}$$

so that $k_3 = 1$. Since $V[Y_{(1)}] = V[Y_{(n)}]$, by symmetry, and since

$$\begin{aligned} \text{cov}\,[Y_{(1)}, Y_{(n)}] &= \frac{n(n-1)}{(n+1+1)} \cdot \frac{\Gamma(1+1)\Gamma(n-1)}{\Gamma(n+1)}\theta^{(1+1)} \\ &\quad - \left[\frac{\theta}{(n+1)}\right]\left[\frac{n\theta}{(n+1)}\right] \\ &= \frac{\theta^2}{(n+2)} - \frac{n\theta^2}{(n+1)^2} \\ &= \frac{\theta^2}{(n+1)^2(n+2)}, \end{aligned}$$

it follows that

$$\begin{aligned} V(\hat{\mu}_3) &= V\left(\frac{1}{2}\left[Y_{(1)} + Y_{(n)}\right]\right) \\ &= \frac{1}{4}[V(Y_{(1)}) + V(Y_{(n)}) + 2\text{cov}(Y_{(1)}, Y_{(n)})] \\ &= \frac{1}{2}\left[\frac{n\theta^2}{(n+1)^2(n+2)} + \frac{\theta^2}{(n+1)^2(n+2)}\right] \\ &= \frac{\theta^2}{2(n+1)(n+2)}. \end{aligned}$$

(d) We have shown that

$$\begin{aligned} V(\hat{\mu}_1) &= \frac{\theta^2}{12n}, \\ V(\hat{\mu}_2) &= \frac{\theta^2}{4n(n+2)}, \\ V(\hat{\mu}_3) &= \frac{\theta^2}{2(n+1)(n+2)}. \end{aligned}$$

Now,

$$4n(n+2) - 12n = 4n^2 - 4n = 4n(n-1) > 0 \quad \text{for } n > 1,$$
$$4n(n+2) - 2(n+1)(n+2) = 2(n+2)(n-1) > 0 \quad \text{for } n > 1,$$

and

$$2(n+1)(n+2) - 12n = 2(n-1)(n-2) > 0 \quad \text{for } n > 2.$$

So, for $n > 2$, we have

$$V(\hat{\mu}_2) < V(\hat{\mu}_3) < V(\hat{\mu}_1).$$

Now,

$$\lim_{n\to\infty} \frac{V(\hat{\mu}_2)}{V(\hat{\mu}_1)} = \lim_{n\to\infty} \frac{V(\hat{\mu}_3)}{V(\hat{\mu}_1)} = 0,$$

so that $\hat{\mu}_1$ has an asymptotic efficiency of 0 relative to $\hat{\mu}_2$ and $\hat{\mu}_3$.

Since

$$\lim_{n\to\infty} \frac{V(\hat{\mu}_2)}{V(\hat{\mu}_3)} = \lim_{n\to\infty} \frac{2(n+1)(n+2)}{4n(n+2)} = \frac{1}{2},$$

this implies that $\hat{\mu}_3$ is asymptotically 50% as efficient as $\hat{\mu}_2$. That $\hat{\mu}_2$ is the estimator of choice should not be surprising since $Y_{(n)}$ is a (complete) sufficient statistic for $\theta$ (and hence for $\mu$), and so $\hat{\mu}_2$ is the minimum variance unbiased estimator (MVUE) of $\mu$.

**Solution 4.41***

(a) The likelihood function $\mathcal{L}$ has the structure

$$\mathcal{L} = \prod_{i=1}^{n} \left\{ \left[ \theta^{x_i}(1-\theta)^{1-x_i} \right] \left[ \frac{1}{\mu(x_i)} \right] e^{-y_i/\mu(x_i)} \right\}$$
$$= \theta^{\sum_{i=1}^{n} x_i}(1-\theta)^{n-\sum_{i=1}^{n} x_i} e^{-\sum_{i=1}^{n}(\alpha+\beta x_i)} e^{-\sum_{i=1}^{n} e^{-(\alpha+\beta x_i)} y_i}.$$

Since $\sum_{i=1}^{n} x_i = n_1$, we have

$$\ln \mathcal{L} = n_1 \ln \theta + n_0 \ln(1-\theta) - n\alpha - \beta n_1 - \sum_{i=1}^{n} e^{-(\alpha+\beta x_i)} y_i.$$

So,

$$\frac{\partial \ln \mathcal{L}}{\partial \theta} = \frac{n_1}{\theta} - \frac{n_0}{(1-\theta)} = 0$$
$$\Rightarrow n_1(1-\theta) - n_0\theta = 0$$
$$\Rightarrow \hat{\theta} = n_1/(n_0+n_1) = n_1/n = \bar{x}.$$

And,

$$\frac{\partial \ln \mathcal{L}}{\partial \alpha} = -n + \sum_{i=1}^{n} e^{-(\alpha+\beta x_i)} y_i = 0$$
$$\Rightarrow -n + e^{-\alpha}\left[ e^{-\beta} n_1 \bar{y}_1 + n_0 \bar{y}_0 \right] = 0. \tag{4.10}$$

Also,

$$\frac{\partial \ln \mathcal{L}}{\partial \beta} = -n_1 + \sum_{i=1}^{n} e^{-(\alpha+\beta x_i)} x_i y_i = 0$$

$$\Rightarrow -n_1 + e^{-\alpha} e^{-\beta} n_1 \bar{y}_1 = 0 \Rightarrow e^{-(\alpha+\beta)} \bar{y}_1 = 1. \tag{4.11}$$

So, using (Equation 4.11) in (Equation 4.10), we obtain

$$-n + n_1 + e^{-\alpha} n_0 \bar{y}_0 = 0 \Rightarrow e^{-\alpha} = \frac{n - n_1}{n_0 \bar{y}_0} = \frac{n_0}{n_0 \bar{y}_0} = \frac{1}{\bar{y}_0}$$

so that

$$-\hat{\alpha} = \ln\left(1/\bar{y}_0\right), \text{ or } \hat{\alpha} = \ln(\bar{y}_0).$$

Finally, since $\hat{\alpha} = \ln(\bar{y}_0)$, it follows from (Equation 4.11) that $e^{-(\hat{\alpha}+\hat{\beta})}\bar{y}_1 = 1$, or $e^{\hat{\beta}} = \bar{y}_1/e^{\hat{\alpha}}$, or $\hat{\beta} = \ln\left(\bar{y}_1/\bar{y}_0\right)$.

So, in summary,

$$\hat{\theta} = \bar{x},\ \hat{\alpha} = \ln(\bar{y}_0), \quad \text{and} \quad \hat{\beta} = \ln\left(\frac{\bar{y}_1}{\bar{y}_0}\right).$$

(b) Now,

$$\frac{\partial^2 \ln \mathcal{L}}{\partial \theta^2} = -\frac{n_1}{\theta^2} - \frac{n_0}{(1-\theta)^2},$$

so that

$$-E\left(\frac{\partial^2 \ln \mathcal{L}}{\partial \theta^2}\right) = \frac{E(n_1)}{\theta^2} + \frac{E(n_0)}{(1-\theta)^2}$$

$$= \frac{n\theta}{\theta^2} + \frac{n(1-\theta)}{(1-\theta)^2} = \frac{n}{\theta} + \frac{n}{(1-\theta)} = \frac{n}{\theta(1-\theta)}.$$

Clearly,

$$\frac{\partial^2 \ln \mathcal{L}}{\partial \theta \partial \alpha} = \frac{\partial^2 \ln \mathcal{L}}{\partial \theta \partial \beta} = 0.$$

Now, with $\boldsymbol{X} = (X_1, X_2, \ldots, X_n)$ and $\boldsymbol{x} = (x_1, x_2, \ldots, x_n)$, we have

$$\frac{\partial^2 \ln \mathcal{L}}{\partial \alpha^2} = -e^{-\alpha} \sum_{i=1}^{n} e^{-\beta x_i} y_i,$$

so that

$$-\mathrm{E}\left(\frac{\partial^2 \ln \mathcal{L}}{\partial \alpha^2}\right) = -\mathrm{E}_x\left\{\mathrm{E}\left[\frac{\partial^2 \ln \mathcal{L}}{\partial \alpha^2}\middle|\mathbf{X}=\boldsymbol{x}\right]\right\}$$

$$= \mathrm{e}^{-\alpha}\mathrm{E}_x\left[\sum_{i=1}^{n} \mathrm{e}^{-\beta x_i}\mathrm{e}^{\alpha+\beta x_i}\right] = n.$$

And,

$$\frac{\partial^2 \ln \mathcal{L}}{\partial \beta^2} = -\mathrm{e}^{-\alpha}\sum_{i=1}^{n} \mathrm{e}^{-\beta x_i} x_i^2 y_i,$$

so that

$$-\mathrm{E}\left(\frac{\partial^2 \ln \mathcal{L}}{\partial \beta^2}\right) = -\mathrm{E}_x\left\{\mathrm{E}\left[\frac{\partial^2 \ln \mathcal{L}}{\partial \beta^2}\middle|\mathbf{X}=\boldsymbol{x}\right]\right\}$$

$$= \mathrm{e}^{-\alpha}\mathrm{E}_x\left[\sum_{i=1}^{n} \mathrm{e}^{-\beta x_i} x_i^2 \mathrm{e}^{\alpha+\beta x_i}\right]$$

$$= \sum_{i=1}^{n} \mathrm{E}_x\left(x_i^2\right) = \sum_{i=1}^{n}\left[\theta(1-\theta)+\theta^2\right] = n\theta.$$

Finally,

$$\frac{\partial^2 \ln \mathcal{L}}{\partial \alpha \partial \beta} = -\mathrm{e}^{-\alpha}\sum_{i=1}^{n} \mathrm{e}^{-(\alpha+\beta x_i)} x_i y_i,$$

so that

$$-\mathrm{E}\left(\frac{\partial^2 \ln \mathcal{L}}{\partial \alpha \partial \beta}\right) = -\mathrm{E}_x\left\{\mathrm{E}\left[\left(\frac{\partial^2 \ln \mathcal{L}}{\partial \alpha \partial \beta}\right)\middle|\mathbf{X}=\boldsymbol{x}\right]\right\}$$

$$= \mathrm{e}^{-\alpha}\mathrm{E}_x\left[\sum_{i=1}^{n} \mathrm{e}^{-\beta x_i} x_i \mathrm{e}^{\alpha+\beta x_i}\right] = \sum_{i=1}^{n} \mathrm{E}_x(x_i) = n\theta.$$

So,

$$\mathcal{I} = \begin{bmatrix} n & n\theta & 0 \\ n\theta & n\theta & 0 \\ 0 & 0 & \dfrac{n}{\theta(1-\theta)} \end{bmatrix},$$

and

$$\mathcal{I}^{-1} = \begin{bmatrix} \frac{1}{n(1-\theta)} & \frac{-1}{n(1-\theta)} & 0 \\ \frac{-1}{n(1-\theta)} & \frac{1}{n\theta(1-\theta)} & 0 \\ 0 & 0 & \frac{\theta(1-\theta)}{n} \end{bmatrix}.$$

So,

$$\mathrm{V}(\hat{\alpha}) \dot{\approx} \frac{1}{n(1-\theta)}, \quad \mathrm{V}(\hat{\beta}) \dot{\approx} \frac{1}{n\theta(1-\theta)}, \quad \mathrm{cov}(\hat{\alpha}, \hat{\beta}) \dot{\approx} \frac{-1}{n(1-\theta)},$$

$$\mathrm{cov}(\hat{\alpha}, \hat{\theta}) = \mathrm{cov}(\hat{\beta}, \hat{\theta}) = 0, \quad \text{and} \quad \mathrm{V}(\hat{\theta}) = \frac{\theta(1-\theta)}{n}.$$

(c) The parameter of interest is $\beta$, and statistical evidence that $\beta \neq 0$ suggests that the true mean time to death differs between adult males with advanced malignant melanoma depending on whether or not these adult males have a family history of skin cancer. An appropriate large-sample 95% CI for $\beta$ is (for large $n$):

$$\hat{\beta} \pm 1.96\sqrt{\frac{1}{n\hat{\theta}(1-\hat{\theta})}}.$$

With $n = 50$, $\hat{\theta} = 0.60$ and $\hat{\beta} = 0.40$, we obtain

$$0.40 \pm 1.96\sqrt{\frac{1}{50(0.60)(1-0.60)}} \quad \text{or} \quad (-0.1658, 0.9658).$$

So, these data provide no evidence that $\beta \neq 0$, since 0 is contained in the computed CI.

**Solution 4.42***

(a) Note that

$$\begin{aligned} \hat{\theta}_x = B_0 + B_1 x &= \frac{1}{N}\sum_{i=1}^{N} Y_{x_i} + x\sum_{i=1}^{N}\left(\frac{x_i}{\sum_{j=1}^{N} x_j^2}\right) Y_{x_i} \\ &= \sum_{i=1}^{N}\left[\frac{1}{N} + \frac{x x_i}{\sum_{j=1}^{N} x_j^2}\right] Y_{x_i} = \sum_{i=1}^{N} c_i Y_{x_i}, \end{aligned}$$

where

$$c_i = \left[\frac{1}{N} + \frac{x x_i}{\sum_{j=1}^{N} x_j^2}\right], \quad i = 1, 2, \ldots, N.$$

So,

$$E(\hat{\theta}_x) = \sum_{i=1}^{N} c_i E(Y_{x_i}) = \sum_{i=1}^{N} c_i(\beta_0 + \beta_1 x_i + \beta_2 x_i^2)$$

$$= \beta_0 \sum_{i=1}^{N} c_i + \beta_1 \sum_{i=1}^{N} c_i x_i + \beta_2 \sum_{i=1}^{N} c_i x_i^2.$$

Now,

$$\sum_{i=1}^{N} c_i = \sum_{i=1}^{N} \left[ \frac{1}{N} + \frac{x x_i}{\sum_{j=1}^{N} x_j^2} \right] = 1 + \left( \frac{x}{\sum_{j=1}^{N} x_j^2} \right) \sum_{i=1}^{N} x_i = 1,$$

since $\mu_1 = 0$. Also,

$$\sum_{i=1}^{N} c_i x_i = \sum_{i=1}^{N} \left[ \frac{1}{N} + \frac{x x_i}{\sum_{j=1}^{N} x_j^2} \right] x_i = x,$$

since $\mu_1 = 0$. Finally,

$$\sum_{i=1}^{N} c_i x_i^2 = \left[ \frac{1}{N} + \frac{x x_i}{\sum_{j=1}^{N} x_j^2} \right] x_i^2 = \mu_2,$$

since $\mu_3 = 0$. So, $E(\hat{\theta}_x) = \beta_0 + \beta_1 x + \beta_2 \mu_2$, where $\mu_2 = N^{-1} \sum_{i=1}^{N} x_i^2$.
And, finally,

$$V(\hat{\theta}_x) = \sum_{i=1}^{N} c_i^2 V(Y_{x_i}) = \sigma^2 \sum_{i=1}^{N} \left[ \frac{1}{N} + \frac{x x_i}{\sum_{j=1}^{N} x_j^2} \right]^2$$

$$= \sigma^2 \sum_{i=1}^{N} \left[ \frac{1}{N^2} + \frac{2 x x_i}{N \sum_{j=1}^{N} x_j^2} + \frac{x^2 x_i^2}{\left(\sum_{j=1}^{N} x_j^2\right)^2} \right]$$

$$= \frac{\sigma^2}{N} + 0 + \sigma^2 x^2 \left[ \sum_{i=1}^{N} x_i^2 \Big/ \left( \sum_{i=1}^{N} x_i^2 \right)^2 \right] = \frac{\sigma^2}{N} \left[ 1 + \frac{x^2}{\mu_2} \right].$$

Finally, $\hat{\theta}_x$ is normally distributed since it is a linear combination of mutually independent and normally distributed random variables.

(b) Since $E(\hat{\theta}_x) = \beta_0 + \beta_1 x + \beta_2 \mu_2$, it follows that $E(\hat{\theta}_x) - \theta_x = \beta_2(\mu_2 - x^2)$. So,

$$Q = \int_{-1}^{1} \left[ \beta_2(\mu_2 - x^2) \right]^2 dx = \beta_2^2 \int_{-1}^{1} (\mu_2^2 - 2\mu_2 x^2 + x^4)\, dx$$

$$= 2\beta_2^2 \left[ \left( \mu_2 - \frac{1}{3} \right)^2 + \frac{4}{45} \right],$$

which is minimized when $\mu_2 = \frac{1}{3}$. So, an optimal design for minimizing the integrated squared bias $Q$ chooses the temperature spacings $x_1, x_2, \ldots, x_N$ such that $\mu_2 = (1/N)\sum_{i=1}^{N} x_i^2 = \frac{1}{3}$, given that $\mu_1 = \mu_3 = 0$. Note that $\mu_1 = \mu_3 = 0$ will be satisfied if the $x_i$ are chosen to be symmetric about zero. For example, when $N = 4$, we can choose $x_1 = -x_4$ and $x_2 = -x_3$. Then, to satisfy $\mu_2 = 2(x_1^2 + x_2^2) = 1/3$ we can choose $x_1$ to be any number in the interval $(0, \sqrt{\frac{1}{6}})$ and then choose $x_2 = \sqrt{\frac{1}{6} - x_1^2}$. For example, $x_1 = \sqrt{\frac{2}{18}}$, $x_2 = \sqrt{\frac{1}{18}}$, $x_3 = -\sqrt{\frac{1}{18}}$, $x_4 = -\sqrt{\frac{2}{18}}$.

**Solution 4.43***

(a)

$$\begin{aligned}\mathrm{cov}(Y_{i0}, Y_{i1}) &= \mathrm{E}(Y_{i0}Y_{i1}) - \mathrm{E}(Y_{i0})\mathrm{E}(Y_{i1}) \\ &= \mathrm{E}_{\alpha_i}[\mathrm{E}(Y_{i0}Y_{i1}|\alpha_i)] - \mathrm{E}_{\alpha_i}[\mathrm{E}(Y_{i0}|\alpha_i)]\mathrm{E}_{\alpha_i}[\mathrm{E}(Y_{i1}|\alpha_i)].\end{aligned}$$

Now,

$$\mathrm{E}(Y_{ij}|\alpha_i) = L_{ij}\mathrm{e}^{(\alpha_i + \beta D_{ij} + \sum_{l=1}^{p} \gamma_l C_{il})}.$$

So, since $\alpha_i \sim \mathrm{N}(0, \sigma_\alpha^2)$, it follows from moment generating function theory that

$$\mathrm{E}(\mathrm{e}^{t\alpha_i}) = \mathrm{e}^{t^2\sigma_\alpha^2/2}, \quad -\infty < t < +\infty.$$

Thus,

$$\mathrm{E}(Y_{ij}) = L_{ij}\mathrm{e}^{(0.50\sigma_\alpha^2 + \beta D_{ij} + \sum_{l=1}^{p} \gamma_l C_{il})}.$$

And, using the assumption that $Y_{i0}$ and $Y_{i1}$ are independent given $\alpha_i$ fixed, we have

$$\begin{aligned}\mathrm{E}(Y_{i0}Y_{i1}) &= \mathrm{E}_{\alpha_i}[\mathrm{E}(Y_{i0}Y_{i1}|\alpha_i)] = \mathrm{E}_{\alpha_i}[\mathrm{E}(Y_{i0}|\alpha_i)\mathrm{E}(Y_{i1}|\alpha_i)] \\ &= \mathrm{E}_{\alpha_i}\left[(L_{i0}\mathrm{e}^{(\alpha_i + \sum_{l=1}^{p} \gamma_l C_{il})})(L_{i1}\mathrm{e}^{(\alpha_i + \beta + \sum_{l=1}^{p} \gamma_l C_{il})})\right] \\ &= L_{i0}L_{i1}\mathrm{e}^{(2\sigma_\alpha^2 + \beta + 2\sum_{l=1}^{p} \gamma_l C_{il})}.\end{aligned}$$

Thus,

$$\mathrm{cov}(Y_{i0}, Y_{i1}) = L_{i0}L_{i1}\mathrm{e}^{(\sigma_\alpha^2 + \beta + 2\sum_{l=1}^{p} \gamma_l C_{il})}\left(\mathrm{e}^{\sigma_\alpha^2} - 1\right).$$

The inclusion of the random effect $\alpha_i$ in the proposed statistical model serves two purposes: (1) to allow for families to have different (baseline) tendencies toward child abuse, and (2) to account for the anticipated positive correlation between $Y_{i0}$ and $Y_{i1}$ for the $i$th family [in particular, note that $\mathrm{cov}(Y_{i0}, Y_{i1}) = 0$ only when $\sigma_\alpha^2 = 0$ and is positive when $\sigma_\alpha^2 > 0$].

(b) Now, $p_{Y_{i1}}(y_{i1}|Y_i = y_i, \alpha_i) = \text{pr}(Y_{i1} = y_{i1}|Y_i = y_i, \alpha_i)$

$$= \frac{\text{pr}[(Y_{i1} = y_{i1}) \cap (Y_i = y_i)|\alpha_i]}{\text{pr}(Y_i = y_i|\alpha_i)}$$

$$= \frac{\text{pr}[(Y_{i1} = y_{i1}) \cap (Y_{i0} = y_i - y_{i1})|\alpha_i]}{\text{pr}(Y_i = y_i|\alpha_i)}$$

$$= \frac{\text{pr}(Y_{i1} = y_{i1}|\alpha_i)\text{pr}[Y_{i0} = (y_i - y_{i1})|\alpha_i]}{\text{pr}(Y_i = y_i|\alpha_i)}$$

$$= \frac{\left[\frac{(L_{i1}\lambda_{i1})^{y_{i1}} e^{-L_{i1}\lambda_{i1}}}{y_{i1}!}\right]\left[\frac{(L_{i0}\lambda_{i0})^{(y_i - y_{i1})} e^{-L_{i0}\lambda_{i0}}}{(y_i - y_{i1})!}\right]}{\left[\frac{(L_{i0}\lambda_{i0} + L_{i1}\lambda_{i1})^{y_i} e^{-(L_{i0}\lambda_{i0} + L_{i1}\lambda_{i1})}}{y_i!}\right]}$$

$$= C_{y_{i1}}^{y_i}\left(\frac{L_{i1}\theta}{L_{i0} + L_{i1}\theta}\right)^{y_{i1}}\left(\frac{L_{i0}}{L_{i0} + L_{i1}\theta}\right)^{(y_i - y_{i1})}, \quad y_{i1} = 0, 1, \ldots, y_i,$$

where

$$\theta = \frac{\lambda_{i1}}{\lambda_{i0}} = \frac{e^{(\alpha_i + \beta + \sum_{l=1}^{p} \gamma_l C_{il})}}{e^{(\alpha_i + \sum_{l=1}^{p} \gamma_l C_{il})}} = e^{\beta}.$$

(c) Since

$$\mathcal{L} = \prod_{i=1}^{n} C_{y_{i1}}^{y_i}\left(\frac{L_{i1}\theta}{L_{i0} + L_{i1}\theta}\right)^{y_{i1}}\left(\frac{L_{i0}}{L_{i0} + L_{i1}\theta}\right)^{(y_i - y_{i1})},$$

it follows that

$$\ln(\mathcal{L}) \propto \sum_{i=1}^{n}[y_{i1}\ln(\theta) - y_i \ln(L_{i0} + L_{i1}\theta)].$$

Thus,

$$\frac{\partial \ln(\mathcal{L})}{\partial \theta} = \sum_{i=1}^{n}\left[\frac{y_{i1}}{\theta} - \frac{y_i L_{i1}}{(L_{i0} + L_{i1}\theta)}\right] = 0,$$

so that the MLE $\hat{\theta}$ of $\theta$ satisfies the equation

$$\hat{\theta}\sum_{i=1}^{n}\left(\frac{y_i L_{i1}}{L_{i0} + L_{i1}\hat{\theta}}\right) = \sum_{i=1}^{n} y_{i1}.$$

(d) Since

$$\frac{\partial^2 \ln(\mathcal{L})}{\partial \theta^2} = \sum_{i=1}^{n}\left[\frac{-y_{i1}}{\theta^2} + \frac{y_i L_{i1}^2}{(L_{i0} + L_{i1}\theta)^2}\right],$$

it follows that

$$-\mathrm{E}\left(\frac{\partial^2 \ln(\mathcal{L})}{\partial\theta^2}|\{y_i\},\{\alpha_i\}\right) = \sum_{i=1}^{n}\left[\frac{y_i\left(\frac{L_{i1}\theta}{L_{i0}+L_{i1}\theta}\right)}{\theta^2} - \frac{y_i L_{i1}^2}{(L_{i0}+L_{i1}\theta)^2}\right]$$

$$= \frac{1}{\theta}\sum_{i=1}^{n}\frac{y_i L_{i0}L_{i1}}{(L_{i0}+L_{i1}\theta)^2}.$$

Thus, a large-sample 95% CI for the rate ratio parameter $\theta$ is

$$\hat{\theta} \pm 1.96\sqrt{\hat{\theta}}\left[\sum_{i=1}^{n}\frac{y_i L_{i0}L_{i1}}{(L_{i0}+L_{i1}\hat{\theta})^2}\right]^{-1/2}.$$

For an important application of this methodology, see Gibbs et al. (2007).

**Solution 4.44*.** With $\boldsymbol{X}^* = (X_1^*, X_2^*, \ldots, X_n^*)$ and $\boldsymbol{x}^* = (x_1^*, x_2^*, \ldots, x_n^*)$, we have

$$\mathrm{E}(\hat{\beta}_1^*|\boldsymbol{X}^* = \boldsymbol{x}^*) = \frac{\sum_{i=1}^{n}(x_i^* - \bar{x}^*)\mathrm{E}(Y_i|\boldsymbol{X}^* = \boldsymbol{x}^*)}{\sum_{i=1}^{n}(x_i^* - \bar{x}^*)^2}.$$

Now, using the nondifferential measurement error assumption, we have, for $i = 1, 2, \ldots, n$,

$$\begin{aligned}\mathrm{E}(Y_i|\boldsymbol{X}^* = \boldsymbol{x}^*) &= \mathrm{E}(Y_i|X_i^* = x_i^*) = \mathrm{E}_{X_i|X_i^*=x_i^*}\left[\mathrm{E}(Y_i|X_i^* = x_i^*, X_i = x_i)\right]\\ &= \mathrm{E}_{X_i|X_i^*=x_i^*}\left[\mathrm{E}(Y_i|X_i = x_i)\right] = \mathrm{E}_{X_i|X_i^*=x_i^*}(\beta_0 + \beta_1 x_i)\\ &= \beta_0 + \beta_1\mathrm{E}(X_i|X_i^* = x_i^*).\end{aligned}$$

And, since $(X_i, X_i^* = X_i + U_i) \sim \mathrm{BVN}[\mu_x, \mu_x, \sigma_x^2, (\sigma_x^2 + \sigma_u^2), \rho]$, where

$$\begin{aligned}\rho &= \frac{\mathrm{cov}(X_i, X_i^*)}{\sqrt{\mathrm{V}(X_i)\mathrm{V}(X_i^*)}} = \frac{\mathrm{cov}(X_i, X_i + U_i)}{\sqrt{\sigma_x^2(\sigma_x^2 + \sigma_u^2)}}\\ &= \frac{\mathrm{V}(X_i)}{\sqrt{\sigma_x^2(\sigma_x^2 + \sigma_u^2)}} = \frac{\sigma_x^2}{\sqrt{\sigma_x^2(\sigma_x^2 + \sigma_u^2)}}\\ &= \frac{1}{\sqrt{1 + \frac{\sigma_u^2}{\sigma_x^2}}} = \frac{1}{\sqrt{1+\lambda}},\end{aligned}$$

it follows from bivariate normal distribution theory that

$$\begin{aligned}
E(X_i|X_i^* = x_i^*) &= \mu_x + \rho\sqrt{\frac{V(X_i)}{V(X_i^*)}}(x_i^* - \mu_x) \\
&= \mu_x + \left(\frac{1}{\sqrt{1+\lambda}}\right)\sqrt{\frac{\sigma_x^2}{(\sigma_x^2 + \sigma_u^2)}}(x_i^* - \mu_x) \\
&= \mu_x + \left(\frac{1}{1+\lambda}\right)(x_i^* - \mu_x).
\end{aligned}$$

Hence, we have

$$E(Y_i|\mathbf{X^*} = \mathbf{x}^*) = \beta_0 + \beta_1\left[\mu_x + \left(\frac{1}{1+\lambda}\right)(x_i^* - \mu_x)\right] = \beta_0^* + \beta_1^* x_i^*,$$

where

$$\beta_0^* = \beta_0 + \beta_1\mu_x\left(\frac{\lambda}{1+\lambda}\right) \quad \text{and} \quad \beta_1^* = \frac{\beta_1}{(1+\lambda)}.$$

Finally, we obtain

$$\begin{aligned}
E(\hat{\beta}_1^*|\mathbf{X^*} = \mathbf{x}^*) &= \frac{\sum_{i=1}^n (x_i^* - \bar{x}^*)(\beta_0^* + \beta_1^* x_i^*)}{\sum_{i=1}^n (x_i^* - \bar{x}^*)^2} \\
&= \beta_1^* = \frac{\beta_1}{(1+\lambda)}.
\end{aligned}$$

So, since $0 < \lambda < \infty$, $|\beta_1^*| < |\beta_1|$, indicating a somewhat common detrimental measurement error effect called *attenuation*. Because the predictor variable is measured with error, the estimator $\hat{\beta}_1^*$ tends, on average, to *underestimate* the true slope $\beta_1$ (i.e., the estimator $\hat{\beta}_1^*$ is said to be attenuated). As $\lambda = \sigma_u^2/\sigma_x^2$ increases in value, the amount of attenuation increases.

For more complicated measurement error scenarios, attenuation should not always be the anticipated measurement error effect; in particular, an estimator could actually have a tendency to *overestimate* a particular parameter of interest [e.g., see Kupper (1984)].

**Solution 4.45***

(a) First,

$$\text{cov}(X, X^*|C) = \text{cov}(\alpha_0 + \alpha_1 X^* + \boldsymbol{\delta}'C + U, X^*|C) = \alpha_1 V(X^*|C).$$

And,

$$V(X|C) = \alpha_1^2 V(X^*|C) + \sigma_u^2.$$

So,

$$\text{corr}(X, X^*|C) = \frac{\text{cov}(X, X^*|C)}{\sqrt{V(X|C)V(X^*|C)}}$$

$$= \frac{\alpha_1 V(X^*|C)}{\sqrt{[\alpha_1^2 V(X^*|C) + \sigma_u^2]V(X^*|C)}}$$

$$= \frac{1}{\sqrt{1 + \sigma_u^2/[\alpha_1^2 V(X^*|C)]}} < 1.$$

(b) With $X = \alpha_0 + \alpha_1 X^* + \boldsymbol{\delta}' C + U$ and given $X^*$ and $C$, it follows directly that $X$ has a *normal* distribution with $E(X|X^*, C) = \alpha_0 + \alpha_1 X^* + \boldsymbol{\delta}' C$ and $V(X|X^*, C) = V(U) = \sigma_u^2$.

(c) Now, appealing to the nondifferential error assumption, we have

$$\begin{aligned}
\text{pr}(Y = 1|X^*, C) &= E(Y|X^*, C) = E_{X|X^*,C}\left[E(Y|X, X^*, C)\right] \\
&= E_{X|X^*,C}\left[E(Y|X, C)\right] = E_{X|X^*,C}\left[\text{pr}(Y = 1|X, C)\right] \\
&= E_{X|X^*,C}\left[e^{(\beta_0 + \beta_1 X + \boldsymbol{\gamma}' C)}\right] = e^{(\beta_0 + \boldsymbol{\gamma}' C)} E_{X|X^*,C}\left(e^{\beta_1 X}\right).
\end{aligned}$$

Thus, from moment generating theory and the results in part (b), we have

$$\begin{aligned}
\text{pr}(Y = 1|X^*, C) &= e^{(\beta_0 + \boldsymbol{\gamma}' C)} e^{[\beta_1(\alpha_0 + \alpha_1 X^* + \boldsymbol{\delta}' C)] + (\beta_1^2 \sigma_u^2)/2} \\
&= e^{(\theta_0 + \theta_1 X^* + \boldsymbol{\xi}' C)},
\end{aligned}$$

where $\theta_0 = (\beta_0 + \beta_1 \alpha_0 + (\beta_1^2 \sigma_u^2)/2), \theta_1 = \beta_1 \alpha_1$, and $\boldsymbol{\xi}' = (\boldsymbol{\gamma}' + \beta_1 \boldsymbol{\delta}')$.

Since $\theta_1/\beta_1 = \alpha_1$, it follows that $0 < \theta_1 \leq \beta_1$ when $0 < \alpha_1 \leq 1$ and that $\theta_1 > \beta_1$ when $\alpha_1 > 1$. So, the use of $X^*$ instead of $X$ will result in biased estimation of the parameter $\beta_1$. In particular, if $0 < \alpha_1 < 1$, the tendency will be to underestimate $\beta_1$; and, if $\alpha_1 > 1$, the tendency will be to overestimate $\beta_1$.

**Solution 4.46***

(a) First, $E(X) = \delta$ and $V(X) = \delta(1 - \delta)$. So,

$$\begin{aligned}
E(X^*) &= E[(X^*)^2] = E\left[E(X^*|X = x)\right] = E\left[\text{pr}(X^* = 1|X = x)\right] \\
&= E(\pi_{x1}) = \pi_{11}\delta + \pi_{01}(1 - \delta),
\end{aligned}$$

and

$$V(X^*) = E(X^*)\left[1 - E(X^*)\right].$$

Also,

$$\begin{aligned}
E(XX^*) &= E\left[E(XX^*|X = x)\right] = E\left[(x)\text{pr}(X^* = 1|X = x)\right] \\
&= E\left[(x)\pi_{x1}\right] = (1)\pi_{11}\delta = \pi_{11}\delta.
\end{aligned}$$

Thus,

$$\text{corr}(X, X^*) = \frac{\text{cov}(X, X^*)}{\sqrt{V(X)V(X^*)}}$$
$$= \frac{E(XX^*) - E(X)E(X^*)}{\sqrt{V(X)V(X^*)}}$$
$$= \frac{\pi_{11}\delta - \delta E(X^*)}{\sqrt{\delta(1-\delta)E(X^*)\,[1 - E(X^*)]}}.$$

So, $\text{corr}(X, X^*) = 1$ when $\pi_{11} = 1$ and $\pi_{01} = 0$ (or, equivalently, when $\pi_{10} = 0$ and $\pi_{00} = 1$), since then $E(X^*) = \delta$. When $\pi_{11} = \text{pr}(X^* = 1|X = 1) < 1$ and/or $\pi_{01} = \text{pr}(X^* = 1|X = 0) > 0$, so that $\text{corr}(X, X^*) < 1$, then $X^*$ is an *imperfect* surrogate for $X$.

(b) Now, using the nondifferential error assumption given earlier, we have

$$\mu^*_{x^*} = \text{pr}(Y = 1|X^* = x^*) = \sum_{x=0}^{1} \text{pr}[(Y = 1) \cap (X = x)|X^* = x^*]$$
$$= \sum_{x=0}^{1} \text{pr}(Y = 1|(X = x) \cap (X^* = x^*)]\text{pr}(X = x|X^* = x^*)$$
$$= \sum_{x=0}^{1} \text{pr}(Y = 1|X = x)\text{pr}(X = x|X^* = x^*)$$
$$= \sum_{x=0}^{1} \mu_x \gamma_{x^*x},$$

where $\gamma_{x^*x} = \text{pr}(X = x|X^* = x^*)$.

So, since $(\gamma_{00} + \gamma_{01}) = (\gamma_{10} + \gamma_{11}) = 1$, we have

$$\theta^* = (\mu_1^* - \mu_0^*) = \sum_{x=0}^{1} \mu_x \gamma_{1x} - \sum_{x=0}^{1} \mu_x \gamma_{0x}$$
$$= (\mu_0\gamma_{10} + \mu_1\gamma_{11}) - (\mu_0\gamma_{00} + \mu_1\gamma_{01})$$
$$= \mu_0(1 - \gamma_{11}) + \mu_1\gamma_{11} - \mu_0(1 - \gamma_{01}) - \mu_1\gamma_{01}$$
$$= (\mu_1 - \mu_0)(\gamma_{11} - \gamma_{01})$$
$$= \theta(\gamma_{11} - \gamma_{01}),$$

so that $|\theta^*| \leq |\theta|$ since $|\gamma_{11} - \gamma_{01}| \leq 1$.

Hence, under the assumption of nondifferential error, the use of $X^*$ instead of $X$ tends, on average, to lead to underestimation of the risk difference parameter $\theta$, a phenomenon known as *attenuation*.

For more detailed discussion about the effects of misclassification error on the validity of analyses of epidemiologic data, see Gustafson (2004) and Kleinbaum et al. (1982).

# 5

# *Hypothesis Testing Theory*

## 5.1 Concepts and Notation

### 5.1.1 Basic Principles

#### *5.1.1.1 Simple and Composite Hypotheses*

A *statistical hypothesis* is an assertion about the distribution of one or more random variables. If the statistical hypothesis completely specifies the distribution (i.e., the hypothesis assigns numerical values to all unknown population parameters), then it is called a *simple* hypothesis; otherwise, it is called a *composite* hypothesis.

#### *5.1.1.2 Null and Alternative Hypotheses*

In the typical statistical hypothesis testing situation, there are two hypotheses of interest: the *null* hypothesis (denoted $H_0$) and the *alternative* hypothesis (denoted $H_1$). The statistical objective is to use the information in a sample from the distribution under study to make a decision about whether $H_0$ or $H_1$ is more likely to be true (i.e., is more likely to represent the true "state of nature").

#### *5.1.1.3 Statistical Tests*

A statistical test of $H_0$ versus $H_1$ consists of a rule which, when operationalized using the available information in a sample, leads to a decision either to *reject*, or *not to reject*, $H_0$ in favor of $H_1$. It is important to point out that a decision *not to reject* $H_0$ does *not* imply that $H_0$ is, in fact, true; in particular, the decision not to reject $H_0$ is often due to data inadequacies (e.g., too small a sample size, erroneous and/or missing information, etc.)

#### *5.1.1.4 Type I and Type II Errors*

For any statistical test, there are two possible decision errors that can be made. A "Type I" error occurs when the decision is made to reject $H_0$ in favor of

$H_1$ when, in fact, $H_0$ is true; the probability of a Type I error is denoted as $\alpha = \text{pr}(\text{test rejects } H_0|H_0 \text{ true})$. A "Type II" error occurs when the decision is made not to reject $H_0$ when, in fact, $H_0$ is false and $H_1$ is true; the probability of a Type II error is denoted as $\beta = \text{pr}(\text{test does not reject } H_0|H_0 \text{ false})$.

### 5.1.1.5 Power

The *power* of a statistical test is the probability of rejecting $H_0$ when, in fact, $H_0$ is false and $H_1$ is true; in particular,

$$\text{POWER} = \text{pr}(\text{test rejects } H_0|H_0 \text{ false}) = (1 - \beta).$$

Type I error rate $\alpha$ is controllable and is typically assigned a value satisfying the inequality $0 < \alpha \leq 0.10$. For a given value of $\alpha$, Type II error rate $\beta$, and hence the power $(1 - \beta)$, will generally vary as a function of the values of population parameters allowable under a composite alternative hypothesis $H_1$.

In general, for a specified value of $\alpha$, the power of any reasonable statistical testing procedure should increase as the sample size increases. Power is typically used as a very important criterion for choosing among several statistical testing procedures in any given situation.

### 5.1.1.6 Test Statistics and Rejection Regions

A statistical test of $H_0$ versus $H_1$ is typically carried out by using a *test statistic*. A test statistic is a random variable with the following properties: (i) its distribution, assuming the null hypothesis $H_0$ is true, is known either exactly or to a close approximation (i.e., for large sample sizes); (ii) its numerical value can be computed using the information in a sample; and, (iii) its computed numerical value leads to a decision either to reject, or not to reject, $H_0$ in favor of $H_1$. More specifically, for a given statistical test and associated test statistic, the set of all possible numerical values of the test statistic under $H_0$ is divided into two disjoint subsets (or "regions"), the *rejection region* $\mathcal{R}$ and the *nonrejection region* $\bar{\mathcal{R}}$. The statistical test decision rule is then defined as follows: if the computed numerical value of the test statistic is in the rejection region $\mathcal{R}$, then reject $H_0$ in favor of $H_1$; otherwise, do not reject $H_0$. The rejection region $\mathcal{R}$ is chosen so that, under $H_0$, the probability that the test statistic falls in the rejection region $\mathcal{R}$ is equal to (or approximately equal to) $\alpha$ (in which case the rejection region and the associated statistical test are both said to be of "size" $\alpha$).

Almost all popular statistical testing procedures use test statistics that, under $H_0$, follow (either exactly or approximately) well-tabulated distributions such as the standard normal distribution, the $t$-distribution, the chi-squared distribution, and the $f$-distribution.

#### 5.1.1.7 P-Values

The *P-value* for a statistical test is the probability of observing a test statistic value at least as rare as the value actually observed under the assumption that the null hypothesis $H_0$ is true. Thus, for a size $\alpha$ test, when the decision is made to reject $H_0$, then the P-value is less than $\alpha$; and, when the decision is made not to reject $H_0$, then the P-value is greater than $\alpha$.

### 5.1.2 Most Powerful (MP) and Uniformly Most Powerful (UMP) Tests

Let $\boldsymbol{X} = (X_1, X_2, \ldots, X_n)$ be a random row vector with likelihood function (or joint distribution) $\mathcal{L}(\boldsymbol{x}; \boldsymbol{\theta})$ depending on a row vector $\boldsymbol{\theta} = (\theta_1, \theta_2, \ldots, \theta_p)$ of $p$ unknown parameters. Let $\mathcal{R}$ denote some subset of all the possible realizations $\boldsymbol{x} = (x_1, x_2, \ldots, x_n)$ of the random vector $\boldsymbol{X}$. Then, $\mathcal{R}$ is the *most powerful* (or MP) rejection region of size $\alpha$ for testing the simple null hypothesis $H_0 : \boldsymbol{\theta} = \boldsymbol{\theta}_0$ versus the simple alternative hypothesis $H_1 : \boldsymbol{\theta} = \boldsymbol{\theta}_1$ if, for every subset $\mathcal{A}$ of all possible realizations $\boldsymbol{x}$ of $\boldsymbol{X}$ for which $\text{pr}(\boldsymbol{X} \in \mathcal{A} | H_0 : \boldsymbol{\theta} = \boldsymbol{\theta}_0) = \alpha$, we have

$$\text{pr}(\boldsymbol{X} \in \mathcal{R} | H_0 : \boldsymbol{\theta} = \boldsymbol{\theta}_0) = \alpha$$

and

$$\text{pr}(\boldsymbol{X} \in \mathcal{R} | H_1 : \boldsymbol{\theta} = \boldsymbol{\theta}_1) \geq \text{pr}(\boldsymbol{X} \in \mathcal{A} | H_1 : \boldsymbol{\theta} = \boldsymbol{\theta}_1).$$

Given $\mathcal{L}(\boldsymbol{x}; \boldsymbol{\theta})$, the determination of the structure of the MP rejection region $\mathcal{R}$ of size $\alpha$ for testing $H_0 : \boldsymbol{\theta} = \boldsymbol{\theta}_0$ versus $H_1 : \boldsymbol{\theta} = \boldsymbol{\theta}_1$ can be made using the *Neyman–Pearson Lemma* (Neyman and Pearson, 1933).

---

**Neyman–Pearson Lemma**

Let $\boldsymbol{X} = (X_1, X_2, \ldots, X_n)$ be a random row vector with likelihood function (or joint distribution) of known form $\mathcal{L}(\boldsymbol{x}; \boldsymbol{\theta})$ that depends on a row vector $\boldsymbol{\theta} = (\theta_1, \theta_2, \ldots, \theta_p)$ of $p$ unknown parameters. Let $\mathcal{R}$ be a subset of all possible realizations $\boldsymbol{x} = (x_1, x_2, \ldots, x_n)$ of $\boldsymbol{X}$. Then, $\mathcal{R}$ is the most powerful (MP) rejection region of size $\alpha$ (and the associated test using $\mathcal{R}$ is the most powerful test of size $\alpha$) for testing the simple null hypothesis $H_0 : \boldsymbol{\theta} = \boldsymbol{\theta}_0$ versus the simple alternative hypothesis $H_1 : \boldsymbol{\theta} = \boldsymbol{\theta}_1$ if, for some $k > 0$, the following three conditions are satisfied:

$$\frac{\mathcal{L}(\boldsymbol{x}; \boldsymbol{\theta}_0)}{\mathcal{L}(\boldsymbol{x}; \boldsymbol{\theta}_1)} < k \quad \text{for every } \boldsymbol{x} \in \mathcal{R};$$

$$\frac{\mathcal{L}(\boldsymbol{x}; \boldsymbol{\theta}_0)}{\mathcal{L}(\boldsymbol{x}; \boldsymbol{\theta}_1)} \geq k \quad \text{for every } \boldsymbol{x} \in \bar{\mathcal{R}};$$

and

$$\text{pr}(\mathbf{X} \in \mathcal{R} | \text{H}_0 : \boldsymbol{\theta} = \boldsymbol{\theta}_0) = \alpha.$$

A rejection region $\mathcal{R}$ is a *uniformly most powerful* (UMP) rejection rejection of size $\alpha$ (and the associated test using $\mathcal{R}$ is a *uniformly most powerful* test of size $\alpha$) for testing a simple null hypothesis $\text{H}_0$ versus a composite alternative hypothesis $\text{H}_1$ if the region $\mathcal{R}$ is a most powerful region of size $\alpha$ *for every* simple alternative hypothesis contained in $\text{H}_1$.

#### 5.1.2.1 Review of Notation

In the subsections to follow, we will utilize the following quantities, which were introduced in Section 4.1.

$\hat{\boldsymbol{\theta}} = (\hat{\theta}_1, \hat{\theta}_2, \ldots, \hat{\theta}_p)$, the MLE of $\boldsymbol{\theta} = (\theta_1, \theta_2, \ldots, \theta_p)$ based on the likelihood $\mathcal{L}(\boldsymbol{x}; \boldsymbol{\theta})$;

$\mathcal{I}(\hat{\boldsymbol{\theta}})$, the estimated expected information matrix based on the likelihood $\mathcal{L}(\boldsymbol{x}; \boldsymbol{\theta})$;

$\mathcal{I}^{-1}(\hat{\boldsymbol{\theta}})$, the estimated large-sample covariance matrix of $\hat{\boldsymbol{\theta}}$ based on expected information for the likelihood $\mathcal{L}(\boldsymbol{x}; \boldsymbol{\theta})$;

$\boldsymbol{I}(\boldsymbol{x}; \hat{\boldsymbol{\theta}})$, the estimated observed information matrix based on the likelihood $\mathcal{L}(\boldsymbol{x}; \boldsymbol{\theta})$;

$\boldsymbol{I}^{-1}(\boldsymbol{x}; \hat{\boldsymbol{\theta}})$, the estimated large-sample covariance matrix of $\hat{\boldsymbol{\theta}}$ based on observed information for the likelihood $\mathcal{L}(\boldsymbol{x}; \boldsymbol{\theta})$.

### 5.1.3 Large-Sample ML-Based Methods for Testing the Simple Null Hypothesis $\text{H}_0 : \boldsymbol{\theta} = \boldsymbol{\theta}_0$ (i.e., $\boldsymbol{\theta} \in \omega$) versus the Composite Alternative Hypothesis $\text{H}_1 : \boldsymbol{\theta} \in \bar{\omega}$

In general, a null hypothesis places a set of restrictions on the *unrestricted parameter space* $\Omega$, where $\Omega$ is the set of all possible values of the parameter vector $\boldsymbol{\theta}$. Let $\omega$ denote the *restricted* parameter space, where $\omega \subset \Omega$. Then, for the simple null hypothesis $\text{H}_0 : \boldsymbol{\theta} = \boldsymbol{\theta}_0$, it follows that $\omega = \{\boldsymbol{\theta} : \boldsymbol{\theta} = \boldsymbol{\theta}_0\}$, and $\Omega = \omega \cup \bar{\omega}$, where $\bar{\omega}$ is the complement of $\omega$.

#### 5.1.3.1 Likelihood Ratio Test

The *likelihood ratio test* statistic $\hat{\lambda}, 0 < \hat{\lambda} < 1$, for testing $\text{H}_0 : \boldsymbol{\theta} = \boldsymbol{\theta}_0$ (i.e., $\boldsymbol{\theta} \in \omega$) versus $\text{H}_1 : \boldsymbol{\theta} \in \bar{\omega}$ is defined as

$$\hat{\lambda} = \frac{\max\limits_{\boldsymbol{\theta} \in \omega} \mathcal{L}(\boldsymbol{x}; \boldsymbol{\theta})}{\max\limits_{\boldsymbol{\theta} \in \Omega} \mathcal{L}(\boldsymbol{x}; \boldsymbol{\theta})} = \frac{\mathcal{L}(\boldsymbol{x}; \boldsymbol{\theta}_0)}{\mathcal{L}(\boldsymbol{x}; \hat{\boldsymbol{\theta}})} \equiv \frac{\mathcal{L}_\omega}{\hat{\mathcal{L}}_\Omega}.$$

Clearly, small values of $\hat{\lambda}$ favor $H_1$, and a size $\alpha$ likelihood ratio test of $H_0$ versus $H_1$ using $\hat{\lambda}$ rejects $H_0$ in favor of $H_1$ when $\hat{\lambda} < k_\alpha$, where $\text{pr}(\hat{\lambda} < k_\alpha | H_0) = \alpha$.

Since the exact distribution of $\hat{\lambda}$ is often difficult to determine (either under $H_0$ or under $H_1$), the following large-sample approximation is typically used (Neyman and Pearson, 1928).

Under certain regularity conditions, for large $n$ and under $H_0 : \boldsymbol{\theta} = \boldsymbol{\theta}_0$,

$$-2 \ln \hat{\lambda} = 2\left[\ln \mathcal{L}(\boldsymbol{x}; \hat{\boldsymbol{\theta}}) - \ln \mathcal{L}(\boldsymbol{x}; \boldsymbol{\theta}_0)\right] \dot{\sim} \chi^2_p.$$

Thus, for a likelihood ratio test of approximate size $\alpha$, one would reject $H_0 : \boldsymbol{\theta} = \boldsymbol{\theta}_0$ in favor of $H_1 : \boldsymbol{\theta} \neq \boldsymbol{\theta}_0$ when $-2 \ln \hat{\lambda} > \chi^2_{p,1-\alpha}$.

### 5.1.3.2 *Wald Test*

The Wald test statistic $\hat{W}, 0 < \hat{W} < +\infty$, for testing $H_0 : \boldsymbol{\theta} = \boldsymbol{\theta}_0$ versus $H_1 : \boldsymbol{\theta} \in \bar{\omega}$ is defined as

$$\hat{W} = (\hat{\boldsymbol{\theta}} - \boldsymbol{\theta}_0)\mathcal{I}(\hat{\boldsymbol{\theta}})(\hat{\boldsymbol{\theta}} - \boldsymbol{\theta}_0)'$$

when using expected information, and is defined as

$$\hat{W} = (\hat{\boldsymbol{\theta}} - \boldsymbol{\theta}_0)\boldsymbol{I}(\boldsymbol{x}; \hat{\boldsymbol{\theta}})(\hat{\boldsymbol{\theta}} - \boldsymbol{\theta}_0)'$$

when using observed information.

Under certain regularity conditions (e.g., see Wald, 1943), for large $n$ and under $H_0 : \boldsymbol{\theta} = \boldsymbol{\theta}_0$, $\hat{W} \dot{\sim} \chi^2_p$. Thus, for a Wald test of approximate size $\alpha$, one would reject $H_0 : \boldsymbol{\theta} = \boldsymbol{\theta}_0$ in favor of $H_1 : \boldsymbol{\theta} \in \bar{\omega}$ when $\hat{W} > \chi^2_{p,1-\alpha}$.

### 5.1.3.3 *Score Test*

With the row vector $\boldsymbol{S}(\boldsymbol{\theta})$ defined as

$$\boldsymbol{S}(\boldsymbol{\theta}) = \left[\frac{\partial \ln \mathcal{L}(\boldsymbol{x}; \boldsymbol{\theta})}{\partial \theta_1}, \frac{\partial \ln \mathcal{L}(\boldsymbol{x}; \boldsymbol{\theta})}{\partial \theta_2}, \ldots, \frac{\partial \ln \mathcal{L}(\boldsymbol{x}; \boldsymbol{\theta})}{\partial \theta_p}\right],$$

the score test statistic $\hat{S}, 0 < \hat{S} < +\infty$, for testing $H_0 : \boldsymbol{\theta} = \boldsymbol{\theta}_0$ versus $H_1 : \boldsymbol{\theta} \in \bar{\omega}$ is defined as

$$\hat{S} = \boldsymbol{S}(\boldsymbol{\theta}_0)\mathcal{I}^{-1}(\boldsymbol{\theta}_0)\boldsymbol{S}'(\boldsymbol{\theta}_0)$$

when using expected information, and is defined as

$$\hat{S} = \boldsymbol{S}(\boldsymbol{\theta}_0)\boldsymbol{I}^{-1}(\boldsymbol{x}; \boldsymbol{\theta}_0)\boldsymbol{S}'(\boldsymbol{\theta}_0)$$

when using observed information. For the simple null hypothesis $H_0 : \boldsymbol{\theta} = \boldsymbol{\theta}_0$, note that the computation of the value of $\hat{S}$ involves no parameter estimation.

Under certain regularity conditions (e.g., see Rao, 1947), for large $n$ and under $H_0 : \boldsymbol{\theta} = \boldsymbol{\theta}_0, \hat{S} \dot{\sim} \chi^2_p$. Thus, for a score test of approximate size $\alpha$, one would reject $H_0 : \boldsymbol{\theta} = \boldsymbol{\theta}_0$ in favor of $H_1 : \boldsymbol{\theta} \in \bar{\omega}$ when $\hat{S} > \chi^2_{p,1-\alpha}$.

For further discussion concerning likelihood ratio, Wald, and score tests, see Rao (1973).

### Example

As an example, let $X_1, X_2, \ldots, X_n$ constitute a random sample of size $n$ from the parent population $p_X(x; \theta) = \theta^x (1 - \theta)^{1-x}, x = 0, 1$ and $0 < \theta < 1$. Consider testing $H_0 : \theta = \theta_0$ versus $H_1 : \theta \neq \theta_0$. Then, with $\hat{\theta} = \bar{X} = n^{-1} \sum_{i=1}^{n} X_i$, it can be shown that

$$-2 \ln \hat{\lambda} = 2n \left[ \bar{X} \ln \left( \frac{\bar{X}}{\theta_0} \right) + (1 - \bar{X}) \ln \left( \frac{1 - \bar{X}}{1 - \theta_0} \right) \right]$$

that

$$\hat{W} = \left[ \frac{(\bar{X} - \theta_0)}{\sqrt{\bar{X}(1 - \bar{X})/n}} \right]^2,$$

and that

$$\hat{S} = \left[ \frac{(\bar{X} - \theta_0)}{\sqrt{\theta_0(1 - \theta_0)/n}} \right]^2.$$

This simple example highlights an important general difference between Wald tests and score tests. Wald tests use parameter variance estimates assuming that $\boldsymbol{\theta} \in \Omega$ is true (i.e., assuming no restrictions on the parameter space $\Omega$), and score tests use parameter variance estimates assuming that $\boldsymbol{\theta} \in \omega$ (i.e., assuming that $H_0$ is true).

### 5.1.4 Large Sample ML-Based Methods for Testing the Composite Null Hypothesis $H_0 : \boldsymbol{\theta} \in \omega$ versus the Composite Alternative Hypothesis $H_1 : \boldsymbol{\theta} \in \bar{\omega}$

Let $R_i(\boldsymbol{\theta}) = 0, i = 1, 2, \ldots, r$, represent $r$ $(<p)$ *independent* restrictions placed on the parameter vector $\boldsymbol{\theta}$, and consider the null hypothesis $H_0 : \boldsymbol{\theta} \in \omega$, where $\omega = \{\boldsymbol{\theta} : R_i(\boldsymbol{\theta}) = 0, i = 1, 2, \ldots, r\}$. For example, with $\boldsymbol{\theta} = (\theta_1, \theta_2, \theta_3, \theta_4)$ for $p = 4$, consider the $r = 3$ linearly independent linear restrictions

$$R_1(\boldsymbol{\theta}) = (\theta_1 - \theta_2) = 0, \quad R_2(\boldsymbol{\theta}) = (\theta_1 - \theta_3) = 0$$

and

$$R_3(\boldsymbol{\theta}) = (\theta_1 - \theta_4) = 0.$$

Then, the null hypothesis $H_0 : R_i(\boldsymbol{\theta}) = 0, i = 1, 2, 3$, is equivalent to the null hypothesis $H_0 : \theta_1 = \theta_2 = \theta_3 = \theta_4$.

In what follows, let $\hat{\boldsymbol{\theta}}_\omega$ denote the *restricted* MLE of $\boldsymbol{\theta}$ under the null hypothesis $H_0 : \boldsymbol{\theta} \in \omega$.

#### 5.1.4.1 Likelihood Ratio Test

The likelihood ratio test statistic $\hat{\lambda}, 0 < \hat{\lambda} < 1$, for testing $H_0 : \boldsymbol{\theta} \in \omega$ versus $H_1 : \boldsymbol{\theta} \in \bar{\omega}$ is defined as

$$\hat{\lambda} = \frac{\max_{\boldsymbol{\theta} \in \omega} \mathcal{L}(\boldsymbol{x}; \boldsymbol{\theta})}{\max_{\boldsymbol{\theta} \in \Omega} \mathcal{L}(\boldsymbol{x}; \boldsymbol{\theta})} = \frac{\mathcal{L}(\boldsymbol{x}; \hat{\boldsymbol{\theta}}_\omega)}{\mathcal{L}(\boldsymbol{x}; \hat{\boldsymbol{\theta}})} \equiv \frac{\hat{\mathcal{L}}_\omega}{\hat{\mathcal{L}}_\Omega}.$$

Under certain regularity conditions, for large $n$ and under $H_0 : \boldsymbol{\theta} \in \omega$,

$$-2 \ln \hat{\lambda} = 2 \left[ \ln \mathcal{L}(\boldsymbol{x}; \hat{\boldsymbol{\theta}}) - \ln \mathcal{L}(\boldsymbol{x}; \hat{\boldsymbol{\theta}}_\omega) \right] \dot{\sim} \chi^2_r.$$

Thus, for a likelihood ratio test of approximate size $\alpha$, one would reject $H_0 : \boldsymbol{\theta} \in \omega$ in favor of $H_1 : \boldsymbol{\theta} \in \bar{\omega}$ when $-2 \ln \hat{\lambda} > \chi^2_{r,1-\alpha}$.

#### 5.1.4.2 Wald Test

Let the $(1 \times r)$ row vector $\boldsymbol{R}(\boldsymbol{\theta})$ be defined as

$$\boldsymbol{R}(\boldsymbol{\theta}) = [R_1(\boldsymbol{\theta}), R_2(\boldsymbol{\theta}), \ldots, R_r(\boldsymbol{\theta})].$$

Also, let the $(r \times p)$ matrix $\boldsymbol{T}(\boldsymbol{\theta})$ have $(i, j)$ element equal to $[\partial R_i(\boldsymbol{\theta})]/\partial \theta_j, i = 1, 2, \ldots, r$ and $j = 1, 2, \ldots, p$.

And, let the $(r \times r)$ matrix $\boldsymbol{\Lambda}(\boldsymbol{\theta})$ have the structure

$$\boldsymbol{\Lambda}(\boldsymbol{\theta}) = \boldsymbol{T}(\boldsymbol{\theta}) \boldsymbol{\mathcal{I}}^{-1}(\boldsymbol{\theta}) \boldsymbol{T}'(\boldsymbol{\theta})$$

when using expected information, and have the structure

$$\boldsymbol{\Lambda}(\boldsymbol{x}; \boldsymbol{\theta}) = \boldsymbol{T}(\boldsymbol{\theta}) \boldsymbol{I}^{-1}(\boldsymbol{x}; \boldsymbol{\theta}) \boldsymbol{T}'(\boldsymbol{\theta})$$

when using observed information.

Then, the Wald test statistic $\hat{W}, 0 < \hat{W} < +\infty$, for testing $H_0 : \boldsymbol{\theta} \in \omega$ versus $H_1 : \boldsymbol{\theta} \in \bar{\omega}$ is defined as

$$\hat{W} = \boldsymbol{R}(\hat{\boldsymbol{\theta}}) \boldsymbol{\Lambda}^{-1}(\hat{\boldsymbol{\theta}}) \boldsymbol{R}'(\hat{\boldsymbol{\theta}})$$

when using expected information, and is defined as

$$\hat{W} = \boldsymbol{R}(\hat{\boldsymbol{\theta}})\boldsymbol{\Lambda}^{-1}(\boldsymbol{x};\hat{\boldsymbol{\theta}})\boldsymbol{R}'(\hat{\boldsymbol{\theta}})$$

when using observed information.

Under certain regularity conditions, for large $n$ and under $\mathrm{H}_0 : \boldsymbol{\theta} \in \omega$, $\hat{W} \dot{\sim} \chi_r^2$. Thus, for a Wald test of approximate size $\alpha$, one would reject $\mathrm{H}_0 : \boldsymbol{\theta} \in \omega$ in favor of $\mathrm{H}_1 : \boldsymbol{\theta} \in \bar{\omega}$ when $\hat{W} > \chi^2_{r,1-\alpha}$.

#### 5.1.4.3 Score Test

The score test statistic $\hat{S}, 0 < \hat{S} < +\infty$, for testing $\mathrm{H}_0 : \boldsymbol{\theta} \in \omega$ versus $\mathrm{H}_1 : \boldsymbol{\theta} \in \bar{\omega}$ is defined as

$$\hat{S} = \boldsymbol{S}(\hat{\boldsymbol{\theta}}_\omega)\mathcal{I}^{-1}(\hat{\boldsymbol{\theta}}_\omega)\boldsymbol{S}'(\hat{\boldsymbol{\theta}}_\omega)$$

when using expected information, and is defined as

$$\hat{S} = \boldsymbol{S}(\hat{\boldsymbol{\theta}}_\omega)\boldsymbol{I}^{-1}(\boldsymbol{x};\hat{\boldsymbol{\theta}}_\omega)\boldsymbol{S}'(\hat{\boldsymbol{\theta}}_\omega)$$

when using observed information.

Under certain regularity conditions, for large $n$ and under $\mathrm{H}_0 : \boldsymbol{\theta} \in \omega, \hat{S} \dot{\sim} \chi_r^2$. Thus, for a score test of approximate size $\alpha$, one would reject $\mathrm{H}_0 : \boldsymbol{\theta} \in \omega$ in favor of $\mathrm{H}_1 : \boldsymbol{\theta} \in \bar{\omega}$ when $\hat{S} > \chi^2_{r,1-\alpha}$.

**Example**

As an example, let $X_1, X_2, \ldots, X_n$ constitute a random sample of size $n$ from a $\mathrm{N}(\mu, \sigma^2)$ parent population. Consider testing the composite null hypothesis $\mathrm{H}_0 : \mu = \mu_0, 0 < \sigma^2 < +\infty$, versus the composite alternative hypothesis $\mathrm{H}_1 : \mu \neq \mu_0, 0 < \sigma^2 < +\infty$. Note that this test is typically called a test of $\mathrm{H}_0 : \mu = \mu_0$ versus $\mathrm{H}_1 : \mu \neq \mu_0$.

It is straightforward to show that the vector $\hat{\boldsymbol{\theta}}$ of MLEs of $\mu$ and $\sigma^2$ for the unrestricted parameter space $\Omega$ is equal to

$$\hat{\boldsymbol{\theta}} = (\hat{\mu}, \hat{\sigma}^2) = \left[\bar{X}, \left(\frac{n-1}{n}\right) S^2\right],$$

where $\bar{X} = n^{-1}\sum_{i=1}^n X_i$ and $S^2 = (n-1)^{-1}\sum_{i=1}^n (X_i - \bar{X})^2$.

Then, it can be shown directly that

$$-2\ln\hat{\lambda} = n\ln\left[1 + \frac{T_{n-1}^2}{(n-1)}\right],$$

where

$$T_{n-1} = \frac{(\bar{X} - \mu_0)}{S/\sqrt{n}} \sim t_{n-1} \text{ under } \mathrm{H}_0 : \mu = \mu_0;$$

thus, the likelihood ratio test is a function of the usual one-sample $t$-test in this simple situation.

In this simple situation, the Wald test is also a function of the usual one-sample $t$-test since

$$\hat{W} = \left(\frac{n}{n-1}\right) T_{n-1}^2.$$

In contrast, the score test statistic has the structure

$$\hat{S} = \left[\frac{(\bar{X} - \mu_0)}{\hat{\sigma}_\omega/\sqrt{n}}\right]^2,$$

where

$$\hat{\sigma}_\omega^2 = n^{-1}\sum_{i=1}^{n}(X_i - \mu_0)^2$$

is the estimator of $\sigma^2$ under the null hypothesis $H_0 : \mu = \mu_0$.

Although all three of these ML-based hypothesis-testing methods (the likelihood ratio test, the Wald test, and the score test) are *asymptotically* equivalent, their use can lead to different conclusions in some actual data-analysis scenarios.

## EXERCISES

**Exercise 5.1.** Consider sampling from the parent population

$$f_X(x;\theta) = \theta x^{\theta-1}, \quad 0 < x < 1, \ \theta > 0.$$

(a) Based on a random sample $X_1$ of size $n = 1$ from this parent population, what is the *power* of the MP test of $H_0 : \theta = 1$ versus $H_1 : \theta = 2$ if $\alpha = \text{pr(Type I error)} = 0.05$?

(b) If $X_1$ and $X_2$ constitute a random sample of size $n = 2$ from this parent population, derive the *exact structure* of the rejection region of size $\alpha = 0.05$ associated with the MP test of $H_0 : \theta = 1$ versus $H_1 : \theta = 2$. Specifically, find the numerical value of the dividing point $k_\alpha$ between the rejection and non-rejection regions.

**Exercise 5.2.** Let $Y_1, Y_2, \ldots, Y_n$ constitute a random sample of size $n$ from the parent density

$$f_Y(y;\theta) = (1+\theta)(y+\theta)^{-2}, \quad y > 1, \ \theta > -1.$$

(a) Develop an explicit expression for the form of the MP rejection region $\mathcal{R}$ for testing $H_0 : \theta = 0$ versus $H_1 : \theta = 1$ when $\text{pr(Type I error)} = \alpha$.

(b) If $n = 1$ and $\alpha = 0.05$, find the numerical value of the dividing point between the rejection and non-rejection regions for this MP test.

(c) If, in fact, $\theta = 1$, what is the exact numerical value of the power of this MP test of $\mathrm{H}_0 : \theta = 0$ versus $\mathrm{H}_1 : \theta = 1$ when $\alpha = 0.05$ and $n = 1$?

**Exercise 5.3.** Let $Y_1, Y_2, \ldots, Y_n$ constitute a random sample of size $n$ from a $\mathrm{N}(0, \sigma^2)$ population. Develop the structure of the rejection region for a uniformly most powerful (UMP) test of $\mathrm{H}_0 : \sigma^2 = 1$ versus $\mathrm{H}_1 : \sigma^2 > 1$. Then, use this result to find a reasonable value for the smallest sample size (say, $n^*$) that is needed to provide a power of at least 0.80 for rejecting $\mathrm{H}_0$ in favor of $\mathrm{H}_1$ when $\alpha = 0.05$ and when the actual value of $\sigma^2$ is no smaller than 2.0 in value.

**Exercise 5.4.** Let $X_1, X_2, \ldots, X_n$ constitute a random sample of size $n$ from

$$p_X(x; \theta_1) = \theta_1^x (1 - \theta_1)^{1-x}, \quad x = 0, 1, \quad \text{and} \quad 0 < \theta_1 < 1;$$

and, let $Y_1, Y_2, \ldots, Y_n$ constitute a random sample of size $n$ from

$$p_Y(y; \theta_2) = \theta_2^y (1 - \theta_2)^{1-y}, \quad y = 0, 1, \quad \text{and} \quad 0 < \theta_2 < 1.$$

(a) If $n = 30$, derive a reasonable numerical value for the power of a size $\alpha = 0.05$ MP test of $\mathrm{H}_0 : \theta_1 = \theta_2 = 0.50$ versus $\mathrm{H}_1 : \theta_1 = \theta_2 = 0.60$.

(b) Now, suppose that it is of interest to test $\mathrm{H}_0 : \theta_1 = \theta_2 = \theta_0$ (where $\theta_0$ is a specified constant, $0 < \theta_0 < 1$) versus $\mathrm{H}_1 : \theta_1 > \theta_2$ at the $\alpha = 0.05$ level using a test statistic that is an explicit function of $(\bar{X} - \bar{Y})$, where $\bar{X} = n^{-1} \sum_{i=1}^n X_i = n^{-1} S_x$ and $\bar{Y} = n^{-1} \sum_{i=1}^n Y_i = n^{-1} S_y$. Provide a reasonable value for the smallest sample size (say, $n^*$) needed so that the power for testing $\mathrm{H}_0$ versus $\mathrm{H}_1$ is at least 0.90 when $\theta_0 = 0.10$ and when $(\theta_1 - \theta_2) \geq 0.20$.

**Exercise 5.5.** An epidemiologist gathers data $(x_i, Y_i)$ on each of $n$ randomly chosen noncontiguous and demographically similar cities in the United States, where $x_i (i = 1, 2, \ldots, n)$ is the known population size (in millions of people) in city $i$, and where $Y_i$ is the random variable denoting the number of people in city $i$ with colon cancer. It is reasonable to assume that $Y_i (i = 1, 2, \ldots, n)$ has a Poisson distribution with mean $\mathrm{E}(Y_i) = \theta x_i$, where $\theta (>0)$ is an unknown parameter, and that $Y_1, Y_2, \ldots, Y_n$ are mutually independent random variables.

(a) Using the available data $(x_i, Y_i), i = 1, 2, \ldots, n$, construct a UMP test of $\mathrm{H}_0\colon \theta = 1$ versus $\mathrm{H}_1\colon \theta > 1$.

(b) If $\sum_{i=1}^n x_i = 0.82$, what is the power of this UMP test for rejecting $\mathrm{H}_0\colon \theta = 1$ versus $\mathrm{H}_1\colon \theta > 1$ when the probability of a Type I error $\alpha \doteq 0.05$ and when, in reality, $\theta = 5$?

**Exercise 5.6.** For $i = 1, 2$, suppose that it is desired to select a random sample $X_{i1}, X_{i2}, \ldots, X_{in_i}$ of size $n_i$ from a $\mathrm{N}(\mu_i, \sigma_i^2)$ population, where $\mu_1$ and $\mu_2$ are *unknown* parameters and where $\sigma_1^2$ and $\sigma_2^2$ are *known* parameters.

For testing $\mathrm{H}_0 : \mu_1 = \mu_2$ versus $\mathrm{H}_1 : \mu_1 - \mu_2 = \delta (> 0)$, the test statistic

$$Z = \frac{(\bar{X}_1 - \bar{X}_2) - 0}{\sqrt{V}}$$

is to be used, where

$$\bar{X}_i = \sum_{j=1}^{n_i} X_{ij} \quad \text{for } i = 1, 2, \text{ and } V = \frac{\sigma_1^2}{n_1} + \frac{\sigma_2^2}{n_2}.$$

(a) If the null hypothesis is to be rejected when $Z > Z_{1-\alpha}$, show that the two conditions pr(Type I error)= $\alpha$ and pr(Type II error)= $\beta$ are simultaneously satisfied when

$$V = \left( \frac{\delta}{Z_{1-\alpha} + Z_{1-\beta}} \right)^2 = \theta, \quad \text{say.}$$

(b) Subject to the constraint $V = \theta$, find (as a function of $\sigma_1^2$ and $\sigma_2^2$) that value of $n_1/n_2$ which minimizes the total sample size $N = (n_1 + n_2)$. Due to logistical constraints, suppose that it is only possible to select a total sample size of $N = (n_1 + n_2) = 100$. If $N = 100, \sigma_1^2 = 9$, and $\sigma_2^2 = 4$, find the appropriate values of $n_1$ and $n_2$.

(c) Again, subject to the constraint $V = \theta$, develop expressions for $n_1$ and $n_2$ (in terms of $\theta, \sigma_1$, and $\sigma_2$) that will minimize the total sampling cost if the cost of selecting an observation from Population 1 is four times the cost of selecting an observation from Population 2. What are the specific sample sizes needed if $\sigma_1 = 5, \sigma_2 = 4, \alpha = 0.05, \beta = 0.10$, and $\delta = 3$?

**Exercise 5.7.** Let $X_1, X_2, \ldots, X_n$ constitute a random sample of size $n$ from the parent population

$$f_X(x;\theta) = \theta^{-1}, \quad 0 < x < \theta,$$

where $\theta$ is an unknown parameter.

Suppose that a statistician proposes the following test of $H_0 : \theta = \theta_0$ versus $H_1 : \theta > \theta_0$: "reject $H_0$ in favor of $H_1$ if $X_{(n)} > c$, where $X_{(n)}$ is the largest observation in the set $X_1, X_2, \ldots, X_n$ and where $c$ is a specified positive constant."

(a) If $\theta_0 = \frac{1}{2}$, find that specified value of $c$, say $c^*$, such that pr(Type I error) = $\alpha$. Note that $c^*$ will be a function of both $n$ and $\alpha$.

(b) If the true value of $\theta$ is actually $\frac{3}{4}$, find the smallest value of $n$ (say, $n^*$) required so that the *power* of the statistician's test is at least 0.98 when $\alpha = 0.05$ and $\theta_0 = \frac{1}{2}$.

**Exercise 5.8.** For the $i$th of $n$ independently selected busy intersections in a certain heavily populated U.S. city, the number $X_i$ of automobile accidents in any given year is assumed to have a Poisson distribution with mean $\mu_i, i = 1, 2, \ldots, n$. It can be assumed that these $n$ intersections are essentially the same with respect to the rate of traffic flow per day. It is of interest to test the null hypothesis $H_0 : \mu_i = \mu, i = 1, 2, \ldots, n$, versus the (unrestricted) alternative hypothesis $H_1$ that the $\mu_i$'s are not necessarily all equal to one another (i.e., that they are completely free to vary in value). In other words, we wish to use the $n$ mutually independent Poisson random variables $X_1, X_2, \ldots, X_n$ to assess whether or not the true average number of accidents in any given year is the same at each of the $n$ intersections. Note that testing $H_0$ versus $H_1$ is equivalent to testing "homogeneity" versus "heterogeneity" among the $\mu_i$'s.

(a) Develop an explicit expression for the likelihood ratio statistic $-2\ln(\hat{\lambda})$ for testing $H_0$ versus $H_1$. If, in a sample of $n = 40$ intersections in a particular year, there were 20 intersections each with a total of 5 accidents, 10 intersections each with a total of 6 accidents, and 10 intersections each with a total of 8 accidents, demonstrate that $H_0$ is not rejected at the $\alpha = 0.05$ level based on these data.

(b) Based on the data and the hypothesis test results for part (a), construct what you deem to be an appropriate 95% CI for $\mu$.

**Exercise 5.9.** It is of interest to compare two cities (say, City 1 and City 2) with regard to their true rates ($\lambda_1$ and $\lambda_2$, respectively) of primary medical care utilization, where these two rates are expressed in units of the number of out-patient doctor visits per person-year of community residence. For $i = 1, 2$, suppose that $n$ adult residents are randomly selected from City $i$; further, suppose that the values of the two variables $X_{ij}$ and $L_{ij}$ are recorded for the $j$th person ($j = 1, 2, \ldots, n$) in this random sample from City $i$, where $X_{ij}$ is the total number of out-patient doctor visits made by this person while residing in City $i$, and where $L_{ij}$ is the length of residency (in years) in City $i$ for this person. Hence, for $i = 1, 2$, the data for City $i$ consist of the $n$ mutually independent pairs $(X_{i1}, L_{i1}), (X_{i2}, L_{i2}), \ldots, (X_{in}, L_{in})$. In what follows, it is to be assumed that the distribution of $X_{ij}$ is $\text{POI}(L_{ij}\lambda_i)$, so that $\text{E}(X_{ij}) = \text{V}(X_{ij}) = L_{ij}\lambda_i$. Furthermore, the $L_{ij}$'s are to be considered as fixed known constants.

(a) Develop an explicit expression for the likelihood function for all $2n$ observations ($n$ from City 1 and $n$ from City 2), and find two statistics which are jointly sufficient for $\lambda_1$ and $\lambda_2$.

(b) Using the likelihood function in part (a), prove that the MLE of $\lambda_i$ is

$$\hat{\lambda}_i = \frac{\sum_{j=1}^{n} X_{ij}}{\sum_{j=1}^{n} L_{ij}}, \quad i = 1, 2.$$

(c) Suppose that it is of interest to test the composite null hypothesis $H_0 : \lambda_1 = \lambda_2$ ($= \lambda$, say) versus the composite alternative hypothesis $H_1 : \lambda_1 \neq \lambda_2$. Assuming that $H_0$ is true, find the MLE $\hat{\lambda}$ of $\lambda$.

(d) Develop an explicit expression for the likelihood ratio statistic which can be used to test $H_0 : \lambda_1 = \lambda_2$ versus $H_1 : \lambda_1 \neq \lambda_2$.

(e) Suppose that $n = 25$, $\hat{\lambda}_1 = 0.02$, $\hat{\lambda}_2 = 0.03$, $\sum_{j=1}^{n} L_{1j} = 200$, and $\sum_{j=1}^{n} L_{2j} = 300$. Use the likelihood ratio statistic developed in part (d) to test $H_0 : \lambda_1 = \lambda_2$ versus $H_1 : \lambda_1 \neq \lambda_2$ at the $\alpha = 0.10$ level. What is the P-value of your test?

**Exercise 5.10.** Suppose that $X$ and $Y$ are continuous random variables representing the survival times (in years) for patients following two different types of surgical procedures for the treatment of advanced colon cancer. Further, suppose that these survival time distributions are assumed to be of the form

$$f_X(x;\alpha) = \alpha e^{-\alpha x}, \quad x > 0, \alpha > 0 \quad \text{and} \quad f_Y(y;\beta) = \beta e^{-\beta y}, \quad y > 0, \ \beta > 0.$$

Let $X_1, X_2, \ldots, X_n$ and $Y_1, Y_2, \ldots, Y_n$ denote random samples of size $n$ from $f_X(x;\alpha)$ and $f_Y(y;\beta)$, respectively. Also, let $\bar{X} = n^{-1}\sum_{i=1}^{n} X_i$ and let $\bar{Y} = n^{-1}\sum_{i=1}^{n} Y_i$.

(a) For the likelihood ratio test of $H_0 : \alpha = \beta$ versus $H_1 : \alpha \neq \beta$, show that the likelihood ratio statistic $\hat{\lambda}$ can be written in the form $\hat{\lambda} = [4u(1-u)]^n$, where $u = \bar{x}/(\bar{x}+\bar{y})$.

(b) If $n = 100, \bar{x} = 1.25$ years, and $\bar{y} = 0.75$ years, use a P-value computation to decide whether or not to reject $H_0$ in favor of $H_1$, and then interpret your finding with regard to these two surgical procedures for the treatment of advanced colon cancer.

**Exercise 5.11.** The number $X$ of speeding tickets issued to a typical teenage driver during a specified two-year period in a certain community (say, Community #1) having mandatory teenage driver education classes is assumed to have the distribution

$$p_X(x;\theta_1) = \theta_1(1-\theta_1)^x, \quad x = 0, 1, \ldots, \infty; \ 0 < \theta_1 < 1.$$

The number $Y$ of speeding tickets issued to a typical teenage driver during that same 2-year period in another community with similar sociodemographic characteristics (say, Community #2), but not having mandatory teenage driver education classes, is assumed to have the distribution

$$p_Y(y;\theta_2) = \theta_2(1-\theta_2)^y, \quad y = 0, 1, \ldots, \infty; \ 0 < \theta_2 < 1.$$

Let $X_1, X_2, \ldots, X_n$ constitute a random sample of size $n$ from $p_X(x;\theta_1)$, and let $x_1, x_2, \ldots, x_n$ denote the corresponding $n$ realizations (i.e., the actual set of observed numbers of speeding tickets) for the set of $n$ randomly chosen teenage drivers selected from Community #1. Further, let $Y_1, Y_2, \ldots, Y_n$ constitute a random sample of size $n$ from $p_Y(y;\theta_2)$, with $y_1, y_2, \ldots, y_n$ denoting the corresponding realizations.

(a) Using the complete set of observed data $\{x_1, x_2, \ldots, x_n; y_1, y_2, \ldots, y_n\}$, develop an explicit expression for the likelihood ratio test statistic $\hat{\lambda}$ for testing the null hypothesis $H_0 : \theta_1 = \theta_2 (= \theta$, say) versus the alternative hypothesis $H_1 : \theta_1 \neq \theta_2$. If $n = 25, \bar{x} = 1.00$, and $\bar{y} = 2.00$, is there sufficient evidence to reject $H_0$ in favor of $H_1$ at the $\alpha = 0.05$ level of significance?

(b) Using *observed* information, use the data in part (a) to compute the numerical value of $\hat{S}$, the score statistic for testing $H_0$ versus $H_1$. How do the conclusions based on the score test compare with those based on the likelihood ratio test?

(c) A highway safety researcher contends that the data do suggest that the teenage driver education classes might actually be beneficial, and he suggests that increasing the sample size $n$ might actually lead to a highly statistically significant conclusion that these mandatory teenage driver education classes do lower the risk of speeding by teenagers. Making use of the available data, comment on the reasonableness of this researcher's contention.

**Exercise 5.12.** Suppose that $n$ randomly selected adult male hypertensive patients are administered a new blood pressure lowering drug during a clinical trial designed to assess the efficacy of this new drug for promoting long-term remission of high blood pressure. Further, once each patient's blood pressure returns to a normal range, suppose that each patient is examined monthly to see if the hypertension returns. For the $i$th patient in the study, let $x_i$ denote the age of the patient at the start of the clinical trial, and let $Y_i$ be the random variable denoting the number of months of follow-up until the hypertension returns for the first time. It is reasonable to assume that $Y_i$ has the geometric distribution

$$p_{Y_i}(y_i; \theta_i) = (1-\theta_i)^{y_i-1}\theta_i, \quad y_i = 1, 2, \ldots, \infty, \; 0 < \theta_i < 1 \text{ and } \; i = 1, 2, \ldots, n.$$

It is well-established that age is a risk factor for hypertension. To take into account the differing ages of the patients at the start of the trial, it is proposed that $\theta_i$ be expressed as the following function of age:

$$\theta_i = \beta x_i/(1+\beta x_i), \quad \beta > 0.$$

Given the $n$ pairs $(x_i, y_i)$, $i = 1, 2, \ldots, n$, of data points, the analysis goal is to obtain the MLE $\hat{\beta}$ of $\beta$, and then to use $\hat{\beta}$ to make statistical inferences about $\beta$.

(a) Prove that the MLE $\hat{\beta}$ of $\beta$ satisfies the equation

$$\hat{\beta} = \frac{n}{\sum_{i=1}^{n} x_i y_i \left(1+\hat{\beta}x_i\right)^{-1}}.$$

(b) Prove that the asymptotic variance of $\hat{\beta}$ is

$$V(\hat{\beta}) = \frac{\beta^2}{\sum_{i=1}^{n}(1+\beta x_i)^{-1}}.$$

(c) If the clinical trial involves 50 patients of age 30 and 50 patients of age 40 at the start of the trial, find a large-sample 95% CI for $\beta$ if $\hat{\beta} = 0.50$.

(d) For the data in part (c), carry out a Wald test of $H_0: \beta = 1$ versus $H_1: \beta \neq 1$ using $\alpha = 0.05$. Do you reject $H_0$ or not? What is the P-value of your test?

(e) To test $H_0: \beta = 1$ versus $H_1: \beta > 1$, consider the test statistic

$$U = \frac{(\hat{\beta}-1)}{\sqrt{V_0(\hat{\beta})}},$$

where

$$V_0(\hat{\beta}) = \frac{1}{\sum_{i=1}^{n}(1+x_i)^{-1}}$$

is the large-sample variance of $\hat{\beta}$ when $H_0: \beta = 1$ is true. Assuming that

$$\frac{(\hat{\beta}-\beta)}{\sqrt{V(\hat{\beta})}} \mathrel{\dot\sim} N(0,1)$$

for large $n$, where $V(\hat{\beta})$ is given in part (b), and using the age data in part (c), what is the approximate *power* of $U$ to reject $H_0 : \beta = 1$ in favor of $H_1 : \beta > 1$ when $\alpha = 0.025$ and when the true value of $\beta$ is equal to 1.10?

**Exercise 5.13.** A random sample of 1000 disease-free heavy smokers is followed for a 20-year period. At the end of this 20-year follow-up period, it is found that exactly 100 of these 1000 heavy smokers developed lung cancer during the follow-up period. It is of interest to make statistical inferences about the population parameter $\psi = \theta/(1-\theta)$, where $\theta$ is the probability that a member of the population from which this random sample came develops lung cancer during this 20-year follow-up period. The parameter $\psi$ is the odds of developing lung cancer during this 20-year follow-up period, namely, the ratio of the probability of developing lung cancer to the probability of not developing lung cancer over this 20-year period.

(a) Using the available numerical information, construct an appropriate 95% CI for the parameter $\psi$.

(b) Carry out Wald and score tests of the null hypothesis $H_0$: $\psi = 0.10$ versus the alternative hypothesis $H_1$: $\psi > 0.10$. What are the P-values of these two tests? Interpret your findings.

**Exercise 5.14.** An environmental scientist postulates that the distributions of the concentrations $X$ and $Y$ (in parts per million) of two air pollutants can be modeled as follows: the conditional density of $Y$, given $X = x$, is postulated to have the structure

$$f_Y(y|X = x; \alpha, \beta) = \frac{1}{(\alpha + \beta)x} e^{-y/(\alpha+\beta)x}, \quad y > 0,\ x > 0,\ \alpha > 0,\ \beta > 0;$$

and, the marginal density of $X$ is postulated to have the structure

$$f_X(x; \beta) = \frac{1}{\beta} e^{-x/\beta}, \quad x > 0,\ \beta > 0.$$

Let $(X_1, Y_1), (X_2, Y_2), \ldots, (X_n, Y_n)$ constitute a random sample of size $n$ from the joint density $f_{X,Y}(x, y; \alpha, \beta)$ of $X$ and $Y$.

(a) Derive explicit expressions for two statistics $U_1$ and $U_2$ that are jointly sufficient for $\alpha$ and $\beta$, and then prove that $\text{corr}(U_1, U_2) = 0$.

(b) Using the random sample $(X_i, Y_i)$ $i = 1, \ldots, n$, derive explicit expressions for the MLEs $\hat{\alpha}$ and $\hat{\beta}$ of the unknown parameters $\alpha$ and $\beta$. Then, if $n = 30, \hat{\alpha} = 2$, and $\hat{\beta} = 1$, find the P-value for a Wald test (based on expected information) of $H_0 : \alpha = \beta$ versus $H_1 : \alpha \neq \beta$. Also, use the available data to compute an appropriate 95% CI for the parameter $(\alpha - \beta)$, and then comment on any numerical connection between the confidence interval result and the P-value.

**Exercise 5.15.** For the $i$th of two large formaldehyde production facilities located in two different southern cities in the United States, the expected amount $E(Y_{ij})$ in pounds of formaldehyde produced by a certain chemical reaction, expressed as a function of the amount $x_{ij}$ (>0) in pounds of catalyst used to promote the reaction, is given by the equation

$$E(Y_{ij}) = \beta_i x_{ij}^2, \quad \text{where } \beta_i > 0 \text{ and } x_{ij} > 0, \quad i = 1,2 \text{ and}$$
$$j = 1,2,\ldots,n.$$

Let $(x_{i1}, Y_{i1}), (x_{i2}, Y_{i2}), \ldots, (x_{in}, Y_{in})$ be $n$ independent pairs of data points from the $i$th production facility, $i = 1,2$. Assume that $Y_{ij}$ has a negative exponential distribution with mean $\alpha_{ij} = E(Y_{ij}) = \beta_i x_{ij}^2$, that the $x_{ij}$'s are known constants, and that the $Y_{ij}$'s are a set of $2n$ mutually independent random variables.

(a) Provide an explicit expression for the joint distribution (i.e., the unconditional likelihood function) of the $2n$ $Y_{ij}$'s, and then provide explicit expressions for two statistics that are jointly sufficient for $\beta_1$ and $\beta_2$.

(b) Under the stated assumptions given earlier, develop an explicit expression (using expected information) for the score statistic $\hat{S}$ for testing $H_0$: $\beta_1 = \beta_2$ versus $H_1$:$\beta_1 \neq \beta_2$. In particular, show that $\hat{S}$ can be expressed solely as a function of $n, \hat{\beta}_1$, and $\hat{\beta}_2$, where $\hat{\beta}_1$ and $\hat{\beta}_2$ are the MLEs of $\beta_1$ and $\beta_2$, respectively, in the unrestricted parameter space. If $n = 25$, $\hat{\beta}_1 = 2$, and $\hat{\beta}_2 = 3$, do you reject $H_0$ in favor of $H_1$ at the $\alpha = 0.05$ level?

**Exercise 5.16.** Consider a clinical trial involving two different treatments for Stage IV malignant melanoma, namely, Treatment A and Treatment B. Let $X_1, X_2, \ldots, X_n$ denote the mutually independent survival times (in months) for the $n$ patients randomly assigned to Treatment A. As a statistical model, consider $X_1, X_2, \ldots, X_n$ to constitute a random sample of size $n$ from

$$f_X(x; \theta) = \theta^{-1} e^{-x/\theta}, \quad x > 0, \ \theta > 0.$$

Further, let $Y_1, Y_2, \ldots, Y_n$ denote the mutually independent survival times (in months) for the $n$ patients randomly assigned to Treatment B. As a statistical model, consider $Y_1, Y_2, \ldots, Y_n$ to constitute a random sample of size $n$ from

$$f_Y(y; \lambda, \theta) = (\lambda\theta)^{-1} e^{-y/\lambda\theta}, \quad y > 0, \ \lambda > 0, \ \theta > 0.$$

Clearly, $E(X) = \theta$ and $E(Y) = \lambda\theta$, so that statistical inferences about the parameter $\lambda$ can be used to decide whether or not the available data provide evidence of a difference in true average survival times for Treatment A and Treatment B.

(a) Find explicit expressions for statistics that are jointly sufficient for making statistical inferences about the unknown parameters $\lambda$ and $\theta$.

(b) Derive explicit expressions for $\hat{\lambda}$ and $\hat{\theta}$, the MLEs of the unknown parameters $\lambda$ and $\theta$.

(c) Using expected information, derive an explicit expression for the score statistic $\hat{S}$ for testing $H_0 : \lambda = 1$ versus $H_1 : \lambda \neq 1$. Also, show directly how a variance estimated under $H_0$ enters into the explicit expression for $\hat{S}$. For a particular data set where $n = 50$, $\bar{x} = n^{-1}\sum_{i=1}^{n} x_i = 30$, and $\bar{y} = n^{-1}\sum_{i=1}^{n} y_i = 40$, what is the approximate P-value when the score statistic $\hat{S}$ is used to test $H_0$ versus $H_1$?

**Exercise 5.17.** An oncologist reasons that the survival time X (in years) for advanced-stage colorectal cancer follows an exponential distribution with *unknown* parameter $\lambda$; that is,

$$f_X(x|\lambda) = \lambda e^{-\lambda x}, \quad x > 0, \ \lambda > 0.$$

Although this oncologist does not know the exact value of $\lambda$, she is willing to assume *a priori* that $\lambda$ also follows an exponential distribution with *known* parameter $\beta$, namely,

$$\pi(\lambda) = \beta e^{-\beta\lambda}, \quad \lambda > 0, \ \beta > 0.$$

In the Bayesian paradigm, $f_X(x|\lambda)$ is called the *likelihood function*, and $\pi(\lambda)$ is called the *prior distribution* of $\lambda$ (i.e., the distribution assigned to $\lambda$ before observing a value for X).

(a) Find the marginal distribution $f_X(x)$ of X (i.e., the distribution of X averaged over all possible values of $\lambda$).

(b) Find the *posterior distribution* $\pi(\lambda|X = x)$ of $\lambda$.

(c) A Bayesian measure of evidence against a null hypothesis ($H_0$), and in favor of an alternative hypothesis ($H_1$), is the *Bayes Factor*, denoted $BF_{10}$. In particular,

$$BF_{10} = \frac{\text{pr}(H_1|X = x)/\text{pr}(H_0|X = x)}{\text{pr}(H_1)/\text{pr}(H_0)} = \frac{\text{pr}(H_1|X = x)\text{pr}(H_0)}{\text{pr}(H_0|X = x)\text{pr}(H_1)},$$

where $\text{pr}(H_k)$ and $\text{pr}(H_k|X = x)$ denote, respectively, the prior and posterior probabilities of hypothesis $H_k, k = 0, 1$. Hence, $BF_{10}$ is the ratio of the posterior odds of $H_1$ to the prior odds of $H_1$. According to Kass and Raftery (1995), $1 < BF_{10} \leq 3$ provides "weak" evidence in favor of $H_1$, $3 < BF_{10} \leq 20$ provides "positive" evidence in favor of $H_1$, $20 < BF_{10} \leq 150$ provides "strong" evidence in favor of $H_1$, and $BF_{10} > 150$ provides "very strong" evidence in favor of $H_1$.

If $\beta = 1$ and $x = 3$, what is the Bayes factor for testing $H_0 : \lambda > 1$ versus $H_1 : \lambda \leq 1$? Using the scale proposed by Kass and Raftery (1995), what is the strength of evidence in favor of $H_1$?

**Exercise 5.18*.** A controlled clinical trial was designed to compare the survival times (in years) of HIV patients receiving once daily dosing of the new drug Epzicom (a combination of 600 mg of Ziagen and 300 mg of Epivir) to the survival times (in years) of HIV patients receiving once daily dosing of the new drug Truvada (a combination of 300 mg of Viread and 200 mg of Emtriva). Randomly chosen HIV patients were paired together based on the values of several important factors, including age, current HIV levels, general health status, and so on. Then, one member of each pair was randomly selected to receive Epzicom, with the other member then receiving Truvada. For the $i$th pair, $i = 1, 2, \ldots, n$, let $X_i$ denote the survival time of the patient receiving Epzicom, and let $Y_i$ denote the survival time of the patient receiving Truvada. Further, assume that $X_i$ and $Y_i$ are independent random variables with respective distributions

$$f_{X_i}(x_i) = (\theta\phi_i)^{-1}e^{-x_i/\theta\phi_i}, \quad x_i > 0,$$

and

$$f_{Y_i}(y_i) = \phi_i^{-1}e^{-y_i/\phi_i}, \quad y_i > 0.$$

Here, $\phi_i$ (>0) is a parameter pertaining to characteristics of the $i$th pair, and $\theta$ (>0) is the parameter reflecting any difference in true average survival times for the two drugs Epzicom and Truvada. Hence, the value $\theta = 1$ indicates no difference between the two drugs with regard to average survival time.

(a) Provide an explicit expression for the joint distribution (i.e., the likelihood) of the $2n$ random variables $X_1, X_2, \ldots, X_n$ and $Y_1, Y_2, \ldots, Y_n$. How many parameters would have to be estimated by the method of ML? Comment on this finding.

(b) A consulting statistician points out that the only parameter of real interest is $\theta$. She suggests that an alternative analysis be based just on the $n$ ratios $R_i = X_i/Y_i$, $i = 1, 2, \ldots, n$. In particular, this statistician claims that the distributions of these ratios depend only on $\theta$ and not on the $\{\phi_i\}$, and that the $\{\phi_i\}$ are so-called *nuisance parameters* (i.e., parameters that appear in assumed statistical models, but that are not of direct relevance to the particular research questions of interest). Prove that this statistician is correct by showing that

$$f_{R_i}(r_i) = \frac{\theta}{(\theta + r_i)^2}, \quad 0 < r_i < +\infty, \ i = 1, 2, \ldots, n.$$

(c) Using the $n$ mutually independent random variables $R_1, R_2, \ldots, R_n$, it is of interest to test $H_0 : \theta = 1$ versus $H_1 : \theta > 1$ at the $\alpha = 0.025$ level. What is the smallest sample size $n^*$ required so that the power of an appropriate large-sample test is at least 0.80 when, in fact, the true value of $\theta$ is 1.50?

**Exercise 5.19***. In many important practical data analysis situations, the statistical models being used involve several parameters, only a few of which are relevant for directly addressing the research questions of interest. The irrelevant parameters, generally referred to as "nuisance parameters," are typically employed to ensure that the statistical models make scientific sense, but are generally unimportant otherwise. One method for eliminating the need to estimate these nuisance parameters, and hence generally to improve both statistical validity and precision, is to employ a *conditional inference* approach, whereby a conditioning argument is used to produce a conditional likelihood function that only involves the relevant parameters. For an excellent discussion of methods of conditional inference, see McCullagh and Nelder (1989).

As an example, suppose that it is of interest to evaluate whether current smokers tend to miss more days of work due to illness than do nonsmokers. For a certain manufacturing industry, suppose that $n$ mutually independent matched pairs of workers, one a current smoker and one a nonsmoker, are formed, where the workers in each pair are chosen (i.e., are matched) to have the same set of general risk factors (e.g., age, current health status, type of job, etc.) for illness-related work absences. These $2n$ workers are then followed for a year, and the number of days missed due to illness during that year is recorded for each worker.

For the $i$th pair of workers, $i = 1, 2, \ldots, n$, let $Y_{ij} \sim \text{POI}(\phi_i \lambda_j), j = 0, 1$, where $j = 0$ pertains to the nonsmoking worker and where $j = 1$ pertains to the worker who currently smokes. Further, assume that $Y_{i1}$ and $Y_{i0}$ are independent random variables. It is of interest to test $\text{H}_0 : \lambda_1 = \lambda_0$ versus $\text{H}_1 : \lambda_1 > \lambda_0$. If $\text{H}_0$ is rejected in favor of $\text{H}_1$, then this finding would supply statistical evidence that current smokers, on average, tend to miss more days of work due to illness than do nonsmokers. The $n$ parameters $\{\phi_1, \phi_2, \ldots, \phi_n\}$ are parameters reflecting inherent differences across the matched pairs with regard to general risk factors for illness-related work absences, and these $n$ nuisance parameters are not of primary interest. The statistical analysis goal is to use a conditional inference approach that eliminates the need to estimate these nuisance parameters and that still produces an appropriate statistical procedure for testing $\text{H}_0$ versus $\text{H}_1$.

(a) Develop an explicit expression for the conditional distribution $p_{Y_{i1}}(y_{i1}|Y_{i1} + Y_{i0} = S_i = s_i)$ of the random variable $Y_{i1}$ given that $(Y_{i1} + Y_{i0}) = S_i = s_i$.

(b) Use the result in part (a) to develop an appropriate ML-based large-sample test of $\text{H}_0$ versus $\text{H}_1$ that is based on the parameter $\theta = \lambda_1/(\lambda_0 + \lambda_1)$. For $n = 50, \sum_{i=1}^{n} s_i = 500$, and $\sum_{i=1}^{n} y_{i1} = 275$, is there statistical evidence for rejecting $\text{H}_0$ in favor of $\text{H}_1$? Can you detect another advantage of this conditional inference procedure?

**Exercise 5.20*.** Let $X_1, X_2, \ldots, X_n$ constitute a random sample of size $n$ from a Poisson distribution with parameter $\lambda_x$. Furthermore, let $Y_1, Y_2, \ldots, Y_n$ constitute a random sample of the same size $n$ from a different Poisson population with parameter $\lambda_y$.

(a) Use these $2n$ mutually independent observations to develop an explicit expression for the score test statistic $\hat{S}$ (based on expected information) for testing the null hypothesis $\text{H}_0 : \lambda_x = \lambda_y$ versus the alternative hypothesis $\text{H}_1 : \lambda_x \neq \lambda_y$. Suppose that $n = 30, \bar{x} = n^{-1} \sum_{i=1}^{n} x_i = 8.00$, and $\bar{y} = n^{-1} \sum_{i=1}^{n} y_i = 9.00$; do you reject $\text{H}_0$ or not using $\hat{S}$?

(b) Now, suppose that $n = 1$, so that only the independent observations $X_1$ and $Y_1$ are available. By considering the conditional distribution of $X_1$ given that $(X_1 + Y_1) = s_1$, develop a method for testing the null hypothesis $\text{H}_0 : \lambda_y = \delta\lambda_x$ versus the alternative hypothesis $\text{H}_1 : \lambda_y > \delta\lambda_x$, where $\delta$ ($> 0$) is a known constant. Suppose that $\delta = 0.60$, $x_1 = 4$, and $y_1 = 10$. What is the exact P-value of your test of $\text{H}_0$ versus $\text{H}_1$?

**Exercise 5.21*.** For older adults with symptoms of Alzheimer's disease, the distribution of the time $X$ (in hours) required to complete a verbal aptitude test designed to measure the severity of dementia is assumed to have the distribution

$$f_X(x) = 1, 0.50 \leq \theta < x < (\theta + 1) < +\infty.$$

Let $X_1, X_2, \ldots, X_n$ constitute a random sample of size $n (> 1)$ from $f_X(x)$. Further, define

$$X_{(1)} = \min\{X_1, X_2, \ldots, X_n\} \quad \text{and} \quad X_{(n)} = \max\{X_1, X_2, \ldots, X_n\}.$$

It is of interest to test $H_0 : \theta = 1$ versus $H_1 : \theta > 1$. Suppose that the following decision rule is proposed: reject $H_0 : \theta = 1$ in favor of $H_1 : \theta > 1$ if and only if the event $A \cup B$ occurs, where A is the event that $X_{(1)} > k$, where B is the event that $X_{(n)} > 2$, and where $k$ is a positive constant.

(a) Find a specific expression for $k$, say $k_\alpha$, such that this particular decision rule has a Type I error rate exactly equal to $\alpha, 0 < \alpha \leq 0.10$.

(b) Find the power function for this decision rule; in particular, consider the power of this decision rule for appropriately chosen disjoint sets of values of $\theta, 1 < \theta < +\infty$.

**Exercise 5.22*.** Suppose that $Y_{11}, Y_{12}, \ldots, Y_{1n}$ constitute a set of $n$ random variables representing the responses to a certain lung function test for $n$ farmers living in the same small neighborhood located very near to a large hog farm in rural North Carolina. Since these $n$ farmers live in the same small neighborhood and so experience roughly the same harmful levels of air pollution from hog waste, it is reasonable to believe that the responses to this lung function test for these $n$ farmers will *not* be independent. In particular, assume that $Y_{1j} \sim N(\mu_1, \sigma^2), j = 1, 2, \ldots, n$, and that $\text{corr}(Y_{1j}, Y_{1j'}) = \rho\,(> 0)$ for every $j \neq j', j = 1, 2, \ldots, n$ and $j' = 1, 2, \ldots, n$.

Similarly, suppose that $Y_{21}, Y_{22}, \ldots, Y_{2n}$ constitute a set of $n$ random variables representing responses to the same lung function test for $n$ farmers living in a different small rural North Carolina neighborhood that experiences only minimal levels of air pollution from hog waste. In particular, assume that $Y_{2j} \sim N(\mu_2, \sigma^2), j = 1, 2, \ldots, n$, and that $\text{corr}(Y_{2j}, Y_{2j'}) = \rho\,(> 0)$ for every $j \neq j', j = 1, 2, \ldots, n$ and $j' = 1, 2, \ldots, n$.

Further, assume that the parameters $\sigma^2$ and $\rho$ have *known values*, that the sets of random variables $\{Y_{1j}\}_{j=1}^{n}$ and $\{Y_{2j}\}_{j=1}^{n}$ are independent of each other, and that the two sample means $\bar{Y}_1 = \sum_{j=1}^{n} Y_{1j}$ and $\bar{Y}_2 = \sum_{j=1}^{n} Y_{2j}$ are each normally distributed.

(a) Find $E(\bar{Y}_1 - \bar{Y}_2)$ and develop an explicit expression for $V(\bar{Y}_1 - \bar{Y}_2)$ that is a function of $n$, $\sigma^2$, and $\rho$.

(b) Given the stated assumptions, provide a hypothesis testing procedure involving the standard normal distribution for testing $H_0 : \mu_1 = \mu_2$ versus $H_1 : \mu_1 > \mu_2$ using a Type I error rate of $\alpha = 0.05$.

(c) Now, suppose that an epidemiologist with minimal statistical training *incorrectly* ignores the positive intra-neighborhood correlation among responses and thus uses a test (based on the standard normal distribution) which incorrectly involves the assumption that $\rho = 0$. If this incorrect test is based on an *assumed* Type I error rate of 0.05, and if $n = 10, \sigma^2 = 2$, and $\rho = 0.50$, compute the exact numerical value of the *actual* Type I error rate associated with the use of this incorrect test. There is an important lesson to be learned here; what is it?

**Exercise 5.23*.** The normally distributed random variables $X_1, X_2, \ldots, X_n$ are said to follow a *first-order autoregressive process* when

$$X_i = \theta X_{i-1} + \epsilon_i, \quad i = 1, 2, \ldots, n,$$

where $X_0 \equiv 0$, where $\theta\,(-\infty < \theta < \infty)$ is an unknown parameter, and where $\epsilon_1, \epsilon_2, \ldots, \epsilon_n$ are *mutually independent* N(0,1) random variables.

(a) Determine the conditional density $f_{X_2}(x_2|X_1 = x_1)$ of $X_2$ given $X_1 = x_1$.

(b) Develop an explicit expression for $f_{X_1,X_2}(x_1, x_2)$, the joint density of $X_1$ and $X_2$.

(c) Let $f^*$ denote the joint density of $X_1, X_2, \ldots, X_n$, where, in general,

$$f^* = f_{X_1}(x_1) \prod_{i=2}^{n} f_{X_i}(x_i|X_1 = x_1, X_2 = x_2, \ldots, X_{i-1} = x_{i-1}).$$

Using a sample $(X_1, X_2, \ldots, X_n)$ from the joint density $f^*$, show that a likelihood ratio test of $H_0 : \theta = 0$ versus $H_1 : \theta \neq 0$ can be expressed explicitly as a function of the statistic

$$\frac{\left(\sum_{i=2}^{n} x_{i-1}x_i\right)^2}{\left(\sum_{i=1}^{n-1} x_i^2\right)\left(\sum_{i=1}^{n} x_i^2\right)}.$$

For $n = 30$, if $\sum_{i=2}^{n} x_{i-1}x_i = 4$, $\sum_{i=1}^{n} x_i^2 = 15$, and $x_n = 2$, would you reject $H_0 : \theta = 0$ at the $\alpha = 0.05$ level using this likelihood ratio test?

**Exercise 5.24*.** For lifetime residents of rural areas in the United States, suppose that it is reasonable to assume that the distribution of the proportion $X$ of a certain biomarker of benzene exposure in a cubic centimeter of blood taken from such a rural resident has a beta distribution with parameters $\alpha = \theta_r$ and $\beta = 1$, namely,

$$f_X(x; \theta_r) = \theta_r x^{\theta_r - 1}, \quad 0 < x < 1, \quad \theta_r > 0.$$

Let $X_1, X_2, \ldots, X_n$ constitute a random sample of size $n$ from $f_X(x; \theta_r)$. Analogously, for lifetime residents of United States urban areas, let the distribution of $Y$, the proportion of this same biomarker of benzene exposure in a cubic centimeter of blood taken from such an urban resident, be

$$f_Y(y; \theta_u) = \theta_u y^{\theta_u - 1}, \quad 0 < y < 1, \quad \theta_u > 0.$$

Let $Y_1, Y_2, \ldots, Y_m$ constitute a random sample of size $m$ from $f_Y(y; \theta_u)$.

(a) Using all $(n + m)$ available observations, find two statistics that are jointly sufficient for $\theta_r$ and $\theta_u$.

(b) Show that a likelihood ratio test of $H_0 : \theta_r = \theta_u (= \theta$, say) versus $H_1 : \theta_r \neq \theta_u$ can be based on the test statistic

$$W = \frac{\sum_{i=1}^{n} \ln(X_i)}{\left[\sum_{i=1}^{n} \ln(X_i) + \sum_{i=1}^{m} \ln(Y_i)\right]}.$$

(c) Find the exact distribution of the test statistic $W$ under $H_0 : \theta_r = \theta_u$ $(= \theta$, say), and then use this result to construct a likelihood ratio test of $H_0$: $\theta_r = \theta_u (= \theta$, say) versus $H_1$: $\theta_r \neq \theta_u$ with an exact Type I error rate of $\alpha = 0.10$ when $n = m = 2$.

**Exercise 5.25***. For two states in the United States with very different distributions of risk factors for AIDS (say, Maine and California), suppose that the number $Y_{ij}$ of new cases of AIDS in county $j$ ($j = 1, 2, \ldots, n$) of state $i$ ($i = 1, 2$) during a particular year is assumed to have the negative binomial distribution

$$p_{Y_{ij}}(y_{ij}; \theta_i) = C_{k-1}^{k+y_{ij}-1} \theta_i^{y_{ij}} (1+\theta_i)^{-(k+y_{ij})}, \quad y_{ij} = 0, 1, \ldots, \infty \text{ and } \theta_i > 0;$$

here, $\theta_1$ and $\theta_2$ are unknown parameters, and $k$ is a known positive constant.

For $i = 1, 2$, let $Y_{i1}, Y_{i2}, \ldots, Y_{in}$ denote $n$ mutually independent random variables representing the numbers of new AIDS cases developing during this particular year in $n$ randomly chosen non-adjacent counties in state $i$. It is desired to use the $2n$ mutually independent observations $\{Y_{11}, Y_{12}, \ldots, Y_{1n}\}$ and $\{Y_{21}, Y_{22}, \ldots, Y_{2n}\}$ to make statistical inferences about the unknown parameters $\theta_1$ and $\theta_2$.

(a) Using these $2n$ mutually independent observations, develop an explicit expression for the likelihood ratio test statistic $-2\ln(\hat{\lambda})$ for testing the null hypothesis $H_0: \theta_1 = \theta_2 (= \theta, \text{say})$ versus the alternative hypothesis $H_1: \theta_1 \neq \theta_2$. For $n = 50$ and $k = 3$, if the observed data are such that $\sum_{j=1}^{n} y_{1j} = 5$ and $\sum_{j=1}^{n} y_{2j} = 10$, use the likelihood ratio statistic to test $H_0$ versus $H_1$ at the $\alpha = 0.05$ significance level. What is the P-value associated with this particular test?

(b) Using the observed data information given in part (a), what is the numerical value of the score statistic $\hat{S}$ for testing $H_0$ versus $H_1$? Use *observed* information in your calculations. What is the P-value associated with the use of $\hat{S}$ for testing $H_0$ versus $H_1$?

(c) For $i = 1, 2$, let $\bar{Y}_i = n^{-1} \sum_{j=1}^{n} Y_{ij}$. A biostatistician suggests that a test of $H_0: \theta_1 = \theta_2$ versus $H_1: \theta_1 \neq \theta_2$ can be based on a test statistic, involving $(\bar{Y}_1 - \bar{Y}_2)$, that is approximately N(0, 1) for large $n$ under $H_0$. Develop the structure of such a large-sample test statistic. For $k = 3$ and $\alpha = 0.05$, if the true parameter values are $\theta_1 = 2.0$ and $\theta_2 = 2.4$, provide a reasonable value for the minimum value of $n$ (say, $n^*$) so that the power of this large-sample test is at least 0.80 for rejecting $H_0$ in favor of $H_1$.

**Exercise 5.26***. Consider an investigation in which each member of a random sample of patients contributes a pair of binary (0–1) outcomes, with the possible outcomes being (1,1), (1,0), (0,1), and (0,0). Data such as these arise when a binary outcome (e.g., the presence or absence of a particular symptom) is measured on the same patient under two different conditions or at two different time points. Interest focuses on statistically testing whether the marginal probability of the occurrence of the outcome of interest differs for the two conditions or time points. To statistically analyze such data appropriately, it is necessary to account for the statistical dependence between the two outcomes measured on the same patient.

For a random sample of $n$ patients, let the discrete random variables $Y_{11}$, $Y_{10}$, $Y_{01}$, and $Y_{00}$ denote, respectively, the numbers of patients having the response patterns (1,1), (1,0), (0,1), and (0,0), where 1 denotes the presence of a particular symptom, 0 denotes the absence of that symptom, and the two outcome measurements are made before and after a particular therapeutic intervention. Assuming that patients respond independently of one another, the observed data $\{y_{11}, y_{10}, y_{01}, y_{00}\}$ may be assumed to arise from a multinomial distribution with corresponding probabilities $\{\pi_{11}, \pi_{10}, \pi_{01}, \pi_{00}\}$, where $\sum_{i=0}^{1}\sum_{j=0}^{1}\pi_{ij} = 1$. Note that the random variable $(Y_{11} + Y_{10})$ is the number of patients who have the symptom prior to the intervention, and that the random variable $(Y_{11} + Y_{01})$ is the number of patients who have the symptom after the intervention. Let

$$\delta = (\pi_{11} + \pi_{10}) - (\pi_{11} + \pi_{01}) = (\pi_{10} - \pi_{01}), \quad -1 < \delta < 1,$$

denote the difference in the probabilities of having the symptom before and after the intervention. Interest focuses on testing $H_0 : \delta = 0$ versus $H_1 : \delta \neq 0$.

(a) Given observed counts $y_{11}, y_{10}, y_{01}$, and $y_{00}$, develop an explicit expression for the MLE $\hat{\delta}$ of $\delta$.

(b) Using *expected* information, derive an explicit expression for the Wald chi-squared test statistic for testing $H_0 : \delta = 0$ versus $H_1 : \delta \neq 0$. What is the P-value of the Wald chi-squared test if $y_{11} = 22$, $y_{10} = 3$, $y_{01} = 7$, and $y_{00} = 13$?

(c) For testing $H_0 : \delta = 0$ versus $H_1 : \delta \neq 0$, the testing procedure known as *McNemar's Test* is based on the test statistic

$$Q_M = \frac{(Y_{01} - Y_{10})^2}{(Y_{01} + Y_{10})}.$$

Under $H_0$, the statistic $Q_M$ follows an asymptotic $\chi_1^2$ distribution, and so a two-sided test at the 0.05 significance level rejects $H_0$ in favor of $H_1$ when $Q_M > \chi^2_{1,0.95}$. Prove that McNemar's test statistic is identical to the score test statistic used to test $H_0 : \delta = 0$ versus $H_1 : \delta \neq 0$. Also, show that the Wald chi-squared statistic is always at least as large in value as the score chi-squared statistic.

(d) For the study in question, the investigators plan to enroll patients until $(y_{10} + y_{01})$ is equal to 10. Suppose that these investigators decide to reject $H_0$ if $Q_M > \chi^2_{1,0.95}$ and decide not to reject $H_0$ if $Q_M \leq \chi^2_{1,0.95}$. What is the exact probability (i.e., the *power*) that $H_0$ will be rejected if $\pi_{11} = 0.80$, $\pi_{10} = 0.10$, $\pi_{01} = 0.05$, and $\pi_{00} = 0.05$?

## SOLUTIONS

### Solution 5.1

(a) To find the form of the MP rejection region, we need to employ the Neyman–Pearson Lemma.
Now, with $\boldsymbol{x} = (x_1, x_2, \ldots, x_n)$, we have $\mathcal{L}(\boldsymbol{x};\theta) = \prod_{i=1}^{n}(\theta x_i^{\theta-1}) = \theta^n\left(\prod_{i=1}^{n} x_i\right)^{\theta-1}$.
In particular, for $n = 1$, $\mathcal{L}(\boldsymbol{x};\theta) = \theta x_1^{\theta-1}$. So,

$$\frac{\mathcal{L}(\boldsymbol{x};1)}{\mathcal{L}(\boldsymbol{x};2)} = \frac{(1)x_1^{1-1}}{(2)x_1^{2-1}} = (2x_1)^{-1} \leq k.$$

Thus, $x_1 \geq k_\alpha$ is the form of the MP rejection region.

Under $H_0: \theta = 1$, $f_{X_1}(x_1; 1) = 1$, $0 < x_1 < 1$, so that $k_\alpha = 0.95$; i.e., we reject $H_0$ if $x_1 > 0.95$.

So,

$$\text{POWER} = \text{pr}\{X_1 > 0.95 | \theta = 2\} = \int_{0.95}^{1} 2x_1^{2-1}\, dx_1 = 0.0975.$$

(b) For $n = 2$, $\mathcal{L}(x; \theta) = \theta^2 (x_1 x_2)^{\theta - 1}$. So,

$$\frac{\mathcal{L}(x;1)}{\mathcal{L}(x;2)} = \frac{1}{4x_1x_2} \leq k.$$

Thus, $x_1 x_2 \geq k_\alpha$ is the form of the MP rejection region.

Under $H_0: \theta = 1$,

$$f_{X_1,X_2}(x_1, x_2; 1) = f_{X_1}(x_1; 1) f_{X_2}(x_2; 1) = (1)(1) = 1,\ 0 < x_1 < 1, \quad 0 < x_2 < 1.$$

So, we need to pick $k_\alpha$ such that $\text{pr}[(X_1, X_2) \in \mathcal{R} | H_0 : \theta = 1] = 0.05$. In other words, we must choose $k_\alpha$ such that

$$\int_{k_\alpha}^{1} \int_{k_\alpha / x_1}^{1} (1)\, dx_2 dx_1 = 0.05 \Rightarrow [x_1 - k_\alpha \ln x_1]_{k_\alpha}^{1} = 0.05$$

$$\Rightarrow 1 - [k_\alpha - k_\alpha \ln k_\alpha] = 0.05 \Rightarrow k_\alpha \mathrel{\dot{\approx}} 0.70.$$

**Solution 5.2**

(a) With $y = (y_1, y_2, \ldots, y_n)$,

$$\mathcal{L}(y; \theta) = (1 + \theta)^n \prod_{i=1}^{n} (y_i + \theta)^{-2}.$$

The MP rejection region has the form

$$\frac{\mathcal{L}(y;0)}{\mathcal{L}(y;1)} = \frac{\prod_{i=1}^{n}(y_i + 0)^{-2}}{2^n \prod_{i=1}^{n}(y_i + 1)^{-2}} = 2^{-n} \prod_{i=1}^{n} \left( \frac{y_i + 1}{y_i} \right)^2 \leq k$$

or, equivalently,

$$\prod_{i=1}^{n} (1 + y_i^{-1})^2 \leq 2^n k.$$

So,

$$\mathcal{R} = \left\{ (y_1, y_2, \ldots, y_n) : \prod_{i=1}^{n} (1 + y_i^{-1})^2 \leq k_\alpha \right\},$$

where $k_\alpha$ is chosen so that $\text{pr}\{(Y_1, Y_2, \ldots, Y_n) \in \mathcal{R} | H_0 : \theta = 0\} = \alpha$.

(b) If $n = 1$, we need to find $k_\alpha$ such that

$$\text{pr}\left\{\left(1 + Y_1^{-1}\right)^2 < k_\alpha \middle| \text{H}_0 : \theta = 0\right\} = 0.05.$$

Since $y_1 > 1$, we have $k_\alpha > 1$, so that

$$\begin{aligned}
&\text{pr}\left\{\left(1 + Y_1^{-1}\right)^2 < k_\alpha \middle| \text{H}_0 : \theta = 0\right\} \\
&= \text{pr}\left\{1 + Y_1^{-1} < \sqrt{k_\alpha} \middle| \text{H}_0 : \theta = 0\right\} \\
&= \text{pr}\{Y_1 > (\sqrt{k_\alpha} - 1)^{-1} | \text{H}_0 : \theta = 0\} \\
&= \int_{(\sqrt{k_\alpha}-1)^{-1}}^{\infty} y_1^{-2}\, dy_1 \\
&= \left[-y_1^{-1}\right]_{(\sqrt{k_\alpha}-1)^{-1}}^{+\infty} \\
&= (\sqrt{k_\alpha} - 1) = 0.05,
\end{aligned}$$

so that $k_\alpha = (1.05)^2 = 1.1025$, or $k'_\alpha = (\sqrt{k_\alpha} - 1)^{-1} = 1/0.05 = 20$.

(c)

$$\begin{aligned}
\text{POWER} &= \text{pr}(Y_1 > 20 | \theta = 1) \\
&= \int_{20}^{\infty} 2(y_1 + 1)^{-2}\, dy_1 = 2\left[-(y_1 + 1)^{-1}\right]_{20}^{\infty} \\
&= \frac{2}{21} = 0.0952.
\end{aligned}$$

The power is very small because $n = 1$.

**Solution 5.3.** For any particular $\sigma_1^2 > 1$, the optimal rejection region for a most powerful (MP) test of $\text{H}_0 : \sigma^2 = 1$ versus $\text{H}_1 : \sigma^2 = \sigma_1^2$ has the form $\mathcal{L}(\boldsymbol{y}; 1)/\mathcal{L}(\boldsymbol{y}; \sigma_1^2) \leq k$, where $\boldsymbol{y} = (y_1, y_2, \ldots, y_n)$. Since

$$\mathcal{L}(\boldsymbol{y}; \sigma) = \prod_{i=1}^{n} \left\{\frac{1}{\sqrt{2\pi}\sigma} \text{e}^{-y_i^2/2\sigma^2}\right\} = (2\pi)^{-n/2}(\sigma^2)^{-n/2}\text{e}^{-(1/2\sigma^2)\sum_{i=1}^n y_i^2},$$

the optimal rejection region has the structure

$$\frac{\mathcal{L}(\boldsymbol{y}; 1)}{\mathcal{L}(\boldsymbol{y}; \sigma_1^2)} = \frac{(2\pi)^{-n/2}\text{e}^{-\frac{1}{2}\sum_{i=1}^n y_i^2}}{(2\pi)^{-n/2}(\sigma_1^2)^{-n/2}\text{e}^{-\frac{1}{2\sigma_1^2}\sum_{i=1}^n y_i^2}} = (\sigma_1^2)^{n/2}\text{e}^{\left(\frac{1}{2\sigma_1^2} - \frac{1}{2}\right)\sum_{i=1}^n y_i^2} \leq k.$$

Since $\left(1/2\sigma_1^2 - 1/2\right) < 0$ when $\sigma_1^2 > 1$, the MP test rejects when $\sum_{i=1}^n y_i^2$ is large, that is when $\sum_{i=1}^n y_i^2 \geq k'$ for some appropriately chosen $k'$. Because we obtain the same optimal rejection region for all $\sigma_1^2 > 1$, we have a UMP test. Under $\mathrm{H}_0 : \sigma^2 = 1$, $\sum_{i=1}^{n} Y_i^2 \sim \chi_n^2$; so, the appropriate critical value $k'$ for an $\alpha$-level test is $k' = \chi_{n,1-\alpha}^2$ because $\mathrm{pr}\left(\sum_{i=1}^{n} Y_i^2 \geq \chi_{n,1-\alpha}^2 \middle| H_0 : \sigma^2 = 1\right) = \alpha$. Now,

$$\begin{aligned}
\text{POWER} &= \mathrm{pr}\left[\sum_{i=1}^{n} Y_i^2 \geq \chi_{n,0.95}^2 \middle| \sigma^2 = 2\right] \\
&= \mathrm{pr}\left[\sum_{i=1}^{n} \left(\frac{Y_i}{\sqrt{2}}\right)^2 \geq \frac{\chi_{n,0.95}^2}{2} \middle| \sigma^2 = 2\right] \\
&= \mathrm{pr}\left[\chi_n^2 \geq \frac{\chi_{n,0.95}^2}{2}\right], \quad \text{since } \frac{Y_i}{\sqrt{2}} \sim \mathrm{N}(0,1) \text{ when } \sigma^2 = 2.
\end{aligned}$$

We want to find the smallest $n$ (say, $n^*$) such that this probability is at least 0.80. By inspection of chi-square tables, we find $n^* = 25$. Also, by the Central Limit Theorem, since $Z_i = Y_i/\sqrt{2} \sim \mathrm{N}(0,1)$ when $\sigma^2 = 2$,

$$\begin{aligned}
\text{POWER} &= \mathrm{pr}\left[\sum_{i=1}^{n} Z_i^2 \geq \frac{\chi_{n,0.95}^2}{2}\right] = \mathrm{pr}\left[\frac{\sum_{i=1}^{n} Z_i^2 - n}{\sqrt{2n}} \geq \frac{\chi_{n,0.95}^2/2 - n}{\sqrt{2n}}\right] \\
&\approx \mathrm{pr}\left[Z \geq \frac{\chi_{n,0.95}^2/2 - n}{\sqrt{2n}}\right],
\end{aligned}$$

where $\mathrm{E}(Z_i^2) = 1$, $\mathrm{V}(Z_i^2) = 2$, and $Z \mathrel{\dot\sim} \mathrm{N}(0,1)$ for large $n$. Since $Z_{0.20} = -0.842$, POWER $\geq 0.80$ when

$$\frac{(\chi_{n,0.95}^2/2) - n}{\sqrt{2n}} \leq -0.842,$$

or, equivalently, when $\chi_{n,0.95}^2 \leq 2[n - 0.842\sqrt{2n}]$, which is satisfied by a minimum value of $n^* = 25$.

**Solution 5.4.** (a) With $\boldsymbol{x} = (x_1, x_2, \ldots, x_n)$ and $\boldsymbol{y} = (y_1, y_2, \ldots, y_n)$, we have

$$\begin{aligned}
\mathcal{L}(\boldsymbol{x}, \boldsymbol{y}; \theta_1, \theta_2) &= \prod_{i=1}^{n} \theta_1^{x_i}(1-\theta_1)^{1-x_i} \cdot \prod_{i=1}^{n} \theta_2^{y_i}(1-\theta_2)^{1-y_i} \\
&= \theta_1^{s_x}(1-\theta_1)^{n-s_x}\theta_2^{s_y}(1-\theta_2)^{n-s_y},
\end{aligned}$$

where $s_x = \sum_{i=1}^{n} x_i$ and $s_y = \sum_{i=1}^{n} y_i$. So, using the Neyman–Pearson Lemma,

$$\begin{aligned}
\frac{\mathcal{L}(\boldsymbol{x},\boldsymbol{y};0.50,0.50)}{\mathcal{L}(\boldsymbol{x},\boldsymbol{y};0.60,0.60)} &= \frac{(0.50)^{s_x}(0.50)^{n-s_x}(0.50)^{s_y}(0.50)^{n-s_y}}{(0.60)^{s_x}(0.40)^{n-s_x}(0.60)^{s_y}(0.40)^{n-s_y}} \\
&= \frac{(0.50)^{2n}}{(0.60)^{(s_x+s_y)}(0.40)^{2n-(s_x+s_y)}} \\
&= \left(\frac{2}{3}\right)^{(s_x+s_y)}\left(\frac{5}{4}\right)^{2n} \leq k \Longrightarrow (s_x+s_y) \geq k'
\end{aligned}$$

is the structure of the MP region. When $\theta_1 = \theta_2 = \theta$, $S = (S_x + S_y) \sim \text{BIN}(2n, \theta)$. So, by the Central Limit Theorem, under $\text{H}_0 : \theta = \frac{1}{2}$

$$\frac{S-n}{\sqrt{n/2}} \stackrel{.}{\sim} \text{N}(0,1)$$

for large $n$. So, for a size $\alpha = 0.05$ test of $\text{H}_0$ versus $\text{H}_1$,

$$\begin{aligned}
\text{POWER} &= \text{pr}\left\{\frac{S-n}{\sqrt{n/2}} > 1.645 \,\middle|\, \text{H}_1\right\} \\
&= \text{pr}\left\{S > 1.645\sqrt{\frac{n}{2}} + n \,\middle|\, \text{H}_1\right\} \\
&= \text{pr}\left\{\frac{S-2n(0.60)}{\sqrt{2n(0.60)(0.40)}} > \frac{1.645\sqrt{\frac{n}{2}} + n - 2n(0.60)}{\sqrt{2n(0.60)(0.40)}} \,\middle|\, \text{H}_1\right\} \\
&\stackrel{.}{\approx} \text{pr}\left\{Z > \frac{1.645\sqrt{\frac{30}{2}} + 30 - 2(30)(0.60)}{\sqrt{2(30)(0.60)(0.40)}} \,\middle|\, \text{H}_1\right\}
\end{aligned}$$

where $Z \sim \text{N}(0,1)$. So

$$\text{POWER} \stackrel{.}{\approx} \text{pr}(Z > 0.0978) = 1 - \Phi(0.0978) = 0.46.$$

(b) Under $\text{H}_0 : \theta_1 = \theta_2 = \theta_0$,

$$\frac{(\bar{X}-\bar{Y})}{\sqrt{\frac{2\theta_0(1-\theta_0)}{n}}} \stackrel{.}{\sim} \text{N}(0,1)$$

for reasonably large $n$. So,

$$\text{POWER} = \text{pr}\left\{\frac{(\bar{X}-\bar{Y})}{\sqrt{\frac{2\theta_0(1-\theta_0)}{n}}} > 1.645 \,\middle|\, (\theta_1-\theta_2) \geq 0.20\right\}$$

$$\geq \text{pr}\left\{\frac{(\bar{X}-\bar{Y})}{\sqrt{\frac{2\theta_0(1-\theta_0)}{n}}} > 1.645 \,\middle|\, (\theta_1-\theta_2) = 0.20\right\}$$

$$= \text{pr}\left\{(\bar{X}-\bar{Y}) > 1.645\sqrt{\frac{2\theta_0(1-\theta_0)}{n}} \,\middle|\, (\theta_1-\theta_2) = 0.20\right\}$$

$$= \text{pr}\left\{\frac{(\bar{X}-\bar{Y})-0.20}{\sqrt{\frac{\theta_1(1-\theta_1)}{n}+\frac{\theta_2(1-\theta_2)}{n}}} > \frac{1.645\sqrt{\frac{2\theta_0(1-\theta_0)}{n}}-0.20}{\sqrt{\frac{\theta_1(1-\theta_1)}{n}+\frac{\theta_2(1-\theta_2)}{n}}}\right\}$$

$$\dot{\approx}\ \text{pr}\left\{Z > \frac{1.645\sqrt{\frac{2\theta_0(1-\theta_0)}{n}}-0.20}{\sqrt{\frac{\theta_1(1-\theta_1)}{n}+\frac{\theta_2(1-\theta_2)}{n}}}\right\},$$

where $Z \sim N(0,1)$. So, for POWER $\geq 0.90$, we require

$$\frac{1.645\sqrt{2\theta_0(1-\theta_0)}-0.20\sqrt{n}}{\sqrt{\theta_1(1-\theta_1)+\theta_2(1-\theta_2)}} \leq -1.282,$$

or

$$n \geq \left[\frac{1.645\sqrt{2\theta_0(1-\theta_0)}+1.282\sqrt{\theta_1(1-\theta_1)+\theta_2(1-\theta_2)}}{0.20}\right]^2.$$

Now, given that $\theta_1 = (\theta_2 + 0.20)$, the quantity $[\theta_1(1-\theta_1)+\theta_2(1-\theta_2)]$ is maximized at $\theta_1 = 0.60$ and $\theta_2 = 0.40$. So, for $\theta_0 = 0.10$ and to cover all $(\theta_1, \theta_2)$ values, choose

$$n \geq \left[\frac{1.645\sqrt{2(0.10)(0.90)}+1.282\sqrt{0.60(0.40)+0.40(0.60)}}{0.20}\right]^2 = 62.88;$$

so, $n^* = 63$.

**Solution 5.5**

(a) Consider the simple null hypothesis $H_0$: $\theta = 1$ versus the simple alternative hypothesis $H_1$: $\theta = \theta_1 (> 1)$, where $\theta_1$ is any specific value of $\theta$ greater than 1.

Then, from the Neyman–Pearson Lemma and with $\boldsymbol{y} = (y_1, y_2, \ldots, y_n)$, the form of the rejection region for a MP test is based on the inequality

$$\frac{\mathcal{L}(\boldsymbol{y};1)}{\mathcal{L}(\boldsymbol{y};\theta_1)} \leq k,$$

or

$$\frac{\left(\prod_{i=1}^{n} x_i^{y_i}\right) e^{-\sum_{i=1}^{n} x_i} / \left(\prod_{i=1}^{n} y_i!\right)}{\theta_1^{\sum_{i=1}^{n} y_i} \left(\prod_{i=1}^{n} x_i^{y_i}\right) e^{-\theta_1 \sum_{i=1}^{n} x_i} / \left(\prod_{i=1}^{n} y_i!\right)} \leq k,$$

or

$$\theta_1^{-\sum_{i=1}^{n} y_i} e^{(\theta_1 - 1)\sum_{i=1}^{n} x_i} \leq k,$$

or

$$\left(-\sum_{i=1}^{n} y_i\right)(\ln \theta_1) \leq k',$$

or

$$\sum_{i=1}^{n} y_i \geq k'', \quad \text{since } \ln \theta_1 > 0.$$

So, the MP rejection region $\mathcal{R} = \{S : S \geq k_\alpha\}$, where $S = \sum_{i=1}^{n} Y_i$. Since $S$ is a discrete random variable, $k_\alpha$ is a positive integer chosen so that $\text{pr}\{S \geq k_\alpha | \text{H}_0\colon \theta = 1\} \doteq \alpha$. Since this same form of rejection region is obtained for any value $\theta_1 > 1$, $\mathcal{R}$ is the UMP region for a test of $\text{H}_0\colon \theta = 1$ versus $\text{H}_1\colon \theta > 1$.

(b) Since the test statistic is $S = \sum_{i=1}^{n} Y_i$, we need to know the distribution of $S$. Since $Y_i \sim \text{POI}(\theta x_i)$, $i = 1, 2, \ldots, n$, and since the $\{Y_i\}$ are mutually independent,

$$\text{M}_S(t) = \text{E}[e^{tS}] = \text{E}[e^{t\sum_{i=1}^{n} Y_i}] = \text{E}\left[\prod_{i=1}^{n} e^{tY_i}\right] = \prod_{i=1}^{n} [\text{E}(e^{tY_i})]$$

$$= \prod_{i=1}^{n} [e^{\theta x_i (e^t - 1)}] = e^{(\theta \sum_{i=1}^{n} x_i)(e^t - 1)} = e^{0.82\theta(e^t - 1)},$$

so that $S \sim \text{POI}(0.82\theta)$. So, under $\text{H}_0\colon \theta = 1$, $S \sim \text{POI}(0.82)$. So, we need to find $k_{.05}$ such that

$$\text{pr}(S \geq k_{.05} | \theta = 1) = 1 - \sum_{s=0}^{(k_{.05} - 1)} \frac{(0.82)^s e^{-0.82}}{s!} \doteq 0.05,$$

or such that

$$\sum_{s=0}^{(k_{.05}-1)} \frac{(0.82)^s e^{-0.82}}{s!} \doteq 0.95.$$

By trial-and-error, $k_{.05} = 3$. So, for $\alpha \doteq 0.05$, we reject $H_0{:}\,\theta = 1$ in favor of $H_1{:}\,\theta > 1$ when $S = \sum_{i=1}^{n} Y_i \geq 3$. Now, when $\theta = 5$, $S \sim \text{POI}(4.10)$, so that

$$\begin{aligned} \text{POWER} &= \text{pr}(S \geq 3|\theta = 5) = 1 - \text{pr}(S < 3|\theta = 5) \\ &= 1 - \sum_{s=0}^{2} \frac{(4.10)^s e^{-4.10}}{s!} = 1 - 0.2238 = 0.7762. \end{aligned}$$

**Solution 5.6.** (a) Now, with $k$ a constant,

$$\alpha = \text{pr}[(\bar{X}_1 - \bar{X}_2) > k|H_0] = \text{pr}\left[Z > \frac{k}{\sqrt{V}}\middle|H_0\right],$$

so that we require $k/\sqrt{V} = Z_{1-\alpha}$, or $k = \sqrt{V}Z_{1-\alpha}$.

And,

$$(1-\beta) = \text{pr}[(\bar{X}_1 - \bar{X}_2) > k|H_1] = \text{pr}\left[Z - \frac{\delta}{\sqrt{V}} > \frac{(k-\delta)}{\sqrt{V}}\middle|H_1\right],$$

which, since $\left(Z - \frac{\delta}{\sqrt{V}}\right) \sim N(0,1)$ when $H_1$ is true, requires that

$$\frac{(k-\delta)}{\sqrt{V}} = -Z_{1-\beta}, \quad \text{or} \quad k = -\sqrt{V}Z_{1-\beta} + \delta.$$

Finally, the equation

$$\sqrt{V}Z_{1-\alpha} = -\sqrt{V}Z_{1-\beta} + \delta$$

gives the requirement $V = \theta$.

(b) Since the goal is to minimize $N = (n_1 + n_2)$ with respect to $n_1$ and $n_2$, subject to the constraint $V = \theta$, consider the function

$$Q = (n_1 + n_2) + \lambda\left(\frac{\sigma_1^2}{n_1} + \frac{\sigma_2^2}{n_2} - \theta\right),$$

where $\lambda$ is a Lagrange multiplier.

Then, simultaneously solving the two equations

$$\frac{\partial Q}{\partial n_1} = 1 - \frac{\lambda\sigma_1^2}{n_1^2} = 0 \quad \text{and} \quad \frac{\partial Q}{\partial n_2} = 1 - \frac{\lambda\sigma_2^2}{n_2^2} = 0$$

gives

$$\frac{n_1^2}{n_2^2} = \frac{\sigma_1^2}{\sigma_2^2} \quad \text{or} \quad \frac{n_1}{n_2} = \frac{\sigma_1}{\sigma_2}.$$

Finally, if $N = 100, \sigma_1^2 = 9$, and $\sigma_2^2 = 4$, then $n_1/n_2 = 1.5$; then, the equation $(n_1 + n_2) = (1.5n_2 + n_2) = 2.5n_2 = 100$ gives $n_2 = 40$ and $n_1 = 60$.

(c) Let $C$ denote the cost of selecting an observation from Population 2, so that the total sampling cost is $(4Cn_1 + Cn_2) = C(4n_1 + n_2)$. So, we want to minimize the function $C(4n_1 + n_2)$ with respect to $n_1$ and $n_2$, subject to the constraint $V = \theta$. Again using Lagrange multipliers, if

$$Q = C(4n_1 + n_2) + \lambda\left(\frac{\sigma_1^2}{n_1} + \frac{\sigma_2^2}{n_2} - \theta\right),$$

then the equation

$$\frac{\partial Q}{\partial n_1} = 0 \quad \text{gives} \quad \lambda = \frac{4Cn_1^2}{\sigma_1^2}$$

and the equation

$$\frac{\partial Q}{\partial n_2} = 0 \quad \text{gives} \quad \lambda = \frac{Cn_2^2}{\sigma_2^2},$$

implying that $n_1/n_2 = \sigma_1/2\sigma_2$.

And, since the equation

$$\frac{\partial Q}{\partial \lambda} = 0 \quad \text{gives} \quad V = \left(\frac{\sigma_1^2}{n_1} + \frac{\sigma_2^2}{n_2}\right) = \theta,$$

so that

$$n_1 = \frac{\left[\sigma_1^2 + (n_1/n_2)\,\sigma_2^2\right]}{\theta},$$

we obtain

$$n_1 = \frac{\left[\sigma_1^2 + (\sigma_1/2\sigma_2)\,\sigma_2^2\right]}{\theta} = \frac{\left(\sigma_1^2 + (\sigma_1\sigma_2)/2\right)}{\theta}$$

and

$$n_2 = \left(\frac{2\sigma_2}{\sigma_1}\right) n_1 = \frac{(2\sigma_1\sigma_2 + \sigma_2^2)}{\theta}.$$

Then, with $\sigma_1 = 5, \sigma_2 = 4, \alpha = 0.05, \beta = 0.10$, and $\delta = 3$, then $Z_{1-\alpha} = Z_{0.95} = 1.645, Z_{1-\beta} = Z_{0.90} = 1.282$, and $V = (3)^2/(1.645 + 1.282)^2 = 1.0505$. Using these values, we obtain $n_1 = 33.3175$ and $n_2 = 53.3079$; in practice, one would use $n_1 = 34$ and $n_2 = 54$.

**Solution 5.7**

(a) Note that the CDF of X is

$$F_X(x) = \text{pr}(X \leq x) = \int_0^x \theta^{-1} dt = \frac{x}{\theta}, \quad 0 < x < \theta.$$

Hence, it follows that

$$\begin{aligned}
\alpha = \text{pr(Type I error)} &= \text{pr}\left[X_{(n)} > c^* \middle| H_0 : \theta = \frac{1}{2}\right] \\
&= 1 - \text{pr}\left\{\bigcap_{i=1}^{n}(X_i \leq c^*) \middle| H_0 : \theta = \frac{1}{2}\right\} \\
&= 1 - \prod_{i=1}^{n}\left[\text{pr}\left(X_i \leq c^* \middle| H_0 : \theta = \frac{1}{2}\right)\right] \\
&= 1 - \left[\frac{c^*}{\left(\frac{1}{2}\right)}\right]^n = 1 - (2c^*)^n.
\end{aligned}$$

$$\Rightarrow \quad (2c^*)^n = (1 - \alpha)$$

$$\Rightarrow \quad c^* = \frac{(1-\alpha)^{1/n}}{2}.$$

For $0 < \alpha < 1$, note that $0 < c^* < \frac{1}{2}$.

(b) When $\alpha = 0.05$, $c^* = (0.95)^{1/n}/2$. So,

$$\begin{aligned}
0.98 \leq \text{POWER} &= \text{pr}\left\{X_{(n)} > c^* \middle| \theta = \frac{3}{4}\right\}. \\
&= 1 - \text{pr}\left\{\bigcap_{i=1}^{n}(X_i \leq c^*) \middle| \theta = \frac{3}{4}\right\}. \\
&= 1 - \prod_{i=1}^{n}\text{pr}\left\{X_i \leq \frac{(0.95)^{1/n}}{2} \middle| \theta = \frac{3}{4}\right\} \\
&= 1 - \left[\frac{(0.95)^{1/n}/2}{\left(\frac{3}{4}\right)}\right]^n
\end{aligned}$$

$$= 1 - (0.95)\left(\frac{2}{3}\right)^n$$

$$\Rightarrow \quad -0.02 \leq -(0.95)\left(\frac{2}{3}\right)^n$$

$$\Rightarrow \quad \left(\frac{2}{3}\right)^n \leq 0.0211 \Rightarrow n^* = 10.$$

**Solution 5.8**

(a) Under $H_1$, and with $\boldsymbol{x} = (x_1, x_2, \ldots, x_n)$ and $\boldsymbol{\mu} = (\mu_1, \mu_2, \ldots, \mu_n)$, the (unrestricted) likelihood and log-likelihood functions are

$$\mathcal{L}(\boldsymbol{x}; \boldsymbol{\mu}) = \prod_{i=1}^{n} \left[\frac{\mu_i^{x_i} e^{-\mu_i}}{x_i!}\right] = \left(\prod_{i=1}^{n} \mu_i^{x_i}\right) e^{-\sum_{i=1}^{n} \mu_i} \left(\prod_{i=1}^{n} x_i!\right)^{-1}$$

and

$$\ln \mathcal{L}(\boldsymbol{x}; \boldsymbol{\mu}) = \sum_{i=1}^{n} x_i \ln \mu_i - \sum_{i=1}^{n} \mu_i - \sum_{i=1}^{n} \ln x_i!.$$

Solving

$$\frac{\partial \ln \mathcal{L}(\boldsymbol{x}; \boldsymbol{\mu})}{\partial \mu_i} = \frac{x_i}{\mu_i} - 1 = 0$$

yields the (unrestricted) MLEs $\hat{\mu}_i = x_i$, $i = 1, 2, \ldots, n$. Thus, with $\hat{\boldsymbol{\mu}} = \boldsymbol{x}$, we have

$$\mathcal{L}(\boldsymbol{x}; \hat{\boldsymbol{\mu}}) = \left(\prod_{i=1}^{n} x_i^{x_i}\right) e^{-\sum_{i=1}^{n} x_i} \left(\prod_{i=1}^{n} x_i!\right)^{-1}.$$

Under $H_0$, the likelihood and log-likelihood functions are

$$\mathcal{L}(\boldsymbol{x}; \mu) = \mu^{\sum_{i=1}^{n} x_i} e^{-n\mu} \left(\prod_{i=1}^{n} x_i!\right)^{-1}$$

and

$$\ln \mathcal{L}(\boldsymbol{x}; \mu) = \left(\sum_{i=1}^{n} x_i\right) \ln \mu - n\mu - \sum_{i=1}^{n} \ln x_i!.$$

Solving

$$\frac{\partial \ln \mathcal{L}(\boldsymbol{x}; \mu)}{\partial \mu} = \frac{\sum_{i=1}^{n} x_i}{\mu} - n = 0$$

yields the (restricted) MLE $\hat{\mu} = \bar{x} = n^{-1}\sum_{i=1}^{n} x_i$. Thus,

$$\mathcal{L}(x;\hat{\mu}) = (\bar{x})^{n\bar{x}} e^{-n\bar{x}} \left(\prod_{i=1}^{n} x_i!\right)^{-1}.$$

So, the likelihood ratio statistic is

$$\hat{\lambda} = \frac{\mathcal{L}(x;\hat{\mu})}{\mathcal{L}(x;\hat{\boldsymbol{\mu}})} = \frac{(\bar{x})^{n\bar{x}} e^{-n\bar{x}} \left(\prod_{i=1}^{n} x_i!\right)^{-1}}{\left(\prod_{i=1}^{n} x_i^{x_i}\right) e^{-n\bar{x}} \left(\prod_{i=1}^{n} x_i!\right)^{-1}} = \frac{(\bar{x})^{n\bar{x}}}{\left(\prod_{i=1}^{n} x_i^{x_i}\right)}.$$

So,

$$\ln\hat{\lambda} = (n\bar{x})\ln\bar{x} - \sum_{i=1}^{n} x_i \ln x_i = (\ln\bar{x})\sum_{i=1}^{n} x_i - \sum_{i=1}^{n} x_i \ln x_i$$
$$= \sum_{i=1}^{n} x_i(\ln\bar{x} - \ln x_i) = \sum_{i=1}^{n} x_i \ln\left(\frac{\bar{x}}{x_i}\right),$$

so that

$$-2\ln\hat{\lambda} = 2\sum_{i=1}^{n} x_i \ln\left(\frac{x_i}{\bar{x}}\right).$$

Under $H_0$: $\mu_1 = \mu_2 = \cdots = \mu_n$, $-2\ln\hat{\lambda} \dot{\sim} \chi^2_{(n-1)}$ for large $n$. For the given data set,

$$\bar{x} = \frac{20(5) + 10(6) + 10(8)}{40} = \frac{240}{40} = 6,$$

so that

$$-2\ln\hat{\lambda} = 2\left\{20\left[5\ln\left(\frac{5}{6}\right)\right] + 10\left[6\ln\left(\frac{6}{6}\right)\right] + 10\left[8\ln\left(\frac{8}{6}\right)\right]\right\}$$
$$= 2[100(-0.1823) + 0 + 80(0.2877)] = 2(23.015 - 18.230)$$
$$= 9.570.$$

Since $\chi^2_{0.95,39} > 50$, we do not reject $H_0$.

(b) Based on the results in part (a), there is no evidence to reject $H_0$. Hence, $\mathcal{L}(x;\mu)$ is the appropriate likelihood to use. Since

$$\frac{\partial \ln \mathcal{L}(x;\mu)}{\partial\mu} = \frac{\sum_{i=1}^{n} x_i}{\mu} - n = n\left(\frac{\bar{x}}{\mu} - 1\right) = \frac{n}{\mu}(\bar{x} - \mu),$$

it follows from exponential family theory that $\bar{X}$ is the MVBUE of $\mu$. Hence, a CI based on $\bar{X}$ would be an appropriate choice.

From ML theory (or Central Limit Theorem theory),

$$\frac{\bar{X} - \mu}{\sqrt{\mathrm{V}(\bar{X})}} = \frac{\bar{X} - \mu}{\sqrt{\mu/n}} \dot{\sim} \mathrm{N}(0, 1)$$

for large $n$. Since $\bar{X}$ is consistent for $\mu$, by Slutsky's Theorem,

$$\frac{\bar{X} - \mu}{\sqrt{\bar{X}/n}} \dot{\sim} \mathrm{N}(0, 1)$$

for large $n$. Thus, an appropriate large-sample $100(1 - \alpha)\%$ CI for $\mu$ is: $\bar{X} \pm Z_{1-\alpha/2}\sqrt{\bar{X}/n}$. For the given data set, and for $\alpha = 0.05$, we have $6 \pm 1.96\sqrt{\frac{6}{40}} = 6 \pm 0.759$, giving $(5.241, 6.759)$ as the computed 95% CI for $\mu$.

**Solution 5.9**

(a) With $\boldsymbol{x} = (x_{11}, x_{12}, \ldots, x_{1n}; x_{21}, x_{22}, \ldots, x_{2n})$,

$$\begin{aligned}
\mathcal{L}(\boldsymbol{x}; \lambda_1, \lambda_2) &= \prod_{i=1}^{2}\prod_{j=1}^{n}\left\{\frac{(L_{ij}\lambda_i)^{x_{ij}}\mathrm{e}^{-L_{ij}\lambda_i}}{x_{ij}!}\right\} \\
&= \left(\prod_{i=1}^{2}\prod_{j=1}^{n}(x_{ij}!)^{-1}\right)\lambda_1^{\sum_{j=1}^{n}x_{1j}}\lambda_2^{\sum_{j=1}^{n}x_{2j}}\left(\prod_{i=1}^{2}\prod_{j=1}^{n}L_{ij}^{x_{ij}}\right) \\
&\quad \times \left(\mathrm{e}^{-\lambda_1\sum_{j=1}^{n}L_{1j}}\mathrm{e}^{-\lambda_2\sum_{j=1}^{n}L_{2j}}\right) \\
&= \left\{\lambda_1^{\sum_{j=1}^{n}x_{1j}}\lambda_2^{\sum_{j=1}^{n}x_{2j}}\mathrm{e}^{-\lambda_1\sum_{j=1}^{n}L_{1j}}\mathrm{e}^{-\lambda_2\sum_{j=1}^{n}L_{2j}}\right\} \\
&\quad \times \left\{\left(\prod_{i=1}^{2}\prod_{j=1}^{n}(x_{ij}!)^{-1}\right)\cdot\left(\prod_{i=1}^{2}\prod_{j=1}^{n}L_{ij}^{x_{ij}}\right)\right\},
\end{aligned}$$

so $\sum_{j=1}^{n} X_{1j}$ and $\sum_{j=1}^{n} X_{2j}$ are jointly sufficient for $\lambda_1$ and $\lambda_2$ by the Factorization Theorem.

(b)

$$\begin{aligned}
\ln \mathcal{L}(\boldsymbol{x}; \lambda_1, \lambda_2) &= \text{constant} + \left(\sum_{j=1}^{n} x_{1j}\right)\ln\lambda_1 + \left(\sum_{j=1}^{n} x_{2j}\right)\ln\lambda_2 \\
&\quad - \lambda_1\sum_{j=1}^{n} L_{1j} - \lambda_2\sum_{j=1}^{n} L_{2j}.
\end{aligned}$$

Solving for $\lambda_i$ in the equation

$$\frac{\partial \ln \mathcal{L}(\boldsymbol{x}; \lambda_1, \lambda_2)}{\partial \lambda_i} = \frac{\sum_{j=1}^{n} x_{ij}}{\lambda_i} - \sum_{j=1}^{n} L_{ij} = 0$$

yields the MLE

$$\hat{\lambda}_i = \sum_{j=1}^{n} X_{ij} \Big/ \sum_{j=1}^{n} L_{ij}, \quad i = 1, 2.$$

(c) Under $\mathrm{H}_0 : \lambda_1 = \lambda_2 (= \lambda, \text{say})$,

$$\ln \mathcal{L}(\boldsymbol{x}; \lambda) = \text{constant} + \left( \sum_{j=1}^{n} x_{1j} + \sum_{j=1}^{n} x_{2j} \right) \ln \lambda - \lambda \left( \sum_{j=1}^{n} L_{1j} + \sum_{j=1}^{n} L_{2j} \right).$$

Solving the equation

$$\frac{\partial \ln L(\boldsymbol{x}; \lambda)}{\partial \lambda} = \frac{\left( \sum_{j=1}^{n} x_{1j} + \sum_{j=1}^{n} x_{2j} \right)}{\lambda} - \left( \sum_{j=1}^{n} L_{1j} + \sum_{j=1}^{n} L_{2j} \right) = 0$$

yields the MLE

$$\hat{\lambda} = \frac{\sum_{j=1}^{n} (x_{1j} + x_{2j})}{\sum_{j=1}^{n} (L_{1j} + L_{2j})}.$$

(d) Now,

$$\frac{\hat{\mathcal{L}}_\omega}{\hat{\mathcal{L}}_\Omega} = \frac{\prod_{i=1}^{2} \left\{ \left( \prod_{j=1}^{n} L_{ij}^{x_{ij}} \right) \hat{\lambda}^{\sum_{j=1}^{n} x_{ij}} e^{-\hat{\lambda} \sum_{j=1}^{n} L_{ij}} \Big/ \prod_{j=1}^{n} x_{ij}! \right\}}{\prod_{i=1}^{2} \left\{ \left( \prod_{j=1}^{n} L_{ij}^{x_{ij}} \right) \hat{\lambda}_i^{\sum_{j=1}^{n} x_{ij}} e^{-\hat{\lambda}_i \sum_{j=1}^{n} L_{ij}} \Big/ \prod_{j=1}^{n} x_{ij}! \right\}}$$

$$= \frac{\hat{\lambda}^{\sum_{j=1}^{n} (x_{1j} + x_{2j})} e^{-\hat{\lambda} \left( \sum_{j=1}^{n} L_{1j} + \sum_{j=1}^{n} L_{2j} \right)}}{\hat{\lambda}_1^{\sum_{j=1}^{n} x_{1j}} \hat{\lambda}_2^{\sum_{j=1}^{n} x_{2j}} e^{-\hat{\lambda}_1 \sum_{j=1}^{n} L_{1j}} e^{-\hat{\lambda}_2 \sum_{j=1}^{n} L_{2j}}}.$$

And, for large $n$,

$$-2 \ln \left( \frac{\hat{\mathcal{L}}_\omega}{\hat{\mathcal{L}}_\Omega} \right) \dot{\sim} \chi_1^2$$

under $\mathrm{H}_0 : \lambda_1 = \lambda_2$.

(e) From part (d),

$$
\begin{aligned}
&-2\ln(\hat{\mathcal{L}}_{\omega}/\hat{\mathcal{L}}_{\Omega})\\
&= -2\left\{\left(\sum_{j=1}^{n} x_{1j} + \sum_{j=1}^{n} x_{2j}\right)\ln\hat{\lambda} - \hat{\lambda}\left(\sum_{j=1}^{n} L_{1j} + \sum_{j=1}^{n} L_{2j}\right)\right.\\
&\quad \left. - \left(\sum_{j=1}^{n} x_{1j}\right)\ln\hat{\lambda}_1 - \left(\sum_{j=1}^{n} x_{2j}\right)\ln\hat{\lambda}_2 + \hat{\lambda}_1\sum_{j=1}^{n} L_{1j} + \hat{\lambda}_2\sum_{j=1}^{n} L_{2j}\right\}\\
&= -2\left\{(4+9)\ln\left(\frac{4+9}{200+300}\right) - \left(\frac{4+9}{200+300}\right)(200+300)\right.\\
&\quad \left. - (4)\ln(0.02) - (9)\ln(0.03) + 0.02(200) + 0.03(300)\right\}\\
&= 0.477.
\end{aligned}
$$

Since $\chi^2_{1,0.90} = 2.706$, we do not reject $H_0 : \lambda_1 = \lambda_2$, and the

$$
\begin{aligned}
\text{P-value} &= \text{pr}\left\{\chi^2_1 > 0.477 | H_0 : \lambda_1 = \lambda_2\right\}\\
&\doteq 0.50.
\end{aligned}
$$

**Solution 5.10**

(a) Under $H_0 : \alpha = \beta (= \gamma, \text{say})$, the restricted likelihood is $\mathcal{L}_{\omega} = \gamma^{2n} e^{-n\gamma(\bar{x}+\bar{y})}$. So,

$$\frac{\partial \ln(\mathcal{L}_{\omega})}{\partial \gamma} = \frac{2n}{\gamma} - n(\bar{x}+\bar{y}) = 0 \quad \text{gives} \quad \hat{\gamma}_{\omega} = 2(\bar{x}+\bar{y})^{-1}.$$

Thus,

$$\hat{\mathcal{L}}_{\omega} = \hat{\gamma}_{\omega}^{2n} e^{-n\hat{\gamma}_{\omega}(\bar{x}+\bar{y})} = \left[\frac{2}{(\bar{x}+\bar{y})}\right]^{2n} e^{-2n}.$$

Under $H_1 : \alpha \neq \beta$, the unrestricted likelihood is $\mathcal{L}_{\Omega} = \alpha^n e^{-n\alpha\bar{x}} \beta^n e^{-n\beta\bar{y}}$. Thus,

$$\frac{\partial \ln(\mathcal{L}_{\Omega})}{\partial \alpha} = \frac{n}{\alpha} - n\bar{x} = 0 \quad \text{gives} \quad \hat{\alpha}_{\Omega} = (\bar{x})^{-1},$$

and

$$\frac{\partial \ln(\mathcal{L}_{\Omega})}{\partial \beta} = \frac{n}{\beta} - n\bar{y} = 0 \quad \text{gives} \quad \hat{\beta}_{\Omega} = (\bar{y})^{-1}.$$

Thus,

$$\hat{\mathcal{L}}_{\Omega} = \left(\bar{x}^{-1}\right)^n e^{-n\bar{x}^{-1}\bar{x}} \left(\bar{y}^{-1}\right)^n e^{-n\bar{y}^{-1}\bar{y}} = (\bar{x}\bar{y})^{-n} e^{-2n}.$$

Finally, the likelihood ratio statistic $\hat{\lambda}$ can be written as

$$\hat{\lambda} = \frac{\hat{\mathcal{L}}_\omega}{\hat{\mathcal{L}}_\Omega} = \left[\frac{4\bar{x}\bar{y}}{(\bar{x}+\bar{y})^2}\right]^n = [4u(1-u)]^n, \quad \text{with} \quad u = \frac{\bar{x}}{(\bar{x}+\bar{y})}.$$

(b) For the given set of data, $\hat{\lambda} = 0.0016$. For large $n$ and under $H_0 : \alpha = \beta$, the random variable $-2\ln(\hat{\lambda}) \dot{\sim} \chi_1^2$. So,

$$\text{P-value} \approx \text{pr}[-2\ln(\hat{\lambda}) > -2\ln(0.0016)] < 0.0005.$$

Since $E(X) = 1/\alpha$ and $E(Y) = 1/\beta$, the available data provide strong statistical evidence that the two surgical procedures lead to different true average survival times for patients with advanced colon cancer.

**Solution 5.11**

(a) With $\boldsymbol{x} = (x_1, x_2, \ldots, x_n)$ and $\boldsymbol{y} = (y_1, y_2, \ldots, y_n)$, the unconstrained likelihood has the form

$$\mathcal{L}_\Omega = \mathcal{L}(\boldsymbol{x}, \boldsymbol{y}; \theta_1, \theta_2) = \prod_{i=1}^{n} \left[\theta_1(1-\theta_1)^{x_i}\theta_2(1-\theta_2)^{y_i}\right]$$
$$= \theta_1^n(1-\theta_1)^{s_x}\theta_2^n(1-\theta_2)^{s_y},$$

where $s_x = \sum_{i=1}^n x_i$ and $s_y = \sum_{i=1}^n y_i$. So, $\ln \mathcal{L}_\Omega = n\ln\theta_1 + s_x\ln(1-\theta_1) + n\ln\theta_2 + s_y\ln(1-\theta_2)$. Solving $\partial \ln \mathcal{L}_\Omega/\partial\theta_1 = 0$ and $\partial \ln \mathcal{L}_\Omega/\partial\theta_2 = 0$ yields the unconstrained MLEs $\hat{\theta}_1 = 1/(1+\bar{x})$ and $\hat{\theta}_2 = 1/(1+\bar{y})$. Substituting these unrestricted MLEs for $\theta_1$ and $\theta_2$ into the expression for $\mathcal{L}_\Omega$ gives

$$\hat{\mathcal{L}}_\Omega = \left(\frac{1}{1+\bar{x}}\right)^n \left(\frac{\bar{x}}{1+\bar{x}}\right)^{n\bar{x}} \left(\frac{1}{1+\bar{y}}\right)^n \left(\frac{\bar{y}}{1+\bar{y}}\right)^{n\bar{y}}.$$

Also, when $\theta_1 = \theta_2 = \theta$, the constrained likelihood has the form $\mathcal{L}_\omega = \theta^{2n}(1-\theta)^{(s_x+s_y)}$, so that $\ln \mathcal{L}_\omega = 2n\ln\theta + (s_x+s_y)\ln(1-\theta)$. Solving $\partial \ln \mathcal{L}_\omega/\partial\theta = 0$ yields the constrained MLE $\hat{\theta} = 2/(2+\bar{x}+\bar{y})$. Then, substituting this restricted MLE for $\theta$ into the expression for $\mathcal{L}_\omega$ gives

$$\hat{\mathcal{L}}_\omega = \left(\frac{2}{2+\bar{x}+\bar{y}}\right)^{2n} \left(\frac{\bar{x}+\bar{y}}{2+\bar{x}+\bar{y}}\right)^{n(\bar{x}+\bar{y})}.$$

So,

$$\hat{\lambda} = \frac{\hat{\mathcal{L}}_\omega}{\hat{\mathcal{L}}_\Omega} = \frac{2^{2n}(1+\bar{x})^{n(1+\bar{x})}(1+\bar{y})^{n(1+\bar{y})}(\bar{x}+\bar{y})^{n(\bar{x}+\bar{y})}}{(\bar{x})^{n\bar{x}}(\bar{y})^{n\bar{y}}(2+\bar{x}+\bar{y})^{n(2+\bar{x}+\bar{y})}}.$$

When $n = 25$, $\bar{x} = 1.00$, and $\bar{y} = 2.00$, $-2\ln\hat{\lambda} = 3.461$. Under $H_0$, $-2\ln\hat{\lambda} \dot{\sim} \chi_1^2$ when $n$ is large. Since $\chi^2_{1,0.95} = 3.841$, we do not reject $H_0$ at the $\alpha = 0.05$ level of significance.

(b) First, with

$$\boldsymbol{\theta} = (\theta_1, \theta_2), \quad \boldsymbol{S}'(\boldsymbol{\theta}) = \begin{bmatrix} \dfrac{\partial \ln \mathcal{L}_\Omega}{\partial \theta_1} \\ \dfrac{\partial \ln \mathcal{L}_\Omega}{\partial \theta_2} \end{bmatrix} = \begin{bmatrix} \dfrac{n}{\theta_1} - \dfrac{s_x}{(1-\theta_1)} \\ \dfrac{n}{\theta_2} - \dfrac{s_y}{(1-\theta_2)} \end{bmatrix}$$

$$= \begin{bmatrix} \dfrac{n - n\theta_1(1+\bar{x})}{\theta_1(1-\theta_1)} \\ \dfrac{n - n\theta_2(1+\bar{y})}{\theta_2(1-\theta_2)} \end{bmatrix}.$$

Now, $\hat{\theta} = 2/(2+\bar{x}+\bar{y}) = 2/(2+1+2) = 0.40$. So, when $n = 25, \bar{x} = 1.00, \bar{y} = 2.00$, and $\hat{\boldsymbol{\theta}}_\omega = (\hat{\theta}, \hat{\theta}) = (0.40, 0.40)$, then $\boldsymbol{S}(\hat{\boldsymbol{\theta}}_\omega) = (20.83333, -20.83333)$. Now,

$$\frac{\partial^2 \ln \mathcal{L}_\Omega}{\partial \theta_1^2} = \frac{-n}{\theta_1^2} - \frac{s_x}{(1-\theta_1)^2} = \frac{-n}{\theta_1^2} - \frac{n\bar{x}}{(1-\theta_1)^2},$$

so that

$$\frac{-\partial^2 \ln \mathcal{L}_\Omega}{\partial \theta_1^2}\bigg|_{\substack{\theta_1=\hat{\theta}=0.40,\\ n=25,\bar{x}=1.0}} = 225.6944.$$

Also,

$$\frac{-\partial^2 \ln \mathcal{L}_\Omega}{\partial \theta_1 \partial \theta_2} = \frac{-\partial^2 \ln \mathcal{L}_\Omega}{\partial \theta_2 \partial \theta_1} = 0.$$

And,

$$\frac{\partial^2 \ln \mathcal{L}_\Omega}{\partial \theta_2^2} = \frac{-n}{\theta_2^2} - \frac{s_y}{(1-\theta_2)^2} = \frac{-n}{\theta_2^2} - \frac{n\bar{y}}{(1-\theta_2)^2},$$

so that

$$\frac{-\partial^2 \ln \mathcal{L}_\Omega}{\partial \theta_2^2}\bigg|_{\substack{\theta_2=\hat{\theta}=0.40,\\ n=25,\bar{y}=2.0}} = 295.1389.$$

Finally,

$$\hat{S} = \boldsymbol{S}(\hat{\boldsymbol{\theta}}_\omega)\boldsymbol{I}^{-1}(\boldsymbol{x}, \boldsymbol{y}; \hat{\boldsymbol{\theta}}_\omega)\boldsymbol{S}'(\hat{\boldsymbol{\theta}}_\omega)$$

$$= (20.8333, -20.8333)\begin{bmatrix} \dfrac{1}{225.6944} & 0 \\ 0 & \dfrac{1}{295.1389} \end{bmatrix}\begin{pmatrix} 20.8333 \\ -20.8333 \end{pmatrix} = 3.394.$$

Under $H_0$, $\hat{S} \mathrel{\dot\sim} \chi_1^2$ for large $n$. Since $\chi_{1,0.95}^2 = 3.841$, we again do not reject $H_0$ at the $\alpha = 0.05$ level of significance. Although the numerical values of $-2\ln\hat{\lambda}$ and $\hat{S}$ agree closely in this particular example, this will not always be the case.

(c) First, the actual P-value for either the likelihood ratio test or the score test satisfies the inequality $0.05 <$ P-value $< 0.10$. Also, since $\bar{X}$ and $\bar{Y}$ are unbiased estimators of E(X) and E(Y), respectively, and since $\bar{x} = 1.00$ is half the size of $\bar{y} = 2.00$, the data do provide some evidence suggesting that the teenage driver education classes are beneficial. So, the suggestion by the highway safety researcher to increase the sample size is very reasonable; power calculations can be used to choose an appropriate sample size.

**Solution 5.12**

(a) The unconditional likelihood $\mathcal{L}(\beta)$ is

$$\mathcal{L}(\beta) = \prod_{i=1}^{n} \left\{ \left(\frac{1}{1+\beta x_i}\right)^{y_i-1} \left(\frac{\beta x_i}{1+\beta x_i}\right) \right\}$$

$$\Rightarrow \quad \ln \mathcal{L}(\beta) = \sum_{i=1}^{n} \left\{ (y_i - 1)\ln\left(\frac{1}{1+\beta x_i}\right) + \ln\left(\frac{\beta x_i}{1+\beta x_i}\right) \right\}$$

$$= \sum_{i=1}^{n} \left[\ln(\beta x_i) - y_i \ln(1+\beta x_i)\right]$$

$$\Rightarrow \quad \frac{d \ln \mathcal{L}(\beta)}{d\beta} = \sum_{i=1}^{n} \left[\frac{1}{\beta} - \frac{x_i y_i}{(1+\beta x_i)}\right] = 0$$

$$\Rightarrow \quad \frac{n}{\hat{\beta}} = \sum_{i=1}^{n} \frac{x_i y_i}{(1+\hat{\beta} x_i)}$$

$$\Rightarrow \quad \hat{\beta} = \frac{n}{\sum_{i=1}^{n} x_i y_i \left(1+\hat{\beta} x_i\right)^{-1}}.$$

(b) From part (a), we know that,

$$\frac{d \ln \mathcal{L}(\beta)}{d\beta} = \frac{n}{\beta} - \sum_{i=1}^{n} \frac{x_i y_i}{(1+\beta x_i)}$$

$$\Rightarrow \quad \frac{d^2 \ln \mathcal{L}(\beta)}{d\beta^2} = \frac{-n}{\beta^2} + \sum_{i=1}^{n} \frac{x_i^2 y_i}{(1+\beta x_i)^2}$$

$$\Rightarrow \quad -\mathrm{E}\left[\frac{d^2 \ln \mathcal{L}(\beta)}{d\beta^2}\right] = \frac{n}{\beta^2} - \sum_{i=1}^{n} \frac{x_i^2 \mathrm{E}(Y_i)}{(1+\beta x_i)^2}.$$

Now, since $\mathrm{E}(Y_i) = \theta_i^{-1} = (1+\beta x_i)/\beta x_i$, it follows that

$$\mathcal{I}(\beta) = -\mathrm{E}\left[\frac{d^2 \ln \mathcal{L}(\beta)}{d\beta^2}\right] = \frac{n}{\beta^2} - \sum_{i=1}^{n} \frac{x_i^2(1+\beta x_i)}{\beta x_i (1+\beta x_i)^2}$$

$$= \frac{n}{\beta^2} - \sum_{i=1}^{n} \frac{x_i}{\beta(1+\beta x_i)} = \frac{1}{\beta^2}\left[\sum_{i=1}^{n}\left(1 - \frac{\beta x_i}{(1+\beta x_i)}\right)\right]$$
$$= \frac{1}{\beta^2}\sum_{i=1}^{n}(1+\beta x_i)^{-1}.$$

So,

$$\mathrm{V}(\hat{\beta}) = \left\{-\mathrm{E}\left[\frac{\mathrm{d}^2 \ln \mathcal{L}(\beta)}{\mathrm{d}\beta^2}\right]\right\}^{-1} = \frac{\beta^2}{\sum_{i=1}^{n}(1+\beta x_i)^{-1}}.$$

(c) For these data,

$$\sum_{i=1}^{n}(1+\hat{\beta}x_i)^{-1} = 50\left[1+\frac{1}{2}(30)\right]^{-1} + 50\left[1+\frac{1}{2}(40)\right]^{-1} = 5.5060,$$

so that

$$\hat{\mathrm{V}}(\hat{\beta}) = \left(\frac{1}{2}\right)^2 / 5.5060 = 0.0454.$$

So, a large-sample 95% CI for $\beta$ is

$$\hat{\beta} \pm 1.96\sqrt{\hat{\mathrm{V}}(\hat{\beta})}$$
$$= 0.50 \pm 1.96\sqrt{0.0454}$$
$$= (0.0824, 0.9176).$$

(d)

$$\mathcal{I}(\beta) = \frac{1}{\beta^2}\sum_{i=1}^{n}(1+\beta x_i)^{-1},$$

so that $\mathcal{I}(\hat{\beta}) = \mathcal{I}(1/2) = 4(5.5060) = 22.0240$.

So, $\hat{W} = \left(\frac{1}{2} - 1\right)^2 (22.0240) = 5.5060$, since $\hat{W} = (\hat{\beta} - \beta_0)\mathcal{I}(\hat{\beta})(\hat{\beta} - \beta_0)$ and $\beta_0 = 1$. Since $\chi^2_{1,0.95} = 3.84$, we reject $\mathrm{H}_0$; P-value $\doteq 0.02$.

(e)

$$\text{POWER} = \mathrm{pr}\{U > 1.96 | \beta = 1.10\}$$
$$= \mathrm{pr}\left\{\frac{\hat{\beta} - 1}{\sqrt{\mathrm{V}_0(\hat{\beta})}} > 1.96 \middle| \beta = 1.10\right\}$$
$$= \mathrm{pr}\left\{\hat{\beta} > 1 + 1.96\sqrt{\mathrm{V}_0(\hat{\beta})} \middle| \beta = 1.10\right\}$$

$$= \text{pr}\left\{\frac{\hat{\beta} - 1.10}{\sqrt{\text{V}(\hat{\beta})}} > \frac{1 + 1.96\sqrt{\text{V}_0(\hat{\beta})} - 1.10}{\sqrt{\text{V}(\hat{\beta})}}\right\}$$

$$\doteq \text{pr}\left\{Z > \frac{1.96\sqrt{\text{V}_0(\hat{\beta})} - 0.10}{\sqrt{\text{V}(\hat{\beta})}}\right\},$$

where $Z \dot{\sim} \text{N}(0, 1)$ for large $n$.

Now, when $\beta = 1$,

$$\text{V}_0(\hat{\beta}) = \frac{1}{50(1+30)^{-1} + 50(1+40)^{-1}} = 0.3531;$$

and, when $\beta = 1.10$,

$$\text{V}(\hat{\beta}) = \frac{\beta^2}{\sum_{i=1}^{n}(1+\beta x_i)^{-1}}$$

$$= \frac{(1.10)^2}{50[1+1.10(30)]^{-1} + 50[1+1.10(40)]^{-1}} = 0.4687.$$

So,

$$\text{POWER} = \text{pr}\left\{Z > \frac{1.96\sqrt{0.3531} - 0.10}{\sqrt{0.4687}}\right\}$$

$$= \text{pr}(Z > 1.5552) \doteq 0.06.$$

**Solution 5.13**

(a) If $X$ is the random variable denoting the number of lung cancer cases developing over this 20-year follow-up period in a random sample of $n = 1000$ heavy smokers, it is reasonable to assume that $X \sim \text{BIN}(n, \theta)$. The maximum likelihood estimator of $\theta$ is

$$\hat{\theta} = \frac{X}{n},$$

with

$$\text{E}(\hat{\theta}) = \theta \quad \text{and} \quad \text{V}(\hat{\theta}) = \frac{\theta(1-\theta)}{n}.$$

Since $n$ is large, by the Central Limit Theorem and by Slutsky's Theorem,

$$0.95 \doteq \text{pr}\left\{-1.96 < \frac{\hat{\theta} - \theta}{\sqrt{\frac{\hat{\theta}(1-\hat{\theta})}{n}}} < 1.96\right\} = \text{pr}\{L < \theta < U\},$$

where

$$L = \hat{\theta} - 1.96\sqrt{\frac{\hat{\theta}(1-\hat{\theta})}{n}} \quad \text{and} \quad U = \hat{\theta} + 1.96\sqrt{\frac{\hat{\theta}(1-\hat{\theta})}{n}}.$$

Since $\psi = \theta/(1-\theta)$, so that $\theta = \psi/(1+\psi)$ and $\theta^{-1} = 1 + \psi^{-1}$, and with $0 < L < U$,

$$\begin{aligned} 0.95 &\doteq \text{pr}\{U^{-1} < \theta^{-1} < L^{-1}\} \\ &= \text{pr}\{U^{-1} < 1 + \psi^{-1} < L^{-1}\} \\ &= \text{pr}\left\{\left(\frac{1}{L} - 1\right)^{-1} < \psi < \left(\frac{1}{U} - 1\right)^{-1}\right\}. \end{aligned}$$

Since

$$L = 0.10 - 1.96\sqrt{\frac{0.10(0.90)}{1000}} = 0.0814$$

and

$$U = 0.10 + 1.96\sqrt{\frac{0.10(0.90)}{1000}} = 0.1186,$$

a large-sample 95% CI for $\psi$ is

$$\left[\left(\frac{1}{0.0814} - 1\right)^{-1}, \left(\frac{1}{0.1186} - 1\right)^{-1}\right] = (0.0886, 0.1346).$$

Or, we can use ML methods directly. Since

$$\mathcal{L}(x;\theta) \propto \theta^{x}(1-\theta)^{n-x},$$

so that

$$\begin{aligned} \ln \mathcal{L}(x;\psi) &\sim \ln\left[\left(\frac{\psi}{1+\psi}\right)^{x}\left(\frac{1}{1+\psi}\right)^{n-x}\right] \\ &= x \ln\left(\frac{\psi}{1+\psi}\right) + (n-x)\ln\left(\frac{1}{1+\psi}\right) \\ &= x\left[\ln\psi - \ln(1+\psi)\right] - (n-x)\ln(1+\psi), \end{aligned}$$

we have

$$\frac{\partial \ln \mathcal{L}(x;\psi)}{\partial \psi} = \frac{x}{\psi} - \frac{x}{(1+\psi)} - \frac{(n-x)}{(1+\psi)}.$$

So,

$$\frac{\partial \ln \mathcal{L}(x;\psi)}{\partial \psi} = 0$$

gives

$$\hat{\psi} = \frac{X}{(n-X)} = \frac{\hat{\theta}}{(1-\hat{\theta})},$$

as expected. Now,

$$\begin{aligned}\frac{\partial^2 \ln \mathcal{L}(x;\psi)}{\partial \psi^2} &= \frac{-x}{\psi^2} + \frac{x}{(1+\psi)^2} + \frac{(n-x)}{(1+\psi)^2}\\ &= \frac{-x}{\psi^2} + \frac{n}{(1+\psi)^2}.\end{aligned}$$

So, using observed information, a large-sample 95% CI for $\psi$ is

$$\begin{aligned}&\hat{\psi} \pm 1.96\left[\frac{x}{\hat{\psi}^2} - \frac{n}{(1+\hat{\psi})^2}\right]^{-1/2}\\ &= 0.1111 \pm 1.96\left[\frac{100}{(0.1111)^2} - \frac{1,000}{(1+0.1111)^2}\right]^{-1/2}\\ &= 0.1111 \pm 1.96(8,101.6202 - 810.0162)^{-1/2}\\ &= 0.1111 \pm 0.0230\\ &= (0.0881, 0.1341).\end{aligned}$$

Or, since

$$-\mathrm{E}_x\left[\frac{\partial^2 \ln \mathcal{L}(x;\psi)}{\partial \psi^2}\right] = \frac{n\theta}{\psi^2} - \frac{n}{(1+\psi)^2} = \frac{n}{\psi(1+\psi)} - \frac{n}{(1+\psi)^2},$$

a 95% CI for $\psi$ using expected information is

$$\begin{aligned}&\hat{\psi} \pm 1.96\left[\frac{n}{\hat{\psi}(1+\hat{\psi})} - \frac{n}{(1+\hat{\psi})^2}\right]^{-1/2}\\ &= 0.1111 \pm 1.96\left[\frac{1,000}{0.1111(1+0.1111)} - \frac{1,000}{(1+0.1111)^2}\right]^{-1/2}\\ &= 0.1111 \pm 0.0230\\ &= (0.0881, 0.1341).\end{aligned}$$

(b) Clearly, testing $H_0$: $\psi = 0.10$ versus $H_1$: $\psi > 0.10$ is equivalent to testing $H_0$: $\theta = 0.0909$ versus $H_1$: $\theta > 0.0909$. Since

$$\frac{\partial \ln \mathcal{L}(x;\theta)}{\partial \theta} = \frac{x}{\theta} - \frac{(n-x)}{(1-\theta)},$$

we have

$$\frac{\partial^2 \ln \mathcal{L}(x;\theta)}{\partial \theta^2} = \frac{-x}{\theta^2} - \frac{(n-x)}{(1-\theta)^2};$$

hence, the estimated observed information is

$$\frac{100}{(0.10)^2} + \frac{900}{(1-0.10)^2} = 10,000 + 1111.111 = 11,111.111.$$

And, since

$$-\mathrm{E}_x\left[\frac{\partial^2 \ln \mathcal{L}(x;\theta)}{\partial \theta^2}\right] = \frac{n\theta}{\theta^2} + \frac{(n-n\theta)}{(1-\theta)^2} = \frac{n}{\theta(1-\theta)},$$

the estimated expected information is $\frac{1,000}{0.10(1-0.10)} = 11,111.111$. So, the Wald statistic is

$$\hat{W} = (0.10 - 0.0909)^2(11,111.111) = 0.9201,$$

with

$$\begin{aligned}\text{P-value} &= \mathrm{pr}(\sqrt{\hat{W}} > \sqrt{0.9201} \mid H_0\colon \theta = 0.0909) \doteq \mathrm{pr}(Z > 0.9592) \\ &\doteq 0.17,\end{aligned}$$

where $Z \sim \mathrm{N}(0,1)$. Since

$$\begin{aligned}\left.\frac{\partial \ln \mathcal{L}(x;\theta)}{\partial \theta}\right|_{\theta=0.0909} &= \frac{(100)}{0.0909} - \frac{(1,000-100)}{(1-0.0909)} \\ &= 1100.11 - 989.9901 = 110.1199,\end{aligned}$$

the score statistic is

$$\hat{S} = \frac{(110.1199)^2}{1000/[0.0909(0.9091)]} = 1.0021,$$

with

$$\text{P-value} = \mathrm{pr}(\sqrt{\hat{S}} > \sqrt{1.0021} | H_0\colon \theta = 0.0909) \doteq \mathrm{pr}(Z > 1.0010) \doteq 0.16,$$

where $Z \sim \mathrm{N}(0,1)$. The results of these Wald and score tests imply that $H_0$: $\theta = 0.0909$ cannot be rejected given the available data.

Of course, we can equivalently work with the parameter $\psi$, and directly test $H_0$: $\psi = 0.10$ versus $H_1$: $\psi > 0.10$ using appropriate Wald and score tests. From part (a), the appropriate estimated observed information is $(8101.6202 - 810.0162) = 7,291.6040$; so, the Wald test statistic is

$$\hat{W} = (0.1111 - 0.10)^2(7,291.6040) = 0.8984,$$

with

$$\text{P-value} = \text{pr}(\sqrt{\hat{W}} > \sqrt{0.8984}|H_0: \psi = 0.10) \doteq \text{pr}(Z > 0.9478) \doteq 0.17,$$

where $Z \sim N(0, 1)$. Since

$$\frac{\partial \ln \mathcal{L}(x; \psi)}{\partial \psi}_{|\psi=0.10} = \frac{100}{0.10} - \frac{1,000}{(1 + 0.10)} = 1000 - 909.0909 = 90.9091,$$

the score test statistic is

$$\hat{S} = \frac{(90.9091)^2}{1000/[.10(1 + .10)^2]} = 1.0000,$$

with

$$\text{P-value} = \text{pr}(\sqrt{\hat{S}} > \sqrt{1.000}|H_0: \psi = 0.10) \doteq \text{pr}(Z > 1.0000) \doteq 0.16,$$

where $Z \sim N(0, 1)$. As before, there is not sufficient evidence to reject $H_0$: $\psi = 0.10$.

**Solution 5.14**

(a) The unrestricted likelihood function $\mathcal{L}_\Omega$ has the structure

$$\mathcal{L}_\Omega = \prod_{i=1}^{n} f_{X,Y}(x_i, y_i; \alpha, \beta) = \prod_{i=1}^{n} f_X(x_i; \beta) f_Y(y_i|X = x_i; \alpha, \beta)$$

$$= \prod_{i=1}^{n} \left\{ \frac{1}{\beta} e^{-x_i/\beta} \frac{1}{(\alpha + \beta)x_i} e^{-y_i/(\alpha+\beta)x_i} \right\}$$

$$= \beta^{-n} e^{-\sum_{i=1}^{n} x_i/\beta} (\alpha + \beta)^{-n} e^{-\sum_{i=1}^{n} \left(\frac{y_i}{x_i}\right)/(\alpha+\beta)} \cdot \left( \prod_{i=1}^{n} x_i \right)^{-1}.$$

By the Factorization Theorem, $U_1 = \sum_{i=1}^{n} X_i$ and $U_2 = \sum_{i=1}^{n} (Y_i/X_i)$ are jointly sufficient for $\alpha$ and $\beta$. If we can show that $X_i$ and $Y_i/X_i$ are uncorrelated, then $U_1$ and $U_2$ will be uncorrelated.

Now, $E(X_i) = \beta$ and $E(Y_i) = E_{x_i}[E(Y_i|X_i = x_i)] = E[(\alpha + \beta)x_i] = (\alpha + \beta)\beta$.

And,

$$\mathrm{E}\left(\frac{Y_i}{X_i}\right) = \mathrm{E}_{x_i}\left[\mathrm{E}\left(\frac{Y_i}{X_i}|X_i = x_i\right)\right]$$
$$= \mathrm{E}_{x_i}\left[\frac{1}{x_i}\mathrm{E}(Y_i|X_i = x_i)\right] = \mathrm{E}\left[\frac{1}{x_i}(\alpha+\beta)x_i\right] = (\alpha+\beta).$$

Since

$$\mathrm{E}\left(X_i \cdot \frac{Y_i}{X_i}\right) = \mathrm{E}(Y_i) = (\alpha+\beta)\beta,$$
$$\mathrm{cov}\left(X_i, \frac{Y_i}{X_i}\right) = \mathrm{E}\left(X_i \cdot \frac{Y_i}{X_i}\right) - \mathrm{E}(X_i)\mathrm{E}\left(\frac{Y_i}{X_i}\right)$$
$$= (\alpha+\beta)\beta - \beta(\alpha+\beta) = 0,$$

and hence $U_1$ and $U_2$ are uncorrelated.

(b) Now,

$$\ln \mathcal{L}_\Omega = -n\ln\beta - \frac{\sum_{i=1}^n x_i}{\beta} - n\ln(\alpha+\beta) - \frac{\sum_{i=1}^n (y_i/x_i)}{(\alpha+\beta)} - \sum_{i=1}^n \ln x_i.$$

So,

$$\frac{\partial \ln \mathcal{L}_\Omega}{\partial \alpha} = \frac{-n}{(\alpha+\beta)} + \frac{\sum_{i=1}^n (y_i/x_i)}{(\alpha+\beta)^2} = 0 \Rightarrow (\hat{\alpha}+\hat{\beta}) = \frac{\sum_{i=1}^n (y_i/x_i)}{n}.$$

And,

$$\frac{\partial \ln \mathcal{L}_\Omega}{\partial \beta} = \frac{-n}{\beta} + \frac{\sum_{i=1}^n x_i}{\beta^2} - \frac{n}{(\alpha+\beta)} + \frac{\sum_{i=1}^n (y_i/x_i)}{(\alpha+\beta)^2} = 0$$
$$\Rightarrow \hat{\beta} = \bar{X} = \frac{\sum_{i=1}^n X_i}{n} \quad \text{and} \quad \hat{\alpha} = \frac{\sum_{i=1}^n (Y_i/X_i)}{n} - \bar{X}.$$

Now,

$$\frac{\partial^2 \ln \mathcal{L}_\Omega}{\partial \alpha^2} = \frac{n}{(\alpha+\beta)^2} - \frac{2\sum_{i=1}^n (y_i/x_i)}{(\alpha+\beta)^3},$$

so that

$$-\mathrm{E}\left(\frac{\partial^2 \ln \mathcal{L}_\Omega}{\partial \alpha^2}\right) = \frac{-n}{(\alpha+\beta)^2} + \frac{2n(\alpha+\beta)}{(\alpha+\beta)^3} = \frac{n}{(\alpha+\beta)^2}.$$

Also,

$$\frac{\partial^2 \ln \mathcal{L}_\Omega}{\partial \beta^2} = \frac{n}{\beta^2} - \frac{2\sum_{i=1}^n x_i}{\beta^3} + \frac{n}{(\alpha+\beta)^2} - \frac{2\sum_{i=1}^n (y_i/x_i)}{(\alpha+\beta)^3},$$

so that

$$-\mathrm{E}\left(\frac{\partial^2 \ln \mathcal{L}_\Omega}{\partial \beta^2}\right) = \frac{-n}{\beta^2} + \frac{2n\beta}{\beta^3} - \frac{n}{(\alpha+\beta)^2} + \frac{2n(\alpha+\beta)}{(\alpha+\beta)^3}$$

$$= \frac{n}{\beta^2} + \frac{n}{(\alpha+\beta)^2}.$$

And,

$$\frac{\partial^2 \ln \mathcal{L}_\Omega}{\partial \alpha \partial \beta} = \frac{n}{(\alpha+\beta)^2} - \frac{2\sum_{i=1}^{n}(y_i/x_i)}{(\alpha+\beta)^3},$$

so that

$$-\mathrm{E}\left(\frac{\partial^2 \ln \mathcal{L}_\Omega}{\partial \alpha \partial \beta}\right) = \frac{-n}{(\alpha+\beta)^2} + \frac{2n(\alpha+\beta)}{(\alpha+\beta)^3} = \frac{n}{(\alpha+\beta)^2}.$$

Thus, the expected information matrix $\mathcal{I}(\alpha, \beta)$ is equal to

$$\mathcal{I}(\alpha,\beta) = \begin{bmatrix} \dfrac{n}{(\alpha+\beta)^2} & \dfrac{n}{(\alpha+\beta)^2} \\ \dfrac{n}{(\alpha+\beta)^2} & \dfrac{n}{\beta^2} + \dfrac{n}{(\alpha+\beta)^2} \end{bmatrix},$$

and so

$$\mathcal{I}^{-1}(\alpha,\beta) = \begin{bmatrix} \dfrac{(\alpha+\beta)^2}{n} + \dfrac{\beta^2}{n} & \dfrac{-\beta^2}{n} \\ \dfrac{-\beta^2}{n} & \dfrac{\beta^2}{n} \end{bmatrix}.$$

For $\mathrm{H}_0 : \alpha = \beta$, or equivalently $\mathrm{H}_0 : R = (\alpha - \beta) = 0$, we have

$$\mathbf{T} = \left[\frac{\partial R}{\partial \alpha}, \frac{\partial R}{\partial \beta}\right] = (1, -1),$$

and so

$$\Lambda = \mathbf{T}\mathcal{I}^{-1}(\alpha,\beta)\mathbf{T}' = \frac{(\alpha+\beta)^2}{n} + \frac{4\beta^2}{n}$$

$$= \mathrm{V}(\hat{R}) = \mathrm{V}(\hat{\alpha} - \hat{\beta}) = \mathrm{V}(\hat{\alpha}) + \mathrm{V}(\hat{\beta}) - 2\mathrm{cov}(\hat{\alpha}, \hat{\beta}).$$

So, the Wald test statistic $\hat{W}$ takes the form

$$\hat{W} = \frac{\hat{R}}{\hat{\Lambda}} = \frac{(\hat{\alpha}-\hat{\beta})^2}{\dfrac{(\hat{\alpha}+\hat{\beta})^2}{n} + \dfrac{4\hat{\beta}^2}{n}} = \left[\frac{(\hat{\alpha}-\hat{\beta}) - 0}{\sqrt{\hat{\mathrm{V}}\left(\hat{\alpha}-\hat{\beta}\right)}}\right]^2, \quad \text{as expected.}$$

For $n = 30, \hat{\alpha} = 2$, and $\hat{\beta} = 1$,

$$\hat{W} = \frac{(2-1)^2}{\frac{(2+1)^2}{30} + \frac{4(1)^2}{30}} = 2.31.$$

Since $\hat{W} \stackrel{.}{\sim} \chi_1^2$ under $H_0 : \alpha = \beta$ for large $n$, the P-value $\doteq \text{pr}(\chi_1^2 > 2.31 | H_0 : \alpha = \beta) \doteq 0.14$. So, for the given data, there is *not* sufficient evidence to reject $H_0 : \alpha = \beta$.

A (large sample) 95% confidence interval for $(\alpha - \beta)$ is

$$(\hat{\alpha} - \hat{\beta}) \pm 1.96\sqrt{\hat{V}(\hat{\alpha} - \hat{\beta})} = (\hat{\alpha} - \hat{\beta}) \pm 1.96\sqrt{\frac{(\hat{\alpha} + \hat{\beta})^2}{n} + \frac{4\hat{\beta}^2}{n}}$$

$$= (2-1) \pm 1.96\sqrt{\frac{(2+1)^2}{30} + \frac{4(1)^2}{30}} = (-0.29, 2.29).$$

The computed 95% CI contains the value 0, which agrees with the conclusion based on the Wald test.

**Solution 5.15**

(a) With $\mathbf{y} = (y_{11}, y_{12}, \ldots, y_{1n}; y_{21}, y_{22}, \ldots, y_{2n})$, we have

$$\mathcal{L}(y; \beta_1, \beta_2) = \prod_{i=1}^{2}\prod_{j=1}^{n}\left\{\frac{1}{\beta_i x_{ij}^2} e^{-y_{ij}/\beta_i x_{ij}^2}\right\}$$

$$= \prod_{i=1}^{2}\left\{\frac{1}{\beta_i^n \prod_{j=1}^{n} x_{ij}^2} e^{-\beta_i^{-1}\sum_{j=1}^{n} x_{ij}^{-2} y_{ij}}\right\}.$$

Hence, by the Factorization Theorem,

$$\sum_{j=1}^{n} \frac{y_{1j}}{x_{1j}^2} \text{ is sufficient for } \beta_1,$$

and

$$\sum_{j=1}^{n} \frac{y_{2j}}{x_{2j}^2} \text{ is sufficient for } \beta_2.$$

(b) Now,

$$\ln \mathcal{L}(y; \beta_1, \beta_2) = \sum_{i=1}^{2}\sum_{j=1}^{n}\left\{-\ln \beta_i - \ln x_{ij}^2 - \frac{y_{ij}}{\beta_i x_{ij}^2}\right\}$$

$$= -n\sum_{i=1}^{2} \ln \beta_i - \sum_{i=1}^{2}\sum_{j=1}^{n} \ln x_{ij}^2 - \sum_{i=1}^{2} \beta_i^{-1} \sum_{j=1}^{n} \frac{y_{ij}}{x_{ij}^2}.$$

So, for $i = 1, 2,$

$$S_i(\beta_1, \beta_2) = \frac{\partial \ln \mathcal{L}(y; \beta_1, \beta_2)}{\partial \beta_i} = \frac{-n}{\beta_i} + \frac{1}{\beta_i^2} \sum_{j=1}^{n} \frac{y_{ij}}{x_{ij}^2} = 0$$

gives

$$\hat{\beta}_i = \frac{1}{n} \sum_{j=1}^{n} \frac{y_{ij}}{x_{ij}^2}, \quad i = 1, 2.$$

And,

$$\frac{\partial^2 \ln \mathcal{L}(y; \beta_1, \beta_2)}{\partial \beta_i^2} = \frac{n}{\beta_i^2} - \frac{2}{\beta_i^3} \sum_{j=1}^{n} \frac{y_{ij}}{x_{ij}^2}, \quad i = 1, 2,$$

so that

$$-\mathrm{E}\left\{ \frac{\partial^2 \ln \mathcal{L}(y; \beta_1, \beta_2)}{\partial \beta_i^2} \right\} = \frac{-n}{\beta_i^2} + \frac{2}{\beta_i^3} \sum_{j=1}^{n} \frac{\mathrm{E}(Y_{ij})}{x_{ij}^2}$$

$$= \frac{-n}{\beta_i^2} + \frac{2}{\beta_i^3} \sum_{j=1}^{n} \frac{\beta_i x_{ij}^2}{x_{ij}^2} = \frac{-n}{\beta_i^2} + \frac{2n}{\beta_i^2} = \frac{n}{\beta_i^2}.$$

Also,

$$\frac{\partial^2 \ln \mathcal{L}(y; \beta_1, \beta_2)}{\partial \beta_1 \partial \beta_2} = \frac{\partial^2 \ln \mathcal{L}(y; \beta_1, \beta_2)}{\partial \beta_2 \partial \beta_1} = 0,$$

so that the expected information matrix is

$$\mathcal{I}(\beta_1, \beta_2) = \begin{bmatrix} n/\beta_1^2 & 0 \\ 0 & n/\beta_2^2 \end{bmatrix}.$$

Under $\mathrm{H}_0$: $\beta_1 = \beta_2$ (= $\beta$, say),

$$\ln \mathcal{L}(y; \beta) = -2n \ln \beta - \sum_{i=1}^{2} \sum_{j=1}^{n} \ln x_{ij}^2 - \frac{1}{\beta} \sum_{i=1}^{2} \sum_{j=1}^{n} \frac{y_{ij}}{x_{ij}^2},$$

so that the equation

$$\frac{\partial \ln \mathcal{L}(y; \beta)}{\partial \beta} = \frac{-2n}{\beta} + \frac{1}{\beta^2} \sum_{i=1}^{2} \sum_{j=1}^{n} \frac{y_{ij}}{x_{ij}^2} = 0$$

gives

$$\hat{\beta} = \frac{\sum_{i=1}^{2}\sum_{j=1}^{n}\frac{y_{ij}}{x_{ij}^2}}{2n} = \frac{1}{2}(\hat{\beta}_1 + \hat{\beta}_2).$$

So, with $\boldsymbol{S}(\hat{\beta}) = [S_1(\hat{\beta},\hat{\beta}), S_2(\hat{\beta},\hat{\beta})]$,

$$\begin{aligned}
\hat{S} &= \boldsymbol{S}(\hat{\beta})\boldsymbol{\mathcal{I}}^{-1}(\hat{\beta},\hat{\beta})\boldsymbol{S}'(\hat{\beta}) \\
&= \left[\frac{-n}{\hat{\beta}} + \frac{n\hat{\beta}_1}{\hat{\beta}^2}, \frac{-n}{\hat{\beta}} + \frac{n\hat{\beta}_2}{\hat{\beta}^2}\right]\begin{bmatrix} \hat{\beta}^2/n & 0 \\ 0 & \hat{\beta}^2/n \end{bmatrix}\begin{bmatrix} \frac{-n}{\hat{\beta}} + \frac{n\hat{\beta}_1}{\hat{\beta}^2} \\ \frac{-n}{\hat{\beta}} + \frac{n\hat{\beta}_2}{\hat{\beta}^2} \end{bmatrix} \\
&= [-\hat{\beta} + \hat{\beta}_1, -\hat{\beta} + \hat{\beta}_2]\begin{bmatrix} \frac{-n}{\hat{\beta}} + \frac{n\hat{\beta}_1}{\hat{\beta}^2} \\ \frac{-n}{\hat{\beta}} + \frac{n\hat{\beta}_2}{\hat{\beta}^2} \end{bmatrix} \\
&= [(\hat{\beta}_1 - \hat{\beta}), (\hat{\beta}_2 - \hat{\beta})]\begin{bmatrix} (\hat{\beta}_1 - \hat{\beta}) \\ (\hat{\beta}_2 - \hat{\beta}) \end{bmatrix}\left(\frac{n}{\hat{\beta}^2}\right) \\
&= \frac{n[(\hat{\beta}_1 - \hat{\beta})^2 + (\hat{\beta}_2 - \hat{\beta})^2]}{\hat{\beta}^2} \\
&= \frac{n\left[\frac{1}{4}(\hat{\beta}_1 - \hat{\beta}_2)^2 + \frac{1}{4}(\hat{\beta}_1 - \hat{\beta}_2)^2\right]}{\frac{1}{4}(\hat{\beta}_1 + \hat{\beta}_2)^2} \\
&= \frac{2n(\hat{\beta}_1 - \hat{\beta}_2)^2}{(\hat{\beta}_1 + \hat{\beta}_2)^2}.
\end{aligned}$$

Under $H_0$: $\beta_1 = \beta_2$, $\hat{S} \dot{\sim} \chi_1^2$ for large $n$. For the given data,

$$\hat{S} = \frac{2(25)(2-3)^2}{(2+3)^2} = \frac{50}{25} = 2.$$

Since $\chi^2_{0.95,1} = 3.841$, we do not reject $H_0$ at the $\alpha = 0.05$ level.

**Solution 5.16**

(a) The unrestricted likelihood function is

$$\mathcal{L}_\Omega = \prod_{i=1}^{n}\theta^{-1}e^{-x_i/\theta} \cdot \prod_{i=1}^{n}(\lambda\theta)^{-1}e^{-y_i/\lambda\theta}$$

$$= \theta^{-n} \exp\left\{-\theta^{-1}\sum_{i=1}^{n} x_i\right\}(\lambda\theta)^{-n} \exp\left\{-(\lambda\theta)^{-1}\sum_{i=1}^{n} y_i\right\};$$

so, by the Factorization Theorem,

$$S_x = \sum_{i=1}^{n} X_i \quad \text{and} \quad S_y = \sum_{i=1}^{n} Y_i$$

are jointly sufficient for $\lambda$ and $\theta$.

(b) From part (a),

$$\ln \mathcal{L}_{\Omega} = -n \ln \theta - \frac{\sum_{i=1}^{n} x_i}{\theta} - n \ln(\lambda\theta) - \frac{\sum_{i=1}^{n} y_i}{\lambda\theta}$$
$$= -2n \ln \theta - n \ln \lambda - \frac{\sum_{i=1}^{n} x_i}{\theta} - \frac{\sum_{i=1}^{n} y_i}{\lambda\theta}.$$

So,

$$\frac{\partial \ln \mathcal{L}_{\Omega}}{\partial \theta} = -\frac{2n}{\theta} + \frac{\sum_{i=1}^{n} x_i}{\theta^2} + \frac{\sum_{i=1}^{n} y_i}{\lambda\theta^2},$$

and

$$\frac{\partial \ln \mathcal{L}_{\Omega}}{\partial \lambda} = -\frac{n}{\lambda} + \frac{\sum_{i=1}^{n} y_i}{\lambda^2\theta}.$$

Now,

$$\frac{\partial \ln \mathcal{L}_{\Omega}}{\partial \lambda} = 0 \implies \frac{-n}{\theta} + \frac{\sum_{i=1}^{n} y_i}{\lambda\theta^2} = 0 \implies \frac{\sum_{i=1}^{n} y_i}{\lambda\theta^2} = \frac{n}{\theta}.$$

Thus,

$$\frac{\partial \ln \mathcal{L}_{\Omega}}{\partial \theta} = 0$$

gives

$$\frac{-2n}{\theta} + \frac{\sum_{i=1}^{n} x_i}{\theta^2} + \frac{n}{\theta} = \frac{-n}{\theta} + \frac{\sum_{i=1}^{n} x_i}{\theta^2} = 0,$$

or

$$\hat{\theta} = \bar{x}.$$

Then,

$$\frac{-n}{\lambda} + \frac{n\bar{y}}{\lambda^2\bar{x}} = 0$$

gives

$$\hat{\lambda} = \frac{\bar{y}}{\bar{x}}.$$

(c) Let $\boldsymbol{\theta} = (\theta, \lambda)$ denote the set of unknown parameters. From part (b),

$$\boldsymbol{S}(\boldsymbol{\theta}) = \left[\frac{-2n}{\theta} + \frac{n\bar{x}}{\theta^2} + \frac{n\bar{y}}{\lambda\theta^2}, \quad \frac{-n}{\lambda} + \frac{n\bar{y}}{\lambda^2\theta}\right].$$

Under $\text{H}_0 : \lambda = 1$, the restricted log likelihood is

$$\ln \mathcal{L}_\omega = -2n \ln \theta - \frac{n(\bar{x} + \bar{y})}{\theta};$$

so,

$$\frac{\partial \ln \mathcal{L}_\omega}{\partial \theta} = \frac{-2n}{\theta} + \frac{n(\bar{x} + \bar{y})}{\theta^2} = 0$$

gives

$$\hat{\theta}_\omega = \frac{(\bar{x} + \bar{y})}{2}.$$

Thus,

$$\hat{\boldsymbol{\theta}}_\omega = \left[(\bar{x} + \bar{y})/2, 1\right].$$

Now,

$$-\frac{2n}{\hat{\theta}_\omega} + \frac{n\bar{x}}{\hat{\theta}^2_\omega} + \frac{n\bar{y}}{(1)\hat{\theta}^2_\omega} = \frac{-4n}{(\bar{x} + \bar{y})} + \frac{4n\bar{x}}{(\bar{x} + \bar{y})^2} + \frac{4n\bar{y}}{(\bar{x} + \bar{y})^2} = 0,$$

and

$$\frac{-n}{(1)} + \frac{2n\bar{y}}{(1)^2(\bar{x} + \bar{y})} = \frac{-n(\bar{x} + \bar{y}) + 2n\bar{y}}{(\bar{x} + \bar{y})} = \frac{n(\bar{y} - \bar{x})}{(\bar{x} + \bar{y})},$$

so that

$$\boldsymbol{S}(\hat{\boldsymbol{\theta}}_\omega) = \left[0, \frac{n(\bar{y} - \bar{x})}{(\bar{x} + \bar{y})}\right].$$

Finally, we need $\mathcal{I}^{-1}(\hat{\boldsymbol{\theta}}_\omega)$. Now,

$$\frac{\partial^2 \ln \mathcal{L}_\Omega}{\partial \theta^2} = \frac{2n}{\theta^2} - \frac{2n\bar{x}}{\theta^3} - \frac{2n\bar{y}}{\lambda\theta^3},$$

so that

$$-\mathrm{E}\left(\frac{\partial^2 \ln \mathcal{L}_\Omega}{\partial \theta^2}\right) = \frac{-2n}{\theta^2} + \frac{2n\theta}{\theta^3} + \frac{2n\lambda\theta}{\lambda\theta^3} = \frac{2n}{\theta^2}.$$

And,

$$\frac{\partial^2 \ln \mathcal{L}_\Omega}{\partial \theta \partial \lambda} = \frac{-n\bar{y}}{\lambda^2\theta^2},$$

so that

$$-\mathrm{E}\left(\frac{\partial^2 \ln \mathcal{L}_\Omega}{\partial \theta \partial \lambda}\right) = \frac{n\lambda\theta}{\lambda^2\theta^2} = \frac{n}{\lambda\theta}.$$

Also,

$$\frac{\partial^2 \ln \mathcal{L}_\Omega}{\partial \lambda^2} = \frac{n}{\lambda^2} - \frac{2n\bar{y}}{\lambda^3\theta},$$

so that

$$-\mathrm{E}\left(\frac{\partial^2 \ln \mathcal{L}_\Omega}{\partial \lambda^2}\right) = \frac{-n}{\lambda^2} + \frac{2n\lambda\theta}{\lambda^3\theta} = \frac{n}{\lambda^2}.$$

So,

$$\mathcal{I}(\boldsymbol{\theta}) = \begin{bmatrix} \frac{2n}{\theta^2} & \frac{n}{\lambda\theta} \\ \frac{n}{\lambda\theta} & \frac{n}{\lambda^2} \end{bmatrix},$$

and hence

$$\mathcal{I}^{-1}(\boldsymbol{\theta}) = \begin{bmatrix} \frac{\theta^2}{n} & \frac{-\lambda\theta}{n} \\ \frac{-\lambda\theta}{n} & \frac{2\lambda^2}{n} \end{bmatrix}.$$

So,

$$\mathcal{I}^{-1}(\hat{\boldsymbol{\theta}}_\omega) = \begin{bmatrix} \frac{(\bar{x}+\bar{y})^2}{4n} & \frac{-(\bar{x}+\bar{y})}{2n} \\ \frac{-(\bar{x}+\bar{y})}{2n} & \frac{2}{n} \end{bmatrix}.$$

Finally,

$$\hat{S} = \left[0, \quad \frac{n(\bar{y}-\bar{x})}{(\bar{x}+\bar{y})}\right] \begin{bmatrix} \frac{(\bar{x}+\bar{y})^2}{4n} & \frac{-(\bar{x}+\bar{y})}{2n} \\ \frac{-(\bar{x}+\bar{y})}{2n} & \frac{2}{n} \end{bmatrix} \begin{bmatrix} 0 \\ \frac{n(\bar{y}-\bar{x})}{(\bar{x}+\bar{y})} \end{bmatrix}$$

$$= \frac{2n(\bar{y}-\bar{x})^2}{(\bar{x}+\bar{y})^2} = \left[\frac{(\bar{y}-\bar{x})}{\sqrt{\hat{V}_0(\bar{y}-\bar{x})}}\right]^2,$$

since

$$V_0(\bar{Y}-\bar{X}) = \frac{\theta^2}{n} + \frac{[(1)\theta]^2}{n} = \frac{2\theta^2}{n} \quad \text{and} \quad \hat{\theta}_\omega = \frac{(\bar{x}+\bar{y})}{2},$$

so that

$$\hat{V}_0(\bar{Y}-\bar{X}) = \frac{2\hat{\theta}_\omega^2}{n} = \frac{(\bar{x}+\bar{y})^2}{2n}.$$

For $n = 50$, $\bar{x} = 30$, and $\bar{y} = 40$,

$$\hat{S} = \frac{2(50)(40-30)^2}{(30+40)^2} = 2.04.$$

So,

$$\text{P-value} = \text{pr}\left(\chi_1^2 > 2.04 \middle| H_0 : \lambda = 1\right) \doteq 0.15.$$

So, there is not sufficient evidence with these data to reject $H_0 : \lambda = 1$.

**Solution 5.17**

(a) The marginal cumulative distribution function (CDF) of $X$, $F_X(x)$, is given by

$$F_X(x) = E_\lambda\left[F_X(x|\lambda)\right] = \int_0^\infty F_X(x|\lambda)\pi(\lambda)\,d\lambda$$

$$= \int_0^\infty \left(1 - e^{-\lambda x}\right)\beta e^{-\beta\lambda} d\lambda$$

$$= 1 - \int_0^\infty \beta e^{-(x+\beta)\lambda} d\lambda$$

$$= 1 - \frac{\beta}{(x+\beta)}\int_0^\infty (x+\beta)e^{-(x+\beta)\lambda} d\lambda$$

$$= 1 - \left(1 + \frac{x}{\beta}\right)^{-1}, \quad x > 0.$$

Thus,

$$f_X(x) = \frac{1}{\beta}\left(1 + \frac{x}{\beta}\right)^{-2}, \quad x > 0, \ \beta > 0,$$

which is a *generalized Pareto distribution* with scale parameter equal to $\beta$, shape parameter equal to 1, and location parameter equal to 0.

(b) Now,

$$\begin{aligned}\pi(\lambda|X = x) &= \frac{f_{X,\lambda}(x,\lambda)}{f_X(x)} = \frac{f_X(x|\lambda)\pi(\lambda)}{f_X(x)} \\ &= \frac{\lambda\beta e^{-(x+\beta)\lambda}}{\frac{1}{\beta}\left(1 + \frac{x}{\beta}\right)^{-2}} \\ &= \lambda\beta^2\left(1 + \frac{x}{\beta}\right)^2 e^{-(x+\beta)\lambda} = \lambda(x+\beta)^2 e^{-(x+\beta)\lambda}, \quad \lambda > 0.\end{aligned}$$

Thus, the posterior distribution for $\lambda$ is GAMMA$[(x+\beta)^{-1}, 2]$. Since $\pi(\lambda)$ is GAMMA$(\beta^{-1}, 1)$, the prior and posterior distributions belong to the same distributional family, and hence $\pi(\lambda)$ is known as a *conjugate prior*.

(c) For a given value of $\lambda$ ($\lambda^*$, say), $\text{pr}(\lambda < \lambda^*) = 1 - e^{-\beta\lambda^*}$ based on the prior distribution $\pi(\lambda)$. And, given an observed value $x$ of $X$,

$$\begin{aligned}\text{pr}(\lambda < \lambda^*|X = x) &= \int_0^{\lambda^*} \pi(\lambda|X = x)\,d\lambda \\ &= \int_0^{\lambda^*} \lambda(x+\beta)^2 e^{-(x+\beta)\lambda}\,d\lambda.\end{aligned}$$

Using integration by parts with $u = \lambda$ and $dv = (x+\beta)^2 e^{-(x+\beta)\lambda}$, we have

$$\begin{aligned}\text{pr}(\lambda < \lambda^*|X = x) &= -\lambda(x+\beta)e^{-(x+\beta)\lambda}\Big|_0^{\lambda^*} + \int_0^{\lambda^*}(x+\beta)e^{-(x+\beta)\lambda}d\lambda \\ &= -\lambda^*(x+\beta)e^{-(x+\beta)\lambda^*} + 1 - e^{-(x+\beta)\lambda^*} \\ &= 1 - \left[\lambda^*(x+\beta) + 1\right]e^{-(x+\beta)\lambda^*}.\end{aligned}$$

With $\lambda^* = 1$, $\beta = 1$ and $x = 3$, $\text{pr}(H_1) = 1 - e^{-1} = 0.6321$, $\text{pr}(H_0) = 1 - \text{pr}(H_1) = 0.3679$, $\text{pr}(H_1|X = x) = 1 - 5e^{-4} = 0.9084$, and $\text{pr}(H_0|X = x) = 1 - \text{pr}(H_1|X = x) = 0.0916$. Thus,

$$BF_{10} = \frac{\text{pr}(H_1|X = x)\text{pr}(H_0)}{\text{pr}(H_0|X = x)\text{pr}(H_1)} = \frac{(0.9084)(0.3679)}{(0.0916)(0.6321)} = 5.77.$$

Hence, observing a survival time of $x = 3$ years yields "positive," but not "strong," evidence in favor of $H_1$.

**Solution 5.18***

(a) With $\boldsymbol{x} = (x_1, x_2, \ldots, x_n)$ and $\boldsymbol{y} = (y_1, y_2, \ldots, y_n)$, the likelihood function has the structure

$$\begin{aligned}\mathcal{L}(\boldsymbol{x}, \boldsymbol{y}; \theta, \phi_1, \phi_2, \ldots, \phi_n) &= \prod_{i=1}^{n} \mathrm{f}_{X_i}(x_i)\mathrm{f}_{Y_i}(y_i) \\ &= \prod_{i=1}^{n} \left\{ (\theta\phi_i)^{-1}\mathrm{e}^{-x_i/\theta\phi_i} \cdot \phi_i^{-1}\mathrm{e}^{-y_i/\phi_i} \right\} \\ &= \theta^{-n}\left(\prod_{i=1}^{n}\phi_i\right)^{-2} \mathrm{e}^{\left(-\sum_{i=1}^{n}\left[\frac{x_i}{\theta\phi_i} + \frac{y_i}{\phi_i}\right]\right)},\end{aligned}$$

$$0 < x_i < \infty, \ 0 < y_i < \infty, \ i = 1, 2, \ldots, n.$$

Using this particular likelihood would entail the estimation of $(n+1)$ parameters, namely, $\theta, \phi_1, \phi_2, \ldots, \phi_n$. Note that there are only $2n$ data points, so that the number of parameters to be estimated is more than half the number of data values; this type of situation often leads to unreliable statistical inferences.

(b) We need to use the method of transformations. We know that

$$\begin{aligned}\mathrm{f}_{X_i,Y_i}(x_i, y_i) &= \mathrm{f}_{X_i}(x_i) \cdot \mathrm{f}_{Y_i}(y_i) \\ &= \frac{1}{\theta\phi_i}\mathrm{e}^{-x_i/\theta\phi_i} \cdot \frac{1}{\phi_i}\mathrm{e}^{-y_i/\phi_i} \\ &= \frac{1}{\theta\phi_i^2}\mathrm{e}^{-\left(\frac{x_i}{\theta\phi_i} + \frac{y_i}{\phi_i}\right)}, \quad 0 < x_i < +\infty, \ 0 < y_i < +\infty.\end{aligned}$$

Let $R_i = X_i/Y_i$ and $S_i = Y_i$, so that $X_i = R_iS_i$ and $Y_i = S_i$. Clearly, $0 < R_i < +\infty$ and $0 < S_i < +\infty$. And,

$$J = \begin{vmatrix} \frac{\partial X_i}{\partial R_i} & \frac{\partial X_i}{\partial S_i} \\ \frac{\partial Y_i}{\partial R_i} & \frac{\partial Y_i}{\partial S_i} \end{vmatrix} = \begin{vmatrix} S_i & R_i \\ 0 & 1 \end{vmatrix} = S_i = |J_i|.$$

So,

$$\mathrm{f}_{R_i,S_i}(r_i, s_i) = \frac{1}{\theta\phi_i^2}\mathrm{e}^{-\left(\frac{r_is_i}{\theta\phi_i} + \frac{s_i}{\phi_i}\right)}(s_i),$$

$0 < r_i < +\infty, 0 < s_i < +\infty$. Finally,

$$f_{R_i}(r_i) = \frac{1}{\theta\phi_i^2}\int_0^{\infty} s_i e^{-\frac{s_i}{\phi_i}\left(\frac{r_i}{\theta}+1\right)} ds_i$$

$$= \frac{1}{\theta\phi_i^2}\int_0^{\infty} s_i e^{-s_i \Big/ \frac{\phi_i}{\left(\frac{r_i}{\theta}+1\right)}} ds_i$$

$$= \frac{1}{\theta\phi_i^2}\left[\frac{\phi_i}{\left(\frac{r_i}{\theta}+1\right)}\right]^2$$

$$= \frac{\theta}{(\theta + r_i)^2}, \quad 0 < r_i < +\infty.$$

(c) With $\boldsymbol{r} = (r_1, r_2, \ldots, r_n)$, we have

$$\mathcal{L}(\boldsymbol{r};\theta) \equiv \mathcal{L} = \prod_{i=1}^{n}\left\{\frac{\theta}{(\theta + r_i)^2}\right\} = \theta^n \prod_{i=1}^{n}(\theta + r_i)^{-2}.$$

So,

$$\ln \mathcal{L} = n \ln \theta - 2\sum_{i=1}^{n} \ln(\theta + r_i),$$

$$\frac{\partial \ln \mathcal{L}}{\partial \theta} = \frac{n}{\theta} - 2\sum_{i=1}^{n}(\theta + r_i)^{-1},$$

$$\frac{\partial^2 \ln \mathcal{L}}{\partial \theta^2} = \frac{-n}{\theta^2} + 2\sum_{i=1}^{n}(\theta + r_i)^{-2}.$$

And,

$$\mathrm{E}[(\theta + r_i)^{-2}] = \int_0^{\infty}(\theta + r_i)^{-2}\frac{\theta}{(\theta + r_i)^2} dr_i$$

$$= \int_0^{\infty}\frac{\theta}{(\theta + r_i)^4} dr_i$$

$$= \theta\left[\frac{-(\theta + r_i)^{-3}}{3}\right]_0^{\infty}$$

$$= \frac{1}{3\theta^2}.$$

So,

$$-\mathrm{E}\left(\frac{\partial^2 \ln \mathcal{L}}{\partial \theta^2}\right) = \frac{n}{\theta^2} - 2\sum_{i=1}^{n}\left(\frac{1}{3\theta^2}\right) = \frac{n}{\theta^2} - \frac{2n}{3\theta^2} = \frac{n}{3\theta^2}.$$

Hence, if $\hat{\theta}$ is the MLE of $\theta$, then

$$\frac{\hat{\theta}-\theta}{\sqrt{3\theta^2/n}} \dot{\sim} \mathrm{N}(0,1)$$

for large $n$.

To test $\mathrm{H}_0$: $\theta = 1$ versus $\mathrm{H}_1 : \theta > 1$, we would reject $\mathrm{H}_0$ if $(\hat{\theta}-1)/\sqrt{3/n} > 1.96$ for a size $\alpha = 0.025$ test; note that this is a score test. So, when $\theta = 1.50$,

$$\begin{aligned}
\text{POWER} &= \text{pr}\left\{\frac{\hat{\theta}-1}{\sqrt{3/n}} > 1.96|\theta = 1.50\right\} \\
&= \text{pr}\left\{\frac{\hat{\theta}-1.50}{\sqrt{3(1.50)^2/n}} > \frac{(1+1.96\sqrt{3/n}-1.50)}{\sqrt{3(1.50)^2/n}}\right\} \\
&\approx \text{pr}\left[Z > \frac{1.96}{1.50} - \frac{\sqrt{n}}{3\sqrt{3}}\right]
\end{aligned}$$

where $Z \sim \mathrm{N}(0,1)$. So, we should choose $n^*$ as the smallest positive integer value of $n$ such that

$$\frac{-\sqrt{n}}{3\sqrt{3}} + \frac{1.96}{1.50} \leq -0.84,$$

or, equivalently,

$$-\sqrt{n} \leq -3\sqrt{3}\left(\frac{1.96}{1.50}+0.84\right) = -11.1546 \implies n^* = 125.$$

**Solution 5.19***

(a) First, we know that $S_i \sim \text{POI}[\phi_i(\lambda_1+\lambda_0)]$. So, for $i = 1, 2, \ldots, n$,

$$\begin{aligned}
p_{Y_{i1}}(y_{i1}|S_i = s_i) &= \frac{\text{pr}[(Y_{i1} = y_{i1}) \cap (S_i = s_i)]}{\text{pr}(S_i = s_i)} \\
&= \frac{\text{pr}[(Y_{i1} = y_{i1}) \cap (Y_{i0} = s_i - y_{i1})]}{\text{pr}(S_i = s_i)} \\
&= \frac{\left[\frac{(\phi_i\lambda_1)^{y_{i1}}e^{-\phi_i\lambda_1}}{y_{i1}!}\right]\left[\frac{(\phi_i\lambda_0)^{(s_i-y_{i1})}e^{-\phi_i\lambda_0}}{(s_i-y_{i1})!}\right]}{\left\{\frac{[\phi_i(\lambda_1+\lambda_0)]^{s_i}e^{-\phi_i(\lambda_1+\lambda_0)}}{s_i!}\right\}} \\
&= C^{s_i}_{y_{i1}}\left(\frac{\lambda_1}{\lambda_1+\lambda_0}\right)^{y_{i1}}\left(\frac{\lambda_0}{\lambda_1+\lambda_0}\right)^{s_i-y_{i1}},
\end{aligned}$$

$y_{i1} = 0, 1, \ldots, s_i$.

So, the conditional distribution of $Y_{i1}$ given $S_i = s_i$ is $\text{BIN}[s_i, \lambda_1/(\lambda_1+\lambda_0)]$.

(b) Based on the result found in part (a), an appropriate (conditional) likelihood function is

$$\mathcal{L}_c = \prod_{i=1}^{n} C_{y_{i1}}^{s_i} \theta^{y_{i1}} (1-\theta)^{s_i - y_{i1}},$$

where $\theta = \lambda_1/(\lambda_1 + \lambda_0)$.
Thus,

$$\ln \mathcal{L}_c \propto \ln\theta \sum_{i=1}^{n} y_{i1} + \ln(1-\theta)\left(\sum_{i=1}^{n} s_i - \sum_{i=1}^{n} y_{i1}\right),$$

so that

$$\frac{\partial \ln \mathcal{L}_c}{\partial\theta} = \frac{\sum_{i=1}^{n} y_{i1}}{\theta} - \frac{(\sum_{i=1}^{n} s_i - \sum_{i=1}^{n} y_{i1})}{(1-\theta)} = 0$$

gives $\hat{\theta} = \sum_{i=1}^{n} y_{i1} / \sum_{i=1}^{n} s_i$.

And, with $\boldsymbol{S} = (S_1, S_2, \ldots, S_n)$ and $\boldsymbol{s} = (s_1, s_2, \ldots, s_n)$,

$$\frac{\partial^2 \ln \mathcal{L}_c}{\partial\theta^2} = \frac{-\sum_{i=1}^{n} y_{i1}}{\theta^2} - \frac{(\sum_{i=1}^{n} s_i - \sum_{i=1}^{n} y_{i1})}{(1-\theta)^2}$$

gives

$$\mathrm{V}(\hat{\theta}|\boldsymbol{S}=\boldsymbol{s}) = -\left[\mathrm{E}\left(\frac{\partial^2 \ln \mathcal{L}_c}{\partial\theta^2}\right)\middle|\boldsymbol{S}=\boldsymbol{s}\right]^{-1}$$

$$= \left[\frac{\theta\sum_{i=1}^{n} s_i}{\theta^2} + \frac{(1-\theta)\sum_{i=1}^{n} s_i}{(1-\theta)^2}\right]^{-1} = \frac{\theta(1-\theta)}{\sum_{i=1}^{n} s_i}.$$

So, given $\boldsymbol{S} = \boldsymbol{s}$, under $\mathrm{H}_0: \theta = 1/2$ (or, equivalently, $\lambda_1 = \lambda_0$) and for large $n$, it follows that

$$U = \frac{\hat{\theta} - \frac{1}{2}}{\sqrt{\frac{(1/2)(1/2)}{(\sum_{i=1}^{n} s_i)}}} \dot{\sim} \mathrm{N}(0,1);$$

so, we would reject $\mathrm{H}_0: \theta = 1/2$ (or, equivalently, $\lambda_1 = \lambda_0$) in favor of $\mathrm{H}_1: \theta > 1/2$ (or, equivalently, $\lambda_1 > \lambda_0$) when the observed value $u$ of $U$ exceeds 1.645. Note that this is a score-type test statistic.

When $n = 50, \sum_{i=1}^{n} s_i = 500$, and $\sum_{i=1}^{n} y_{i1} = 275$, then $\hat{\theta} = 275/500 = 0.55$, so that

$$u = \frac{0.55 - 0.50}{\left(\frac{1}{2}\right)\sqrt{1/500}} = 2.236;$$

so, these data provide strong evidence (P-value = 0.0127) for rejecting $H_0 : \theta = 1/2$ in favor of $H_1 : \theta > 1/2$. Another advantage of this conditional inference procedure is that its use avoids the need to estimate the parameters $\lambda_1$ and $\lambda_0$ separately.

**Solution 5.20***

(a) With $\boldsymbol{x} = (x_1, x_2, \ldots, x_n)$ and $\boldsymbol{y} = (y_1, y_2, \ldots, y_n)$,

$$\mathcal{L} \equiv \mathcal{L}(\boldsymbol{x}, \boldsymbol{y}; \lambda_x, \lambda_y) = \prod_{i=1}^{n} \left[ \frac{\lambda_x^{x_i} \mathrm{e}^{-\lambda_x}}{x_i!} \cdot \frac{\lambda_y^{y_i} \mathrm{e}^{-\lambda_y}}{y_i!} \right]$$

$$= \frac{\lambda_x^{n\bar{x}} \mathrm{e}^{-n\lambda_x}}{\prod_{i=1}^{n} x_i!} \cdot \frac{\lambda_y^{n\bar{y}} \mathrm{e}^{-n\lambda_y}}{\prod_{i=1}^{n} y_i!}.$$

So,

$$\ln \mathcal{L} \sim n\bar{x} \ln \lambda_x + n\bar{y} \ln \lambda_y - n(\lambda_x + \lambda_y).$$

Thus,

$$\frac{\partial \ln \mathcal{L}}{\partial \lambda_x} = \frac{n\bar{x}}{\lambda_x} - n, \quad \frac{\partial \ln \mathcal{L}}{\partial \lambda_y} = \frac{n\bar{y}}{\lambda_y} - n,$$

$$\frac{\partial^2 \ln \mathcal{L}}{\partial \lambda_x^2} = \frac{-n\bar{x}}{\lambda_x^2}, \quad \frac{\partial^2 \ln \mathcal{L}}{\partial \lambda_y^2} = \frac{-n\bar{y}}{\lambda_y^2},$$

and

$$\frac{\partial^2 \ln \mathcal{L}}{\partial \lambda_x \partial \lambda_y} = \frac{\partial^2 \ln \mathcal{L}}{\partial \lambda_y \partial \lambda_x} = 0.$$

Hence,

$$-\mathrm{E}\left[ \frac{\partial^2 \ln \mathcal{L}}{\partial \lambda_x^2} \right] = \frac{n\lambda_x}{\lambda_x^2} = \frac{n}{\lambda_x},$$

and

$$-\mathrm{E}\left[ \frac{\partial^2 \ln \mathcal{L}}{\partial \lambda_y^2} \right] = \frac{n\lambda_y}{\lambda_y^2} = \frac{n}{\lambda_y}.$$

So,

$$\mathcal{I}(\lambda_x, \lambda_y) = \begin{bmatrix} \dfrac{n}{\lambda_x} & 0 \\ 0 & \dfrac{n}{\lambda_y} \end{bmatrix} \quad \text{and} \quad \mathcal{I}^{-1}(\lambda_x, \lambda_y) = \begin{bmatrix} \dfrac{\lambda_x}{n} & 0 \\ 0 & \dfrac{\lambda_y}{n} \end{bmatrix}.$$

Under $H_0 : \lambda_x = \lambda_y$ ($= \lambda$, say), $\ln \mathcal{L}_\omega \sim n\bar{x}\ln\lambda + n\bar{y}\ln\lambda - 2n\lambda$.

Solving

$$\frac{\partial \ln \mathcal{L}_\omega}{\partial \lambda} = \frac{n\bar{x}}{\lambda} + \frac{n\bar{y}}{\lambda} - 2n = 0$$

gives

$$\hat{\lambda} = \frac{(\bar{x}+\bar{y})}{2}.$$

So,

$$\hat{S} = \begin{bmatrix} \frac{n\bar{x}}{\hat{\lambda}} - n \\ \frac{n\bar{y}}{\hat{\lambda}} - n \end{bmatrix}' \begin{bmatrix} \hat{\lambda}/n & 0 \\ 0 & \hat{\lambda}/n \end{bmatrix} \begin{bmatrix} \frac{n\bar{x}}{\hat{\lambda}} - n \\ \frac{n\bar{y}}{\hat{\lambda}} - n \end{bmatrix}.$$

Since $\hat{\lambda} = (8.00 + 9.00)/2 = 8.500$ and $n = 30$,

$$\hat{S} = \begin{bmatrix} \frac{30(8)}{8.5} - 30 \\ \frac{30(9)}{8.5} - 30 \end{bmatrix}' \begin{bmatrix} \frac{8.5}{30} & 0 \\ 0 & \frac{8.5}{30} \end{bmatrix} \begin{bmatrix} \frac{30(8)}{8.5} - 30 \\ \frac{30(9)}{8.5} - 30 \end{bmatrix} = 1.7645.$$

Since,

$$\text{P-value} = \text{pr}(\chi_1^2 > 1.7645) \geq 0.15,$$

we would not reject $H_0$ at any conventional $\alpha$-level.

(b) Now, since $(X_1 + Y_1) \sim \text{POI}(\lambda_x + \lambda_y)$, we have

$$\begin{aligned}
p_{X_1}(x_1|X_1+Y_1=s_1) &= \text{pr}(X_1 = x_1|X_1+Y_1=s_1) \\
&= \frac{\text{pr}\left[(X_1=x_1)\cap(X_1+Y_1=s_1)\right]}{\text{pr}\left[X_1+Y_1=s_1\right]} \\
&= \frac{\text{pr}(X_1=x_1)\text{pr}(Y_1=s_1-x_1)}{\text{pr}(X_1+Y_1=s_1)} \\
&= \frac{\left[\frac{\lambda_x^{x_1}e^{-\lambda_x}}{x_1!}\right]\left[\frac{\lambda_y^{s_1-x_1}e^{-\lambda_y}}{(s_1-x_1)!}\right]}{\left[\frac{(\lambda_x+\lambda_y)^{s_1}e^{-(\lambda_x+\lambda_y)}}{s_1!}\right]} \\
&= \frac{s_1!}{x_1!(s_1-x_1)!}\left(\frac{\lambda_x}{\lambda_x+\lambda_y}\right)^{x_1}\left(1-\frac{\lambda_x}{\lambda_x+\lambda_y}\right)^{s_1-x_1} \\
&= C_{x_1}^{s_1}\,\pi^{x_1}(1-\pi)^{s_1-x_1}
\end{aligned}$$

for $x_1 = 0, 1, \ldots, s_1$ and $\pi = \lambda_x/(\lambda_x + \lambda_y)$. So, given $(X_1 + Y_1) = s_1$,

$$X_1 \sim \text{BIN}\left[n = s_1, \pi = \frac{\lambda_x}{(\lambda_x + \lambda_y)}\right].$$

If $\delta = 0.60$, then $\text{H}_0 : \lambda_y = 0.60\lambda_x$ is equivalent to testing

$$\text{H}_0' : \pi = \frac{\lambda_x}{\lambda_x + 0.60\lambda_x} = \frac{1}{1.60} = 0.625,$$

and $\text{H}_1 : \lambda_y > 0.60\lambda_x$ is equivalent to testing

$$\text{H}_1' : \pi < \frac{\lambda_x}{\lambda_x + 0.60\lambda_x} = 0.625.$$

So, for the given data, the exact P-value is

$$\begin{aligned} \text{P-value} &= \text{pr}(X_1 \leq 4 | S_1 = 14, \theta = 0.625) \\ &= \sum_{x_1=0}^{4} \text{C}_{x_1}^{14} (0.625)^{x_1} (0.375)^{14-x_1} \doteq 0.0084. \end{aligned}$$

So, given the observed values of $x_1 = 4$ and $y_1 = 10$, one would reject $\text{H}_0 : \lambda_y = 0.60\lambda_x$ in favor of $\text{H}_1 : \lambda_y > 0.60\lambda_x$ using this conditional test.

**Solution 5.21***

(a) From standard order statistics theory, it follows directly that

$$\text{f}_{X_{(1)}}(x_{(1)}) = n[1 - x_{(1)} + \theta]^{n-1}, \quad 0.50 \leq \theta < x_{(1)} < (\theta + 1) < +\infty,$$

that

$$\text{f}_{X_{(n)}}(x_{(n)}) = n[x_{(n)} - \theta]^{n-1}, \quad 0.50 \leq \theta < x_{(n)} < (\theta + 1) < +\infty,$$

and that

$$\text{f}_{X_{(1)},X_{(n)}}(x_{(1)}, x_{(n)}) = n(n-1)[x_{(n)} - x_{(1)}]^{n-1},$$
$$0.50 \leq \theta < x_{(1)} < x_{(n)} < (\theta + 1) < +\infty.$$

Now, $\text{pr}(\text{B}|\text{H}_0 : \theta = 1) = \text{pr}(X_{(n)} > 2|\text{H}_0 : \theta = 1) = 0$ since $1 < X_{(n)} < 2$ when $\theta = 1$. Thus, it follows that the probability of a Type I error is equal to

$$\begin{aligned} \text{pr}(\text{A}|\text{H}_0 : \theta = 1) &= \text{pr}(X_{(1)} > k|\theta = 1) = \int_k^2 n(2 - x_{(1)})^{n-1} \text{d}x_{(1)} \\ &= \left[-(2 - x_{(1)})^n\right]_k^2 = (2 - k)^n; \end{aligned}$$

thus, solving the equation $(2-k)^n = \alpha$ gives

$$k_\alpha = 2 - \alpha^{1/n}, \quad 0 < \alpha \leq 0.10.$$

(b) First, consider the power for values of $\theta$ satisfying $\theta > k_\alpha > 1$. In this situation, $X_{(1)} > k_\alpha$, so that $\text{pr}(\text{A}|\theta > k_\alpha) = \text{pr}(X_{(1)} > k_\alpha|\theta > k_\alpha) = 1$, so that the power is 1 for $\theta > k_\alpha$.

For values of $\theta$ satisfying $1 < \theta \leq k_\alpha$,

$$\text{POWER} = \text{pr}(\text{A}|1 < \theta \leq k_\alpha) + \text{pr}(\text{B}|1 < \theta \leq k_\alpha) - \text{pr}(\text{A} \cap \text{B}|1 < \theta \leq k_\alpha).$$

Now, with $k_\alpha = 2 - \alpha^{1/n}$,

$$\begin{aligned}\text{pr}(\text{A}|1 < \theta \leq k_\alpha) &= \int_{k_\alpha}^{\theta+1} n[1 - x_{(1)} + \theta]^{n-1} dx_{(1)} \\ &= (1 - k_\alpha + \theta)^n = \left[\theta - 1 + \alpha^{1/n}\right]^n.\end{aligned}$$

And,

$$\begin{aligned}\text{pr}(\text{B}|1 < \theta \leq k_\alpha) &= \int_{2}^{\theta+1} n[x_{(n)} - \theta]^{n-1} dx_{(n)} \\ &= 1 - (2-\theta)^n.\end{aligned}$$

Finally,

$$\begin{aligned}\text{pr}(\text{A} \cap \text{B}|1 < \theta \leq k_\alpha) &= \int_{2}^{\theta+1} \int_{k_\alpha}^{x_{(n)}} n(n-1)(x_{(n)} - x_{(1)})^{n-2} dx_{(1)} dx_{(n)} \\ &= \int_{2}^{\theta+1} n\left[-(x_{(n)} - x_{(1)})^{n-1}\right]_{k_\alpha}^{x_{(n)}} dx_{(n)} \\ &= \int_{2}^{\theta+1} n(x_{(n)} - k_\alpha)^{n-1} dx_{(n)} = \left[(x_{(n)} - k_\alpha)^n\right]_2^{\theta+1} \\ &= (\theta + 1 - k_\alpha)^n - (2 - k_\alpha)^n = \left(\theta - 1 + \alpha^{1/n}\right)^n - \alpha.\end{aligned}$$

So, for $1 < \theta \leq k_\alpha = 2 - \alpha^{1/n}$,

$$\begin{aligned}\text{POWER} &= \left[\theta - 1 + \alpha^{1/n}\right]^n + \left[1 - (2-\theta)^n\right] - \left[\theta - 1 + \alpha^{1/n}\right]^n + \alpha \\ &= 1 + \alpha - (2-\theta)^n.\end{aligned}$$

As required, the above expression equals $\alpha$ when $\theta = 1$ and equals 1 when $\theta = k_\alpha = 2 - \alpha^{1/n}$.

**Solution 5.22***

(a) Clearly, $E(\bar{Y}_1 - \bar{Y}_2) = E(\bar{Y}_1) - E(\bar{Y}_2) = (\mu_1 - \mu_2)$. Now,

$$V(\bar{Y}_i) = V\left(\frac{1}{n}\sum_{j=1}^{n} Y_{ij}\right) = \frac{1}{n^2}\left\{\sum_{j=1}^{n} V(Y_{ij}) + 2\sum_{\text{all} j<j'} \text{cov}(Y_{ij}, Y_{ij'})\right\}$$

$$= \frac{1}{n^2}\left\{n\sigma^2 + 2\frac{n(n-1)}{2}\rho\sigma^2\right\} = \frac{\sigma^2}{n}[1 + (n-1)\rho], \quad i = 1, 2.$$

So,

$$V(\bar{Y}_1 - \bar{Y}_2) = V(\bar{Y}_1) + V(\bar{Y}_2) = \frac{2\sigma^2}{n}[1 + (n-1)\rho].$$

(b) Under the stated assumptions,

$$(\bar{Y}_1 - \bar{Y}_2) = N\left\{(\mu_1 - \mu_2), \frac{2\sigma^2}{n}[1 + (n-1)\rho]\right\}.$$

So,

$$Z = \frac{(\bar{Y}_1 - \bar{Y}_2) - (\mu_1 - \mu_2)}{\sqrt{\frac{2\sigma^2}{n}[1 + (n-1)\rho]}} \sim N(0, 1).$$

Thus, to test $H_0 : \mu_1 = \mu_2$ versus $H_1 : \mu_1 > \mu_2$ at the $\alpha = 0.05$ level, we reject $H_0$ in favor of $H_1$ when

$$\frac{(\bar{Y}_1 - \bar{Y}_2) - 0}{\sqrt{\frac{2\sigma^2}{n}[1 + (n-1)\rho]}} > 1.645.$$

(c) If one incorrectly assumes that $\rho = 0$, one would use (under the stated assumptions) the test statistic

$$\frac{(\bar{Y}_1 - \bar{Y}_2)}{\sqrt{2\sigma^2/n}},$$

and reject $H_0 : \mu_1 = \mu_2$ in favor of $H_1 : \mu_1 > \mu_2$ when

$$\frac{(\bar{Y}_1 - \bar{Y}_2)}{\sqrt{\frac{2\sigma^2}{n}}} > 1.645.$$

Thus, the *actual* Type I error rate using this incorrect testing procedure (when $n = 10, \sigma^2 = 2$, and $\rho = 0.50$) is:

$$\text{pr}\left[\frac{(\bar{Y}_1 - \bar{Y}_2) - 0}{\sqrt{\frac{2\sigma^2}{n}}} > 1.645 \Big| n = 10, \sigma^2 = 2, \rho = 0.50\right]$$

$$= \text{pr}\left[\frac{(\bar{Y}_1 - \bar{Y}_2) - 0}{\sqrt{\frac{2\sigma^2}{n}[1 + (n-1)\rho]}} > \frac{1.645\sqrt{2\sigma^2/n}}{\sqrt{\frac{2\sigma^2}{n}[1 + (n-1)\rho]}} \Big| n = 10, \sigma^2 = 2, \rho = 0.50\right]$$

$$= \text{pr}(Z > 0.7014] \doteq 0.24.$$

This simple example illustrates that ignoring positive "intra-cluster" (in our case, intra-neighborhood) response correlation can lead to inflated Type I error rates, and more generally, to invalid statistical inferences.

**Solution 5.23***

(a) Given $X_1 = x_1$, where $x_1$ is a fixed constant, $X_2 = \theta x_1 + \epsilon_2 \sim N(\theta x_1, \sigma^2)$.

(b)

$$\begin{aligned} f_{X_1,X_2}(x_1, x_2) &= f_{X_1}(x_1) f_{X_2}(x_2 | X_1 = x_1) \\ &= \frac{1}{\sqrt{2\pi\sigma^2}} e^{-x_1^2/2\sigma^2} \cdot \frac{1}{\sqrt{2\pi\sigma^2}} e^{-(x_2-\theta x_1)^2/2\sigma^2}, \\ & -\infty < x_1 < \infty, \ -\infty < x_2 < \infty. \end{aligned}$$

(c) For $i = 2, 3, \ldots, n$, since $X_i = \theta X_{i-1} + \epsilon_i$, it follows from part (a) that

$$f_{X_i}(x_i | X_j = x_j, j = 1, 2, \ldots, i-1) = f_{X_i}(x_i | X_{i-1} = x_{i-1}),$$

where the conditional density of $X_i$ given $X_{i-1} = x_{i-1}$ is $N(\theta x_{i-1}, \sigma^2)$. So,

$$f^* = f_{X_1}(x_1) \prod_{i=2}^{n} f_{X_i}(x_i | X_{i-1} = x_{i-1})$$

$$= \frac{1}{\sqrt{2\pi\sigma^2}} e^{-x_1^2/2\sigma^2} \cdot \prod_{i=2}^{n} \left\{ \frac{1}{\sqrt{2\pi\sigma^2}} e^{-(x_i-\theta x_{i-1})^2/2\sigma^2} \right\}$$

$$= (2\pi)^{-n/2}(\sigma^2)^{-n/2}\exp\left\{-\frac{1}{2\sigma^2}\left[x_1^2 + \sum_{i=2}^{n}(x_i - \theta x_{i-1})^2\right]\right\},$$

$$-\infty < x_i < \infty,\ i = 1, 2, \ldots, n.$$

So,

$$\ln f^* = -\frac{n}{2}\ln(2\pi) - \frac{n}{2}\ln\sigma^2 - \frac{1}{2\sigma^2}\left[x_1^2 + \sum_{i=2}^{n}(x_i - \theta x_{i-1})^2\right].$$

So, in the unrestricted parameter space $\Omega$,

$$\frac{\partial \ln f^*}{\partial\theta} = \frac{1}{\sigma^2}\sum_{i=2}^{n} x_{i-1}(x_i - \theta x_{i-1}) = 0 \Rightarrow \hat{\theta}_\Omega = \frac{\sum_{i=2}^{n} x_{i-1}x_i}{\sum_{i=1}^{n-1} x_i^2}.$$

And,

$$\frac{\partial \ln f^*}{\partial(\sigma^2)} = \frac{-n}{2\sigma^2} + \frac{1}{2\sigma^4}\left[x_1^2 + \sum_{i=2}^{n}(x_i - \theta x_{i-1})^2\right]$$

$$\Rightarrow \hat{\sigma}_\Omega^2 = \frac{1}{n}\left[x_1^2 + \sum_{i=2}^{n}(x_i - \hat{\theta}_\Omega x_{i-1})^2\right]$$

$$= \frac{1}{n}\sum_{i=1}^{n}(x_i - \hat{\theta}_\Omega x_{i-1})^2 \quad \text{since} \quad x_0 \equiv 0.$$

So,

$$\hat{\mathcal{L}}_\Omega = f^*_{|\theta=\hat{\theta}_\Omega, \sigma^2=\hat{\sigma}^2_\Omega} = (2\pi)^{-n/2}\left(\hat{\sigma}^2_\Omega\right)^{-n/2} e^{-n/2}.$$

And, in the restricted parameter space $\omega$ (i.e., where $\theta = 0$),

$$\ln f^*_{|\theta=0} = -\frac{n}{2}\ln(2\pi) - \frac{n}{2}\ln\sigma^2 - \frac{1}{2\sigma^2}\sum_{i=1}^{n} x_i^2,$$

$$\frac{\partial \ln f^*}{\partial(\sigma^2)} = \frac{-n}{2\sigma^2} + \frac{1}{2\sigma^4}\sum_{i=1}^{n} x_i^2 = 0$$

$$\Rightarrow \hat{\sigma}^2_\omega = \frac{\sum_{i=1}^{n} x_i^2}{n},$$

so that

$$\hat{\mathcal{L}}_{\omega} = f^{*}_{|\theta=0,\sigma^2=\hat{\sigma}^2_{\omega}} = (2\pi)^{-n/2}\left(\hat{\sigma}^2_{\omega}\right)^{-n/2} e^{-n/2}.$$

Thus,

$$\hat{\lambda} = \frac{\hat{\mathcal{L}}_{\omega}}{\hat{\mathcal{L}}_{\Omega}} = \left(\frac{\hat{\sigma}^2_{\Omega}}{\hat{\sigma}^2_{\omega}}\right)^{n/2}$$

$$\Rightarrow \hat{\lambda}^{2/n} = \frac{\hat{\sigma}^2_{\Omega}}{\hat{\sigma}^2_{\omega}} = \frac{\sum_{i=1}^{n}(x_i - \hat{\theta}_{\Omega}x_{i-1})^2}{\sum_{i=1}^{n} x_i^2}$$

$$= \frac{\sum_{i=1}^{n} x_i^2 - 2\hat{\theta}_{\Omega}\sum_{i=1}^{n} x_{i-1}x_i + \hat{\theta}^2_{\Omega}\sum_{i=1}^{n} x_{i-1}^2}{\sum_{i=1}^{n} x_i^2}$$

Note that $\sum_{i=1}^{n} x_{i-1}x_i = \sum_{i=2}^{n} x_{i-1}x_i$ and $\sum_{i=1}^{n} x_{i-1}^2 = \sum_{i=1}^{n-1} x_i^2$ since $x_0 \equiv 0$. Thus, we have

$$\hat{\lambda}^{2/n} = 1 - \left\{\frac{2\left[\left(\sum_{i=2}^{n} x_{i-1}x_i\right)^2 / \sum_{i=1}^{n-1} x_i^2\right] - \left[\left(\sum_{i=2}^{n} x_{i-1}x_i\right)^2 / \sum_{i=1}^{n-1} x_i^2\right]}{\sum_{i=1}^{n} x_i^2}\right\}$$

$$= 1 - \frac{\left(\sum_{i=2}^{n} x_{i-1}x_i\right)^2}{\left(\sum_{i=1}^{n-1} x_i^2\right)\left(\sum_{i=1}^{n} x_i^2\right)}.$$

For the given data,

$$\hat{\lambda}^{\frac{2}{30}} = 1 - \frac{(4)^2}{(15-4)(15)} = 0.9030 \Rightarrow \hat{\lambda} = (0.9030)^{15} = 0.2164$$

$$\Rightarrow -2\ln\hat{\lambda} = 3.0610.$$

Since $\chi^2_{1,0.95} = 3.84$, these data do *not* provide sufficient evidence to reject $H_0 : \theta = 0$.

**Solution 5.24***

(a) Now, with $\boldsymbol{x} = (x_1, x_2, \ldots, x_n)$ and $\boldsymbol{y} = (y_1, y_2, \ldots, y_m)$, we have

$$\mathcal{L}(\boldsymbol{x}, \boldsymbol{y}; \theta_r, \theta_u) = \left\{\prod_{i=1}^{n}\left(\theta_r x_i^{\theta_r - 1}\right)\right\}\left\{\prod_{i=1}^{m}\left(\theta_u y_i^{\theta_u - 1}\right)\right\}$$

$$= \theta_r^n\left(\prod_{i=1}^{n} x_i\right)^{\theta_r - 1}\theta_u^m\left(\prod_{i=1}^{m} y_i\right)^{\theta_u - 1}.$$

So, by the Factorization Theorem, $\prod_{i=1}^{n} X_i$ and $\prod_{i=1}^{m} Y_i$ are jointly sufficient for $\theta_r$ and $\theta_u$.

(b) From part (a), the unrestricted log likelihood is

$$\ln \mathcal{L}_\Omega = n \ln \theta_r + (\theta_r - 1) \sum_{i=1}^{n} \ln x_i + m \ln \theta_u + (\theta_u - 1) \sum_{i=1}^{m} \ln y_i.$$

So,

$$\frac{\partial \ln \mathcal{L}_\Omega}{\partial \theta_r} = \frac{n}{\theta_r} + \sum_{i=1}^{n} \ln x_i = 0$$

gives

$$\hat{\theta}_r = -n \left( \sum_{i=1}^{n} \ln x_i \right)^{-1}.$$

Similarly, by symmetry,

$$\frac{\partial \ln \mathcal{L}_\Omega}{\partial \theta_u} = \frac{m}{\theta_u} + \sum_{i=1}^{m} \ln y_i = 0$$

gives

$$\hat{\theta}_u = -m \left( \sum_{i=1}^{m} \ln y_i \right)^{-1}.$$

So,

$$\hat{\mathcal{L}}_\Omega = \hat{\theta}_r^n \left( \prod_{i=1}^{n} x_i \right)^{\hat{\theta}_r - 1} \hat{\theta}_u^m \left( \prod_{i=1}^{m} y_i \right)^{\hat{\theta}_u - 1}.$$

Now, under $H_0 : \theta_r = \theta_u (= \theta$, say), the restricted log likelihood is

$$\ln \mathcal{L}_\omega = (n + m) \ln \theta + (\theta - 1) \left( \sum_{i=1}^{n} \ln x_i + \sum_{i=1}^{m} \ln y_i \right),$$

so that

$$\frac{\partial \ln \mathcal{L}_\omega}{\partial \theta} = \frac{(n+m)}{\theta} + \left( \sum_{i=1}^{n} \ln x_i + \sum_{i=1}^{m} \ln y_i \right) = 0$$

gives

$$\hat{\theta} = -(n+m)\left(\sum_{i=1}^{n} \ln x_i + \sum_{i=1}^{m} \ln y_i\right)^{-1}.$$

So,

$$\hat{\mathcal{L}}_{\omega} = \hat{\theta}^{(n+m)}\left(\prod_{i=1}^{n} x_i \cdot \prod_{i=1}^{m} y_i\right)^{\hat{\theta}-1}.$$

So,

$$\begin{aligned}
\hat{\lambda} = \frac{\hat{\mathcal{L}}_{\omega}}{\hat{\mathcal{L}}_{\Omega}} &= \frac{\hat{\theta}^{(n+m)}\left(\prod_{i=1}^{n} x_i \cdot \prod_{i=1}^{m} y_i\right)^{\hat{\theta}-1}}{\hat{\theta}_r^n \hat{\theta}_u^m \left(\prod_{i=1}^{n} x_i\right)^{\hat{\theta}_r-1}\left(\prod_{i=1}^{m} y_i\right)^{\hat{\theta}_u-1}} \\
&= \left(\frac{\hat{\theta}}{\hat{\theta}_r}\right)^n \left(\frac{\hat{\theta}}{\hat{\theta}_u}\right)^m \left(\prod_{i=1}^{n} x_i\right)^{\hat{\theta}-\hat{\theta}_r} \left(\prod_{i=1}^{m} y_i\right)^{\hat{\theta}-\hat{\theta}_u} \\
&= \left(\frac{n+m}{n}\right)^n W^n \left(\frac{n+m}{m}\right)^m (1-W)^m \left(\prod_{i=1}^{n} x_i\right)^{\hat{\theta}-\hat{\theta}_r} \left(\prod_{i=1}^{m} y_i\right)^{\hat{\theta}-\hat{\theta}_u}.
\end{aligned}$$

Thus,

$$\begin{aligned}
\ln \hat{\lambda} &= n \ln\left(\frac{n+m}{n}\right) + m \ln\left(\frac{n+m}{m}\right) + n \ln W + m \ln(1-W) \\
&\quad + (\hat{\theta} - \hat{\theta}_r)\sum_{i=1}^{n} \ln x_i + (\hat{\theta} - \hat{\theta}_u)\sum_{i=1}^{m} \ln y_i \\
&= n \ln\left(\frac{n+m}{n}\right) + m \ln\left(\frac{n+m}{m}\right) + n \ln W + m \ln(1-W) \\
&\quad - (n+m)W \\
&\quad + n - (n+m)(1-W) + m \\
&= n \ln\left(\frac{n+m}{n}\right) + m \ln\left(\frac{n+m}{m}\right) + \ln[W^n(1-W)^m].
\end{aligned}$$

Finally,

$$-2\ln \hat{\lambda} = -2n \ln\left(\frac{n+m}{n}\right) - 2m \ln\left(\frac{n+m}{m}\right) - 2\ln[W^n(1-W)^m].$$

Under $H_0: \theta_r = \theta_u$, we know that $-2\ln\hat{\lambda} \dot{\sim} \chi_1^2$ for large $n$ and $m$. Since $0 < W < 1$, $-2\ln\hat{\lambda}$ will be large (and hence favor rejecting $H_0$) when either $W$ is close to 0 or $W$ is close to 1.

(c) Under $H_0 : \theta_r = \theta_u$ $(= \theta$, say) $f_X(x;\theta) = \theta x^{\theta-1}, 0 < x < 1$, and $f_Y(y;\theta) = \theta y^{\theta-1}$, $0 < y < 1$. Now, let $U = -\ln X$, so that $X = e^{-U}$ and $dX = -e^{-U}dU$. Hence,

$$f_U(u;\theta) = \theta(e^{-u})^{\theta-1}e^{-u} = \theta e^{-\theta u}, \quad 0 < u < \infty,$$

so that $U = -\ln X \sim \text{GAMMA}(\alpha = \theta^{-1}, \beta = 1)$. Thus,

$$\sum_{i=1}^{n}(-\ln X_i) = -\sum_{i=1}^{n}\ln X_i \sim \text{GAMMA}(\alpha = \theta^{-1}, \beta = n).$$

Analogously,

$$\sum_{i=1}^{m}(-\ln Y_i) = -\sum_{i=1}^{m}\ln Y_i \sim \text{GAMMA}(\alpha = \theta^{-1}, \beta = m).$$

Thus,

$$W = \frac{\sum_{i=1}^{n}\ln X_i}{\sum_{i=1}^{n}\ln X_i + \sum_{i=1}^{m}\ln Y_i} = \frac{-\sum_{i=1}^{n}\ln X_i}{-\sum_{i=1}^{n}\ln X_i - \sum_{i=1}^{m}\ln Y_i} = \frac{R}{(R+S)}$$

where $R \sim \text{GAMMA}(\alpha = \theta^{-1}, \beta = n)$, $S \sim \text{GAMMA}(\alpha = \theta^{-1}, \beta = m)$, and $R$ and $S$ are independent random variables. So,

$$\begin{aligned} f_{R,S}(r,s;\theta) &= \left[\frac{\theta^n r^{n-1}e^{-\theta r}}{\Gamma(n)}\right]\left[\frac{\theta^m s^{m-1}e^{-\theta s}}{\Gamma(m)}\right] \\ &= \theta^{(n+m)}r^{n-1}s^{m-1}e^{-\theta(r+s)}/\Gamma(n)\Gamma(m), \quad r > 0, \ s > 0. \end{aligned}$$

So, let $W = R/(R+S)$ and $P = (R+S)$; hence, $R = PW$ and $S = (P - PW) = P(1 - W)$. Clearly, $0 < W < 1$ and $0 < P < +\infty$. Also,

$$J = \begin{vmatrix} \frac{\partial R}{\partial P} & \frac{\partial R}{\partial W} \\ \frac{\partial S}{\partial P} & \frac{\partial S}{\partial W} \end{vmatrix} = \begin{vmatrix} W & P \\ (1-W) & -P \end{vmatrix} = -P,$$

so that $|J| = P$. Finally,

$$\begin{aligned} f_{W,P}(w,p;\theta) &= \frac{\theta^{(n+m)}(pw)^{n-1}[p(1-w)]^{m-1}e^{-\theta[pw+p(1-w)]}(p)}{\Gamma(n)\Gamma(m)} \\ &= \left[\frac{\Gamma(n+m)}{\Gamma(n)\Gamma(m)}w^{n-1}(1-w)^{m-1}\right]\left[\frac{\theta^{n+m}p^{(n+m)-1}e^{-\theta p}}{\Gamma(n+m)}\right], \end{aligned}$$

$0 < w < 1$, $0 < p < \infty$. So, $W \sim \text{BETA}(\alpha = n, \beta = m)$, $P \sim \text{GAMMA}(\alpha = \theta^{-1}, \beta = n + m)$, and $W$ and $P$ are independent random variables. When $n = m = 2$,

$$f_W(w) = \frac{\Gamma(4)}{\Gamma(2)\Gamma(2)} w(1-w) = 6w(1-w), \quad 0 < w < 1,$$

when $H_0 : \theta_r = \theta_u$ is true. So, we want to choose $k_{.05}$ such that

$$\int_0^{k_{.05}} 6t(1-t)dt = 0.05,$$

or $(3k_{.05}^2 - 2k_{.05}^3) = 0.05$, or (by trial-and-error) $k_{.05} \doteq 0.135$. So, for $n = m = 2$, reject $H_0 : \theta_r = \theta_u$ when either $W < 0.135$ or $W > 0.865$ for $\alpha = 0.10$.

**Solution 5.25***

(a) For the unrestricted parameter space,

$$\mathcal{L}_\Omega = \prod_{i=1}^{2}\prod_{j=1}^{n}\left\{C_{k-1}^{k+y_{ij}-1}\theta_i^{y_{ij}}(1+\theta_i)^{-(k+y_{ij})}\right\},$$

and

$$\ln \mathcal{L}_\Omega = \sum_{i=1}^{2}\sum_{j=1}^{n}\left\{\ln C_{k-1}^{k+y_{ij}-1} + y_{ij}\ln\theta_i - (k+y_{ij})\ln(1+\theta_i)\right\},$$

so that

$$\frac{\partial \ln \mathcal{L}_\Omega}{\partial \theta_i} = \sum_{j=1}^{n}\left[\frac{y_{ij}}{\theta_i} - \frac{(k+y_{ij})}{(1+\theta_i)}\right] = \frac{n\bar{y}_i}{\theta_i} - \frac{n(k+\bar{y}_i)}{(1+\theta_i)} = 0,$$

where

$$\bar{y}_i = n^{-1}\sum_{j=1}^{n} y_{ij}.$$

Thus,

$$\hat{\theta}_i = \frac{\bar{y}_i}{k} = \frac{\sum_{j=1}^{n} y_{ij}}{nk}, \quad i = 1, 2.$$

So,

$$\hat{\mathcal{L}}_\Omega = \prod_{i=1}^{2}\prod_{j=1}^{n}\left\{C_{k-1}^{k+y_{ij}-1}\hat{\theta}_i^{y_{ij}}(1+\hat{\theta}_i)^{-(k+y_{ij})}\right\}.$$

For the restricted parameter space,

$$\mathcal{L}_\omega = \prod_{i=1}^{2}\prod_{j=1}^{n}\left\{C_{k-1}^{k+y_{ij}-1}\theta^{y_{ij}}(1+\theta)^{-(k+y_{ij})}\right\}$$

$$= \left(\prod_{i=1}^{2}\prod_{j=1}^{n}C_{k-1}^{k+y_{ij}-1}\right)\theta^{s}(1+\theta)^{-(2nk+s)}, \quad \text{where } s = \sum_{i=1}^{2}\sum_{j=1}^{n} y_{ij}.$$

So,

$$\frac{\partial \mathcal{L}_\omega}{\partial\theta} = \frac{s}{\theta} - \frac{(2nk+s)}{(1+\theta)} = 0 \quad \text{gives} \quad \hat{\theta} = \frac{\bar{y}}{k}, \quad \text{where } \bar{y} = \frac{1}{2}(\bar{y}_1 + \bar{y}_2).$$

Thus,

$$\hat{\mathcal{L}}_\omega = \left(\prod_{i=1}^{2}\prod_{j=1}^{n}C_{k-1}^{k+y_{ij}-1}\right)\hat{\theta}^{s}(1+\hat{\theta})^{-(2nk+s)}.$$

Hence,

$$-2\ln\hat{\lambda} = -2\ln\left\{\frac{\hat{\theta}^{s}(1+\hat{\theta})^{-(2nk+s)}}{\prod_{i=1}^{2}\prod_{j=1}^{n}\hat{\theta}_i^{y_{ij}}(1+\hat{\theta}_i)^{-(k+y_{ij})}}\right\}$$

$$= -2\Bigg\{s\ln\hat{\theta} - (2nk+s)\ln(1+\hat{\theta})$$

$$- \sum_{i=1}^{2}[n\bar{y}_i\ln\hat{\theta}_i - n(k+\bar{y}_i)\ln(1+\hat{\theta}_i)]\Bigg\}.$$

Now, $s = 15$, $n = 50$, $k = 3$, $\bar{y}_1 = \frac{5}{50} = 0.10$, $\bar{y}_2 = \frac{10}{50} = 0.20$, $\hat{\theta} = \frac{\bar{y}}{k} = \frac{(\bar{y}_1+\bar{y}_2)}{2k} = \frac{0.10+0.20}{2(3)} = 0.05$, $\hat{\theta}_1 = \frac{0.10}{3} = 0.0333$, and $\hat{\theta}_2 = \frac{0.20}{3} = 0.0667$.
So, $-2\ln\hat{\lambda} = 1.62$. Since $\chi^2_{1,0.95} = 3.841$, we do *not* reject $H_0$; the P-value $\doteq 0.22$.

(b) From part (a),

$$\mathbf{S}'(\theta_1,\theta_2) = \begin{bmatrix} \dfrac{\partial \ln \mathcal{L}_\Omega}{\partial\theta_1} \\[2ex] \dfrac{\partial \ln \mathcal{L}_\Omega}{\partial\theta_2} \end{bmatrix} = \begin{bmatrix} \dfrac{n\bar{y}_1}{\theta_1} - \dfrac{n(k+\bar{y}_1)}{1+\theta_1} \\[2ex] \dfrac{n\bar{y}_2}{\theta_2} - \dfrac{n(k+\bar{y}_2)}{1+\theta_2} \end{bmatrix}.$$

Under $H_0$: $\theta_1 = \theta_2 (= \theta$, say$)$, $\hat{\theta} = \frac{(\bar{y}_1+\bar{y}_2)}{2k} = 0.05$. So,

$$\mathbf{S}'(\hat{\theta},\hat{\theta}) = \begin{bmatrix} \dfrac{(50)(0.10)}{0.05} - \dfrac{(50)(3+0.10)}{(1+0.05)} \\[2ex] \dfrac{(50)(0.20)}{0.05} - \dfrac{(50)(3+0.20)}{(1+0.05)} \end{bmatrix} = \begin{bmatrix} -47.6190 \\ +47.6190 \end{bmatrix}.$$

Now,

$$\frac{\partial^2 \ln \mathcal{L}_\Omega}{\partial \theta_1^2} = \frac{-n\bar{y}_1}{\theta_1^2} + \frac{n(k+\bar{y}_1)}{(1+\theta_1)^2},$$

$$\frac{\partial^2 \ln \mathcal{L}_\Omega}{\partial \theta_2^2} = \frac{-n\bar{y}_2}{\theta_2^2} + \frac{n(k+\bar{y}_2)}{(1+\theta_2)^2},$$

and

$$\frac{\partial^2 \ln \mathcal{L}_\Omega}{\partial \theta_1 \partial \theta_2} = \frac{\partial^2 \ln \mathcal{L}_\Omega}{\partial \theta_2 \partial \theta_1} = 0.$$

So, with $\boldsymbol{y} = (y_{11}, y_{12}, \ldots, y_{1n}; y_{21}, y_{22}, \ldots, y_{2n})$, we have

$$\begin{aligned}
\mathbf{I}(\boldsymbol{y}; \hat{\theta}) &= \begin{bmatrix} \dfrac{n\bar{y}_1}{\hat{\theta}^2} - \dfrac{n(k+\bar{y}_1)}{(1+\hat{\theta})^2} & 0 \\ 0 & \dfrac{n\bar{y}_2}{\hat{\theta}^2} - \dfrac{n(k+\bar{y}_2)}{(1+\hat{\theta})^2} \end{bmatrix} \\
&= \begin{bmatrix} \dfrac{(50)(0.10)}{(0.05)^2} - \dfrac{(50)(3+0.10)}{(1+0.05)^2} & 0 \\ 0 & \dfrac{(50)(0.20)}{(0.05)^2} - \dfrac{(50)(3.20)}{(1+0.05)^2} \end{bmatrix} \\
&= \begin{bmatrix} 1,859.4104 & 0 \\ 0 & 3,854.8753 \end{bmatrix}.
\end{aligned}$$

So,

$$\hat{S} = \mathbf{S}(\hat{\theta}, \hat{\theta})\mathbf{I}^{-1}(\boldsymbol{y}; \hat{\theta})\mathbf{S}'(\hat{\theta}, \hat{\theta}) = \frac{(-47.6190)^2}{1859.4104} + \frac{(47.6190)^2}{3854.8753} = 1.81.$$

Since $\chi^2_{1,0.95} = 3.84$, we do *not* reject $\mathrm{H}_0$; the P-value $\doteq 0.18$.

(c) With $X_{ij} = k + Y_{ij}$, then

$$\mathrm{p}_{X_{ij}}(x_{ij}; \theta) = \mathrm{C}_{k-1}^{x_{ij}-1}\left(\frac{1}{1+\theta_i}\right)^k \left(\frac{\theta_i}{1+\theta_i}\right)^{x_{ij}-k}, \quad x_{ij} = k, k+1, \ldots, \infty.$$

So,

$$\mathrm{E}(Y_{ij}) = \mathrm{E}(X_{ij}) - k = k(1+\theta_i) - k = k\theta_i$$

and

$$\mathrm{V}(Y_{ij}) = k\left(\frac{\theta_i}{1+\theta_i}\right)(1+\theta_i)^2 = k\theta_i(1+\theta_i).$$

In general,

$$\frac{(\bar{Y}_1 - \bar{Y}_2) - k(\theta_1 - \theta_2)}{\sqrt{\frac{k\theta_1(1+\theta_1)}{n} + \frac{k\theta_2(1+\theta_2)}{n}}} \dot{\sim} \mathrm{N}(0,1)$$

for large $n$ by the Central Limit Theorem. Under $\mathrm{H}_0$: $\theta_1 = \theta_2 (= \theta$, say), then

$$\frac{(\bar{Y}_1 - \bar{Y}_2) - 0}{\sqrt{\frac{2k\theta(1+\theta)}{n}}} \dot{\sim} \mathrm{N}(0,1) \quad \text{for large } n.$$

Thus, via Slutsky's Theorem, we could reject $\mathrm{H}_0$: $\theta_1 = \theta_2 (= \theta$, say) at the $\alpha$-level for large $n$ when

$$\left| \frac{(\bar{Y}_1 - \bar{Y}_2) - 0}{\sqrt{\frac{2k\hat{\theta}(1+\hat{\theta})}{n}}} \right| > Z_{1-\alpha/2}.$$

Now, for large $n$,

$$\begin{aligned}
\text{POWER} &\doteq \mathrm{pr}\left\{ \left| \frac{\bar{Y}_1 - \bar{Y}_2}{\sqrt{\frac{2k\theta(1+\theta)}{n}}} \right| > Z_{1-\alpha/2} \middle| \theta_1 \neq \theta_2 \right\} \\
&= \mathrm{pr}\left\{ \frac{(\bar{Y}_1 - \bar{Y}_2)}{\sqrt{\frac{2k\theta(1+\theta)}{n}}} < -Z_{1-\alpha/2} \middle| \theta_1 \neq \theta_2 \right\} \\
&\quad + \mathrm{pr}\left\{ \frac{(\bar{Y}_1 - \bar{Y}_2)}{\sqrt{\frac{2k\theta(1+\theta)}{n}}} > Z_{1-\alpha/2} \middle| \theta_1 \neq \theta_2 \right\}.
\end{aligned}$$

For $\theta_1 = 2.0$ and $\theta_2 = 2.4$, the contribution of the second term will be negligible. So,

$$\begin{aligned}
\text{POWER} &\doteq \mathrm{pr}\left\{ \frac{(\bar{Y}_1 - \bar{Y}_2)}{\sqrt{\frac{2k\theta(1+\theta)}{n}}} < -Z_{1-\alpha/2} \middle| \theta_1 = 2.0, \theta_2 = 2.4 \right\} \\
&= \mathrm{pr}\left\{ \frac{(\bar{Y}_1 - \bar{Y}_2) - k(\theta_1 - \theta_2)}{\sqrt{\frac{k\theta_1(1+\theta_1)}{n} + \frac{k\theta_2(1+\theta_2)}{n}}} < \right.
\end{aligned}$$

$$\left. \frac{-Z_{1-\alpha/2}\sqrt{\frac{2k\theta(1+\theta)}{n}} - k(\theta_1 - \theta_2)}{\sqrt{\frac{k\theta_1(1+\theta_1)}{n} + \frac{k\theta_2(1+\theta_2)}{n}}} \right| \theta_1 = 2.0, \theta_2 = 2.4 \Bigg\}$$

$$= \mathrm{pr}\left\{ Z < \left. \frac{-Z_{1-\alpha/2}\sqrt{\frac{2k\theta(1+\theta)}{n}} - k(\theta_1 - \theta_2)}{\sqrt{\frac{k\theta_1(1+\theta_1)}{n} + \frac{k\theta_2(1+\theta_2)}{n}}} \right| \theta_1 = 2.0, \theta_2 = 2.4 \right\},$$

where $Z \dot{\sim} \mathrm{N}(0,1)$ for large $n$. So, with $\theta_1 = 2.0, \theta_2 = 2.4, \alpha = 0.05, k = 3, (1-\beta) = 0.80$, and $\theta = (\theta_1 + \theta_2)/2 = 2.2$, we require the smallest $n$ (say, $n^*$) such that

$$\frac{-1.96\sqrt{2(3)(2.2)(1+2.2)} - \sqrt{n}(3)(2.0-2.4)}{\sqrt{3(2.0)(1+2.0) + 3(2.4)(1+2.4)}} \geq 0.842,$$

giving $n^* = 231$.

**Solution 5.26***

(a) The multinomial likelihood function $\mathcal{L}$ is given by

$$\mathcal{L} == \frac{n!}{y_{11}!y_{10}!y_{01}!y_{00}!}\pi_{11}^{y_{11}}\pi_{10}^{y_{10}}\pi_{01}^{y_{01}}\pi_{00}^{y_{00}},$$

and so

$$\ln \mathcal{L} \propto \sum_{i=0}^{1}\sum_{j=0}^{1} y_{ij} \ln \pi_{ij}.$$

To maximize $\ln \mathcal{L}$ subject to the constraint $(\pi_{11} + \pi_{10} + \pi_{01} + \pi_{00}) = 1$, we can use the method of Lagrange multipliers. Define

$$U = \sum_{i=0}^{1}\sum_{j=0}^{1} y_{ij} \ln \pi_{ij} + \lambda \left(1 - \sum_{i=0}^{1}\sum_{j=0}^{1} \pi_{ij}\right).$$

The equations

$$\frac{\partial U}{\partial \pi_{ij}} = \frac{y_{ij}}{\pi_{ij}} - \lambda = 0, \quad i = 0,1 \text{ and } j = 0,1,$$

imply that

$$\frac{y_{11}}{\pi_{11}} = \frac{y_{10}}{\pi_{10}} = \frac{y_{01}}{\pi_{01}} = \frac{y_{00}}{\pi_{00}} = \lambda,$$

or equivalently,

$$\frac{y_{ij}}{\lambda} = \pi_{ij}, \quad i = 0, 1 \text{ and } j = 0, 1.$$

Additionally,

$$\sum_{i=0}^{1}\sum_{j=0}^{1} \pi_{ij} = 1 \implies \sum_{i=0}^{1}\sum_{j=0}^{1} \frac{y_{ij}}{\lambda} = 1$$
$$\implies \lambda = \sum_{i=0}^{1}\sum_{j=0}^{1} y_{ij} = n.$$

Hence,

$$\hat{\pi}_{ij} = \frac{y_{ij}}{n}, \quad i = 0, 1 \text{ and } j = 0, 1.$$

By the invariance property for MLEs, it follows that the MLE of $\delta$ is equal to

$$\hat{\delta} = (\hat{\pi}_{11} + \hat{\pi}_{10}) - (\hat{\pi}_{11} + \hat{\pi}_{01}) = (\hat{\pi}_{10} - \hat{\pi}_{01}) = \frac{(y_{10} - y_{01})}{n}.$$

(b) Using the equality $\pi_{00} = (1 - \pi_{11} - \pi_{10} - \pi_{01})$, the log likelihood function can be written as

$$\ln \mathcal{L} \propto y_{11} \ln \pi_{11} + y_{10} \ln \pi_{10} + y_{01} \ln \pi_{01} + y_{00} \ln(1 - \pi_{11} - \pi_{10} - \pi_{01}).$$

Now,

$$\frac{\partial \ln \mathcal{L}}{\partial \pi_{11}} = \frac{y_{11}}{\pi_{11}} - \frac{y_{00}}{(1 - \pi_{11} - \pi_{10} - \pi_{01})},$$
$$\frac{\partial \ln \mathcal{L}}{\partial \pi_{10}} = \frac{y_{10}}{\pi_{10}} - \frac{y_{00}}{(1 - \pi_{11} - \pi_{10} - \pi_{01})},$$

and

$$\frac{\partial \ln \mathcal{L}}{\partial \pi_{01}} = \frac{y_{01}}{\pi_{01}} - \frac{y_{00}}{(1 - \pi_{11} - \pi_{10} - \pi_{01})}.$$

So, for $(i, j)$ equal to $(1, 1), (1, 0)$, or $(0, 1)$, we have

$$\frac{\partial^2 \ln \mathcal{L}}{\partial \pi_{ij}^2} = -\frac{y_{ij}}{\pi_{ij}^2} - \frac{y_{00}}{(1 - \pi_{11} - \pi_{10} - \pi_{01})^2},$$

and hence

$$-\mathrm{E}\left[\frac{\partial^2 \ln \mathcal{L}}{\partial \pi_{ij}^2}\right] = \left(\frac{n}{\pi_{ij}} + \frac{n}{\pi_{00}}\right).$$

In addition,

$$\frac{\partial^2 \ln \mathcal{L}}{\partial \pi_{11} \partial \pi_{10}} = \frac{\partial^2 \ln \mathcal{L}}{\partial \pi_{11} \partial \pi_{01}} = \frac{\partial^2 \ln \mathcal{L}}{\partial \pi_{10} \partial \pi_{01}} = -\frac{y_{00}}{(1 - \pi_{11} - \pi_{10} - \pi_{01})^2},$$

and so

$$-\mathrm{E}\left[\frac{\partial^2 \ln \mathcal{L}}{\partial \pi_{11} \partial \pi_{10}}\right] = -\mathrm{E}\left[\frac{\partial^2 \ln \mathcal{L}}{\partial \pi_{11} \partial \pi_{01}}\right] = -\mathrm{E}\left[\frac{\partial^2 \ln \mathcal{L}}{\partial \pi_{10} \partial \pi_{01}}\right] = \frac{n}{\pi_{00}}.$$

Hence, with $\boldsymbol{\pi} = (\pi_{11}, \pi_{10}, \pi_{01})$, the expected Fisher information matrix $\mathcal{I}(\boldsymbol{\pi})$ is

$$\mathcal{I}(\boldsymbol{\pi}) = n \begin{bmatrix} \left(\frac{1}{\pi_{11}} + \frac{1}{\pi_{00}}\right) & \frac{1}{\pi_{00}} & \frac{1}{\pi_{00}} \\ \frac{1}{\pi_{00}} & \left(\frac{1}{\pi_{10}} + \frac{1}{\pi_{00}}\right) & \frac{1}{\pi_{00}} \\ \frac{1}{\pi_{00}} & \frac{1}{\pi_{00}} & \left(\frac{1}{\pi_{01}} + \frac{1}{\pi_{00}}\right) \end{bmatrix},$$

with $\pi_{00} = (1 - \pi_{11} - \pi_{10} - \pi_{01})$.

So,

$$\mathcal{I}^{-1}(\boldsymbol{\pi}) = \frac{1}{n} \begin{bmatrix} \pi_{11}(1 - \pi_{11}) & -\pi_{11}\pi_{10} & -\pi_{11}\pi_{01} \\ -\pi_{11}\pi_{10} & \pi_{10}(1 - \pi_{10}) & -\pi_{10}\pi_{01} \\ -\pi_{11}\pi_{01} & -\pi_{10}\pi_{01} & \pi_{01}(1 - \pi_{01}) \end{bmatrix}.$$

The null hypothesis of interest is $H_0 : R(\boldsymbol{\pi}) \equiv R = (\pi_{10} - \pi_{01}) = 0$. Hence, with $\boldsymbol{T}(\boldsymbol{\pi}) \equiv \boldsymbol{T} = [0, 1, -1]$,

$$\begin{aligned} \Lambda(\boldsymbol{\pi}) \equiv \Lambda &= \boldsymbol{T}\mathcal{I}^{-1}(\boldsymbol{\pi})\boldsymbol{T}' \\ &= \frac{\pi_{10}(1 - \pi_{10})}{n} + \frac{\pi_{01}(1 - \pi_{01})}{n} + \frac{2\pi_{10}\pi_{01}}{n} \\ &= \frac{(\pi_{10} + \pi_{01})}{n} - \frac{(\pi_{10} - \pi_{01})^2}{n}. \end{aligned}$$

So, the Wald test statistic $\hat{W}$ takes the form

$$\begin{aligned} \hat{W} &= \frac{\hat{R}^2}{\hat{\Lambda}} \\ &= \frac{(\hat{\pi}_{10} - \hat{\pi}_{01})^2}{\frac{(\hat{\pi}_{10} + \hat{\pi}_{01})}{n} - \frac{(\hat{\pi}_{10} - \hat{\pi}_{01})^2}{n}} \end{aligned}$$

$$= \frac{(y_{10} - y_{01})^2/n^2}{\left[\frac{y_{10}}{n^2} + \frac{y_{01}}{n^2} - \left(\frac{y_{10} - y_{01}}{n}\right)^2\right] \Big/ n}$$

$$= \frac{(y_{10} - y_{01})^2}{(y_{10} + y_{01}) - \frac{(y_{10} - y_{01})^2}{n}}.$$

A simpler way to derive this test statistic is to note that

$$\hat{W} = \frac{\hat{\delta}^2}{\hat{V}(\hat{\delta})},$$

where $\hat{V}(\hat{\delta})$ denotes the MLE of $V(\hat{\delta})$. Now,

$$V(\hat{\delta}) = V\left[\frac{(Y_{10} - Y_{01})}{n}\right]$$

$$= \frac{V(Y_{10}) + V(Y_{01}) - 2\text{cov}(Y_{10}, Y_{01})}{n^2}$$

$$= \frac{n\pi_{10}(1 - \pi_{10}) + n\pi_{01}(1 - \pi_{01}) + 2n\pi_{10}\pi_{01}}{n^2}$$

$$= \frac{(\pi_{10} + \pi_{01}) - (\pi_{10} - \pi_{01})^2}{n}.$$

By the invariance property, it follows that

$$\hat{V}(\hat{\delta}) = \frac{(\hat{\pi}_{10} + \hat{\pi}_{01}) - (\hat{\pi}_{10} - \hat{\pi}_{01})^2}{n}$$

$$= \frac{\frac{(y_{10} + y_{01})}{n} - \frac{(y_{10} - y_{01})^2}{n^2}}{n}.$$

Finally, the Wald test statistic is given by

$$\hat{W} = \frac{(\hat{\delta})^2}{\hat{V}(\hat{\delta})}$$

$$= \frac{\frac{(y_{10} - y_{01})^2}{n^2}}{\left[\frac{(y_{10} + y_{01})}{n} - \frac{(y_{10} - y_{01})^2}{n^2}\right] \Big/ n}$$

$$= \frac{(y_{10} - y_{01})^2}{(y_{10} + y_{01}) - \frac{(y_{10} - y_{01})^2}{n}}.$$

When $y_{11} = 22$, $y_{10} = 3$, $y_{01} = 7$, and $y_{00} = 13$, so that $n = 45$, the Wald test statistic is equal to

$$\hat{W} = \frac{(3-7)^2}{(3+7) - \dfrac{(3-7)^2}{45}} = 1.6590.$$

An approximate P-value is

$$\text{P-value} = \text{pr}\left(\chi_1^2 > 1.6590 | \text{H}_0\right) \doteq 0.1977.$$

(c) The score vector has the form $\boldsymbol{S}(\boldsymbol{\pi}) = (s_1, s_2, s_3)$, where

$$s_1 = \frac{\partial \ln \mathcal{L}}{\partial \pi_{11}} = \frac{y_{11}}{\pi_{11}} - \frac{y_{00}}{\pi_{00}},$$

$$s_2 = \frac{\partial \ln \mathcal{L}}{\partial \pi_{10}} = \frac{y_{10}}{\pi_{10}} - \frac{y_{00}}{\pi_{00}},$$

and

$$s_3 = \frac{\partial \ln \mathcal{L}}{\partial \pi_{01}} = \frac{y_{01}}{\pi_{01}} - \frac{y_{00}}{\pi_{00}}.$$

Under $H_0 : \pi_{10} = \pi_{01}$ ($= \pi$, say), the restricted log likelihood is

$$\ln \mathcal{L}_\omega \propto y_{11} \ln \pi_{11} + y_{10} \ln \pi + y_{01} \ln \pi + y_{00} \ln \pi_{00}.$$

Using the LaGrange multiplier method with

$$U = y_{11} \ln \pi_{11} + y_{10} \ln \pi + y_{01} \ln \pi + y_{00} \ln \pi_{00} + \lambda(1 - \pi_{11} - 2\pi - \pi_{00}),$$

we have

$$\frac{\partial U}{\partial \pi_{11}} = \frac{y_{11}}{\pi_{11}} - \lambda = 0,$$

$$\frac{\partial U}{\partial \pi} = \frac{(y_{10} + y_{01})}{\pi} - 2\lambda = 0,$$

and

$$\frac{\partial U}{\partial \pi_{00}} = \frac{y_{00}}{\pi_{00}} - \lambda = 0.$$

Since $\lambda = n$, the restricted MLEs are

$$\hat{\pi}_{\omega 11} = \frac{y_{11}}{n},$$

$$\hat{\pi}_\omega = \frac{(y_{10} + y_{01})}{(2n)} \quad (= \hat{\pi}_{\omega 10} = \hat{\pi}_{\omega 01}),$$

and

$$\hat{\pi}_{\omega 00} = \frac{y_{00}}{n}.$$

Thus, with $\hat{\boldsymbol{\pi}}_{\omega} = \left(\hat{\pi}_{\omega 11}, \hat{\pi}_{\omega 10}, \hat{\pi}_{\omega 01}\right) = \left(\hat{\pi}_{\omega 11}, \hat{\pi}_{\omega}, \hat{\pi}_{\omega}\right)$, we have

$$\hat{\mathcal{I}}^{-1}(\hat{\boldsymbol{\pi}}_{\omega}) = \frac{1}{n^3}$$

$$\times \begin{bmatrix} y_{11}(n - y_{11}) & \frac{-y_{11}(y_{10}+y_{01})}{2} & \frac{-y_{11}(y_{10}+y_{01})}{2} \\ \frac{-y_{11}(y_{10}+y_{01})}{2} & \frac{(y_{10}+y_{01})(n+y_{11}+y_{00})}{4} & \frac{-(y_{10}+y_{01})^2}{4} \\ \frac{-y_{11}(y_{10}+y_{01})}{2} & \frac{-(y_{10}+y_{01})^2}{4} & \frac{(y_{10}+y_{01})(n+y_{11}+y_{00})}{4} \end{bmatrix}.$$

Now,

$$\frac{\partial \ln \mathcal{L}}{\partial \pi_{11}}_{|\boldsymbol{\pi}=\hat{\boldsymbol{\pi}}_{\omega}} = 0,$$

$$\frac{\partial \ln \mathcal{L}}{\partial \pi_{10}}_{|\boldsymbol{\pi}=\hat{\boldsymbol{\pi}}_{\omega}} = n\left(\frac{y_{10} - y_{01}}{y_{10} + y_{01}}\right)$$

and

$$\frac{\partial \ln \mathcal{L}}{\partial \pi_{01}}_{|\boldsymbol{\pi}=\hat{\boldsymbol{\pi}}_{\omega}} = n\left(\frac{y_{01} - y_{10}}{y_{10} + y_{01}}\right).$$

So,

$$\boldsymbol{S}(\hat{\boldsymbol{\pi}}_{\omega}) = \left[0, \hat{s}, -\hat{s}\right],$$

where

$$\hat{s} = n\left(\frac{y_{10} - y_{01}}{y_{10} + y_{01}}\right).$$

Finally, it can be shown with some algebra that

$$\boldsymbol{S}(\hat{\boldsymbol{\pi}}_{\omega})\hat{\mathcal{I}}^{-1}(\hat{\boldsymbol{\pi}}_{\omega})\boldsymbol{S}'(\hat{\boldsymbol{\pi}}_{\omega}) = Q_{M}.$$

When comparing the Wald and score test statistics, the nonnegative numerators of the two test statistics are identical. Since the nonnegative denominator of the score statistic is always at least as large as the denominator of the Wald statistic, it follows that the Wald statistic will always be at least as large in value as the score statistic.

(d) Since $\chi^2_{1,0.95} = 3.84$, $H_0$ will be rejected if $Q_M > 3.84$ and will not be rejected if $Q_M \leq 3.84$. Let $Q(y_{10}; 10)$ denote the value of $Q_M$ when $Y_{10} = y_{10}$ and when $(y_{10} + y_{01}) = 10$. Note that

$$Q(0;10) = Q(10;10) = 10.0;$$

$$Q(1;10) = Q(9;10) = 6.4;$$

$$Q(2;10) = Q(8;10) = 3.6;$$

$$Q(3;10) = Q(7;10) = 1.6;$$

$$Q(4;10) = Q(6;10) = 0.4;$$

$$Q(5;10) = 0.0.$$

Thus, the null hypothesis will be rejected if $Y_{10}$ takes any of the four values 0,1,9, or 10, and will not be rejected otherwise. For each randomly selected subject who has a discordant response pattern [i.e., (0,1) or (1,0)], the conditional probability of a (1,0) response [given that the response is either (1,0) or (1,0)] is equal to $\pi_{10}/(\pi_{10} + \pi_{01})$. This probability remains constant and does not depend on the number of subjects who have a concordant [(0,0) or (1,1)] response, and so the binomial distribution applies. Under the assumption that $\pi_{10} = 0.10$ and $\pi_{10} = 0.05$, the probability of rejecting the null hypothesis is equal to

$$\begin{aligned} \text{POWER} &= \sum_{y \in \{0,1,9,10\}} C_y^{10} \left( \frac{0.10}{0.10 + 0.05} \right)^y \left( \frac{0.05}{0.10 + 0.05} \right)^{10-y} \\ &= 0.0000169 + 0.000339 + 0.0867 + 0.01734 \\ &= 0.1044. \end{aligned}$$

Thus, there is roughly a 10% chance that the null hypothesis will be rejected. A larger sample size is needed in order to achieve reasonable power for testing $H_0\colon \delta = 0$ versus $H_1\colon \delta \neq 0$ when $\pi_{10} = 0.10$ and $\delta = 0.05$.

# Appendix

## Useful Mathematical Results

### A.1 Summations

a. *Binomial:* $\sum_{j=0}^{n} C_j^n a^j b^{(n-j)} = (a+b)^n$, where $C_j^n = \frac{n!}{j!(n-j)!}$.

b. *Geometric:*

i. $\sum_{j=0}^{\infty} r^j = \frac{1}{1-r}$, $|r| < 1$.

ii. $\sum_{j=1}^{\infty} r^j = \frac{r}{1-r}$, $|r| < 1$.

iii. $\sum_{j=0}^{n} r^j = \frac{1-r^{(n+1)}}{1-r}$, $-\infty < r < +\infty, r \neq 0$.

c. *Negative Binomial:* $\sum_{j=0}^{\infty} C_k^{j+k} \pi^j = (1-\pi)^{-(k+1)}$, $0 < \pi < 1$,

$k$ a positive integer.

d. *Exponential:* $\sum_{j=0}^{\infty} \frac{x^j}{j!} = e^x$, $-\infty < x < +\infty$.

e. *Sums of Integers:*

i. $\sum_{i=1}^{n} i = \frac{n(n+1)}{2}$.

ii. $\sum_{i=1}^{n} i^2 = \frac{n(n+1)(2n+1)}{6}$.

iii. $\sum_{i=1}^{n} i^3 = \left[\frac{n(n+1)}{2}\right]^2$.

### A.2 Limits

a. $\lim_{n\to\infty} \left(1+\frac{a}{n}\right)^n = e^a$, $-\infty < a < +\infty$.

## A.3 Important Calculus-Based Results

a. *L'Hôpital's Rule:* For differentiable functions f($x$) and g($x$) and an "extended" real number $c$ (i.e., $c \in \Re_1$ or $c = \pm\infty$), suppose that $\lim_{x\to c} f(x) = \lim_{x\to c} g(x) = 0$, or that $\lim_{x\to c} f(x) = \lim_{x\to c} g(x) = \pm\infty$. Suppose also that $\lim_{x\to c} f'(x)/g'(x)$ exists [in particular, $g'(x) \neq 0$ near $c$, except possibly at $c$]. Then,

$$\lim_{x\to c} \frac{f(x)}{g(x)} = \lim_{x\to c} \frac{f'(x)}{g'(x)}.$$

L'Hôpital's Rule is also valid for one-sided limits.

b. *Integration by Parts:* Let $u = f(x)$ and $v = g(x)$, with differentials $du = f'(x)\,dx$ and $dv = g'(x)\,dx$. Then,

$$\int u\,dv = uv - \int v\,du.$$

c. *Jacobians for One- and Two-Dimensional Change-of-Variable Transformations:* Let X be a scalar variable with support $\mathcal{A} \subseteq \Re^1$. Consider a one-to-one transformation $U = g(X)$ that maps $\mathcal{A} \to \mathcal{B} \subseteq \Re^1$. Denote the inverse of $U$ as $X = h(U)$. Then, the corresponding one-dimensional Jacobian of the transformation is defined as

$$J = \frac{d[h(U)]}{dU},$$

so that

$$\int_{\mathcal{A}} f(X)\,dX = \int_{\mathcal{B}} f[h(U)]|J|\,dU.$$

Similarly, consider scalar variables $X$ and $Y$ defined on a two-dimensional set $\mathcal{A} \subseteq \Re^2$, and let $U = g_1(X, Y)$ and $V = g_2(X, Y)$ define a one-to-one transformation that maps $\mathcal{A}$ in the $xy$-plane to $\mathcal{B} \subseteq \Re^2$ in the $uv$-plane. Define $X = h_1(U, V)$ and $Y = h_2(U, V)$. Then, the Jacobian of the (two-dimensional) transformation is given by the second-order determinant

$$J = \begin{vmatrix} \dfrac{\partial h_1(U,V)}{\partial U} & \dfrac{\partial h_1(U,V)}{\partial V} \\ \dfrac{\partial h_2(U,V)}{\partial U} & \dfrac{\partial h_2(U,V)}{\partial V} \end{vmatrix},$$

so that

$$\iint_{\mathcal{A}} f(X,Y)\, dX\, dY = \iint_{\mathcal{B}} f[h_1(U,V), h_2(U,V)]|J|\, dU\, dV.$$

## A.4 Special Functions

a. *Gamma Function:*

i. For any real number $t > 0$, the Gamma function is defined as

$$\Gamma(t) = \int_0^\infty y^{t-1} e^{-y}\, dy.$$

ii. For any real number $t > 0$, $\Gamma(t+1) = t\Gamma(t)$.

iii. For any positive integer $n$, $\Gamma(n) = (n-1)!$.

iv. $\Gamma(1/2) = \sqrt{\pi}$; $\Gamma(3/2) = \sqrt{\pi}/2$; $\Gamma(5/2) = (3\sqrt{\pi})/4$.

b. *Beta Function:*

i. For $\alpha > 0$ and $\beta > 0$, the Beta function is defined as

$$B(\alpha,\beta) = \int_0^1 y^{\alpha-1}(1-y)^{\beta-1}\, dy.$$

ii. $B(\alpha,\beta) = \dfrac{\Gamma(\alpha)\Gamma(\beta)}{\Gamma(\alpha+\beta)}$.

c. *Convex and Concave Functions:* A real-valued function $f(\cdot)$ is said to be *convex* if, for any two points $x$ and $y$ in its domain and any $t \in [0,1]$, we have

$$f[tx + (1-t)y] \le tf(x) + (1-t)f(y).$$

Likewise, $f(\cdot)$ is said to be *concave* if

$$f[tx + (1-t)y] \ge tf(x) + (1-t)f(y).$$

Also, $f(x)$ is concave on $[a,b]$ if and only if $-f(x)$ is convex on $[a,b]$.

## A.5 Approximations

a. *Stirling's Approximation:*

For $n$ a nonnegative integer, $n! \approx \sqrt{2\pi n}\left(\frac{n}{e}\right)^n$.

b. *Taylor Series Approximations:*

i. *Univariate Taylor Series:* If $f(x)$ is a real-valued function of $x$ that is infinitely differentiable in a neighborhood of a real number $a$, then a Taylor series expansion of $f(x)$ around $a$ is equal to

$$f(x) = \sum_{k=0}^{\infty} \frac{f^{(k)}(a)}{k!}(x-a)^k,$$

where

$$f^{(k)}(a) = \left[\frac{d^k f(x)}{dx^k}\right]_{|x=a}, \quad k = 0, 1, \ldots, \infty.$$

When $a = 0$, the infinite series expansion above is called a *Maclaurin* series.

As examples, a *first-order* (or linear) Taylor series approximation to $f(x)$ around the real number $a$ is equal to

$$f(x) \approx f(a) + \left[\frac{df(x)}{dx}\right]_{|x=a} (x-a),$$

and a *second-order* Taylor series approximation to $f(x)$ around the real number $a$ is equal to

$$f(x) \approx f(a) + \left[\frac{df(x)}{dx}\right]_{|x=a} (x-a) + \frac{1}{2!}\left[\frac{d^2 f(x)}{dx^2}\right]_{|x=a} (x-a)^2.$$

ii. *Multivariate Taylor series*: For $p \geq 2$, if $f(x_1, x_2, \ldots, x_p)$ is a real-valued function of $x_1, x_2, \ldots, x_p$ that is infinitely differentiable in a neighborhood of $(a_1, a_2, \ldots, a_p)$, where $a_i, i = 1, 2, \ldots, p$, is a real number, then a multivariate Taylor series expansion of $f(x_1, x_2, \ldots, x_p)$ around $(a_1, a_2, \ldots, a_p)$ is equal to

$$f(x_1, x_2, \ldots, x_p) = \sum_{k_1=0}^{\infty} \sum_{k_2=0}^{\infty} \cdots \sum_{k_p=0}^{\infty} \frac{f^{(k_1+k_2+\cdots+k_p)}(a_1, a_2, \ldots, a_p)}{k_1! k_2! \cdots k_p!} \times \prod_{i=1}^{p} (x_i - a_i)^{k_i},$$

where

$$f^{(k_1+k_2+\cdots+k_p)}(a_1, a_2, \ldots, a_p) = \left[\frac{\partial^{(k_1+k_2+\cdots+k_p)} f(x_1, x_2, \ldots, x_p)}{\partial x_1^{k_1} \partial x_2^{k_2} \cdots \partial x_p^{k_p}}\right]_{|(x_1, x_2, \ldots, x_p)=(a_1, a_2, \ldots, a_p)}.$$

As examples, when $p = 2$, a *first-order* (or *linear*) multivariate Taylor series approximation to $\mathrm{f}(x_1, x_2)$ around $(a_1, a_2)$ is equal to

$$\mathrm{f}(x_1, x_2) \approx f(a_1, a_2) + \sum_{i=1}^{2} \left[ \frac{\partial \mathrm{f}(x_1, x_2)}{\partial x_i} \right]_{|(x_1,x_2)=(a_1,a_2)} (x_i - a_i),$$

and a *second-order* multivariate Taylor series approximation to $\mathrm{f}(x_1, x_2)$ around $(a_1, a_2)$ is equal to

$$\begin{aligned} \mathrm{f}(x_1, x_2) \approx \mathrm{f}(a_1, a_2) &+ \sum_{i=1}^{2} \left[ \frac{\partial \mathrm{f}(x_1, x_2)}{\partial x_i} \right]_{|(x_1,x_2)=(a_1,a_2)} (x_i - a_i) \\ &+ \frac{1}{2!} \sum_{i=1}^{2} \left[ \frac{\partial^2 \mathrm{f}(x_1, x_2)}{\partial x_i^2} \right]_{|(x_1,x_2)=(a_1,a_2)} (x_i - a_i)^2 \\ &+ \left[ \frac{\partial^2 \mathrm{f}(x_1, x_2)}{\partial x_1 \partial x_2} \right]_{|(x_1,x_2)=(a_1,a_2)} (x_1 - a_1)(x_2 - a_2). \end{aligned}$$

## A.6 Lagrange Multipliers

The method of *Lagrange multipliers* provides a strategy for finding stationary points $\boldsymbol{x}^*$ of a differentiable function $\mathrm{f}(\boldsymbol{x})$ subject to the constraint $\mathbf{g}(\boldsymbol{x}) = \boldsymbol{c}$, where $\boldsymbol{x} = (x_1, x_2, \ldots, x_p)'$, where $\mathbf{g}(\boldsymbol{x}) = [\mathrm{g}_1(\boldsymbol{x}), \mathrm{g}_2(\boldsymbol{x}), \ldots, \mathrm{g}_m(\boldsymbol{x})]'$ is a set of $m(< p)$ constraining functions, and where $\boldsymbol{c} = (c_1, c_2, \ldots, c_m)'$ is a vector of known constants. The stationary points $\boldsymbol{x}^* = (x_1^*, x_2^*, \ldots, x_p^*)'$ can be (local) maxima, (local) minima, or saddle points. The Lagrange multiplier method involves consideration of the Lagrange function

$$\Lambda(\boldsymbol{x}, \boldsymbol{\lambda}) = \mathrm{f}(\boldsymbol{x}) - \left[\mathbf{g}(\boldsymbol{x}) - \boldsymbol{c}\right]' \boldsymbol{\lambda},$$

where $\boldsymbol{\lambda} = (\lambda_1, \lambda_2, \ldots, \lambda_m)'$ is a vector of scalars called "Lagrange multipliers." In particular, the stationary points $\boldsymbol{x}^*$ are obtained as the solutions for $\boldsymbol{x}$ using the $(p + m)$ equations

$$\frac{\partial \Lambda(\boldsymbol{x}, \boldsymbol{\lambda})}{\partial \boldsymbol{x}} = \frac{\partial \mathrm{f}(\boldsymbol{x})}{\partial \boldsymbol{x}} - \left\{ \frac{\partial \left[\mathbf{g}(\boldsymbol{x}) - \boldsymbol{c}\right]'}{\partial \boldsymbol{x}} \right\} \boldsymbol{\lambda} = \mathbf{0}$$

and

$$\frac{\partial \Lambda(\boldsymbol{x}, \boldsymbol{\lambda})}{\partial \boldsymbol{\lambda}} = -\left[\mathbf{g}(\boldsymbol{x}) - \boldsymbol{c}\right] = \mathbf{0},$$

where $\partial f(x)/\partial x$ is a $(p \times 1)$ column vector with $i$th element equal to $\partial f(x)/\partial x_i$, $i = 1, 2, \ldots, p$, where $\partial[\mathbf{g}(x) - \mathbf{c}]'/\partial x$ is a $(p \times m)$ matrix with $(i, j)$th element equal to $\partial g_j(x)/\partial x_i$, $i = 1, 2, \ldots, p$ and $j = 1, 2, \ldots, m$, and where $\mathbf{0}$ denotes a column vector of zeros.

Note that the second matrix equation gives $\mathbf{g}(x) = \mathbf{c}$.

As an example, consider the problem of finding the stationary points $(x^*, y^*)$ of the function $f(x, y) = (x^2 + y^2)$ subject to the constraint $\mathbf{g}(x, y) = g_1(x, y) = (x + y) = 1$. Here, $p = 2, m = 1$, and the Lagrange multiplier function is given by

$$\Lambda(x, y, \lambda) = (x^2 + y^2) - \lambda(x + y - 1).$$

The stationary points $(x^*, y^*)$ are obtained by solving the system of equations

$$\frac{\partial \Lambda(x, y, \lambda)}{\partial x} = 2x - \lambda = 0,$$

$$\frac{\partial \Lambda(x, y, \lambda)}{\partial y} = 2y - \lambda = 0,$$

$$\frac{\partial \Lambda(x, y, \lambda)}{\partial \lambda} = x + y - 1 = 0.$$

Solving these three equations yields the solution $x^* = y^* = 1/2$. Since

$$\frac{\partial \Lambda^2(x, y, \lambda)}{\partial x^2} = \frac{\partial \Lambda^2(x, y, \lambda)}{\partial y^2} > 0 \quad \text{and} \quad \frac{\partial \Lambda^2(x, y, \lambda)}{\partial x \partial y} = 0,$$

this solution yields a *minimum* subject to the constraint $x + y = 1$.

# References

Berkson J. 1950. "Are there two regressions?," *Journal of the American Statistical Association*, 45(250), 164–180.

Birkett NJ. 1988. "Evaluation of diagnostic tests with multiple diagnostic categories," *Journal of Clinical Epidemiology*, 41(5), 491–494.

Blackwell D. 1947. "Conditional expectation and unbiased sequential estimation," *Annals of Mathematical Statistics*, 18, 105–110.

Bondesson L. 1983. "On uniformly minimum variance unbiased estimation when no complete sufficient statistics exist," *Metrika*, 30, 49–54.

Breslow NE and Day NE. 1980. *Statistical Methods in Cancer Research, Volume I: The Analysis of Case–Control Studies*, International Agency for Research on Cancer (IARC) Scientific Publications.

Casella G and Berger RL. 2002. *Statistical Inference*, Second Edition, Duxbury, Thomson Learning, Belmont, CA.

Cramér H. 1946. *Mathematical Methods of Statistics*, Princeton University Press, Princeton, NJ.

Dempster AP, Laird NM, and Rubin DB. 1977. "Maximum likelihood from incomplete data via the EM algorithm," *Journal of the Royal Statistical Society, Series B*, 39, 1–22.

Feller W. 1968. *An Introduction to Probability Theory and Its Applications, Volume I*, Third Edition, John Wiley and Sons, Inc., Hoboken, NJ.

Fuller WA. 2006. *Measurement Error Models*, paperback, John Wiley and Sons, Inc., Hoboken, NJ.

Gibbs DA, Martin SL, Kupper LL, and Johnson RE. 2007. "Child maltreatment in enlisted soldiers' families during combat-related deployments," *Journal of the American Medical Association*, 298(5), 528–535.

Gustafson P. 2004. *Measurement Error and Misclassification in Statistics and Epidemiology: Impacts and Bayesian Adjustments*, Chapman & Hall/CRC Press, London, UK.

Halmos PR and Savage LJ. 1949. "Applications of the Radon-Nikodym theorem to the theory of sufficient statistics," *Annals of Mathematical Statistics*, 20, 225–241.

Hogg RV, Craig AT, and McKean JW. 2005. *Introduction to Mathematical Statistics*, Sixth Edition, Prentice-Hall, Upper Saddle River, NJ.

Hosmer DW and Lemeshow S. 2000. *Applied Logistic Regression*, Second Edition, John Wiley and Sons, Inc., Hoboken, NJ.

Hosmer DW, Lemeshow S, and May S. 2008. *Applied Survival Analysis: Regression Modeling of Time to Event Data*, Second Edition, John Wiley and Sons, Inc., Hoboken, NJ.

Houck N, Weller E, Milton DK, Gold DR, Ruifeng L, and Spiegelman D. 2006. "Home endotoxin exposure and wheeze in infants: correction for bias due to exposure measurement error," *Environmental Health Perspectives*, 114(1), 135–140.

Kalbfleisch JG. 1985. *Probability and Statistical Inference, Volume 1: Probability*, Second Edition, Springer, New York, NY.

Kalbfleisch JG. 1985. *Probability and Statistical Inference, Volume 2: Statistical Inference*, Second Edition, Springer, New York, NY.

Kass RE and Raftery AE. 1995. "Bayes factors," *Journal of the American Statistical Association*, 90, 773–795.

Kleinbaum DG and Klein M. 2002. *Logistic Regression: A Self-Learning Text*, Second Edition, Springer, New York, NY.

Kleinbaum DG and Klein M. 2005. *Survival Analysis: A Self-Learning Text*, Second Edition, Springer, New York, NY.

Kleinbaum DG, Kupper LL, and Morgenstern H. 1982. *Epidemiologic Research: Principles and Quantitative Methods*, John Wiley and Sons, Inc., Hoboken, NJ.

Kleinbaum DG, Kupper LL, Nizam A, and Muller KE. 2008. *Applied Regression Analysis and Other Multivariable Methods*, Fourth Edition, Duxbury Press, Belmont, CA.

Kupper LL. 1984. "Effects of the use of unreliable surrogate variables on the validity of epidemiologic research studies," *American Journal of Epidemiology*, 120(4), 643–648.

Kupper LL and Hafner KB. 1989. "How appropriate are popular sample size formulas?," *The American Statistician*, 43(2), 101–105.

Kupper LL and Haseman JK. 1978. "The use of a correlated binomial model for the analysis of certain toxicological experiments," *Biometrics*, 34, 69–76.

Kutner MH, Nachtsheim CJ, and Neter J. 2004. *Applied Linear Regression Models*, Fourth Edition, McGraw-Hill/Irwin, Burr Ridge, IL.

Lehmann EL. 1983. *Theory of Point Estimation*, Springer, New York, NY.

Makri FS, Philippou AN, and Psillakis ZM. 2007. "Shortest and longest length of success runs in binary sequences," *Journal of Statistical Planning and Inference*, 137, 2226–2239.

McCullagh P and Nelder JA. 1989. *Generalized Linear Models*, Second Edition, Chapman & Hall/CRC Press, London, UK.

Neyman J and Pearson ES. 1928. "On the use and interpretation of certain test criteria for purposes of statistical inference," *Biometrika*, 20A, 175–240 and 263–294.

Neyman J and Pearson ES. 1933. "On the problem of the most efficient tests of statistical hypotheses," *Philosophical Transactions, Series A*, 231, 289–337.

Rao CR. 1945. "Information and accuracy attainable in the estimation of statistical parameters," *Bulletin of the Calcutta Mathematical Society*, 37, 81–91.

Rao CR. 1947. "Large sample tests of statistical hypotheses concerning several parameters with applications to problems of estimation," *Proceedings of the Cambridge Philosophical Society*, 44, 50–57.

Rao CR. 1973. *Linear Statistical Inference and Its Applications*, Second Edition. John Wiley and Sons, Inc., Hoboken, NJ.

Ross S. 2006. *A First Course in Probability*, Seventh Edition, Prentice-Hall, Inc., Upper Saddle River, NJ.

Samuel-Cahn E. 1994. "Combining unbiased estimators," *The American Statistician*, 48(1), 34–36.

Serfling RJ. 2002. *Approximation Theorems of Mathematical Statistics*, John Wiley and Sons, Inc., Hoboken, NJ.

Taylor DJ, Kupper LL, Rappaport SM, and Lyles RH. 2001. "A mixture model for occupational exposure mean testing with a limit of detection," *Biometrics*, 57(3), 681–688.

Wackerly DD, Mendenhall III W, and Scheaffer RL. 2008. *Mathematical Statistics With Applications*, Seventh Edition, Duxbury, Thomson Learning, Belmont, CA.
Wald A. 1943. "Tests of statistical hypotheses concerning several parameters when the number of observations is large," *Transactions of the American Mathematical Society*, 54, 426–482.

# Index

Printed in the United States
by Baker & Taylor Publisher Services